T0324108

Polymer Electrolyte Fuel Cell Degradation

Polymer Electrolyte Fuel Cell Degradation

Matthew M. Mench

Emin Caglan Kumbur

T. Nejat Veziroglu

AMSTERDAM • BOSTON • HEIDELBERG • LONDON
NEW YORK • OXFORD • PARIS • SAN DIEGO
SAN FRANCISCO • SINGAPORE • SYDNEY • TOKYO

Academic Press is an Imprint of Elsevier

Academic Press is an imprint of Elsevier
The Boulevard, Langford Lane, Kidlington, Oxford, OX5 1GB
225 Wyman Street, Waltham, MA 02451, USA

First published 2012

British Library Cataloguing in Publication Data
A catalogue record for this book is available from the British Library

Library of Congress Number: 2011931448

ISBN: 978-0-12-810359-3

For information on all Academic Press publications
visit our website at www.elsevierdirect.com

Printed and bound in the United States

12 13 14 15   10 9 8 7 6 5 4 3 2 1

Contents

Preface

The field of fuel cell science and technology continues to evolve rapidly, with conversion to the commercialization stage already well underway. For example, in 2009 over 24,000 fuel cell units from various manufacturers were shipped worldwide. This represents a 40% increase compared to 2008 [1]. In the last decade alone, the cost of mass production (e.g. 500,000 units) of transportation fuel cell systems has dropped by nearly an order of magnitude, to an estimated cost of $45 per kW in 2010 [1]. All other metrics of performance have continued to improve as well, including operational stability, cold-start capability and survivability, and operating efficiency. The cost of hydrogen production and storage has also continually dropped, and is now already at levels that are considered economically competitive with existing sources in certain markets. Even as budget constraints and political considerations have limited expansion in the United States, the rest of the world has continued to accelerate hydrogen and fuel cell development, with numerous plans for continued growth and commercialization in Asia and Europe.

While real commercialization has already begun, there are still significant development challenges, such as long-term durability, that require advanced research and engineering solutions. The subject of long-term durability of fuel cell systems, and in particular polymer electrolyte fuel cells (PEFCs), is the subject of this book. It should be noted that this book does not cover system-related issues such as blowers or pumps, but instead focuses on the fundamental and practical issues of the stack itself. All materials and layers in the stack from the electrolyte to the bipolar plates are covered, as well as computational and experimental methods, and the protocol used to evaluate and predict degradation. Presently, degradation is observed in all materials involved, so that none can be overlooked, and no one breakthrough will solve all of the industrial needs. Although advances in durability have been rapid, they have often been founded on trial-and-error, and the fundamental science has lagged behind. Where significant progress has been made through more Edisonian approaches, more academic studies are needed. Conversely, the academic studying these issues needs to understand the underlying applied issues and the big picture. To respond to the underlying need for a combined exchange of applied and academic information, this book is designed as a compilation of the state-of-the-art fundamental knowledge and applied practice of some of the top international experts in the field, both in industry and academia. The focus, wherever possible, is on the fundamental understanding and underlying physics of the degradation processes involved inside the fuel cell stack, so that it is timeless, and not a function of a particular design. The combined applied and academic nature of the book makes it suitable as a reference in a graduate class studying the topic, or as an educational summary and reference for practicing engineers.

Chapter 1 presents an overview of the durability status and United States Department of Energy future targets for PEFCs, and is authored by Professor E.C. Kumbur and his students at Drexel University. This chapter is provided to give the reader a general overview of the present state-of-the-art and the future requirements needed to ensure competitive systems in a variety of applications, before diving into the rest of the more specific material. Of course, the moment it was written, the state-of-the-art became out of date, so the reader should refer to the online material available from the Department of Energy or other resources for a more timely reference.

Chapter 2 is written by top researchers at General Motors Fuel Cell Activities, and covers physical and chemical degradation of the membrane. The membrane is perhaps the most difficult and complex medium to design for high durability. Coupled physicochemical degradation occurs, and there is always a push to move to higher temperature, lower humidity environments that exacerbate these mechanisms. This important chapter summarizes the multidisciplinary understanding that overlaps fields of electrochemistry, transport, mechanical engineering, and polymer science, all of which are needed to fully understand the underlying physics of this material.

Chapter 3 is written by Dr Shyam Kocha, formerly of Nissan Motor Company and presently with the National Renewable Energy Laboratory (NREL) in the United States. Based on his dozen years of experience in the research and development of PEMFCs in industry (UTC Fuel Cells, General Motors and Nissan), Dr Kocha's chapter summarizes the fundamental understanding of the modes, mechanisms, and mitigation strategies involved in degradation and durability of the electrocatalyst and carbon-based support structure. This should be of great interest to a large cross-section of engineers, since the driving forces for the various degradation modes affect many aspects of the system and material design.

Chapter 4 of this book is believed to be a unique contribution to literature in this field, as it explores our understanding of degradation in the fuel cell diffusion media layer. Written by Professor Bruno Pollet of the University of Birmingham in the United Kingdom, it summarizes results from industry and academics in this relatively unexplored topic, and should be of great value to the community.

Chapter 5 moves to the edge of the fuel cell, and examines the bipolar plate degradation mechanisms. Written by Professor H. Tawfik, of Farmingdale State University, Dr Y. Hung of Brookhaven National Laboratory, and Professor D. Mahajan of Stony Brook University, this chapter is perhaps the first such summary available in the literature. A special focus on metallic bipolar plates is given, as they represent the least expensive alternative available and are therefore more likely to be implemented on a massive scale, especially in automotive based applications.

In Chapter 6, the subject matter moves away from a specific material layer and instead reviews the phenomenon of freeze-related degradation in polymer electrolyte fuel cells. Written by Professor M. M. Mench of the University of Tennessee, Knoxville and his Ph.D. student A. K. Srouji, the field of freeze-related damage and mitigation strategies is reviewed. This topic is extremely rich in fundamental science and complex transport phenomena, and should be of interest to engineers involved in the design of any system exposed to extremely cold environments.

Chapter 7 is an extremely unique contribution, covering the practical experimental diagnostics and durability testing protocols used at one of the leading fuel cell developers; UTC Power. Written by Rob Darling, Ryan Balliet and Mike Perry of UTC Power, this chapter is completely fresh, and not found in other literature to date. Since it covers protocols developed to experimentally understand degradation phenomena, it should be of particular interest to practicing engineers and industrial development programs.

Chapter 8 was developed to cover some unique areas of experimental diagnostics that can be used to identify and understand degradation mechanisms in different fuel cell materials. Written by well-established experts in this field at The University of Delaware, this chapter should be of great use to graduate students and faculty, as well as practicing engineers seeking new tools for analysis.

The final chapter of this book also represents a unique contribution in the challenging field of computational modeling of PEFC durability. Due to the disparate timescale between real-time performance changes and various degradation modes, this area is one of the least explored and difficult areas in the field of fuel cell science. Written by Dr Yu Morimoto of Toyota Motor Company, this chapter shows new contributions toward establishing a computational framework for prediction of various modes of fuel cell degradation, and should be of great interest to modelers and experimentalists alike.

The editors of this book hope that the reader finds this compilation both relevant and enlightening, and welcome any feedback to help us further contribute to this evolving field.

[1] United States Department of Energy, September 2010.

Chapter 1

Durability of Polymer Electrolyte Fuel Cells: Status and Targets

E.A. Wargo, C.R. Dennison and E.C. Kumbur
Electrochemical Energy Systems Laboratory, Department of Mechanical Engineering and Mechanics, Drexel University, Philadelphia, PA, USA

1. BACKGROUND

Growing concerns over energy supply and surrounding security and environmental issues have placed considerable emphasis on the development of alternatives to conventional energy sources. Much of the world's energy is derived from fossil fuels; coal, petroleum, and natural gas. In the United States, for example, in 2009 over 50% of the country's electricity was generated by coal-fired power plants, which consequently contributed 40% of the nation's carbon dioxide (CO_2) emissions [1]. Figure 1.1(a) shows recent world wide consumption of the top five energy sources, with non-renewables (such as petroleum, coal, and natural gas) displaying a more dramatic increase than renewables (such as hydroelectricity). The demand for energy – petroleum in particular – has been on the rise since the early 1980s, following the 1970s oil crisis. In 2009 over 84 million barrels of oil were consumed internationally each day, compared to 58 million in 1983 [2]. All regions of the world have seen significant growth in oil demand, but this trend has been most notable in Asia (Fig. 1.1(b)).

The increase in fossil fuel consumption over the past 40 years and the accompanying rise in carbon emissions have provided a substantial driving force for the development of alternative energy systems. Fuel cells are an alternative energy technology which show great potential, as they have the ability to alleviate both consumption and emissions concerns. They can be utilized for transportation, residential and commercial power, electronic devices, and so forth.

Due to their numerous benefits and wide range of application areas, they have attracted significant research and investment; specifically in polymer electrolyte fuel cells (PEFCs). This forms the main emphasis of this book. PEFC technology is highly efficient (83% theoretical, at room temperature [4]), produces near-zero detrimental greenhouse gas emissions, operates at low

Polymer Electrolyte Fuel Cell Degradation. DOI: 10.1016/B978-0-12-386936-4.10001-6

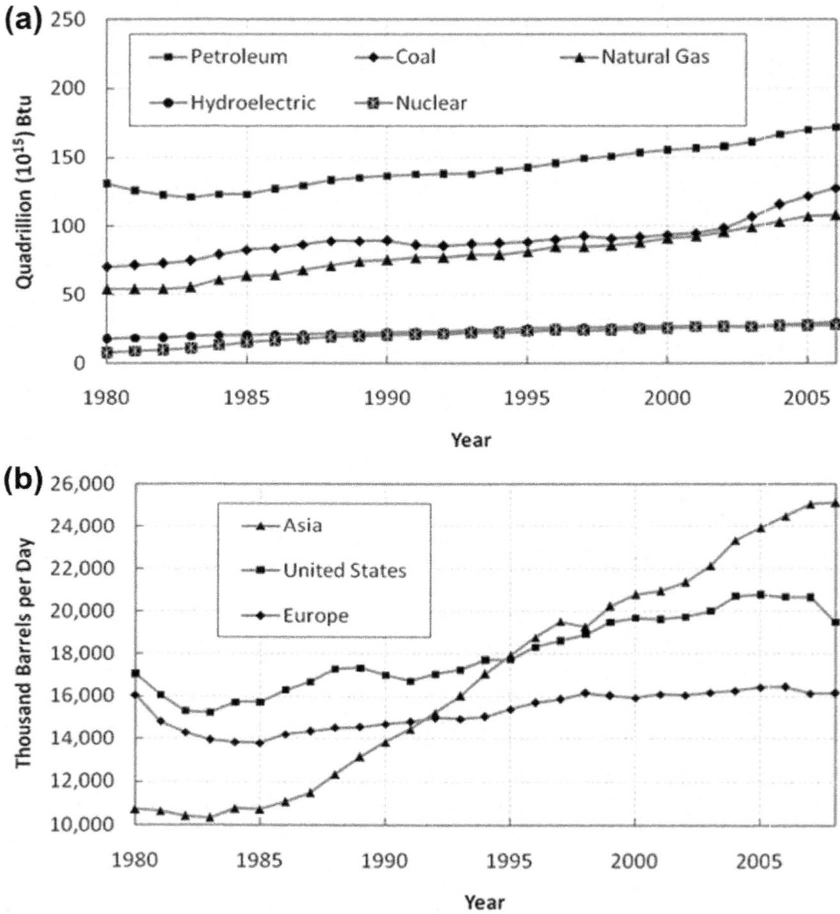

FIGURE 1.1 (a) World consumption of primary energy by energy type, 1980–2006 [2]. (b) Petroleum consumption by region, 1980–2008 [2].

temperatures (generally less than 100°C), and features rapid start-up and transient response characteristics. In addition to stationary and portable power applications, PEFCs are a very promising alternative to internal combustion engines for transportation applications. All of these reasons mean that the technology has attracted much attention from governments, industrial developers, and research institutions, resulting in a focused, collective effort aimed at the development and commercialization of PEFC systems [3].

In order to be truly competitive, PEFCs must meet or exceed the technological advantages of heat engines and other conventional power systems, on both the small and large scale. This includes balancing power output, system lifetime, and cost for a specific application of interest. Current research efforts are focused on improving each of these aspects, while recognizing the complex and coupled relationships between these characteristics. For example, increasing the membrane thickness will increase its durability and lifetime, but this requires more membrane material and lowers the cell's performance via higher protonic resistance, both of which elevate the overall cost [3]. A relatively limited lifetime is currently one of the greatest shortcomings of PEFCs compared to heat engines and other competing technologies. In order to improve cell lifetime without sacrificing cost and/or performance, much attention is now focused on identifying and investigating the factors that impact PEFC durability.

2. DURABILITY TARGETS FOR PEFC TECHNOLOGY

Despite the present cost and durability challenges of PEFC technology, there is a strong consensus that it is a practical, marketable, alternative energy technology. This factor, along with the impressive advantages of PEFCs, has led to the establishment of several government programs which have been designed to facilitate the development and implementation of PEFC power systems. Most notable are the efforts of the United States Department of Energy, the European Commission, and Japan's Ministry of Economy, Trade and Industry, each of which has established targets and timelines for the commercialization of PEFCs for a range of applications. The technical targets established by each of these organizations will be presented in Sections 2.1–2.3 and compared in Section 3 at the conclusion of the chapter. While the focus of this book is PEFC durability, due to the coupling between durability, cost, and performance, other relevant targets are also presented to provide the reader with the complete context in which these targets must be achieved.

2.1. United States Office of Energy Efficiency and Renewable Energy

The United States Hydrogen Energy Program was initially authorized in 1976 under the management of the National Science Foundation, and transferred to the Department of Energy (DOE) in 1990. In 2005, the Office of Energy Efficiency and Renewable Energy (EERE) released the Multi-Year Research, Development and Demonstration Plan which established goals for the Hydrogen, Fuel Cells and Infrastructure Technologies Program, and outlined a research and development plan for attaining those goals by 2015 [5].

The EERE has identified specific long-term objectives for PEFC development in various application areas, namely: transportation, stationary, consumer electronics, and auxiliary power units (APUs). Transportation systems are expected to operate in environments between −40 and 40°C, while stationary systems are expected to operate in environments between −35 and 40°C. Recent (2005) performance milestones, as identified by the EERE, for efficiency, durability, and cost are summarized in Table 1.1 [5].

TABLE 1.1 2005 Performance Milestones, US Department of Energy, Office of Energy Efficiency and Renewable Energy [5]

Application	Efficiency	Durability (Hours)	Cost
Transportation	50%	~1,000	$110/kW
Stationary	32%	20,000	$2,500/kW
Consumer Electronics (<50W)	–	>500	$40/W
Auxiliary Power Unit (3–30kW)	15%	100	>$2,000/kW

The EERE's 2010 and 2015 targets for overall efficiency, durability, and cost are summarized in Table 1.2 [5]. As previously noted, these characteristics are inherently coupled, and must be considered together. The EERE projects its targets out to the year 2015, which agrees with the fuel cell development timelines of other regional organizations throughout the world. It should be noted that for transportation applications, the 2010 targets allow for external humidification. However, the 2015 targets specify that the membrane must function without external humidification, with an end of life (EOL) performance loss of less than 5% beginning of life (BOL) performance [5].

In order to achieve the objectives summarized in Table 1.2, the EERE identifies targets for individual fuel cell components. The EERE is specifically concerned with bipolar plate/membrane electrode assembly (MEA) development for general applications, as well as membranes/electrocatalysts for

TABLE 1.2 Long-term Targets by Application, US Department of Energy, Office of Energy Efficiency and Renewable Energy [5]

Application	Efficiency	Durability (Hours)	Cost, 2010	Cost, 2015
Transportation	60%	5,000	$45/kW	$30/kW
Stationary	40%	40,000	$750/kW[1]	–
Consumer Electronics (<50W)	–	5,000	$3/W	–
Auxiliary Power Unit (3–30kW)	40%	35,000	$400/kW	$400/kW

[1]Milestone delayed from 2010 to 2011 due to appropriations shortfall and Congressionally directed activities [5].

TABLE 1.3 Technical Targets for Membranes for Transportation Applications, US Department of Energy, Office of Energy Efficiency and Renewable Energy [5]

Characteristic	Units	2005 Status	2010	2015
Cost	$/m^2	25	20	20
Durability with cycling[1]				
At operating temperature of ≤ 80°C	Hours	~2,000	5,000	5,000
At operating temperature of > 80°C	Hours	N/A	2,000	5,000

[1]*Based on the DOE stress test protocol [6].*

transportation use. The EERE technical targets for membranes for transportation applications are summarized in Table 1.3 [5]. It is worth noting that, to date, the EERE has not acknowledged any representative data available for high temperature membranes (> 80°C). However, high temperature membranes are still expected to meet the same durability criteria as low temperature membranes by 2015.

In addition to membranes, electrocatalysts for automotive applications have been identified for development by the EERE (Table 1.4) [5]. The targets for durability are based on an accelerated stress test protocol issued by the DOE [6]. This protocol is intended to accelerate component degradation by varying

TABLE 1.4 Technical Targets for Electrocatalysts for Transportation Applications, US Department of Energy, Office of Energy Efficiency and Renewable Energy [5]

		2005 Status		Stack Targets	
Characteristic	Units	Cell	Stack	2010	2015
Cost	$/kW	9	55	5	3
Durability with cycling					
Operating temp ≤ 80°C	Hours	>2,000	~2,000	5,000	5,000
Operating temp > 80°C	Hours	N/A	N/A	2,000	5,000
Electrochemical area loss	%	90	90	<40	<40
Electrocatalyst loss	mV after 100 hours at 1.2V	>30	N/A	<30	<30

TABLE 1.5 Technical Targets for Membrane Electrode Assemblies, US Department of Energy, Office of Energy Efficiency and Renewable Energy [5]

Characteristic	Units	2005 Status	2010	2015
Cost	\$/kW	60	10	5
Durability with cycling				
At operating temp of ≤ 80°C	Hours	~2,000	5,000	5,000
At operating temp of > 80°C	Hours	N/A	2,000	5,000
End of life (EOL) power density	% BOL	95	90	95

voltage conditions for 30,000 cycles. Cell diagnostics are performed at specified intervals to ensure that the cell still meets certain criteria. The EERE determines electrochemical area and electrocatalyst support loss according to protocols developed by General Motors [5]. As with high temperature membranes, high temperature electrocatalysts are still in the development stage, and reliable durability data were not available in 2005. However, they are still expected to achieve 5,000 hours of operation by 2015. Presently, electrocatalysts require significant advances in durability and reductions in cost to achieve EERE targets by 2015 [5].

Beyond transportation applications, the EERE has also identified targets for fuel cell components in general applications. Table 1.5 details the EERE technical targets for non-automotive MEAs [5]. These targets are also based on the DOE accelerated stress test [6]. Similar to high temperature membranes and electrocatalysts, high temperature MEAs are still in the development stage, which limits the availability of historical data and the determination of specific targets. The end of life (EOL) power density targets are determined as a function of the BOL power density, and account for degradation due to detrimental impurities in the air and fuel supply. It appears that the 2015 EOL power density target was already achieved in 2005. However, this data may be misleading, as it is based on a single data point. It is possible that if more data is considered, the present status for EOL power density will actually be below the 2005 value and no longer satisfy the 2015 target of 95% BOL.

Bipolar plates have been identified as another key component affecting the cost, efficiency, and durability of PEFCs. Table 1.6 details the EERE technical targets for bipolar plates [5]. Reductions in weight and cost are desired, although the current targets for weight have already been satisfied. The trend toward greater flexibility and lower flexural strength is due to the desire for improved cell sealing, lower contact resistance, and survivability under freeze/thaw conditions. The EERE notes that 'if all corrosion product ions remain in ionomer', the corrosion target may have to be decreased to 1 nA/cm^2 [5].

TABLE 1.6 Technical Targets for Bipolar Plates, US Department of Energy, Office of Energy Efficiency and Renewable Energy [5]

Characteristic	Units	2005 Status	2010	2015
Cost	\$/kW	10	5	3
Corrosion	$\mu A/cm^2$	<1	<1	<1
Flexural Strength	MPa	>34	>25	>25
Flexibility	% deflection at mid-span	1.5–3.5	3–5	3–5
Weight	kg/kW	0.36	<0.4	<0.4

2.2. European Hydrogen and Fuel Cell Technology Platform

In May 2003, the European Commission began preparations for the establishment of a European Hydrogen and Fuel Cell Technology Platform (HFP), geared towards the R&D and commercialization of hydrogen and fuel cell infrastructure in Europe [7]. To address this mission, the HFP developed a Strategic Research Agenda and Deployment Strategy in March 2005, and published an Implementation Plan in January 2007. The plan includes four Innovation and Development Actions, or application areas, namely:

i. Hydrogen Vehicles and Refueling Stations;
ii. Sustainable Hydrogen Production and Supply;
iii. Fuel Cells for CHP and Power Generation; and
iv. Fuel Cells for Early Markets.

The relevant durability goals established in the 2007 plan are summarized in Table 1.7 [8]. Cost and efficiency targets are also included, since they are ultimately coupled with durability. In May 2008, a new organization, the Fuel Cells and Hydrogen Joint Undertaking (FCH JU), was launched to further accelerate the HFP's goals. The FCH JU is a public-private partnership to support research, development, and demonstration activities with the overall goals of lowering CO_2 emissions and hydrocarbon dependence, while contributing to economic growth [7]. The FCH JU places additional driving force behind the four application areas established in the 2007 Implementation Plan with updated titles, including:

i. Transportation & Refueling Infrastructure;
ii. Hydrogen Production, Storage & Distribution;
iii. Stationary Power Generation & Combined Heat and Power; and
iv. Early Markets.

TABLE 1.7 Long-term Targets by Application, the European Hydrogen and Fuel Cell Technology Platform [8]

Market	Characteristic	Units	Target
			2015 Target
Road propulsion fuel cell system	Efficiency (NEDC)	%	>40
	Specific cost	€/kW	100
	Durability: Passenger car Bus	hours	 5,000 10,000
Road APU fuel cell system	Efficiency (P_{max})	%	>35
	Specific cost	€/kW	<500
	Durability: Passenger car Heavy goods-vehicle	hours	 5,000 40,000
Aircraft fuel cell power unit with reformer	Efficiency (full load incl. head and power)	% LHV	>60
	Specific cost	€/kW	500
	Durability	hours	~30,000
Maritime fuel cell power unit	Efficiency (full load)	%	>55
	Specific cost	€/kW	<1,000
	Durability	hours	50,000
Rail fuel cell propulsion and power unit (PEFC)	Efficiency (full load)	%	>45
	Specific cost	€/kW	500
	Durability	hours	50,000
Low-power system consumer electronic devices	Gravimetric power density	W/kg	80–200
	Specific cost	€/W	3–5
	Durability	hours	1,000–5,000
			2009–2012 Target
Stationary applications 1–10 kW (residential)	Electrical efficiency at BOL, including DC/AC conversion	%	34–40
	Total fuel efficiency BOL; at best point	%	80

TABLE 1.7 Long-term Targets by Application, the European Hydrogen and Fuel Cell Technology Platform [8]—cont'd

Market	Characteristic	Units	Target
	System cost	€/kW	6,000
	Stack durability (90% BOL performance)	hours	>12,000
Stationary applications ≥100 kW (community/ industrial)	Electrical efficiency at BOL, including DC/AC conversion	%	50
	Total fuel efficiency BOL; at best point	%	90
	System cost	€/kW	1,500–5,000
	90% BOL performance	hours	>30,000

These application areas were allocated funding of 26.4, 5.7, 25.9, and 10.3 million euros (M€), respectively, indicating the significant interest of European governments in fuel cell technology. The FCH JU began publishing Annual Implementation Plans, and the 2008 and 2009 plans both issued calls for research proposals. The proposal targets relevant to PEFC durability are summarized in Table 1.8 [7,9].

2.3. Japanese New Energy and Industrial Technology Development Organization

The Japanese New Energy and Industrial Technology Development Organization (NEDO) was established in 1980 [10] by the Japanese Ministry of Economy, Trade and Industry, and was originally charged with organizing and coordinating R&D efforts for new petroleum-alternative energy technologies. Since then, NEDO's activities have grown in scope. NEDO regulates and supports the development of fuel cell technologies, specifically PEFCs and solid oxide fuel cells. From 2008 to 2009, NEDO budgeted over 21 billion yen for their Fuel Cell and Hydrogen Technologies program [11]. Beyond fuel cell technology, this program funds many other aspects of hydrogen energy technology, such as safety, standardization and infrastructure.

Accordingly, NEDO has developed several R&D road maps to guide efforts in the area of fuel cells. Each road map is specific to a particular fuel cell application or technology, and details specific long-term targets extending to 2030. For PEFC technology, the road map identifies targets for fuel cell

TABLE 1.8 Proposal Targets, the European Fuel Cells and Hydrogen Joint Undertaking [7,9]

Topic	Program Duration	Application	Characteristic	Target
Large-scale demonstration of road vehicles and refueling infrastructure	At least 5 years	Cars	Durability	>2,000 hrs initially, min 3,000 hrs program target
			MTBF	>1,000 km
			Efficiency	>40% (NEDC)
		Buses	Durability	>4,000 hrs initially, min 6,000hrs program target
			Fuel Consumption	<11 – 13 kg H_2 / 100 km depending on drive cycle
Development and optimization of PEMFC electrodes and GDLs	Up to 3 years	Electrodes	Pt-loadings	<0.15 g/kW at > 55% efficiency (LHV=lower heating value)
		Electrodes and GDLs	Durability	>5 000 hrs at dynamic operation (car)[1]
Fundamentals of fuel cell degradation for stationary power applications	Up to 3 years		Durability	40,000 hrs
Demonstration of fuel cell-powered materials handling vehicles including infrastructure	Up to 3 years	Example: Forklifts	Durability	>5,000 hrs
			Cost	<4,000 €/kW
			Efficiency (tank to wheel)	>40%
Portable fuel cell generators, back-up and UPS power systems	Up to 3 years		Durability	>5 years
			Cost	5,000 €/kW
			Reliability	100%

[1] *Includes start/stop and freeze/thaw cycles.*

TABLE 1.9 Long-term Targets and Recent Milestones for Fuel Cell Vehicles, the Japanese New Energy and Industrial Technology Development Organization [12]

Characteristic	Units	2007 Status	2010	2015	2020–2030
Vehicle efficiency[1,2]	%	50	>50	60	>60
Durability[3]	Hours	1,000	3,000	5,000	>5,000
Operating temperature	°C	80	−30 to 90	−30 to 90 −100	−40 to 100 −120
Cost	Yen/kW ($/kW)[4]	–	50,000–60,000 (539–647)	10,000 (108)	<4,000 (43)

[1] *Vehicle efficiency measured using chassis dynamometer.*
[2] *Values referenced to LHV.*
[3] *Durability targets include consideration of start/stop cycles at required operation conditions.*
[4] *Currency exchange rate used: 92.7716 yen/$.*

vehicles and stationary applications [12]. The targets and recent milestones for PEFC vehicle efficiency, durability, operating temperature, and cost, are summarized in Table 1.9, for varying operating conditions [12]. For instance, the targets for 2015 anticipate operation ranging from 90 to 100°C under less than 30% relative humidity (RH), while the targets for 2020 to 2030 project operation from 100 to 120°C, under non-humidified conditions.

In terms of stationary applications, the recent milestones and targets for efficiency, durability, operating temperature, and cost are summarized in Table 1.10 [12]. Similar to the targets for automotive applications, the targets noted in Table 1.10 are specified for varying operating conditions. The targets for 2010 anticipate operation from 70 to 90°C under less than 65% RH, while the targets for 2015 onwards anticipate operation at or above 90°C under less than 30% RH.

3. CONCLUDING REMARKS

There is a strong international consensus that PEFCs are a promising technology which demand considerable investment in research and development. Numerous countries, including China, India, South Korea, Russia, among others, are funding R&D and commercialization efforts [13]. The United States, Europe, and Japan are of particular interest and have each commissioned organizations to guide these efforts. Each of these organizations has established targets for the development and commercialization of PEFCs. Strong agreements can be seen between the targets established by each of these

TABLE 1.10 Long-term Targets and Recent Milestones for Stationary Fuel Cells, the Japanese New Energy and Industrial Technology Development Organization [12]

Characteristic	Units	2007 Status	2010	2015	2020–2030
Electrical efficiency[1]	%	36	37	37	>40
Durability[2]	Hours	20,000	40,000	40,000 −90,000	90,000
Operating temperature	°C	70	70	70 to 90	90
Cost	Million yen/kW (thousand $/kW)[3]	4.8 (52)	0.7–1.2[4] (7.5–13)	5–0.7[5] (5–7.5)	<0.4 (4)

[1]*Values referenced to LHV.*
[2]*Durability targets include consideration of start/stop cycles at required operation conditions.*
[3]*Currency exchange rate used: 92.7716 yen/$.*
[4]*Assumes production rate of 10,000 systems/company·year.*
[5]*Assumes production rate of 100,000 systems/company·year.*

organizations. As expected, the intent of these targets is to improve PEFC performance by increasing efficiency and durability while decreasing cost. The targets of each organization are also indicative of the general vision of their respective region. The US and Europe have established targets for PEFCs in portable applications, while Japan has not. Europe has also established goals for various transportation applications beyond automobiles, including aircraft, maritime vessels, and rail vehicles.

Each organization targets 5,000 hours of operation in automotive applications by 2015, with a cost of ~$110/kW. The efficiency targets for automotive applications range from 40 to 60%, with Japan setting the most aggressive target at 60% for total vehicle efficiency (i.e. 'tank-to-wheels'). The targets for the US and Japan indicate a trend away from external reactant humidification for automotive applications. The US, Europe, and Japan each target a minimum of 30,000 to 40,000 hours of durability for PEFCs in stationary applications. Europe has set the most aggressive target for stationary PEFC efficiency at 50% by 2012, including DC/AC conversion. Japan and the US have set more conservative targets of 37% and 40% respectively, by 2015. The US has set a very aggressive cost target of $750/kW by 2010, while Europe and Japan have targeted ~$6,000/kW and $7,500/kW or less, respectively.

It is important to note that the information summarized in this chapter represents regional targets at the time of publication. These targets are subject to change with improvements in technology and increases in allocated resources for R&D efforts. Readers are encouraged to refer to [5,7,8,9,12] for more detailed information regarding regional targets.

In conclusion, durability currently represents one of the greatest technological challenges for PEFCs, with significant opportunities for development. Each component has its own durability considerations associated with its role in the operation of the fuel cell. The mechanisms of degradation within PEFCs and areas for further investigation will be discussed in detail in the following chapters. It is important to recognize the fact that fuel cell development is a balance between the coupled characteristics of durability, efficiency, and cost. While immediate improvements in durability are possible at the expense of efficiency and cost, the true challenge of R&D efforts in this area is to advance PEFC durability without adversely affecting cost and/or performance.

ACRONYMS AND ABBREVIATIONS

€	Euro
AC	Alternating current
APU	Auxiliary power unit
BOL	Beginning of life
CHP	Combined heat and power (generation)
CO_2	Carbon dioxide
DC	Direct current
DOE	Department of Energy (United States)
EERE	The Office of Energy Efficiency and Renewable Energy (United States)
EOL	End of life
FCH JU	Fuel Cells and Hydrogen Joint Undertaking (Europe)
GDL	Gas diffusion layer
HFP	The European Hydrogen and Fuel Cell Technology Platform
LHV	Lower heating value
MEA	Membrane electrode assembly
MTBF	Mean time between failures
NEDC	New European Driving Cycle
NEDO	The Japanese New Energy and Industrial Technology Development Organization
PEFC	Polymer electrolyte fuel cell
R&D	Research and development
RH	Relative humidity
UPS	Uninterruptable power supply
US	United States

REFERENCES

[1] T.C. Merkel, H. Lin, X. Wei, R. Baker, Power plant post-combustion carbon dioxide capture: An opportunity for membranes, Journal of Membrane Science volume 359 1–2 (1 September

2010). Membranes and CO_2 Separation, PP. 126–139, ISSN 0376-7388, DOI: 10.1016/j.memsci.2009.10.041.

[2] United States Energy Information Administration, World Petroleum Consumption. http://www.eia.doe.gov/ipm/, 2009.

[3] R. Borup, J. Meyers, B. Pivovar, Y.S. Kim, R. Mukundan, N. Garland, D. Myers, M. Wilson, F. Garzon, D. Wood, P. Zelenay, K. More, K. Stroh, T. Zawodzinski, J. Boncella, J.E. McGrath, M. Inaba, K. Miyatake, M. Hori, K. Ota, Z. Ogumi, S. Miyata, A. Nishikata, Z. Siroma, Y. Uchimoto, K. Yasuda, K.-i. Kimijima, N. Iwashita, Scientific Aspects of Polymer Electrolyte Fuel Cell Durability and Degradation, Chemical Reviews 10 (2007) 3904–3951.

[4] World Energy Council, Fuel Cell Efficiency. http://www.worldenergy.org, 2010.

[5] Hydrogen, Fuel Cells & Infrastructure Technologies Program, US Department of Energy, Office of Energy Efficiency and Renewable Energy, Multi-Year Research, Development and Demonstration Plan (2009). http://www1.eere.energy.gov/hydrogenandfuelcells/mypp/.

[6] United States Department of Energy, DOE Cell Component Accelerated Stress Test Protocols For PEM Fuel Cells. http://www1.eere.energy.gov/hydrogenandfuelcells/fuelcells/pdfs/component_durability_profile.pdf, 2007.

[7] Fuel Cell and Hydrogen Joint Undertaking, Annual Implementation Plan. http://ec.europa.eu/research/fch/pdf/calls-2009/fch_ju_annual_implementation_plan_2009.pdf, 2009.

[8] European Hydrogen & Fuel Cell Technology Platform, Implementation Plan - Status 2006. http://ec.europa.eu/research/fch/pdf/hfp_ip06_final_20apr2007.pdf, 2007.

[9] Fuel Cell and Hydrogen Joint Undertaking, Annual Implementation Plan. http://ec.europa.eu/research/fch/pdf/fch_ju_aip2008.6.10.08.fin.pdf, 2008.

[10] New Energy and Industrial Technology Development Organization, What is NEDO? http://www.nedo.go.jp/english/introducing/what.html.

[11] New Energy and Industrial Technology Development Organization, Outline of NEDO 2008–2009, http://www.nedo.go.jp/kankobutsu/pamphlets/kouhou/2008gaiyo_e/index.html.

[12] New Energy and Industrial Technology Development Organization, Polymer Electrolyte Fuel Cell (PEFC) Technology Road Map, From personal correspondence with the NEDO Washington, DC Office (2007).

[13] Fuel Cells, 2000, International Hydrogen and Fuel Cell Policy and Funding. http://www.fuelcells.org/InternationalH2-FCpolicyfunding.pdf, 2008.

Chapter 2

Membrane Durability: Physical and Chemical Degradation

Craig S. Gittleman, Frank D. Coms and Yeh-Hung Lai
General Motors Electrochemical Energy Research Lab, NY, USA

1. INTRODUCTION

1.1. Background

One of the key challenges facing the commercialization of fuel cells is in developing membrane electrode assemblies (MEAs) that can meet industry durability targets. Proton exchange membranes (PEMs) are the most promising membranes for automotive applications because of their relatively high proton conductivity at low temperatures. These membranes serve to conduct protons from the anode electrode to the cathode electrode of the fuel cell, while simultaneously insulating electronic current from passing across the membrane and preventing crossover of the reactant gases, H_2 and O_2. State-of-the-art PEM fuel cells for high power density operation utilize perfluorosulfonic acid (PFSA) membranes that are typically no more than 25 microns thick. The most common example of a PFSA PEM is Nafion® from DuPont™, for which the chemical structure is shown below.

$$[CF_2\text{-}CF_2]_x\text{-}[CF\text{-}CF_2]_y$$
$$[O\text{-}CF_2\text{-}CF]_z\text{-}O\text{-}CF_2\text{-}CF_2\text{-}SO_3H$$
$$CF_3$$

To be viable for automotive applications, these membranes must survive 10 years in a vehicle and 5,500 hours of operation including transient operation with start/stop and freeze/thaw cycles. The requirements for the chemical and mechanical stability of these thin membranes are significantly more demanding than the thick membranes (100–200 μm) used in the past. Fuel cells cannot operate effectively if even small amounts of gas are able to permeate the membrane through microscopic pinholes. Ultimately, fuel cells fail because such pinholes develop and propagate within the polymer membranes. Fuel cells can also fail if electronic current passes through the membranes causing the

Polymer Electrolyte Fuel Cell Degradation. DOI: 10.1016/B978-0-12-386936-4.10002-8

system to short. It is critical that these membranes are sufficiently robust with respect to cracking and shorting over the range of conditions experienced during fuel cell operation.

This chapter addresses the three primary root causes of membrane failure in automotive fuel cell systems. These failure modes are:

1. Chemical degradation: direct attack of the polymer membrane by radical species generated as by-products or side reactions of the fuel cell electrochemical reactions causing polymer decomposition.
2. Mechanical degradation: membrane fracture caused by cyclic or fatigue stresses imposed on the membrane via humidity and thermal fluctuations in a constrained cell.
3. Shorting: electronic current passing though the membrane caused by cell over-compression and topographical irregularities of the neighboring components (electrodes, gas diffusion layers (GDL)) leading to local over-compression and creep.

Each of these failure modes is discussed in detail in the sections below. Each section will include the underlying theory of the specific membrane failure mode, the *ex-situ* and *in-situ* tests used to evaluate membranes for robustness against that failure mode, and mitigation strategies employed to prevent membrane failure. Additionally, this chapter will address the impact of the combination of stresses which lead to the membrane failure modes described above. For example, when membranes experience both chemical and mechanical stressors simultaneously, the effect is that membrane failure by each individual failure mode is accelerated by the presence of the other.

It should be noted that, while PEMs can fail by thermal decomposition, typical PEM thermal decomposition temperatures are sufficiently above the operating temperature of automotive fuel cell systems and, thus, thermal degradation is not addressed in this chapter. For example, thermal decomposition of Nafion® occurs near 400°C [1,2], whereas automotive fuel cells are not expected to operate above 120°C. Such high temperatures can occur subsequent to membrane failure by gas crossover, which leads to exothermic combustion of H_2 on the fuel cell electrocatalysts, or membranes shorting, which leads to extreme resistive heating. Both cases can lead to thermal runaway and membrane thermal decomposition, but in neither case is membrane thermal degradation the primary failure mode.

1.2. Performance-Durability Trade-offs

Current polymer electrolyte fuel cell (PEFC) systems run at a maximum temperature of 80°C because operation above that temperature requires too much system support (e.g. pressure, humidification) to maintain the high humidity required for acceptable proton conduction. However, in order to effectively design a system capable of rejecting the heat generated at maximum vehicle power requires that fuel cells can operate, at least for relatively short times, at temperatures up to 95°C. The pressure and humidification constraints referred to above further require PEFC operation at these elevated temperatures at less than 50% relative humidity (RH). The US Department of Energy (DOE) requirements for membrane performance require them to effectively conduct protons at 120°C without external humidification [3]. The complete DOE requirements for PEMs for automotive applications are given in Table 2.1. These membranes will also experience temperatures as low as −40°C during freezing conditions and will be expected to withstand exposure to liquid water over the entire temperature range. The push to meet these challenging performance requirements has led researchers to focus on developing materials with reduced proton transport resistance in both wet and dry conditions. The two primary means to achieve said reduced proton transport resistance is to (a) increase polymer proton conductivity and (b) reduce membrane thickness.

There exists a great challenge to simultaneously address the performance and durability demands of PEMs for automotive systems. This is because many of the approaches to reduce membrane proton transport resistance tend to increase the risk of membrane failure. For example, membranes with lower equivalent weight (EW) or a higher concentration of acid sites typically exhibit higher proton conductivity. However, lower EW membranes also exhibit greater swelling upon exposure to liquid water. Higher swelling leads to greater stresses imposed on the membrane during fuel cell operation and, subsequently, reduces the membrane lifetime in accelerated humidity cycling stress tests [4]. Another relatively simple way to reduce proton transport resistance is to use thinner membranes. Thinner membranes, however, exhibit higher gas crossover, which lowers the efficiency of the fuel cell system and can also lead to increased voltage decay rates [5] and earlier membrane failure. Thinner membranes also have lower electron transport resistance across the membrane and are more susceptible to shorting.

Other approaches to improve PEM performance at low relative humidities involve incorporating additives or composites into the membrane. Researchers have added inorganic particles into the membrane to enable it to retain water at low RH, and thus enhance conductivity. These inorganics include metal oxides, such as silica and titania, as well as zeolites and zirconium phosphate. Others have incorporated functionalized inorganic species such as heteropolyacids (HPAs) or acid functionalized metal oxides. These HPAs include phosphotungstic acid (PWA) [6], phosphomolybdic acid (PMA) [7], 12-silicotungstic acid (SiWA) [8], sulfonated zeolites [9] and sulfonated polyhedral oligomeric silsesquioxanes (POSS) [10] among others. Use of these additives has produced varying degrees of success for improving membrane performance. However, addition of such inorganic materials can embrittle the membrane, making it more susceptible to mechanical degradation [11]. These additives can also migrate out of the membrane into the electrodes causing voltage decay issues.

TABLE 2.1 US Department of Energy Targets for Membranes for Automotive Fuel Cell Applications

Characteristic		Units	2015 target
Maximum operating temperature		°C	120
Area specific proton transport resistance	Maximum operating temp. and water partial pressures from 40 to 80 kPa	$\Omega\cdot cm^2$	0.02
	80°C and water partial pressures from 25–45 kPa	$\Omega\cdot cm^2$	0.02
	30°C and water partial pressures up to 4 kPa	$\Omega\cdot cm^2$	0.03
	−20°C	$\Omega\cdot cm^2$	0.2
Maximum Oxygen cross over[a]		mA/cm^2	2
Maximum Hydrogen cross over[a]		mA/cm^2	2
Minimum Electrical resistance		$\Omega\cdot cm^2$	1000
Cost (500K vehicles/yr)		$\$/m^2$	20
Mechanical Durability		RH Cycles	20,000
Chemical Durability		hours	500

[a] *Tested on MEA at 1 atm O_2 or H_2 at nominal stack operating temperature*

Similarly, many of the approaches to improve membrane durability can cause a decrease in performance. One of the common approaches to improve membrane mechanical durability is to incorporate a reinforcing or support layer into it, such as expanded poly [tetrafluoroethylene] (ePTFE) [12,13]. While supported PEMs have led to longer life in accelerated humidity cycling stress tests, the addition of a non-conductive support increases the proton transport resistance of the composite membranes. Similar concerns about performance loss exist for other approaches that use composite membranes to increase membrane mechanical stability. Such approaches include the use of two- and three-dimensional porous polymer support structures to reduce swelling in the plane of the membrane [14], blending of polyelectrolytes with elastomers such as polyvinylidene difluoride (PVDF) [15–18], and nano-capillary network membranes where a mat of electrospun ionomer fibers are embedded with an inert stiff polymer [19]. In these cases, the non-conductive component invariably causes an increase of the proton transport resistance across the membrane.

Several approaches have been developed in order to prevent chemical degradation of PEMs by attack from radical species. Significant work has been done on PFSA ionomers to eliminate the presence of non-fluorinated end groups, which have been reported to be initiation sites for polymer chemical degradation [20]. While elimination of such sites has led to improved lifetimes in *ex-situ* Fenton's degradation tests, this approach alone has not enabled PFSA membranes to meet *in-situ* durability targets. The most effective approaches utilize stabilization additives [21–23]. These additives may be metal ions, such as Ce^{3+} or Mn^{2+}, and are typically incorporated into the membrane either as salts or oxides. However, such ions can bind to the cation exchange sites in the membrane or the ionomer in the electrodes, thus lowering the proton conductivity of these respective layers. These cationic additives can also migrate out of the membrane into the electrodes leading to accelerated voltage degradation. In such systems, trade-off studies must be done to find the optimum concentration of stabilization additives to sufficiently stabilize the PEM without significant performance loss.

Two relatively simple approaches can be applied to reduce the risk of membrane shorting caused by local over-compression. Fuel cell stacks are normally compressed at up to 1.5 MPa in order to minimize the electronic contact resistances in the cells, particularly between the bipolar plates and the GDL. While lower stack compression can significantly reduce the risk of shorting, the increased associated contact resistances lead to unacceptable performance losses. Another approach is to utilize a thick microporous layer (MPL) between the GDL and the electrodes. This layer can cushion any topographical irregularities of the GDL that lead to local high compression. However, addition of a thick MPL adds to both the electronic and gas transport resistances in the cell, thus creating undesirable performance losses. Shorting can also be mitigated by careful fuel cell system control to prevent voltage

TABLE 2.2 US Department of Energy Membrane Chemical Durability Test Protocol and Metrics (test using an MEA)

Test condition	Steady state OCV, single cell 25–50 cm^2	
Total time	500 h	
Temperature	90°C	
Relative humidity	Anode/Cathode 30/30% (inlet)	
Fuel/Oxidant	Hydrogen/Air at stoichs of 10/10 at 0.2 A/cm^2 equivalent flow	
Pressure, inlet kPa abs (bara)	Anode 150 (1.5), Cathode 150 (1.5)	
METRIC	FREQUENCY	TARGET
F^- release or equivalent for non-fluorine membranes	At least every 24 h	No target – for monitoring
Hydrogen crossover	Every 24 h	$\leq$2 mA/cm^2
OCV	Continuous	$\leq$20% loss in OCV
High-frequency resistance	Every 24 h at 0.2 A/cm^2	No target – for monitoring
Shorting resistance	Every 24 h	>1,000 $\Omega \cdot cm^2$

[a]*Crossover current per USFCC 'Single Cell Test Protocol' Section A3-2, electrochemical hydrogen crossover method [25]*

spikes or cell reversal. In order to develop appropriate mitigation strategies, it is critical to understand the underlying mechanisms by which membranes fail, as well as develop accelerated tests by which mitigation strategies can be readily evaluated.

1.3. Accelerated Durability Testing and Failure Analysis

The US DOE has published updated recommended *in-situ* test protocols for membrane mechanical and chemical durability [24]. These protocols are given in Table 2.2 and Table 2.3, respectively. The accelerated chemical degradation protocol (Table 2.2) involves holding a single cell at open circuit voltage (OCV) for several hundred hours at 30% RH and 90°C. The health of the membranes is monitored by periodically checking for H_2 crossover and membrane shorting resistance. Degradation of fluorinated PEMs can be monitored in real time by measuring the fluoride release rate (FRR) of the cell's effluent water. General Motors has developed a more stressful accelerated chemical degradation protocol, which will be discussed in Section 2, because

TABLE 2.3 US Department of Energy Membrane Mechanical Durability Test Protocol and Metrics (test using an MEA)

Cycle	Cycle 0% RH (2 min) to 90°C dew point (2 min), single cell 25–50 cm^2	
Total time	Until crossover >2 mA/cm^2 or 20,000 cycles	
Temperature	80°C	
Relative humidity	Cycle from 0% RH (2 min) to 90°C dew point (2 min)	
Fuel/Oxidant	Air/Air at 2 SLPM on both sides	
Pressure	Ambient or no back-pressure	
METRIC	FREQUENCY	TARGET
Hydrogen Crossover[a]	Every 24 h	≤2 mA/cm^2
Shorting resistance	Every 24 h	>1,000 Ω·cm^2

[a] *Crossover current per USFCC 'Single Cell Test Protocol' Section A3-2, electrochemical hydrogen crossover method [25]*

the DOE protocol cannot effectively discriminate PEMs incorporating novel, relatively effective mitigation strategies. The accelerated mechanical degradation protocol (Table 2.3) involves cycling a single cell between dry and liquid-contacting conditions at 80°C in the absence of reactive gases or an electric potential. In this test, which was adapted from the General Motors mechanical-durability humidity-cycling test protocol [26], the health of the membranes is monitored by periodically checking for a physical leak and measuring the membrane shorting resistance. The DOE has not developed a test for accelerated membrane shorting, but the General Motors recommended test will be discussed in Section 5.

For any accelerated durability test, it is essential that the test accelerates the actual failure mode that occurs in real systems and does not generate different, yet potentially irrelevant, failure modes. Careful postmortem analysis is required to validate the exact mode of membrane failure. Figures 2.1 and 2.2 show membrane cross-sections of microtomed samples illustrating examples of membrane failure by the three primary failure modes discussed herein. Figure 2.1 shows scanning electron micrograph (SEM) images whereas Fig. 2.2 shows optical micrographs. Figures 2.1(a) and 2.1(b) show cross-sections of new MEAs made with commercially available PFSA membranes. Figure 2.1(a) is a DuPont Nafion® NRE-211 SEM cross-section showing the 25 μm thick homogeneous membrane between the electrode layers. Figure 2.1(b) is a Primea® Series 57 MEA from WL Gore™, where the three-layer composite

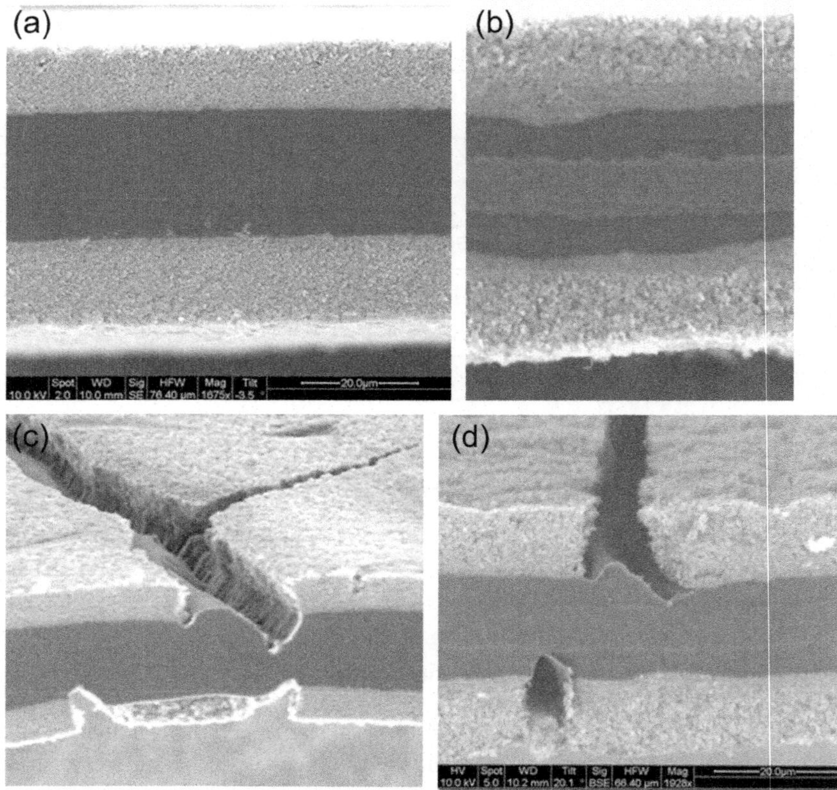

FIGURE 2.1 Scanning electron micrographs of MEA cross-sections: (a) new DuPont Nafion® NRE-211; (b) new Gore Primea® Series 57; (c) DuPont Nafion® NRE-211 after several thousand RH cycles; (d) Gore Primea® Series 57 after several thousand RH cycles.

membrane consists of two layers of neat PFSA ionomer on either side of an ePTFE support filled with ionomer (here after called the support layer). These images are typical of virgin, defect-free membranes. Figures 2.1(c) and 2.1(d) are SEM cross-sections of NRE-211 and Series 57 MEAs, respectively, after being subject to several thousand mechanical stress cycles using the DOE accelerated humidity cycling protocol. For both membranes, cracks in the membrane coincide with cracks in the electrode layers, and there is evidence of membrane creep into those electrode cracks that leads to crazing and fracture of the membrane adjacent to the electrode crack edge. For the reinforced Series 57, the crack appears to be arrested by the support layer. These cross-sections are typical of the onset of a purely mechanical fatigue-driven membrane failure. Figure 2.2(a) is an optical micrograph of a Series 57 MEA from a stack that underwent an automotive drive cycle durability test. Here the crack is seen

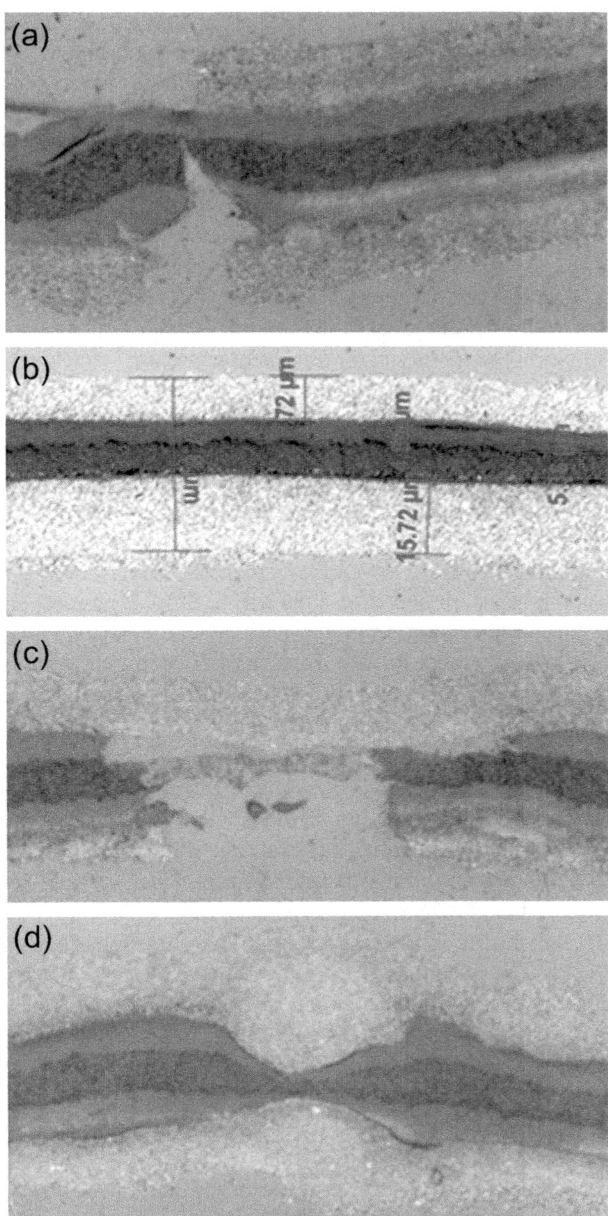

FIGURE 2.2 Optical micrographs of Gore Primea® Series 57 MEAs showing: (a) mechanical failure; (b) global chemical failure; (c) local chemical failure; (d) potential shorting failure.

extending through the support layer, suggesting that an ePTFE support is insufficient to prevent mechanical membrane failure leading to gas crossover.

Figures 2.2(b) and 2.2(c) show evidence of membrane chemical degradation of series 57 MEAs after automotive stack durability testing. Figure 2.2(b) shows that the neat ionomer layer on the cathode side of (below) the support layer has completely disappeared. Chemical degradation often manifests itself as membrane thinning. In this case, where a multilayer composite membrane is used, it is possible to identify from which side of the membrane degradation began. In Fig. 2.2(c), one can see a local region where the ionomer across the entire thickness of the membrane has degraded. All that remains is the inert ePTFE support from wherein most of the ionomer has disappeared and through which reactant gases can readily pass. It is quite common to see local regions of chemical degradation where significant areas of electrode (in this case the cathode) are missing. Finally, Fig. 2.2(d) shows an optical micrograph of a cross-section of a Series 57 MEA, highlighting a scenario than can lead to membrane shorting. In this image catalyst agglomerates or nuggets in both electrodes are aligned so that the membrane thickness has decreased to about 3 μm. Not much compression would be needed for complete membrane penetration and direct contact between the electrode layers, leading to a direct short when a potential is applied across the MEA. Similar phenomena can also occur when a carbon fiber from the GDL penetrates the MEA. Once an actual short occurs, local overheating follows very quickly, melting the membrane and rapidly damaging the MEA beyond the point where the short initiation can be distinguished. Thus, direct evidence of the root cause of a short after failure occurs is not presented.

Different types of membranes will behave differently under the stresses experienced during fuel cell operation. For example, hydrocarbon-based PEMs such as sulfonated polyphenylene (S-Parmax) and sulfonated poly-ether-ether-ketone (SPEEK) tend to exhibit a brittle fracture when subject to humidity cycling [27], rather than the ductile failure observed for PFSA membranes (Figs. 2.1c and 2.1d). This brittle failure leads to significantly shorter lifetimes in RH cycling tests [28]. The accelerated tests described above were developed for PFSA membranes, and validation of their relevance was based on comparing failure modes from the accelerated tests and actual fuel cell stacks and systems. Unfortunately, actual system data and postmortem analysis is not readily available for novel PEMs, and membrane developers must assume that the accelerated tests developed for PFSA membranes are appropriate for these new materials.

The existing ASTs serve a useful purpose in evaluating the effectiveness of mitigation strategies in preventing specific membrane failure modes. These tests are not mature enough to be used for predicting membrane life in field applications. This is mainly due to the limited amount of available field lifetime data, the difficulty in isolating the damage caused by the individual stressors in field tests, and the fact that any acceleration factors determined on the basis of the small set of materials for which there is field lifetime data are not likely to apply for novel materials. Significant further study beyond accelerated test results is required to enable lifetime predictions. For example, a fuel cell system will likely not spend significant time at OCV conditions. Thus any predictive model for membrane chemical degradation must consider the effect of cell potential on PEM decay rate. Similarly, although an actual system will not experience many RH cycles between completely wet and dry conditions, the dependence of both the rate and magnitude of the RH swing on membrane damage must be incorporated in any mechanical durability lifetime prediction method. These issues, and others regarding membrane life prediction, will be discussed in more detail below.

The following sections of this chapter will focus of the individual failure modes of chemical degradation, mechanical degradation and membrane shorting. The theories and experimental approaches discussed are primarily based on PFSA membranes, although perspective will be included on their application toward non-fluorinated and partially fluorinated PEMs. There is also a section dedicated to the impact of simultaneous exposure to chemical and mechanical membrane stressors and the effectiveness of combining chemical and mechanical stabilization mitigation strategies.

2. CHEMICAL DEGRADATION

2.1. Background

From the earliest applications of PEM fuel cells in the 1960s, chemical degradation of membranes has been recognized as a primary life limiting process. The first proton exchange membranes, based on sulfonated polystyrenes, offered constant current lifetimes of no more than a few hundred hours before developing cracks and pinholes leading to gas crossover and shorting failures [29,30]. This membrane chemical degradation is attributed to the action of aggressive radical species that form during fuel cell operation and attack the abundance of kinetically vulnerable bonds of the polystyrene structure. By contrast, PFSA membranes are widely used today due to their relatively high chemical stability. The much improved stability of PFSA materials is attributed to the chemically inert Teflon®-like fluorocarbon structure. Although PFSA membranes are much more stable than the early sulfonated polystyrenes, these perfluorinated materials are by no means chemically inert during fuel cell operation, especially when subjected to strenuous duty cycles involving large amplitude humidity and voltage cycling. This section will address the details of PFSA chemical degradation and mitigation strategies from a chemically rigorous, mechanistic perspective, providing insights into this long-standing obstacle to the widespread use and commercialization of this technology. While this section does not specifically address chemical degradation of the polyelectrolyte in the electrode layers, the fundamentals governing membrane ionomer degradation and the mitigation strategies to prevent it can also be applied to the electrode layers. It should be noted that there is no conclusive evidence for chemical degradation of the ionomer in the electrode layers during fuel cell operation.

Chemical degradation of PFSA membranes can be characterized by membrane thinning and the emission of HF, CO_2 and H_2SO_4 in the fuel cell exhaust streams. The thinning leads to increased rates of gas crossover and mechanical weakening that ultimately lead to cell failure. The rate of chemical degradation can be conveniently monitored *in-situ* by quantifying the fluoride emission using a variety of analytical techniques or via *in-situ* electrochemical monitoring of hydrogen gas crossover. The degree of chemical degradation is often expressed in terms of total membrane fluoride inventory loss, although this value alone does not always provide a meaningful indication of membrane health. For example, if the region of degradation is highly localized, MEAs can fail at very low inventory losses (ca. 1%) [31]. On the other hand, MEAs undergoing near uniform degradation can tolerate losses of 50% or greater before failure. Over the past decade, many investigations have contributed greatly to understanding the factors that influence the rate of chemical degradation of PFSA membranes. Accelerating factors include high temperature, low relative humidity of incoming gas streams [32,33], low electrochemical load [31,34] and high reactant gas (both H_2 and O_2) pressures [32,33]. Because the requisite lifetimes of PFSA membranes are many thousands of hours, tests of chemical stability necessarily are performed under accelerated conditions. Therefore, investigations of pure chemical degradation often consist of OCV tests performed at high temperature and low humidity. High temperature OCV testing causes largely uniform degradation conditions across the MEA active area and accordingly leads to near uniform thinning of the membrane [31]. Unmitigated 25 μm PFSA PEMs can withstand little more than 200 hours of DOE specified OCV testing (Table 2.2) before cell failure occurs [31].

Figure 2.3 shows the voltage degradation and fluoride release rate profiles of a Nafion® 1000EW based 20 μm membrane run at OCV conditions at 95°C, and 50% RH of the air and hydrogen feed streams. An MEA was constructed with standard Pt/C electrodes (0.4 mg/cm^2 Pt) on both the anode and cathode. Over the course of the 170 hour test, the voltage dropped by about 75 mV (ca. 450 μV/h) and the FRR accelerated from an initial value of 2×10^{-8} gF/h·cm^2 to 4×10^{-6} gF/h·cm^2. At the end of the test, the membrane had lost a calculated 20% of the original fluoride inventory. The rapid increase in the fluoride release rate during the test is consistently observed when unmitigated PFSA MEAs are tested in this manner [35] and provides valuable information about the degradation mechanism.

It is generally accepted that the chemical degradation of PFSA membranes is caused by the action of aggressive, highly oxidative species

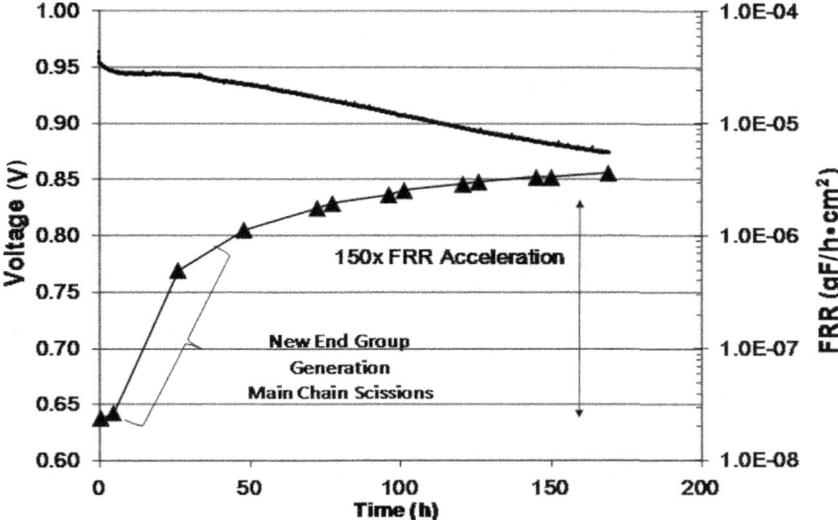

FIGURE 2.3 50 cm^2 OCV durability test of cast 20 μm Nafion® 1000EW membrane: H_2/Air, 95°C, 50% RH inlets, 50 kPag. Solid line is voltage, triangles are fluoride release rate.

generated during fuel cell operation [35]. These reactive species include hydroxyl radical (•OH), hydroperoxyl radical (•OOH) and hydrogen peroxide (H_2O_2) and, indeed all of these species have been detected directly or indirectly in operating fuel cells [36,37]. The electrochemical and thermodynamic properties of these species, along with those of other PFSA fuel cell related materials, are summarized in Tables 2.4 and 2.5 [38]. The data in the tables clearly show that hydroxyl radical is by far the most reactive of the three oxidizing species, possessing a very high propensity to abstract hydrogen atoms from O-H and C-H bonds and form the water molecule. This high activity is thermodynamically driven by the formation of the very

TABLE 2.4 Reduction Potentials of Reactive Oxygen Species

Half Cell Reaction	Reduction Potential at pH = 0 (SHE)
$\cdot OH + H^+ + e^- \rightarrow H_2O$	2.59
$\cdot H + H^+ + e^- \rightarrow H_2$	2.30
$\cdot OOH + H^+ + e^- \rightarrow H_2O_2$	1.48
$H_2O_2 + 2H^+ + 2e^- \rightarrow 2H_2O$	1.74

TABLE 2.5 Experimental Bond Enthalpies (ΔH_{298}) of Species Involved in PFSA Fuel Cell Degradation

Bond	kJ/mol	Bond	kJ/mol
H_3C-H	439.3	HO-H	497.1
F_3C-F	546.4	HOO-H	369.0
F_3C-H	449.4	F-H	570.7
RCO_2-H	442.7	HO-OH	210.9
H-H	436.0	HO-F	215.1

strong (497 kJ/mol) O-H bond. It is worth noting that •OOH, while present in operating fuel cells, is not capable of abstracting hydrogen atoms from fuel cell degradation intermediates and, while it may participate in chain termination events, it can be considered benign from an initiation and propagation perspective [35]. Hydrogen peroxide is certainly an active participant in fuel cell degradation processes, as will be discussed in subsequent sections. However, acting alone in solution, it is not capable of damaging PFSAs.

A mechanism for the decomposition of PFSA membranes (Eqns 2.1–2.3) was proposed several years ago by Curtin *et al.* [20]. The decomposition process begins with hydrogen atom abstraction from a perfluorocarboxylic acid group by •OH to form the perfluorocarbon radical, CO_2 and H_2O. The fluorocarbon radical is trapped as the fluoroalcohol, which loses two equivalents of HF to ultimately become another carboxylic acid shortened by one CF_2 unit. This degradation mechanism, which is thermodynamically driven by the formation of very strong H-F bonds (Table 2.5), can continue many times over in a so-called unzipping process, wherein the average molecular weight of the polymer decreases with the equivalent weight remaining constant as the degradation proceeds from the end groups. Note that this mechanism assumes the presence of vulnerable end groups and provides no means to create new end groups, implying that the rate of degradation remains relatively constant. As shown in Fig. 2.3, OCV chemical degradation rates can accelerate with time, a phenomenon not addressed by the Curtin scheme. Note also that this scheme implies that a perfect PFSA polymer with no vulnerable end groups would be quite stable to OCV conditions. This, too, is not consistent with fuel cell results that have shown poor correlation between degradation rates and the number of vulnerable end groups present in fresh MEAs [39]. Taken together, these findings suggest that end group creating, chain scission processes contribute to chemical degradation during fuel cell operation. Chain scission processes will be discussed in detail in a subsequent section.

$$R_fCF_2CF_2CO_2H + \bullet OH \rightarrow R_fCF_2CF_2\bullet + H_2O + CO_2 \quad (2.1)$$

$$R_fCF_2CF_2\bullet + \bullet OH \rightarrow R_fCF_2CF_2OH \rightarrow R_fCF_2COF + HF \quad (2.2)$$

$$R_fCF_2COF + H_2O \rightarrow R_fCF_2CO_2H + HF \quad (2.3)$$

2.2. Initiation: Oxidants

As mentioned above, •OH is the only oxidant that is capable of abstracting vulnerable, but thermodynamically stable, O-H bonds of perfluorocarboxylic acids present in chemically degrading PFSA PEM fuel cells. The sources of •OH and its major reaction paths within the fuel cell environment are therefore of paramount importance for understanding chemical degradation, and for developing mitigation strategies. Previous studies indicate that there are several sources of the hydroxyl radical, including both chemical and electrochemical paths. The chemical pathways of •OH formation involve crossover gases interacting with a platinum catalyst to form the radical directly. Single sided MEA studies have shown that this process can occur at either the anode or cathode with comparable degradation rates [40,41]. Chemical formation of •OH is also thought to occur within the membrane through reaction of crossover gases with nanocrystals of Pt deposited within the membrane [42]. The Pt in the membrane stems from dissolution of Pt in the cathode and varies in density and position depending on gas compositions and partial pressures [43,44]. The yield of •OH from membrane-deposited Pt may actually be enhanced in regions of low Pt density compared to the high density regions associated with the so called Pt line often observed in degraded membranes [42]. Recent *in-situ* ESR spin trapping studies indicate that •OH radical is also formed on the Pt black catalyst surface under closed circuit conditions [36]. Hydroxyl radical can also be formed from H_2O_2 by reaction with a Fenton active metal like Fe^{2+} or with membrane deposited Pt. Hydrogen peroxide can be formed either chemically via crossover gases or electrochemically at potentials less than 0.6V [37]. Given the multiple potential sources of •OH both at the electrode surfaces and within the membrane, the fuel cell chemistry of •OH must be addressed.

As mentioned above, hydroxyl radicals have a high propensity to abstract hydrogen atoms wherever they are found. In a fuel cell, the two most abundant sources of abstractable hydrogen atoms are hydrogen gas and hydrogen peroxide. Hydroxyl radicals are known to undergo facile reactions with both of these hydrogen atom donors at 95°C to produce H_2O and radicals, as shown in Eqns 2.4 and 2.5 [45,46]. Together, these two reactions will limit the average lifetime of hydroxyl radicals in a fuel cell. The rate constants of reaction, when used in combination with reasonable estimates of

TABLE 2.6 Kinetic Parameters of H_2 and H_2O_2 at 95°C

Reactant	Rate Constant ($M^{-1}s^{-1}$)	Aqueous Conc. (M)	k_{obs} (s^{-1})
H_2	2×10^8	8×10^{-4}	1.6×10^5
H_2O_2	1×10^8	3×10^{-4}	3×10^4

aqueous concentrations of H_2 [45] and H_2O_2 [37] and the diffusion coefficient of •OH (approximated by that of H_2O, 2.5×10^{-5} cm^2s^{-1}), can be used to estimate the average lifetime and diffusion distance of •OH in an operating fuel cell. Using the kinetic estimates in Table 2.6, the apparent pseudo first order rate constant for •OH reactions is approximately 2×10^5 s^{-1}, giving an average lifetime of 5 μs and an average diffusion distance given by $(D/k)^{1/2}$ of 0.1 μm. Note that the average •OH lifetime is dominated by its reaction with H_2 gas, and therefore this value will be a strong function of the concentration of H_2 under various conditions. For example, the H_2 concentration used in Table 2.6 is derived from aqueous solution values at 95°C and 100 kPa, [45] which is a reasonable approximation of a fully hydrated membrane. If, however, the H_2 concentration, like permeability, decreases under dry conditions, •OH lifetime will increase and likely lead to greater membrane damage.

$$\bullet OH + H_2 \rightarrow \bullet H + H_2O \quad (2.4)$$

$$\bullet OH + H_2O_2 \rightarrow \bullet OOH + H_2O \quad (2.5)$$

$$\bullet H + O_2 \rightarrow \bullet OOH \quad (2.6)$$

The dominant reaction of •OH with H_2 deserves further comment, because the radical product of this reaction, •H, also has high oxidizing power [47,48], as shown in Table 2.4. Hydrogen abstraction reactions by hydrogen atoms form H_2 gas, which has a bond strength of 436 kJ/mol. The H-H bond formed is weaker than the abstractable hydrogen atoms of PFSAs and, as a result, such abstraction reactions are not kinetically significant. Hydrogen radicals, on the other hand, do have a strong thermodynamic driving force to abstract fluorine atoms, owing to the extremely high bond strength of HF (571 kJ/mol). Despite the strong driving force for these reactions, density functional theory calculations [35] indicate that these reactions are slow compared to the diffusion controlled reaction with oxygen (Eqn 2.6) to form •OOH [48]. Thus, the two fastest reactions available to •OH in an operating fuel cell (H_2 and H_2O_2) both ultimately produce the relatively benign •OOH radical. Therefore, chemical degradation of PFSA membranes occurs from that fraction of •OH that escapes deactivation by H_2 and H_2O_2.

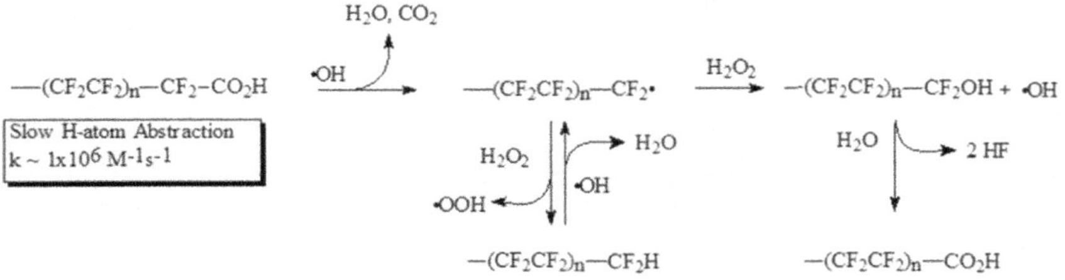

H_2O, CO_2

$-(CF_2CF_2)_n-CF_2-CO_2H \xrightarrow{\bullet OH} -(CF_2CF_2)_n-CF_2\bullet \xrightarrow{H_2O_2} -(CF_2CF_2)_n-CF_2OH + \bullet OH$

Slow H-atom Abstraction
k ~ 1x10^6 M^{-1}s^{-1}

H_2O_2 H_2O $\bullet OOH$ $\bullet OH$ H_2O 2 HF

$-(CF_2CF_2)_n-CF_2H$ $-(CF_2CF_2)_n-CO_2H$

SCHEME 2.1 Modified Curtin Mechanism.

2.3. End Chain Degradation Pathways

In the Curtin degradation scheme, the fluorocarbon radical intermediate is trapped by a hydroxyl radical to form the fluoroalcohol, which then further degrades. While there is little doubt concerning the formation of the fluoroalcohol, this pathway as depicted is very unlikely, as it requires the combination of two highly reactive species present in very low concentrations. Rather than combine with the ephemeral hydroxyl radical, the almost equally reactive fluorocarbon radical will more likely react with a high concentration hydrogen donor, like H_2O_2 or H_2, to form the hydrofluorocarbon at a near-diffusion-controlled rate [49]. Thus, rather than direct conversion to the fluoroalcohol, the fluorocarboxylic acid will be largely converted to the hydrofluorocarbon, which is not an end product, since it is converted back to the fluorocarbon radical via reaction with •OH. This modified Curtin mechanism is shown in Scheme 2.1. Table 2.7 shows the best estimate rate constants for the hydrogen abstraction reactions of the carboxylic acid and hydrofluorocarbon at 25°C, along with other relevant fuel cell reactions at this temperature. It is worth noting that, of the two weak end groups, the carboxylic acid is by far the least reactive. Regarding the

TABLE 2.7 Rate Constant Data (25°C) for Fuel Cell Reactions

Reaction	Rate Constant ($M^{-1}s^{-1}$)
$\bullet OH + H_2O_2 \rightarrow \bullet OOH + H_2O$	2.7×10^7
$\bullet OH + H_2 \rightarrow \bullet H + H_2O$	4.3×10^7
$\bullet OH + R_fCF_2CO_2H \rightarrow H_2O + CO_2 + R_fCF_2\bullet$	$<1.0 \times 10^6$
$\bullet OH + R_fCF_2H \rightarrow H_2O + R_fCF_2\bullet$	$<1.0 \times 10^7$
$\bullet OH + Ce^{3+} + H^+ + \rightarrow H_2O + Ce^{4+}$	3×10^8
$\bullet OH + Mn^{2+} + H^+ + \rightarrow H_2O + Mn^{3+}$	4×10^7

pathway from the fluororadical to fluoroalcohol, a new reaction has been proposed involving a second mode of reaction with hydrogen peroxide wherein the weak O-O bond (211 kJ/mol, Table 2.5) is cleaved. In addition to the fluoroalcohol, an equivalent of hydroxyl radical is generated. Based on experimental heats of formation for $\bullet CF_3$, this reaction is highly exothermic (−284 kJ/mol) and can be rationalized by the formation of a strong C-O bond (ca. 470 kJ/mol) while cleaving a weak O-O bond. While the O-O cleavage reaction is considerably slower than the H_2O_2 hydrogen transfer detour reaction, it is for more likely than the radical recombination and it serves as the siphon for the fluororadical to chain shortening degradation. Furthermore, this reaction provides another source of •OH which, importantly, is independent of a Pt catalytic surface.

2.4. Acid Site Degradation Pathways

As mentioned in Section 2.1, it has long been recognized that PEM fuel cell operation at low relative humidities and high potentials often leads to high rates of chemical degradation. For example, OCV operation can produce rapidly accelerating fluoride release rates that can be 100-fold higher than those obtained from steady state constant current operation [31]. A similar impact of low humidity operation has been observed in *ex-situ* hydrogen peroxide vapor tests of PFSA membranes. In these studies, membranes subjected to hydrogen peroxide vapor exhibit high rates of fluoride release, along with increasing carboxylic acid end group concentration and decreasing molecular weight with time, all of which provide strong evidence for main chain scission reactions [50]. High rates of PFSA degradation are observed in the vapor tests even without Fe^{2+} doping. The chemical degradation behaviors observed in the vapor tests are in stark contrast to results obtained with non-Fe^{2+} doped PFSA membranes treated with aqueous 30% H_2O_2 solution at 80°C, where negligible levels of degradation occur over many days of treatment [51]. Several hypotheses have been proposed to account for the accelerated degradation under low humidity conditions, including increased concentration of H_2O_2, owing to both its low volatility and enhanced electrochemical yield [52], increased activity of H_2O_2 [51], and increased •OH activity and lifetime. Another hypothesis involving changes in the sulfonic acid group ionization at low relative humidities has also been proposed [35].

PFSAs are sometimes referred to as super acids because of their extremely low pKa values which have been estimated to be near −6 for Nafion®[53]. In a fully humidified PFSA membrane, the sulfonic acid groups are virtually 100% deprotonated, with the acidic proton residing on water molecules, forming H_3O^+, the proposed proton carrier in a hydrated PEM fuel cell. Thus, under wet conditions, the sulfonic acid group exists in the rather chemically inert sulfonate salt form -SO_3^-. As the level of hydration decreases, the pKa value of the sulfonic acid will necessarily increase, due to the diminished solvation of the charged products of ionization. In the dry limit, the acidic proton must reside on the sulfonic acid group as -SO_3H. At 20% RH, PFSA membranes are hydrated to the extent of about two water molecules per SO_3H group (λ=2) [54]. Modeling studies of PFSA ionization indicate that at least three water molecules per SO_3H group are required to support significant ionization [55,56]. It is therefore reasonable to conclude that, as the membrane becomes drier, the proton will tend to increase its residence time on the sulfonic acid group. Even at these low hydration levels, the acid will be largely deprotonated, but the fraction of associated acid will increase by several orders of magnitude relative to the fully hydrated case. As discussed below, the location of the proton has a significant impact on the degradation chemistry of the dry membrane of a fuel cell.

When the acidic proton resides on the -SO_3 group as -SO_3H, it is susceptible to abstraction by •OH to form the -SO_3• radical [35]. Again, this abstraction is thermodynamically driven by the formation of the strong O-H bond of water. By analogy to the carboxyl radical, $R_fCF_2CO_2$•, the perfluorosulfonyl radical, $R_fCF_2SO_3$• can fragment to give SO_3 and a fluoroalkyl radical (Eqns 2.7 and 2.8). Density functional calculations show that, unlike the carboxyl radical fragmentation which is exothermic (ca. −40 kJ/mol) and virtually instantaneous, the perfluorosulfonyl radical fragmentation is slightly endothermic [35]. Given the small calculated fragmentation energy (ca. 40 kJ/mol), the rupture of the C-S bond of the perfluorosulfonyl radical is anticipated to occur on the sub-microsecond time scale at 95°C. Significantly, the perfluorosulfonyl radical is unique among other sulfonyl radicals with respect to its facile fragmentation. As shown in Table 2.8, alkyl and aryl sulfonyl radicals have stronger C-S bonds and are calculated to fragment 10^7 and 10^{14} times slower that the perfluorosulfonyl radical, respectively. The increased fragmentation energies of the alkyl and aryl sulfonyl radicals indicate that the sulfonyl radicals, once formed, will abstract hydrogen from good donors such as hydrogen peroxide and return to the sulfonic acid faster than the C-S bond ruptures. The weakness of the C-S bond of perfluorosulfonates is a general phenomenon [35], and is

TABLE 2.8 Estimated Fragmentation Rates of Sulfonyl Radicals[a]

Radical	M05-2X/6-31G(d) Calc Frag Energy (kJ/mol)	Rounded Fragmentation Energy (kJ/mol)	Calculated Fragmentation 368 K Rate (s^{-1})	Relative Rate
CF_3SO_3•	43	45	$5e10^6$	1
CH_3SO_3•	91	90	$2e^0$	$5e10^{-7}$
$C_6H_5SO_3$•	145	140	$1e10^{-7}$	$2e10^{-14}$

[a]*Data taken from reference [36]*

a consequence of the low electron density of the C-S bond resulting from the presence of the strong electron withdrawing effects on the carbon of the fluorine atoms as well as the SO_3 group itself, which is also strongly electron withdrawing. Thus, the strong electron withdrawing groups responsible for the desired high acidity of perfluorosulfonic acids are also responsible for the fragility of the C-S bond in $R_fCF_2SO_3H$.

$$R_fCF_2CO_2\bullet \rightarrow R_fCF_2\bullet + CO_2 \quad E_a = 0\ \mathrm{kJ/mol} \tag{2.7}$$

$$R_fCF_2SO_3\bullet \rightarrow R_fCF_2\bullet + SO_3 \quad E_a = 40\ \mathrm{kJ/mol} \tag{2.8}$$

A second pathway by which the $R_fCF_2SO_3\bullet$ side chain radical can be formed under dry conditions involves a reaction between the protonated acid and hydrogen peroxide to form bissulfonyl peroxide as shown in Eqns 2.9 and 2.10. Bissulfonyl peroxides of perfluorosulfonic acids are unstable compounds that rapidly decompose via a radical mechanism to give SO_3 and radical recombination products [57]. Importantly, the reaction between hydrogen peroxide and the sulfonic acid can only occur with the protonated form and therefore can occur only under very dry conditions such as found during OCV tests and hydrogen peroxide vapor tests.

$$2R_fCF_2SO_3H + H_2O_2 \rightarrow (R_fCF_2SO_3)_2 + 2H_2O \tag{2.9}$$

$$(R_fCF_2SO_3)_2 \rightarrow 2RCF_2\bullet + 2SO_3 \tag{2.10}$$

The side chain perfluorocarbon radical, which has been detected by ESR spectroscopy [58], can further degrade via the unzipping mechanism up to the main chain whereupon chain scission occurs, producing two new ends groups. Thus, the side chain $R_fCF_2SO_3$ radical pathway is a specific and highly probable mechanism of PFSA main chain scission. It is important to note that only a relatively small number of these side chain radicals are required to generate additional chain breaks that lead to the degradation rates observed during fuel cell chemical degradation tests [39,50]. Furthermore, by this mechanism, the side chain damage can be initiated only during very dry conditions, whereas the damage can be propagated via the unzipping process during relatively wet conditions. After great efforts to prevent chemical degradation through creating a pure PFSA, it is rather ironic that the inherent weakness or Achilles heel of PFSA ionomers could potentially be based on the highly acidic proton conducting group itself.

The sulfonyl radical initiation mechanism described above is consistent with a number of other experimental observations. First, this mechanism indirectly provides a rationale for the inert nature of perfluorosulfonic acid model compounds in the Fenton's solution test [59,60]. Since the model compounds are dissolved in aqueous solution, the sulfonic acid groups are fully ionized and thus unable to form a sulfonyl radical via reaction with either hydroxyl radicals or hydrogen peroxide. Fenton and others have reported that PFSA membranes in which all protons are exchanged for alkali metal ions are stable in a fuel cell running at OCV conditions [41]. Presumably, even though these membranes cannot conduct protons, the reactive species (hydroxyl radical/hydrogen peroxide) are still generated, yet are unable to induce damage because the sulfonic acid group is deprotonated.

2.5. Mitigation of Chemical Degradation

Given the central role of •OH in the degradation of PFSA ionomers, it is a high value target for arresting the damaging chemistry. There are two main strategies for significantly reducing chemical degradation processes in a PEM fuel cell. The first, and most direct approach, involves prevention of •OH formation. Unfortunately, because some of the primary generation mechanisms of •OH involve gas diffusion through the membrane, it is unlikely that this process can be completely arrested in PFSA systems with thin membranes. Some have attempted to reduce the amount of H_2 and/or O_2 gas crossover by platinizing the membrane [61]. In fact, there have been several recent studies regarding the role of Pt from the cathode that deposits in the membrane, forming the aforementioned Pt line, on PEM chemical degradation [42,62,63]. These studies present contradicting theories as to whether Pt in the membrane contributes to membrane degradation, or if the Pt line is actually formed because of the same initiators (i.e. high cell potential, gas crossover) that can drive PEM degradation. Reactant gas crossover is also dependent on the polyelectrolyte chemistry. Thus, •OH generation could be significantly attenuated in other families of membranes that exhibit lower gas permeability, such as hydrocarbon-based materials.

The second approach requires the deactivation of •OH before it induces damage to the polymer chain via the hydrogen containing end groups. From a chemical degradation perspective, the most effective deactivation reaction of •OH involves its one-electron reduction to give water. As discussed in Section 2.3, the majority of •OH is deactivated in an operating fuel cell through its reactions with H_2 and H_2O_2 to generate the relatively benign •OOH. In order for the •OH quenching system to be effective, favorable thermodynamics of deactivation must exist but, importantly, several *kinetic* requirements must also be fulfilled. First, the deactivation chemistry must be faster than the reaction of •OH with the reactive end groups of the degrading PFSA polymer. The rate constant of hydrogen atom abstraction from a perfluorocarboxylic acid intermediate of Scheme 2.1 is estimated to be no greater than $1\times10^{6}\ M^{-1}s^{-1}$ (Table 2.7). This reaction rate constant is quite slow for the highly reactive hydroxyl radical and is accordingly the slow step in the chemical degradation process. Studies by Hommura *et al.* indicate that the level of chemical degradation of PFSA membranes correlates with the concentration of carboxylic acid groups [50], supporting the idea that hydrogen atom abstraction is the rate determining step in PFSA chemical degradation. Secondly, because fuel cell MEAs are continually bombarded with hydroxyl radicals throughout their

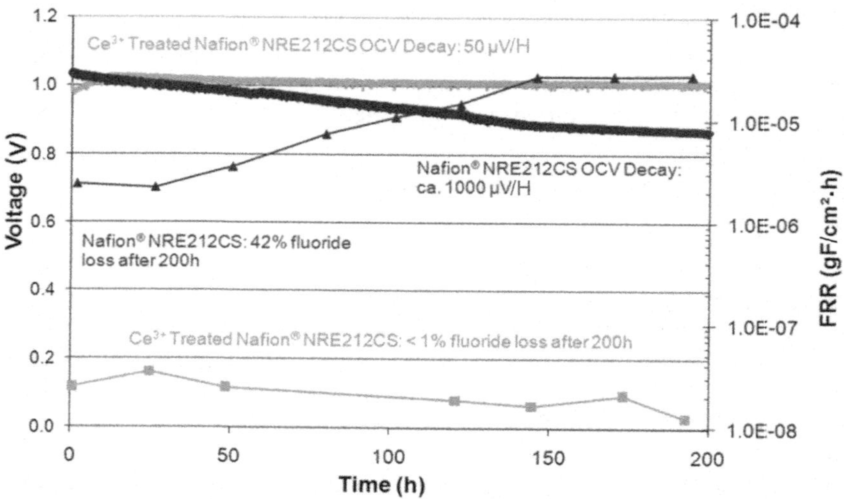

FIGURE 2.4 50 cm^2 OCV durability tests of cerium stabilized and as-received 50 μm Nafion® NRE212CS MEAs: H_2/Air, 95°C, 50% RH inlets, 50 kPag. Solid line is voltage, symbols are fluoride release rate.

operational lifetimes, any practical deactivation scheme requires catalytic rather than sacrificial •OH reduction chemistry. That is, an effective mitigation scheme requires the oxidized form of the •OH quencher be recycled to its reduced form for repeated quenching chemistry. Importantly, the recycling chemistry must also be faster than the degradation chemistry, so that there is no significant decrease in the concentration of the active form of the •OH quencher. Finally, the quenching agent must remain in the membrane for its required lifetime and not degrade significantly with time nor produce additional •OH or other highly aggressive species.

It has recently been reported that Ce^{3+}, Mn^{2+} and their metal oxides are very effective mitigants of PFSA membrane chemical degradation [21–23]. Because Ce^{3+} is highly effective at stabilizing PFSA membranes to chemical degradation, the majority of the discussion will focus on its use and mechanism. Mn^{2+} use will be discussed by analogy. The metal salts can be introduced to the MEA either by membrane doping or via the electrode ink or ionomer overcoat. In the latter two cases, EPMA studies reveal that the metal ions partition into the membrane during MEA processing [22]. The remarkable impact of Ce^{3+} mitigation is illustrated by Fig. 2.4 which compares the 95°C/50% RH OCV degradation rates of unmitigated and mitigated Nafion® NRE212CS-based MEAs. In this example, the mitigated MEA contains 0.4 mg (2.8 μmol) Ce^{3+}, which complexes about 1% of the sulfonic acid sites within the membrane. The Ce^{3+} mitigated MEA leads to a 20-fold reduction in the voltage degradation rate and a nearly 1,000-fold reduction in the FRR dropping to a very low

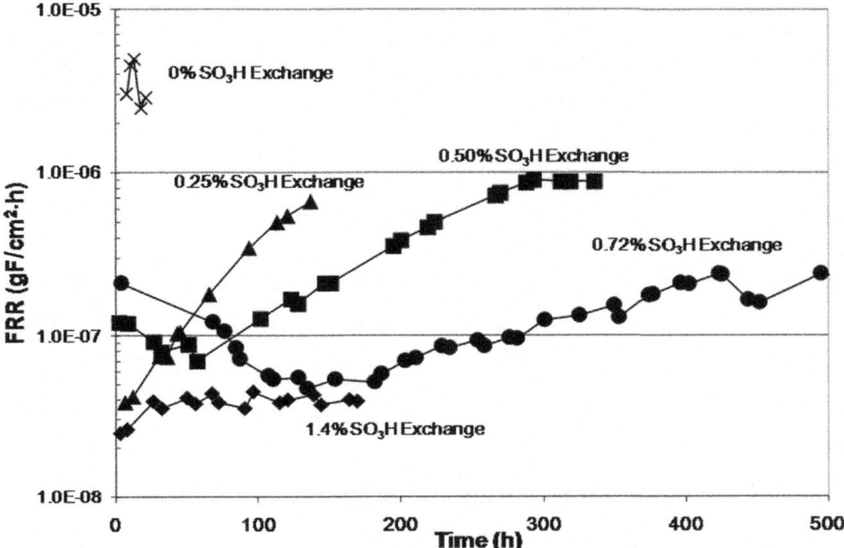

FIGURE 2.5 Durability impact of cerium loading. Fluoride release rates obtained during 50 cm^2 OCV durability testing of NRE212CS 50 cm^2 MEAs with various Ce^{3+} doping levels: H_2/Air, 95°C, 50% RH inlets, 50 kPag.

2×10^{-8} $gF/cm^2 \cdot h$. The effectiveness of Ce^{3+} mitigation is further emphasized by noting that at the end of a 200 hour OCV test, stabilized MEA lost less than 0.2% of its fluoride inventory while the unmitigated membrane thinned from an original 55 μm to 35 μm during the course of losing over 40% of its original fluoride inventory. Similarly, mitigated MEAs can withstand more than 1,000 hours of 95°C/50% RH OCV testing without thinning or membrane failure. Compared to Ce^{3+} mitigation, equivalent levels of Mn^{2+} doping are 4–5 times less effective, producing FRR values near 1×10^{-7} $gF/cm^2 \cdot h$ [22].

Ce^{3+} is a highly effective mitigant of PFSA membrane chemical degradation because it functions as an efficient, highly reversible and stable redox quencher of •OH. As shown in Table 2.7, Ce^{3+} undergoes a facile one-electron redox reaction with the hydroxyl radical to produce H_2O and the corresponding oxidized cation [22,48]. Note, importantly, that the rate constant of •OH reduction by Ce^{3+} is more than two orders of magnitude greater than the estimated rate of hydrogen atom abstraction by •OH. This rapid reaction with •OH fulfills the first crucial requirement of an effective scavenging system, as the mitigation system relies on its ability to out compete the damaging chemical reactions. The successful competition can be accomplished via combinations of high inherent quenching reactivity and quencher concentration. This relative rate concept is illustrated in Fig. 2.5, which shows the impact of Ce^{3+} doping level on FRR during OCV testing. Increasing levels of Ce^{3+} doping leads to increasingly effective •OH scavenging (decreased FRR) and improved chemical durability. The relatively low effectiveness of Mn^{2+} as a mitigant may be due to its lower rate constants of reaction with •OH radical (Table 2.7). Note that the rate constant for the reaction of Mn^{2+} with •OH is still considerably faster than the rate of hydrogen atom abstraction from carboxylic acid end groups.

As mentioned above, reduction of the hydroxyl radical constitutes only one part of the catalytic cycle needed for efficient mitigation chemistry. The oxidized cations must be reduced in order to capture additional hydroxyl radicals. Ce^{4+} is a very strong oxidizing agent as indicated by its reduction potential of 1.72 V. Indeed, Ce^{4+} oxidizes H_2O_2 to oxygen as shown in Scheme 2.2 [64–66]. In addition to the uncatalyzed hydrogen peroxide mediated reduction of Ce^{4+} in the fuel cell environment, Ce^{4+} can also be catalytically reduced by both H_2 gas and H_2O. The last two pathways of Ce^{4+} reduction require a Pt catalyst in the membrane and, although the H_2 reaction is facile, the oxygen evolution reaction is quite slow. The metal ion reduction pathways available to Ce^{4+} are crucial to the effectiveness of this mitigation system because they enable the hydroxyl radical quenching reactions to be accomplished with very small amounts of Ce^{3+}.

The effectiveness of chemical degradation stabilizers such as Ce^{3+} has rendered standard, single condition OCV accelerated degradation protocols (i.e. 90°C/30% RH, Table 2.2) inefficient because mitigated MEAs can have very long lifetimes under such conditions. Harsher tests are now required in order to screen novel, chemically robust materials more quickly, and to discriminate between mitigation options that might otherwise appear quite similar using less punishing protocols. One option is to adopt very aggressive single condition testing protocols such as 120°C/18% RH [67]. While such conditions will accelerate chemical degradation, the extremely harsh conditions can lead to very early failures and, thus, loss of useful information. An alternative approach involves employing a multistep degradation protocol with escalating degrees of aggressiveness. One such example is shown in Fig. 2.6 wherein the conditions are ramped from an initial OCV condition of 95°C/50% RH to 95°C/25% RH and finally to 110°C/25% RH. The example in

$$Ce^{4+} + H_2O_2 \underset{k_{-1}}{\overset{k_1}{\rightleftharpoons}} Ce^{3+} + \bullet OOH + H^+ \qquad k_1 = 1 \times 10^6\ M^{-1}s^{-1}$$

$$Ce^{4+} + \bullet OOH \xrightarrow{k_2} Ce^{3+} + O_2 + H^+ \qquad k_2/k_{-1} = 13 \pm 2$$

$$2Ce^{4+} + H_2 \xrightarrow{Pt} 2Ce^{3+} + 2H^+ \qquad \text{fast}$$

$$4Ce^{4+} + 2H_2O \xrightarrow{Pt} 4Ce^{3+} + 4H^+ + O_2 \qquad \text{very slow}$$

SCHEME 2.2 Fuel cell chemistry of Ce^{4+} reduction.

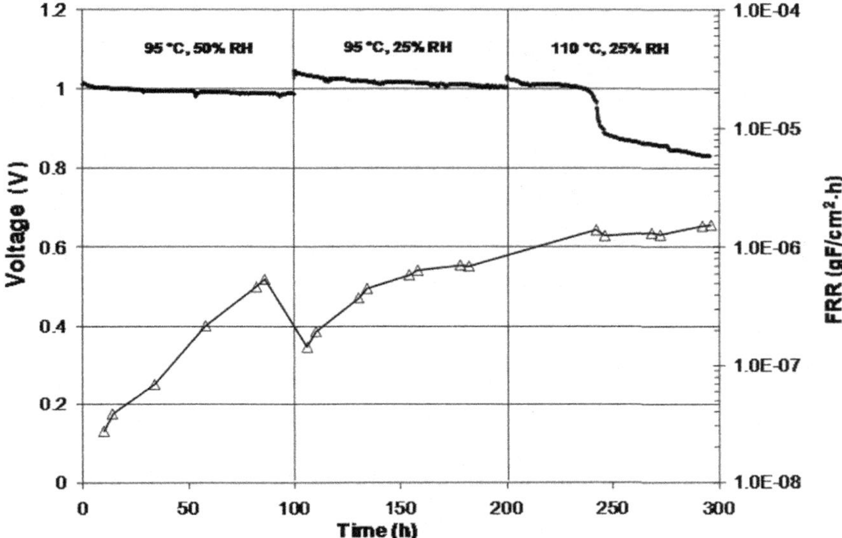

FIGURE 2.6 Escalating 50 cm^2 OCV chemical durability test. Voltage decay curve and FRR profile of a NRE212CS MEA containing a low level of Ce^{3+} mitigation (0.25 mol% SO_3H substitution) : H_2/Air, 50 kPag, variable temperature and RH.

Fig. 2.6 demonstrates the behavior of a Nafion® 212CS MEA stabilized with a low level of Ce^{3+} (ca. 0.25% SO_3H exchange). The FRR rose steadily during the 95°C/50% step and then dropped by a factor of three upon decreasing the RH to 25%. This may signify that hydrolysis of the acyl fluoride intermediate becomes partially rate limiting. Over 100 hours at 25% RH the FRR increases by 4 times. Upon changing conditions to 110°C/25% RH, the FRR increases again and the MEA fails within about 40 hours at this condition. Using this type of protocol in conjunction with voltage degradation rate, FRR and other diagnostic data such as polarization curves and electrochemical crossover measurements reveals considerable detail about the degradation behaviors of a wide variety of chemically mitigated MEAs that can be quite useful for materials optimization.

Because incorporation of metal ions into an MEA displaces acidic protons from the proton exchange membrane, there exists the potential for performance, or power density losses as metal ion load increases. Large active area stacks are excellent platforms for assessing the performance-durability tradeoffs of MEAs with different stabilization levels. Figure 2.7 shows four selected polarization curves obtained for the seven sets of MEAs that differed only in cerium loading. The membranes employed in the experiment were 25 μm thick, extruded, homogeneous PFSAs with an equivalent weight of 1,100. Extruded membranes were used in order to reduce the risk of mechanical membrane failure (as will be discussed in Section 3). The Ce^{3+} doping levels varied from

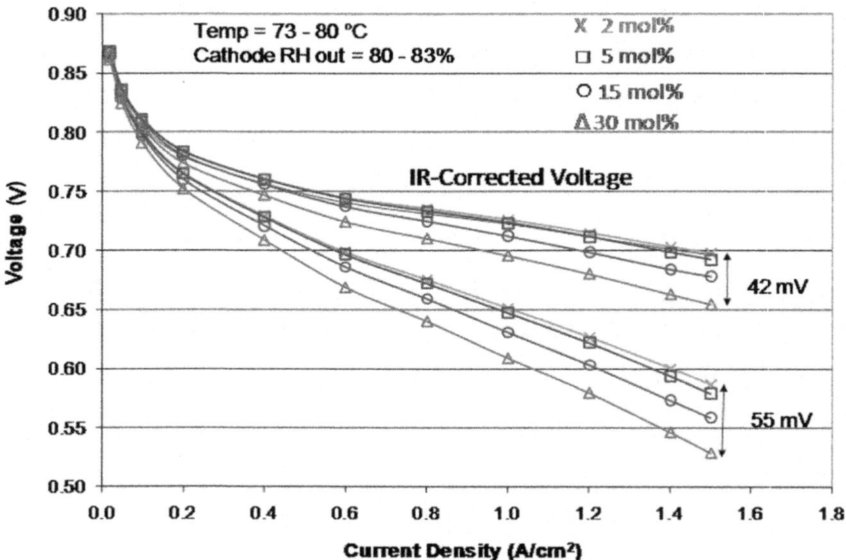

FIGURE 2.7 Performance impact of cerium loading. Polarization curves from a fuel cell short stack of extruded 1100EW PFSA MEAs with varying levels (2, 5, 15, 30 mol%) of Ce^{3+} (see Table 2.9). Measured (lower) and IR-corrected (upper) voltages shown.

zero to 30 mol% (0, 2, 5, 10, 15, 20, and 30%) based on the fraction of sulfonic acid sites complexed. Because each Ce^{3+} ion can complex three sulfonic acid groups (displacing three protons), the highest level of Ce^{3+} represents 10% of the total number of sulfonic acid sites in the membrane. Figure 2.7 and Table 2.9 clearly show that as the cerium loading increases, cell performance decreases. While the performance losses relative to the unmitigated MEAs are quite small for up to 5% loading, they become very significant as the loading increases to 10% and beyond. Note, as indicated by the IR-corrected cell voltages, the performance losses are not dominated by increasing cell resistance at the high loadings. Investigations indicate that the majority of observed losses in cation contaminated MEAs are thermodynamic in nature wherein the contaminant cations migrate toward the cathode under load, displacing protons and creating a proton gradient across the MEA between the anode and cathode [68]. The impact of cerium mitigation on chemical stability is clearly seen in Fig. 2.8, which shows the stack fluoride release rate as a function of operating time during durability testing. The accelerated automotive durability protocol was conducted at 80–85°C and contained frequent voltage and humidity cycling. Initially, the stack FRR was quite high while the unmitigated series of MEAs were present. After 280 hours of durability testing, all of the unmitigated MEAs failed due to hydrogen crossover. After removal of the unmitigated cells, the FRR rapidly dropped by three orders of magnitude as durability testing was

TABLE 2.9 Summary of Cerium Stack Performance Parameters for 80°C/ 85% RH Out Polarization Curve

% Ce level	Voltage at 1.5 A/cm^2	Performance Penalty (mV)	IR-corrected Voltage at 1.5 A/cm^2	Performance Penalty (mV)
0	0.584	0	0.697	0
2	0.587	−3	0.698	−2
5	0.579	5	0.693	4
10	0.568	16	0.684	13
15	0.559	25	0.679	18
20	0.547	37	0.668	29
30	0.529	55	0.655	42

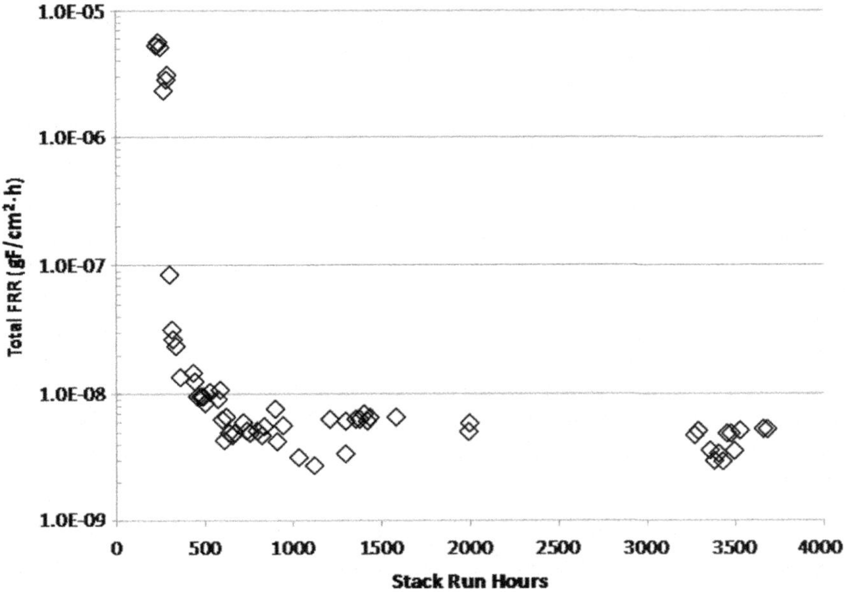

FIGURE 2.8 Fluoride release rate profile from the Ce level loading stack. Initially, stack contained a series of cells without Ce^{3+} mitigation. After 280 hours of operation, unmitigated cells were removed.

continued for total of 3,800 hours without additional crossover failures. Furthermore, voltage degradation rates did not vary across the doping level series over the entire durability testing time. Thus, the durability data obtained for the Ce^{3+} loading stack clearly indicate that the 2 mol% loading level provides excellent protection against chemical degradation. Significantly, this level of cerium doping induces no detectable performance penalty. As suggested above, higher levels of cerium mitigation may be necessary for harsher operating conditions such as those outlined in Table 2.2.

2.6. Chemical Durability of Hydrocarbon-based PEMs

Hydrocarbon-based PEMs have considerable possibilities for use in a variety of fuel cell applications owing to their improved design flexibility and cost advantages over existing PFSA materials [29–30]. Typically, hydrocarbon-based membranes consist of a polyaromatic backbone which most often is sulfonated on the aromatic rings of the main chain, although side chain sulfonated materials are also known. Representative hydrocarbon PEMs include sulfonated polystyrenes, polyethersulfones, polyetheretherketones, polyphenylenes, polyimides and polybenzimidazoles. As discussed below, the chemistry of aromatic-rich hydrocarbon membranes in the highly oxidative environment of an operating fuel cell is quite different from that of PFSAs, which accounts for the differences in behavior of these two classes in *ex-situ* and *in-situ* chemical degradation testing. Because hydrocarbon membranes encompass a wide variety of structures and, accordingly, many potential degradation mechanisms, no attempt to present a comprehensive discussion of hydrocarbon membrane chemical degradation will be made here. Instead, the discussion will be based on the general modes of chemical reactivity that can, in large part, explain the qualitatively different behaviors of hydrocarbons and PFSAs in both *in-situ* and *ex-situ* degradation testing.

As discussed in Section 2.2, the hydroxyl radical is an extremely reactive, electrophilic chemical intermediate of virtually unsurpassed aggressiveness. Due to the chemically inert, Teflon®-like backbone of PFSAs, the reactivity of hydroxyl radical is restricted to kinetically sluggish ($K_{abs} = 10^6$-$10^7 M^{-1}s^{-1}$) abstractions of hydrogen atoms present in low concentration [48]. By contrast, hydrocarbon membranes present a target rich environment for this radical. In particular, addition of hydroxyl radical to the highly abundant aromatic rings of hydrocarbon membranes occurs at near diffusion controlled rates. That is, the second order rate constants of addition to an aromatic ring are on the order of $10^{10} M^{-1}s^{-1}$ [48]. This huge reactivity difference between the addition and abstraction reactions accounts for the extremely poor stability of hydrocarbon membranes in Fenton's solution tests [29,69]. Multiple additions of hydroxyl radical lead to chain scissions [70], reductions in molecular weight and loss of mechanical integrity. Abstractions of the aromatic C-H bonds of aromatic rings are not kinetically significant, but aliphatic hydrogen atoms are susceptible to this mode of reactivity. In particular, the weak benzylic hydrogen atoms (ca. 335 kJ/mol) present in the backbone of polystyrene systems are particularly vulnerable to abstraction.

In stark contrast to the poor stability of hydrocarbon membranes in Fenton's tests, some hydrocarbon membranes have shown surprisingly good stability in fuel cell environments. Sethuraman *et al.* showed that a sulfonated biphenyl sulfone (BPSH-35) membrane survived 340 h at OCV, 100°C and 25% RH whereas Nafion® N112 failed after 60 h under these conditions [69]. In contrast, the BPSH lost 100% of its weight after 24 h in Fenton's solution while Nafion® lost less than 20% of its weight after 96 h. In another example, a side chain sulfonated polyimide membrane ran for 5,000 h at a current density of 0.2 A/cm^2 at both 60 and 90% RH [71]. At the end of the test, the polyimide and Nafion® 112 membranes were found to have thinned to about the same extent (ca. 20–30%) despite the 200-fold stability difference in Fenton's solution tests. While it is important to note that the constant current test cited above is not nearly as demanding as standard chemical durability tests such as high temperature OCV, this result suggests that understanding the chemical durability of hydrocarbon membranes requires a paradigm that is distinct from that used for PFSAs.

There are numerous plausible explanations for the surprising stability of hydrocarbon membranes in operating fuel cells. First, hydrocarbon membranes, in general, have significantly lower H_2 and O_2 permeabilities than PFSAs [54]. Given that much of the hydroxyl radical and H_2O_2 formation is thought to originate from reactant gas crossover (Section 2.2), the decreased gas permeabilities should decrease the yield of the damaging oxidants in hydrocarbon PEM systems. Thus, the Fenton's test could grossly overestimate the relative flux of hydroxyl radical present in the fuel cell environment across hydrocarbon and PFSA classes. Second, the hydroperoxyl radical, which is largely benign against PFSA systems [35], may have greater damage potential for some hydrocarbons. Third, hydroxyl radical additions to aromatics, the most likely mode of reactivity, will change the physical properties of the membranes, as numerous sites for hydrogen bonding are introduced. This could explain the increase in brittleness that is often reported after fuel cell testing of hydrocarbon membranes [72]. Fourth, as stated above, it is generally believed that a series of hydroxyl radical additions are required to induce chain breakages but, unlike PFSAs, chain breaks within hydrocarbon membranes do not lead to rapid increases in reactive sites of attack and accompanying acceleration of degradation, because polyaromatic hydrocarbon membranes always possess very high reactivity toward the hydroxyl radical.

Two other differences between the fuel cell degradation chemistry of hydrocarbon and PFSA membranes are of significance. The sulfonic acid mechanism of initiation is not operative in hydrocarbon systems for two reasons. First, the abstraction of the hydrogen atom from Ar-SO_3H is orders of magnitude slower than the hydroxyl radical additions and is therefore highly disfavored. Second, if the Ar-SO_3• radical does form, via direct abstraction or a hydrogen peroxide mechanism, it will not fragment, due to enhanced C-S bond strength (Table 2.8), but rather, will abstract a hydrogen atom from a good donor and return to the acid. Finally, Ce^{3+} (or Mn^{2+}) mitigation is not anticipated to be effective with hydrocarbon systems because the rate of •OH addition to aromatic rings is about two orders of magnitude faster than the •OH radical quenching redox reactions (Table 2.7).

In summary, relative to PFSAs, hydrocarbon membrane fuel cell systems are much more reactive toward oxidants like hydroxyl radicals but likely benefit from lower production rates of these oxidants owing to their low gas permeabilities. Clearly, much mechanistic work must be done on hydrocarbon membranes in order to understand the chemical stability landscape of this structurally diverse class of materials. Another difficulty in studying the chemical durability of hydrocarbon PEMs is the inability to use FRR as an *in-situ* diagnostic for membrane degradation. Appropriate degradation products must be identified and monitored, assuming such degradation products are present in the cell effluent water streams. Some recent studies have focused on monitoring sulfate ion release rates during OCV tests [63]. Sulfate ion release can provide insight into acid site loss, but may not be sufficient to monitor degradation of a hydrocarbon PEM polymer backbone. In addition to the chemical degradation issues discussed here, hydrocarbon membranes are more sensitive than PFSAs to hydration levels and associated mechanical stresses as discussed in the following section.

3. MECHANICAL DEGRADATION

3.1. Background

Since proton exchange membranes are at least partially constrained in the fuel cell by the electrode, gas diffusion layers, and bipolar plates, expansion and contraction of the membrane from temperature/hydration change can induce mechanical stresses. As the fuel cell operating condition fluctuates to meet the changing power demand, the membrane experiences hydration and temperature fluctuations, effectively subjecting the membrane to hygrothermal fatigue loading. As in other engineering materials, fatigue in the membrane, whether resulting from cyclic mechanical, thermal, or hygral stresses, can lead to mechanical degradation and failure through the initiation and propagation of microscopic cracks that result in the crossover of reactant gas. The postmortem analyses of mechanically degraded membrane samples from fuel cell stacks from field applications as well as from single fuel cells subjected to accelerated humidity cycling testing have revealed tell-tale signs of hygrothermal fatigue failure by cracks propagating through the membrane thickness [73].

Driven by hygrothermal stresses, mitigation of membrane mechanical degradation is a challenging task, which requires delicate balancing acts among the water uptake, modulus, and strength in developing high performance membranes. For example, a hydrocarbon membrane such as SPEEK [74] was reported to have superior tensile break strength in a simple tensile test compared to PFSA membranes such as Nafion®. However, the high modulus coupled with high mechanical strains induced by hygral swelling of the constrained hydrocarbon membranes in fuel cells can result in high enough stresses to overcome their higher mechanical strength. The imbalance between the hygrothermal stress and strength, plus the increased brittleness in these hydrocarbon membranes, have often resulted in poor fuel cell durability in humidity cycling tests [26,28]. Because the complex interplay among mechanical stress, strength, and fracture toughness can ultimately determine the membrane's mechanical degradation, it is desirable to assess these three parameters independently in *ex-situ* tests as well as jointly in *in-situ* tests.

In recent years, the humidity cycling test originally developed at General Motors has formed the basis for the recommended US DOE membrane mechanical durability test for evaluating membrane *in-situ* mechanical durability (Table 2.3). To independently characterize the stress, strength and fracture toughness of membranes under the conditions in the humidity cycling test as well as in an operating fuel cell, a series of characterization methods based on a viscoelastic framework have been jointly developed at General Motors and Virginia Tech [75–82]. These methods have been successfully used to understand the mechanical degradation of several model PFSA membranes that were studied extensively using the humidity cycling tests. In this section, we will review these *ex-situ* methods and the insights gained by comparing three commercial membranes of distinct humidity cycling durability – Gore-Select® Series 57, Nafion® NRE-211, and Ion Power N111-IP.

Nafion® NRE-211 (E. I. du Pont de Nemours and Company, Wilmington, DE) is a 25 μm-thick, re-cast, homogeneous, PFSA membrane. The N111-IP (Ion Power, Inc., New Castle, DE) membrane, also nominally 25 μm thick, is an extruded version of NRE-211. The 18 μm thick Gore-Select® Series 57 (W. L. Gore & Associates Inc., Elkton, MD) membrane has three layers, including a central reinforcing layer consisting of a composite network of ePTFE imbibed with PFSA ionomer, and two outer pure PFSA layers (Fig. 2.1(b)). All three membranes are made with ionomers with EWs of ca. 1,100. NRE-211 has the shortest lifetime (~4,500 cycles) in the humidity cycling test, Gore-Select 57 shows slight improvement (~6,000 cycles), and N111-IP has a significantly longer humidity cycling lifetime (>20,000 cycles without failure) [26,73]. In the postmortem analysis of the test samples, it was found that NRE-211 and Gore-Select 57 membranes had microcracks of various lengths in the membrane's thickness direction (Fig. 2.1 (c), (d)), while the N111-IP membrane was free of cracks. Additionally, it was observed that the Gore-Select® 57 membrane had numerous cracks in the outer PFSA layers that appeared to arrest when reaching the reinforcement layer (Fig. 2.1(d)).

The postmortem analysis also revealed that most of the cracks are located under the flow channel regions of the flowfield, suggesting lower compression pressure could exacerbate membrane cracking.

3.2. Initiation: Hygrothermal Mechanical Stress

Proton exchange membranes, which are required to function over a wide range of operating conditions from freezing to near boiling, and from dry to fully wet, have mechanical properties which are strongly dependent on temperature and humidity. Constrained membranes that nominally maintain a total strain of zero, with mechanical strains offsetting hygrothermal strains in the order of 10–20%, could experience a mechanical deformation that exceeds their elastic limit. Several membrane stress models that incorporate thermal and hygral dependency have been proposed, with some incorporating time-independent constitutive material behavior based on linear elasticity [83] or elasticity-plasticity [84–88] and others incorporating time-dependent constitutive models of viscoelasticity [4,77], viscoplasticity [89], and nonlinear visco-elasticity [90]. In these studies, the stress in the constrained membrane was determined using simple configurations such as a half cell including lands and channels or by applying constraints around the membrane's in-plane perimeter. Furthermore, the thermal and hygral conditions were prescribed rather than being determined from more rigorous fuel cell models. These simplified models employing hygrothermal loading and boundary conditions have been expanded through more rigorous, multiphysics, fuel cell models that incorporate the electrochemistry, fluid dynamics, and solid mechanics to determine the membrane stress distribution [91–93]. Although these models only account for the hygrothermal elastic behavior of membranes, they provide a platform that can be further developed to incorporate more comprehensive material models.

Following the framework of linear viscoelasticity established by Lai and Dillard [73] that considers the material dependency of temperature, water content, and time, the simplified form of membrane stress under the constrained condition can be expressed as:

$$\sigma_{ij}(\mathrm{t},\mathrm{T},\lambda) = \int_0^{\mathrm{t}} \mathrm{E}(\mathrm{t}-\xi,\mathrm{T},\lambda)\,\frac{d[-\delta_{ij}\beta(\mathrm{T},\lambda)]\Delta\lambda}{d\xi}d\xi \tag{2.11}$$

where σ_{ij} is the stress tensor; E is the uniaxial relaxation modulus; β is the linear coefficient of hygral expansion; δ_{ij} is the Kronecker delta; T is temperature; t and ξ are time; and λ is the water content of the membrane in terms of the number of water molecules per sulfonic acid. Note that the thermal expansion has been assumed to be negligible compared to the hygral expansion for these extremely hydrophilic membranes, where the hygral strain is in the order of 20% for typical humidity swings encountered in operating fuel cells. Two material properties are required to use this stress model: uniaxial relaxation modulus E and linear coefficient of hygral expansion, β. The uniaxial relaxation modulus E can be measured by dynamic mechanical analysis (DMA) in the relaxation test mode [75]. Alternatively, creep compliance or storage/loss modulus measurements from the creep and dynamic test modes, respectively, can also be converted into relaxation modulus [94]. Hygrothermal relaxation modulus master curves can be established by testing under various temperature and RH combinations and double-shifting the modulus data with both temperature and water content along the time axis to represent the constitutive properties over many decades in time. In forming the hygrothermal relaxation modulus master curve, both thermal and hygral shift factors, a_T and a_H, can be obtained individually. Furthermore, Patankar *et al.* have shown that the total hygrothermal shift factor a_{TH} is the product of separable thermal and hygral shift factors and is given by $\log a_{TH} = \log a_T + \log a_H$ [95]. Figure 2.9 shows the hygrothermal master curves for the three commercial PFSA membranes plotted at a reference condition of 80°C and 30% RH with NRE-211 showing slightly higher modulus than the others at longer times. The thermal and hygral shift factors determined from the double-shifting process are shown in Figs 2.10 and 2.11, respectively. The three membranes show similar shift factors. To determine the coefficient of hygral expansion, β,

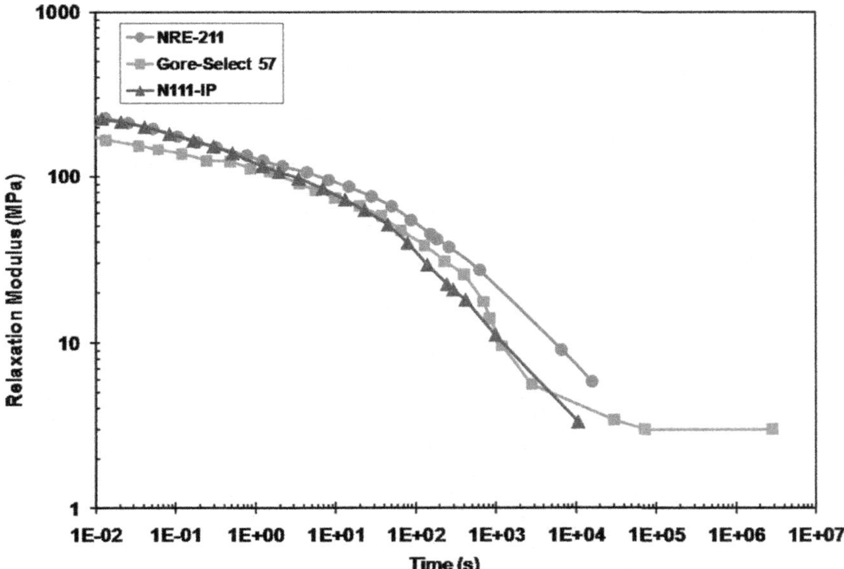

FIGURE 2.9 Relaxation modulus master curves of the three commercial membranes, Nafion® NRE-211, Gore-Select® Series 57, and Ion Power N111-IP, determined from the DMA stress relaxation test. The reference conditions are 80°C and 30% RH.

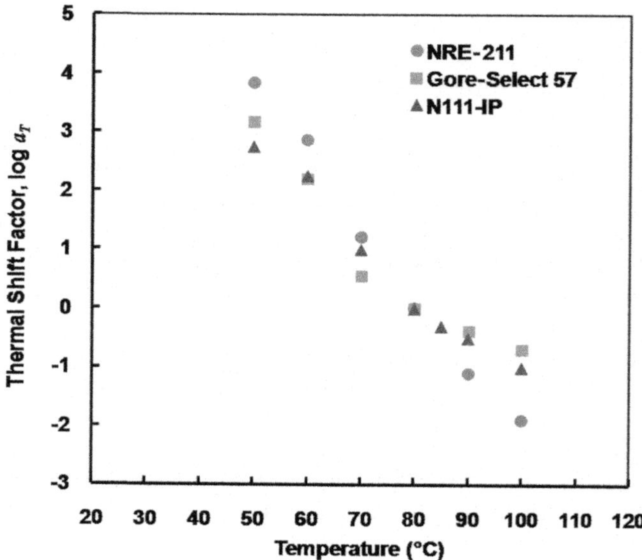

FIGURE 2.10 Thermal shift factors of the three commercial membranes for the relaxation modulus master curves in Fig. 2.9.

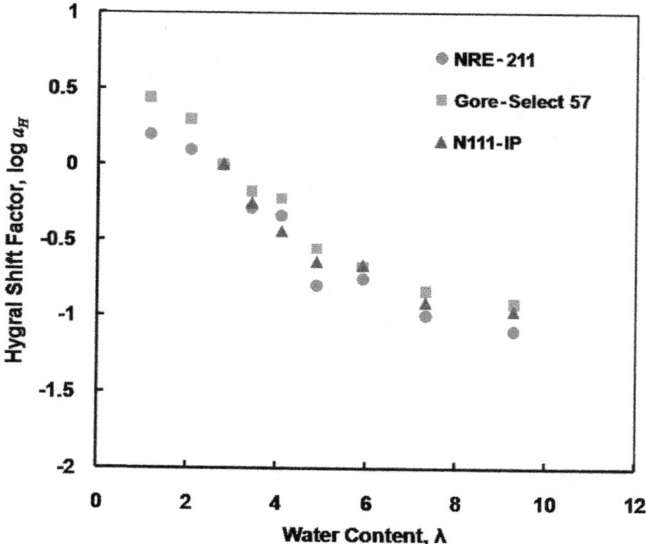

FIGURE 2.11 Hygral shift factors of the three commercial membranes for the relaxation modulus master curves in Fig. 2.9.

a variety of techniques may be used [95–98]. Among them, the DMA method [95,97] appears to be the most promising and convenient since the equipment, sample preparation, and test control are similar to those required in determining the viscoelastic properties of membrane materials. Testing nominally in creep mode at very small stress levels (< 0.01 MPa), the hygral strain after an RH change at a constant temperature can be readily measured. The coefficient of hygral expansion can then be determined by a linear curve fit through the hygral strain vs. water content data. Figure 2.12 shows the hygral strains and the coefficients of hygral expansion at 80°C for the three commercial membranes as-received [95].

To compare the hygrothermal stresses of the three commercial membranes, analyses based on a numerical scheme such as the finite element method were conducted to determine the stress generated when biaxially constrained membranes are subject to humidity cycles from 0% to 100% RH at 80°C [77]. Humidity cycles were simulated as trapezoidal waves in terms of the water content λ, and cycled between $\lambda = 0$ and $\lambda = 12$ with an initial (cell build state) $\lambda = 3$. The cycle period was 480 s and the hydration/dehydration rate was 0.2 λ/s. During the hydration portion of the cycles, the membrane attempts to swell, but because of the in-plane constraint, compressive stress develops as the

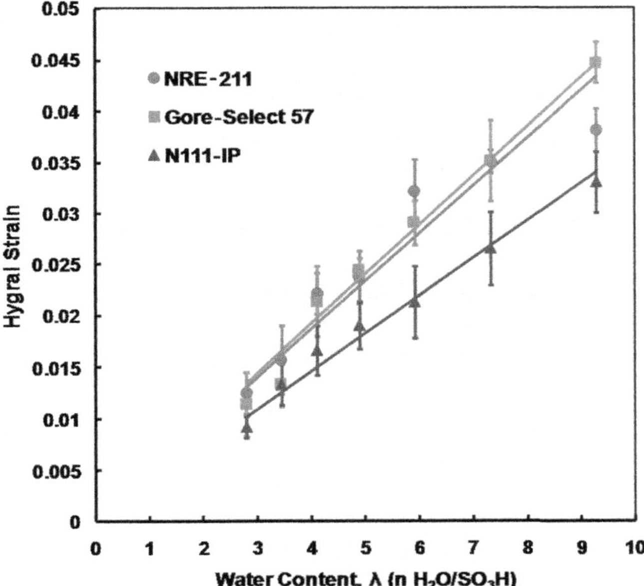

FIGURE 2.12 Hygral strain of the three commercial membranes measured from DMA test in creep condition under a stress lower than 0.01 MPa. The linear coefficients of hygral expansion β, as determined from the linear regression fit, are 0.0047, 0.0048, and 0.0037 for NRE-211, Gore-Select® 57, and N111-IP, respectively.

hygral strains are offset by equal and opposite mechanical strains. As the membrane continues to absorb water, the modulus decreases and, after the water content reaches its maximum value, the magnitude of the compressive stress begins to decrease due to relaxation. In this analysis, 218 s after the water content reaches $\lambda = 12$, the membrane stress appears to approach a plateau value of about -0.5 MPa, the negative value indicating a compressed state. During the dehydration portion of the cycles, the membrane attempts to contract. Because the hygral strains are again offset by equal and opposite mechanical strains from an already relaxed and only slightly compressive state the contracting hygral strain can result in a positive (tensile) stress state. When the membrane returns to its initial hydration state of $\lambda=0$ and zero hygral strain, the membrane reaches a maximum tensile stress. Beyond this point the stress begins to decay as the water content and the hygral strain are held constant. In the subsequent hydration cycles, similar behavior is repeated. Steady state, cycle to cycle, behavior is reached after only a few cycles. From this analysis it is seen that net negative strains (or shrinkage in membrane after a complete hydration cycle) are not required for the membrane to develop tensile stresses. This stress ratcheting effect [99] in a viscoelastic membrane is enough to induce significant tensile stress when the membrane experiences cyclic expansion. When swollen by moisture, viscoelastic compressive flow is accelerated in the plasticized material, resulting in more tensile stresses when the membrane is subsequently redried and viscoelastic flow is slowed.

The stress profiles for the three commercial PFSA membranes over 2 RH cycles, as shown in Fig. 2.13, are very similar, which is not surprising since their constitutive material properties such as relaxation modulus and hygrothermal shift factors are nearly the same, as shown in Figs 2.9–2.12. The most notable difference is in the lower maximum stress in the N111-IP membrane. Compared to the maximum stresses of 4.9 MPa for N111-IP, the NRE-211 and Gore-Select® 57 have nearly identical maximum stresses of 6.6 and 6.3 MPa, respectively. This analysis suggests that the N111-IP membrane has the advantage of slightly lower maximum tensile stress during humidity cycling. The lower stress in the N111-IP membrane is mostly due to the lower coefficient of hygral expansion – about 25% lower than the NRE-211 and Gore-Select 57 membranes.

3.3. Membrane Strength

Compared to the stress analysis and constitutive characterization of the PEMs, the characterization of strength and fracture toughness is relatively sparse. A great deal about membrane strength and fracture toughness has been learned through tensile tests on unnotched [83,88] and notched [76] specimens. When loaded in uniaxial tension, however, extensive plasticity occurs and failure strains can be an order of magnitude larger than strains anticipated in operating fuel cells. This large scale deformation is also not observed in failed

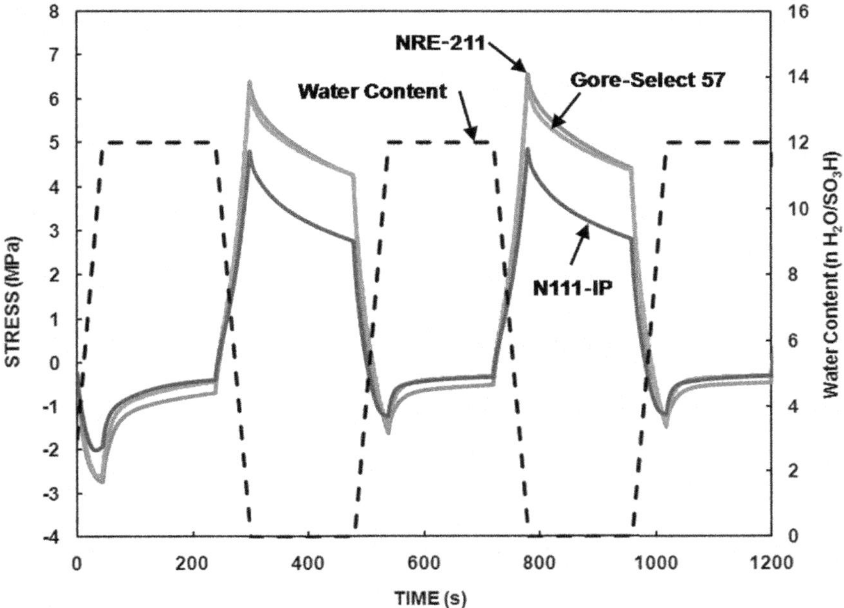

FIGURE 2.13 Hygrothermal stress response of the three commercial membranes subjected to an idealized trapezoidal-shaped water content history of 240-s hydration cycling and a hydration/dehydration rate of 0.2 λ/s.

membranes from fuel cell tests. The unnotched uniaxial tensile specimens typically fail catastrophically, which does not represent membrane failure within the fuel cell where slow gas crossover leaks through slowly propagating cracks and pinholes appears to be the main failure mode. This failure via slow crack propagation and gradually increasing membrane gas crossover rates has created a challenge to developing representative *ex-situ* strength tests to evaluate the membrane durability. To address this challenge, Dillard *et al.* have proposed the use of a circular pressure-loaded blister configuration, schematically shown in Fig. 2.14, to evaluate the membrane's strength under a variety of loading and environmental conditions [78]. The biaxial stress state mimics that of the constrained PEM in a fuel cell, resulting in significantly smaller strains at failure than seen in uniaxial tension specimens. The use of a pressurized gas also provides a very convenient and informative method to characterize the formation of pinholes or other flaws that would lead to gas crossover. By measuring how leakage initiates and increases, one has a more meaningful assessment of relevant membrane failure than from stressing a specimen to mechanical separation, as is typically done in tensile tests. The blister test is easily implemented under a wide range of pressure profiles to achieve ramp-to-burst, static creep-to-leak, or cyclic fatigue-to-leak loading.

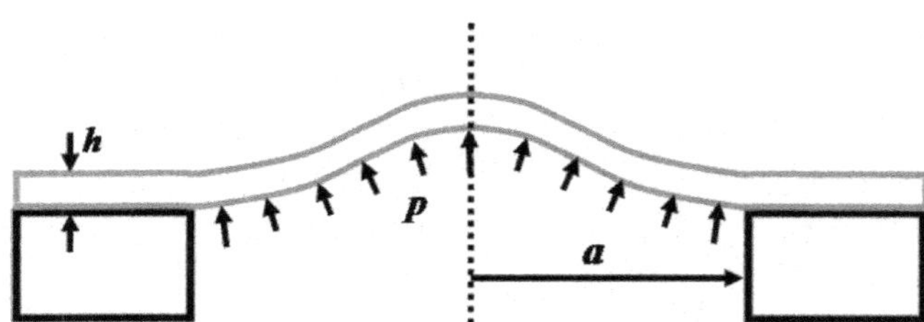

FIGURE 2.14 Schematic of a circular blister specimen with radius *a* and thickness *h*. The applied pressure is *p*.

Testing under various temperatures and RHs using the blister specimens, Li *et al.* [79] and Grohs *et al.* [80] have established master curves for burst, static fatigue, and cyclic fatigue strengths by shifting the strength curves according to the time-temperature-humidity superposition method described above for DMA relaxation tests.

Figures 2.15 and 2.16 illustrate the strength vs. lifetime curves for the three commercial membranes in the static creep and cyclic fatigue test modes, respectively, at 90°C and 2% RH. A clear difference is seen in the stress/life behavior of the three membranes. At higher stress levels, the solution cast NRE-211 has a shorter lifetime than both the reinforced Gore-Select® 57 and the extruded N111-IP in both test modes, while N111-IP has a longer lifetime than the other two membranes over the entire stress range. An important

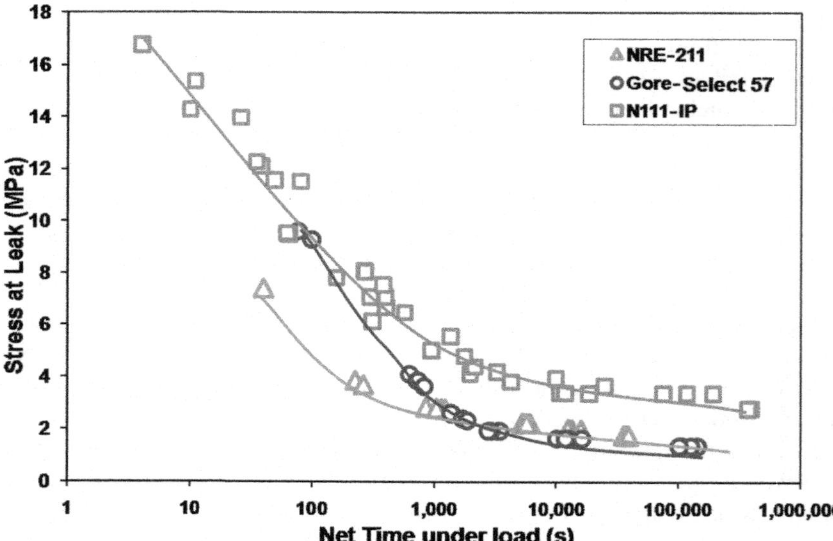

FIGURE 2.15 Strength vs. lifetime curves for the three commercial membranes in the static creep test modes at 90°C and 2% RH.

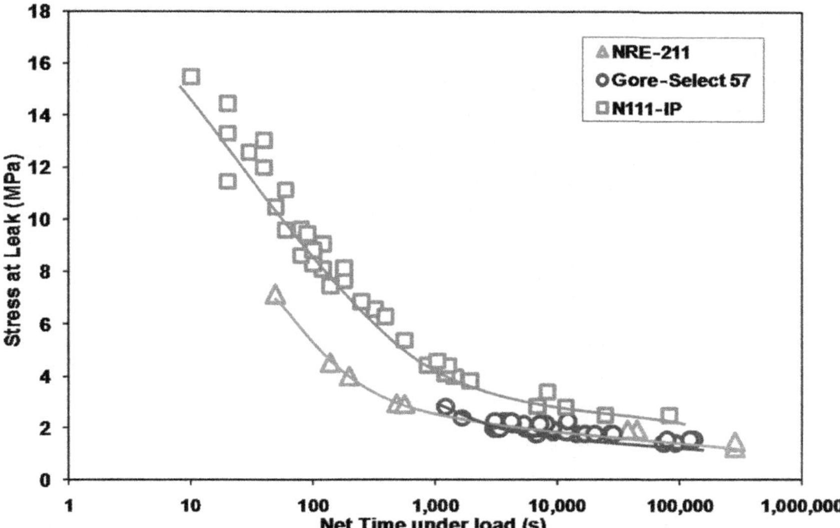

FIGURE 2.16 Strength vs. lifetime curves for the three commercial membranes in the cyclic fatigue test modes at 90°C and 2% RH.

observation in this study is that the static creep and cyclic fatigue-to-leak results are almost identical as the membrane's lifetime is determined by the total time under stress. This result suggests that the lifetime is predominantly determined by the magnitude of stress and the accumulated time under stress rather than the cycle count or frequency. It also suggests that it is possible to track membrane damage, and therefore predict the lifetime of the membrane during fuel cell operation if the membrane stress history is known and incorporated into a cumulative damage model.

3.4. Membrane Fracture Toughness

Recognizing that pinholes typically result from cracking within the membrane, it is essential to understand the membrane fracture process and to quantify the fracture toughness of the PEM. Several techniques have been reported for measuring the fracture toughness of PEMs, including the double-edge notch test (DENT) [100], trouser tear test [12], and knife slit test [76]. Among these, the knife slit test, shown schematically in Fig. 2.17, is the most informative because of the greatly reduced crack tip plasticity and the capability of measuring very slow crack propagation that has the potential to elucidate the intrinsic fracture energy of polymer material. An additional advantage of the knife slit test is the capability of forming doubly-shifted (hygrothermal) master curves that allow the study of the fracture behavior over a wide range of conditions and timescales.

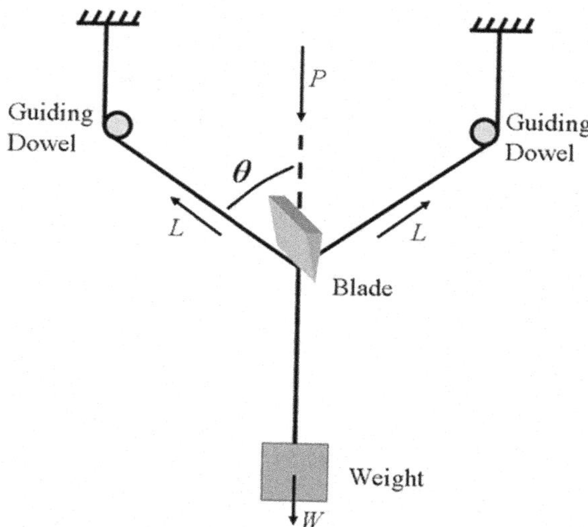

FIGURE 2.17 Schematic of the knife slit test setup where P is the applied cutting force on the knife blade; L is the pulling force in each leg of the specimen; θ is the cutting angle; and W is the tearing weight hanging under the uncut ligament.

Figure 2.18 shows the fracture toughness master curves for the three commercial PFSA membranes, where the master curves were determined from tests at 23, 40, 50, 60, 70, 80, and 90°C. The data were collected at two RH levels: 0% and 50% RH. For a given fracture energy (or crack driving force), NRE-211 has a significantly higher crack propagation rate than N111-IP: from about 30 times at lower fracture energies to about 1,000 times at higher levels, suggesting that NRE-211 would have poorer durability once initial cracks have formed. This is in agreement with the humidity cycling durability test results. On the other hand, Gore-Select® 57 appears to have higher fracture energies or lower crack propagation rates than both NRE-211 and N111-IP. Although this is not consistent with humidity cycling durability test results [26], the increased fracture resistance may be attributed to the tough ePTFE layer used in the composite Gore membranes [12]. This discrepancy may be explained by the fact that the knife slit test measures crack propagation along the length of the membrane, whereas, in a fuel cell, cracks tend to propagate through the thickness of the membrane (see Fig. 2.1(c,d)).

3.5. Mitigation of Mechanical Degradation

The comparison of the three commercial PFSA membranes revealed that N111-IP experiences 25% lower hygrothermal stress during humidity cycling due to its lower coefficient of hygral expansion and has significantly higher blister strength and fracture toughness. These results agree very well with the humidity cycling test results and are believed to mechanistically explain N111-IP's outstanding mechanical durability. N111-IP and NRE-211 are both 25 μm

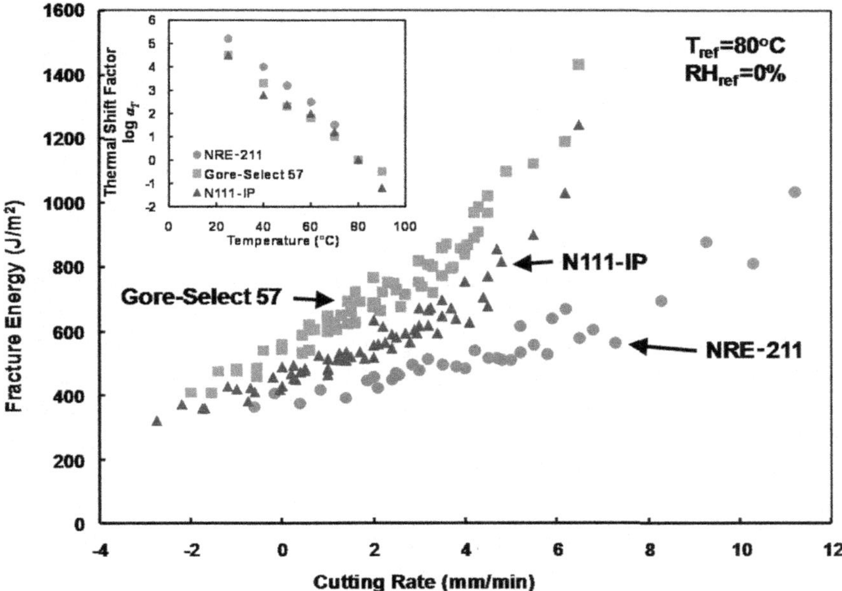

FIGURE 2.18 Fracture energy master curves of the three commercial membranes for various cutting rates at 23, 40, 50, 60, 70, 80, and 90°C. The data set was collected at 0% and 50% RH. The inset shows the thermal shift factors.

thick, 1100EW PFSA membranes, with the key difference being that NRE-211 is prepared by solution casting whereas the N111-IP is extruded. The results discussed above suggest that extruded polymeric membranes have potential to enable enhanced mechanical durability. It should be noted that nearly all thin PEMs developed for automotive applications are solution cast because acid functionalization of extruded membranes typically requires complex and costly post-treatment. Further study is required to determine why extrusion processes lead to more mechanically robust membranes.

High humidity cycling durability by simultaneously reducing mechanical stress via hygral expansion reduction and increasing strength has also been achieved recently in an updated Gore composite membrane, Gore-Select® 5720 [24]. The Gore-Select® 5720 membrane has an improved mechanical reinforcement layer and an in-plane hygral expansion 10 times smaller than, the Gore-Select® 57 membrane. In addition, the Gore-Select® 5720 membrane has significantly improved blister strength, with an increase of about 100% over the entire net-time-to-burst range compared with the Gore-Select® 57 membrane. The simultaneous improvement in both hygral expansion and strength has resulted in a humidity cycling lifetime that exceeds 40,000 cycles [24]. Other forms of composite membranes have been used in an attempt to improve PEM mechanical durability. Approaches that have proven to extend lifetimes in humidity cycling tests include the use of stiff two- and three-dimensional porous polymer support structures to reduce swelling in the plane of the membrane [14] and blending of polyelectrolytes with elastomers such as PVDF [15–18]. Other approaches to improve mechanical durability still need to be validated by *in-situ* humidity cycling tests, such as blending with ceramics (such as silica [101,102], alumina [103], titania [104], zeolites [105], and zirconium phosphate [106]), and nanocapillary network membranes where a mat of electrospun ionomer fibers are embedded with an inert stiff polymer [19].

While hydrocarbon membranes have been reported to perform relatively poorly in humidity cycling durability experiments [26–28,72], they also present unique opportunities for mechanical durability enhancements. Gross *et al.* have shown that mechanical durability can be improved by tailoring the morphology of sulfonated poly(arylene ether) copolymers without sacrificing proton conductivity [72]. This was achieved by using and optimizing block copolymers. For example, a random sulfonated poly(arylene ether ketone) PEM failed the DOE humidity cycling test in less than 40 RH cycles, while a triblock copolymer PEM with long range order (hundreds of nm) lasted for 800 cycles, and one with short range order (tens of nm) lasted for nearly 3,000 cycles. More work is required to understand the fundamental correlation between membrane morphology and mechanical durability. One hypothesis is that swelling behavior and dimensional stability can be controlled by tailoring membrane morphology. McGrath and Baird have shown that BPSH/BPS multi-block copolymers show anisotropic swelling, with PEMs swelling preferably in the thickness direction, whereas random copolymers swell isotropically [107]. Membranes with higher block lengths are more anisotropic than those with lower block lengths. Such anisotropic swelling behavior is preferential for reducing stresses induced by humidity cycling (Section 3.2). In general, the block copolymers also tend to adsorb less total water than random polymers [72,108], leading to lower induced stresses during RH cycling. For a given polymer system, membrane mechanical durability can also be improved by increasing the molecular weight. Another approach has been to improve membrane strength and reduce swelling of hydrocarbon PEMs by using cross-linked polymers [18,109,110]. While cross-linking does reduce membrane swelling and increase tensile strength, it can also make the membrane more brittle. Such approaches have yet to be proven using *in-situ* humidity cycling tests.

Guided by the viscoelastic framework described above, mitigation methods to reduce membrane mechanical degradation can also be developed by tailoring the fuel cell operating conditions and design strategies to reduce membrane stresses or to increase strength. One such strategy, referring to Eqn 2.11, is to reduce the magnitude and rate of change of the membrane's water content during the fuel cell operation. In a parametric stress analysis of Nafion® NR-111, Lai *et al.* [77] found that maximum stress of the constrained membrane during the humidity cycling can be reduced significantly with reduced hydration/dehydration rate ($d\lambda/dt$) or reduced hydration swing ($\Delta\lambda$). For example, the maximum stress of 8.0 MPa at a humidity swing between $\lambda=0$ and 12 (0–100% RH) at $d\lambda/dt=0.6\ s^{-1}$ can be reduced to 5.1 MPa at $d\lambda/dt=0.2\ s^{-1}$, and further to 1.1 MPa if the humidity swing is held between $\lambda=7.3$ and 12 (80–100% RH) at this lower dehydration rate. These calculations are consistent with humidity cycling tests of Gore-Select® 57, where humidity cycling lifetimes can be extended from about 6,000 cycles to more than 40,000 cycles when tested between 80–150% RH compared with 0–150% RH [73]. In addition to mitigating through hydration amplitudes and rates, temperature can also be an effective control parameter. In general, operating a fuel cell at a lower temperature can increase the mechanical durability. This is due to an increase in mechanical strength, as lower temperatures shift the strength curves toward longer lifetimes (see Figs 2.15, 2.16, and 2.18). It should be noted that the membrane modulus does increase at lower temperatures and therefore increases the stress generated at those conditions. However, the effect of increased strength appears to outweigh the increase in stress, as humidity cycling results for Gore-Select® 57 show that the lifetimes at 70°C, 80°C, and 90°C are 12,000, 5,000, and 4,500 cycles, respectively [73].

Another means to mitigate membrane mechanical degradation is through appropriate fuel cell hardware design and build conditions. Postmortem analyses of humidity cycled membranes indicate that most of the micro-cracks and pinholes are located under the channel regions of the flowfield. This suggests that the fuel cell compression, particularly over the channel region, may also affect membrane mechanical durability. A series of humidity cycling tests was conducted to study the effect of fuel cell compression. 50 cm^2 single cells with Gore Primea® Series 57 MEAs were built at various nominal compression pressures that were held constant throughout the tests. Flowfields with 1 or 2 mm wide lands and channels were used. Figure 2.19 illustrates the number of humidity cycles to failure vs. the contact pressure between the GDL and MEA over the mid-point of channels. The GDL/MEA contact pressure over the channel was determined through finite element analyses described by Lai *et al.* [111]. The three-number notation next to each symbol in the figure represents the land width, channel width, and the nominal cell compression pressure. The result shows that the humidity cycling lifetime increases as the GDL/MEA compression over the channel increases. The increased durability with increasing compression pressure may be due to the suppression of in-plane stress from the increasing through-plane compressive stress [83], the reduced crack driving force in propagating through-plane cracks when the crack opening displacement is reduced by the increasing constraining force from the GDL, and/or the reduced propensity for buckling with the increasing lateral (through-plane direction) support [112]. The results suggest that narrower flowfield channels, GDL with higher shear and bending stiffness, and high overall fuel cell compression can help mitigate membrane mechanical degradation. Similar benefits of increased mechanical durability upon exposure to humidity cycling

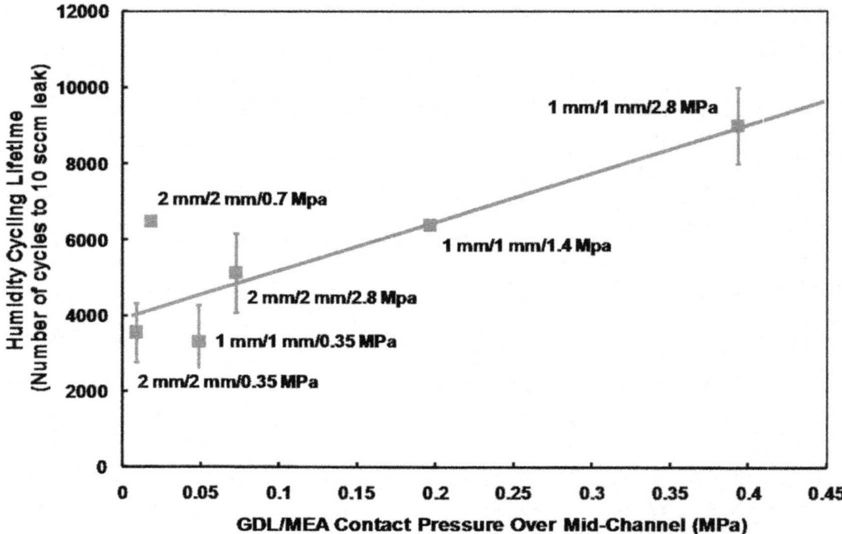

FIGURE 2.19 Humidity cycling lifetime versus compression between GDL and MEA over the mid-point of flow channels in the 50 cm^2 fuel cells built at various land/channel widths and cell compression. The US DOE membrane mechanical durability test protocol (shown in Table 2.3) was used.

can be achieved by increasing the adhension between the layers (membrane/electrode, electrode/GDL) in the MEA [113].

4. COMBINED CHEMICAL AND MECHANICAL DEGRADATION

Up until this point we have discussed only the isolated effects of independent stressors that can lead to chemical or mechanical degradation of the PEM. It is critical to investigate the effect of combining chemical and mechanical stressors and to verify if the approaches to mitigate the individual degradation modes can be combined to provide a comprehensive mitigation strategy that can withstand all the stressors imposed by an automotive drive cycle. Very little work has been done to focus on the impact of the synergistic effects of chemical and mechanical degradation of PEMs. The work done to date falls into two categories. One relies on *ex-situ* measurements of membranes after they have been subject to accelerated *in-situ* degradation. Huang et al., for example, did *ex-situ* stress-strain measurements on membranes after either humidity cycling to cause mechanical degradation or OCV testing to cause chemical degradation [87]. They found that the strain-to-break decreases after both humidity cycles and OCV testing. In another study, Escobedo *et al.* found that the strain-to-break for cast Nafion® decreases after increasing exposure to Fenton's reagent [114]. However, as discussed earlier, neither tensile tests nor Fenton's tests are necessarily the most appropriate means of determining mechanical and chemical durability [24]. Other studies have investigated *in-situ* PEM durability where the MEA was exposed to alternating chemical and mechanical stressors. For example, Escobedo *et al.* subjected membranes to 24 h of humidity cycling (N_2/N_2, 80°C; RH of inlet gases was cycled between 0 and 100% every 30 minutes) followed by 24 h of load cycling (H_2/O_2; 50% RH, 80°C; the load was cycled between 10 and 800 mA/cm^2 (7 min/3 min)) [114]. They found that incorporating a mechanical reinforcement could more than double the life of cast Nafion® in that test. The life was further extended, and the amount of fluoride released was reduced by a factor of 14, by adding a non-disclosed chemical stabilization. The studies described above, however, do not address the simultaneous application of mechanical and chemical stresses on the membrane.

To study the effects of simultaneous exposure to both chemical and mechanical stressors, membranes were subject to humidity cycling in an operating fuel cell. The DOE humidity cycling test was modified to run in a H_2/air cell operating at a constant current density of 0.1 A/cm^2. The humidity of both H_2 and air feed streams was cycled between dry for 2 minutes and 95°C dew point for 2 minutes, with constant flow stoichiometries of 20. Periodically, the cells were held at the wet conditions for several hours in order to collect product water to measure the fluoride ion release. Product water was also collected during the cycling stages to get the fluoride ion release during humidity cycling. The cell temperature was fixed at 80°C with no back pressure. Tests were run on a 50 cm^2 cell with straight 2 mm wide channels separated by 2 mm lands in counter flow operation. Recast NRE-211 Nafion® and extruded N111-IP Nafion® were evaluated using this protocol. As-received membranes were tested, as were membranes that were chemically stabilized by doping with Ce^{3+}. MEAs were prepared with 0.4 mg Pt/cm^2 electrodes on both the anode and cathode.

The lifetimes until membrane crossover failure, defined as a 10 sccm crossover leak when one side of the membrane is exposed to air at 3 psi, are shown in Table 2.10. As discussed above, the recast Nafion® fails the purely mechanical RH cycling test in less than 5,000 cycles, whereas the extruded Nafion® is still healthy after 20,000 cycles. When exposing the membranes to similar mechanical stresses in an operating fuel cell rather than in an inert atmosphere, the recast Nafion® fails in less than 1,000 cycles, exhibiting an acceleration factor of >4×. The extruded Nafion® fails in less than 2,000 cycles, exhibiting an acceleration factor of >10×. The addition of Ce^{3+} to the Nafion® membranes completely eliminates any accelerated failure caused by chemical degradation during fuel cell operation. Ce^{3+} doped NRE-211 survives 5,300 RH cycles in an operating fuel cell, which is slightly more than the 4,000–4,500 cycles to failure in an inert atmosphere. The longer lifetime in the fuel cell test can be attributed to the fact that at 0.1 A/cm^2 the cell does not dry out to the same extent as in the air/air test, where no current is generated and no

TABLE 2.10 RH Cycling Results in an Inert Atmosphere and in an Operating Fuel Cell

Membrane	Cycles to Failure w/o load (Air/Air)	Cycles to Failure @ 0.1 A/cm² (H_2/Air)
DuPont™ recast Nafion® (NRE-211)	4,000–4,500	800–1,000
Ce^{3+} doped NRE-211		5,300
Ion Power™ extruded Nafion® (N111-IP)	20,000+	1,800
Ce^{3+} doped N111-IP		26,000+

water is produced during the dry portion of the humidity cycle. Thus, the magnitude of the humidity swing and, accordingly, the hygrothermal tensile stress on the membrane, is lower in the operating fuel cell than in the air/air RH cycling test. The Ce^{3+} doped N111-IP survives 26,000 RH cycles in an operating fuel cell with no signs of failure, essentially replicating the results of the mechanical-only RH cycling tests. These results indicate that combining mechanical and chemical degradation mitigation strategies (i.e. extrusion and Ce^{3+} doping) can effectively prevent membrane degradation upon exposure to simultaneous stressors. Micrographs of cross-sections of the Ce^{3+} doped N111-IP after 26,000 RH cycles show no evidence of membrane cracking or thinning.

The fluoride release rate measured in the product water for the four membranes tested are compared in Fig. 2.20. FRRs are shown for both steady state wet and humidity cycling conditions. The as-received membranes exhibit much higher FRRs than their Ce^{3+} doped counterparts. During humidity cycling, the FRR is reduced by two and three orders of magnitude, respectively, for NRE-211 and N111-IP by doping the membranes with Ce^{3+}. This is consistent with accelerated chemical degradation experiments described in Section 2, where Ce^{3+} doping dramatically reduces FRR, thus significantly reducing chemical degradation. Another interesting observation is that, for the as-received membranes, the FRR is about 10 times higher during humidity cycling than it is during steady state wet operation, whereas for the Ce^{3+} doped membranes there is no discernable difference in the FRR measured under the two different conditions. A higher FRR during current cycling conditions compared to steady current conditions was also observed by Liu and Case, who saw a 30-fold increase in FRR from Nafion® 112 during a 16-step current cycle ranging from OCV to 1.06 A/cm^2 compared to a cell held at a constant current of 1.06 A/cm^2 [115]. The difference in the FRR of as-received Nafion® at the

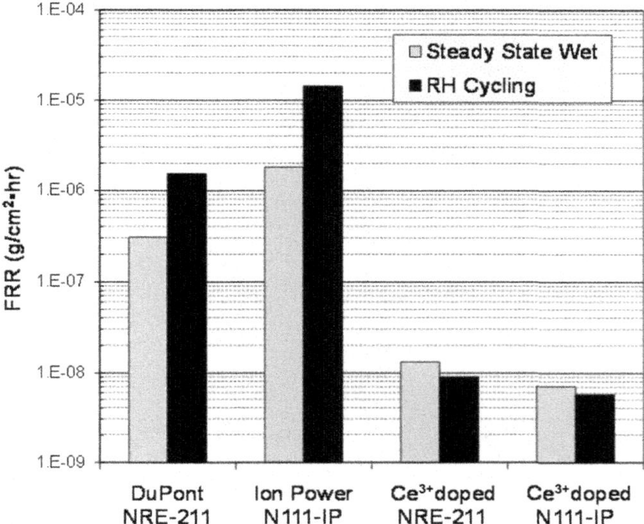

FIGURE 2.20 Fluoride release rates measured during 50 cm^2 RH cycling tests at 0.1 A/cm^2 of as-received and Ce^{3+} mitigated Nafion® membranes.

wet and cycling conditions can be explained by the higher chemical decay rate at dry conditions. Liu *et al.* showed that FRRs are 15 to 20 times greater for Nafion® 112 at 25% RH (H_2/O_2, open circuit, 95°C), than they are at fully humidified conditions [32]. Still, one cannot rule out that the mechanical stresses imposed by humidity cycling could accelerate membrane chemical decomposition and, thus, further study is required.

Another accelerated durability test has been developed that simultaneously combines chemical and mechanical stressors, while varying the magnitude of the stressors over the cell active area. In this test, rather than cycling the inlet RH to create a humidity cycle, the humidity fluctuations are created by cycling the current density. In a co-flow cell, the inlet RH is fixed at 40% and the current density is cycled between 0.05 and 0.8 A/cm^2. By holding the feed flow rate constant, the cell will experience humidity cycling toward the outlet region of the cell, whereas the cell inlet region will see a nearly constant RH. Thus, the cell inlet will see only predominantly chemical stressors, while the cell outlet will experience both mechanical and chemical stressors. A modeled humidity profile along the length of a channel during the durability cycle is shown in Fig. 2.21. The RH at the channel inlet stays at 40% on both the anode and cathode sides during the entire cycle, while the RH near the channel outlet cycles between 48% and liquid water. The magnitude of the RH swing increases monotonically from the cell inlet to the outlet.

An advantage of this protocol is that the location of membrane failure during the accelerated durability test can provide information about the

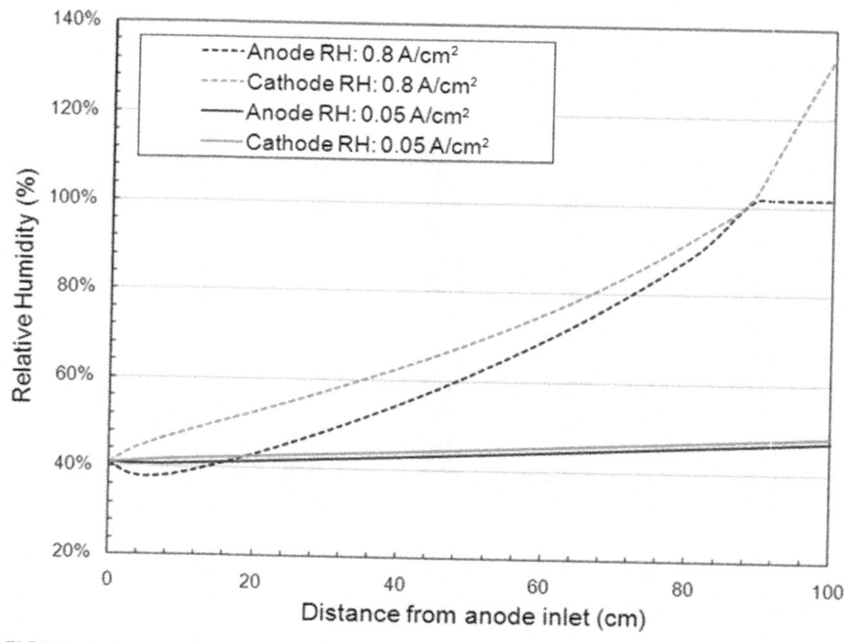

FIGURE 2.21 Modeled RH profiles along the cell channel length for a 50 cm^2 accelerated durability test where RH cycling is induced by cycling the cell current density between 0.05 and 0.8 A/cm^2.

dominant stressors. A useful non-destructive failure analysis tool involves using a segmented cell to generate a hydrogen crossover map of the active area [116]. Figure 2.22(a) shows a H_2 crossover current map of an NRE-211 membrane after 13 h of durability testing in a 50 cm^2 serpentine cell. The H_2 crossover current increases steadily from the inlet (top) to the outlet (bottom) of the cell. Thus the degree of membrane damage coincides with the magnitude of the RH cycle and, thus, the mechanical stress. This is not surprising because, as described above, NRE-211 is susceptible to RH-cycling induced failure. Fig. 2.22(b) shows a similar H_2 crossover distribution map of N111-IP after 29 h of accelerated durability testing. The N111-IP shows a clear rapid increase in degree of crossover near the cell outlet. This is not intuitive because, as described above, N111-IP does not fail when subject to only RH-cycling induced mechanical stresses. As also described above, chemical degradation tends to be accelerated at drier conditions. These two facts would suggest that, if there were no interaction between chemical and mechanical stressors, the N111-IP would fail faster near the cell inlet where the steady state chemical stressors are higher. However, the fact that the amount of damage is greater at the outlet, indicates that, once membrane chemical degradation is initiated, RH cycling can accelerate failure of an inherently mechanically robust membrane to the point where mechanical stressors become dominant.

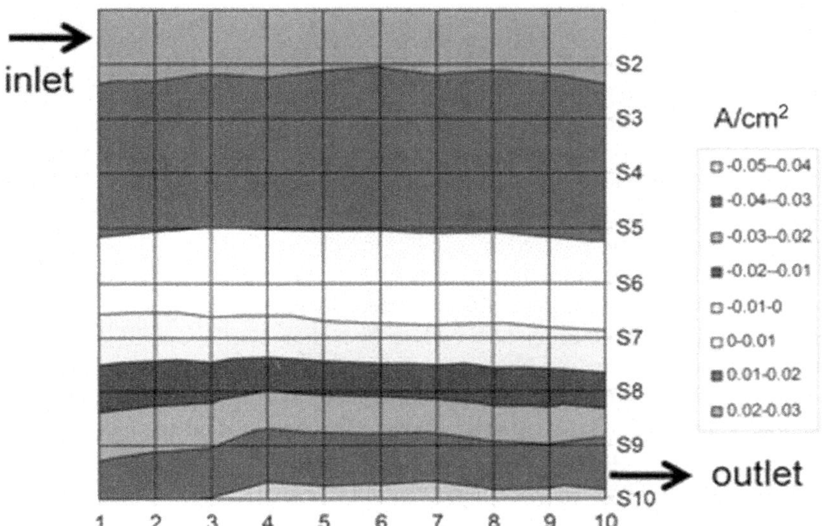

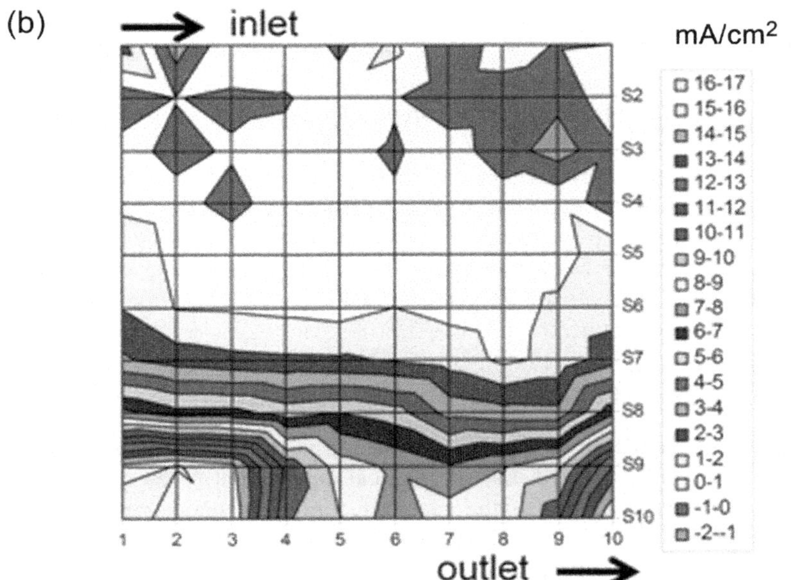

FIGURE 2.22 H_2 crossover current maps of membranes after accelerated durability testing in a 50 cm^2 serpentine cell using a protocol where RH cycling is induced by cycling the cell current density between 0.05 and 0.8 A/cm^2. (a) NRE-211 after 13 h, (b) N111-IP after 29 h.

5. MEMBRANE SHORTING

5.1. Background

Ohmic shorting through the membrane has been identified as one of the major failure modes in PEM fuel cells [29,117]. Shorting occurs when electrons flow directly from the anode to the cathode instead of through the device being powered. Ohmic shorting not only can reduce the performance of fuel cells, but can also lead to local heat generation in the vicinity of the short, causing membrane damage that can ultimately result in gas crossover failure in fuel cells. Although acknowledged as a failure mode, the lack of in-depth study of this subject in the open literature is conspicuous. Several challenges exist in the study of PEM shorting failure: the extremely localized shorting sites are very difficult to find (of the order of a carbon fiber diameter: about 10 μm); when occuring, the shorting morphology is often lost because of the destructive postmortem process that involves the removal of gas diffusion layers along with any likely shorting initiators; and the shorting failures is often accompanied and obscured by other failure modes such as membrane melting, thinning and pinholes that are more visible in the postmortem analyses.

During the course of fuel cell durability testing at General Motors, two types of membrane shorts have been observed: soft shorts and hard shorts. A *soft short* is a sub-critical short that does not immediately lead to fuel cell failure. However, significant accumulation of soft shorts can reduce the overall cell performance, and compromises fuel cell durability through cell voltage degradation. A *hard short* is a critical short that is the result of thermal runaway from an existing soft short. A hard short can directly lead to membrane crossover and cell failure. Hard shorts can occur suddenly in an operating fuel cell stack where one cell develops a significantly higher ohmic resistance compared to the cells in the rest of the stack. As the stack continues to draw current from all cells, the cell with the abnormally high resistance can develop an excessive voltage drop and even reverse its cell potential, creating a negative cell voltage much lower than −1 V. As will be seen in the upcoming discussion, −3 V is sufficient to create a hard short through a localized thermal runaway from an existing soft short.

5.2. Compression Induced Soft Shorts

Although the exact cause of membrane shorting is still under investigation, it is generally acknowledged that mechanical penetration of electronically conductive objects external to the membrane plays a major role. These shorting objects may include the carbon fibers and binders of gas diffusion layers, aggregates in the catalyst layers, or conductive debris trapped between membrane and electrode/GDL during MEA fabrication. Shorting induced by penetration of electronically conductive objects is strongly dependent on cell compression. Shorts driven solely by compression tend to be soft shorts, with resistances above 100 $\Omega \cdot cm^2$. Mittelsteadt and Liu [54] developed a shorting resistance test using a non-operating single fuel cell fixture to investigate the membrane shorting by diffusion media compression. An MEA or membrane was sandwiched between two carbon fiber GDLs, all of which were then compressed between two 50 cm^2 fuel cell flowfields. The cell was built with 600 kPa compression in the land regions. Specific resistance of the specimen was monitored with a voltage of 0.5 V applied across the cell at 95°C and 100% RH with inert nitrogen gas flowing on both sides of the membrane. The compression was then increased stepwise up to 6,200 kPa using a pneumatic cylinder and the resistance was monitored at each compression level. They investigated the effect of membrane thickness, membrane type, and catalyst layers, and found that increased membrane thickness and the use of catalyst layers and membrane reinforcement layers can increase the shorting resistance.

This test procedure was used to investigate the impact of GDL type on membrane shorting. Figure 2.23 shows the resistance of Nafion® NRE-211 as a function of compression pressure for three GDL types. Cells built with SGL 21BC GDL start to see a drop in electronic resistance at relatively low compression levels – below 1 MPa. GDLs can be tailored for improved shorting

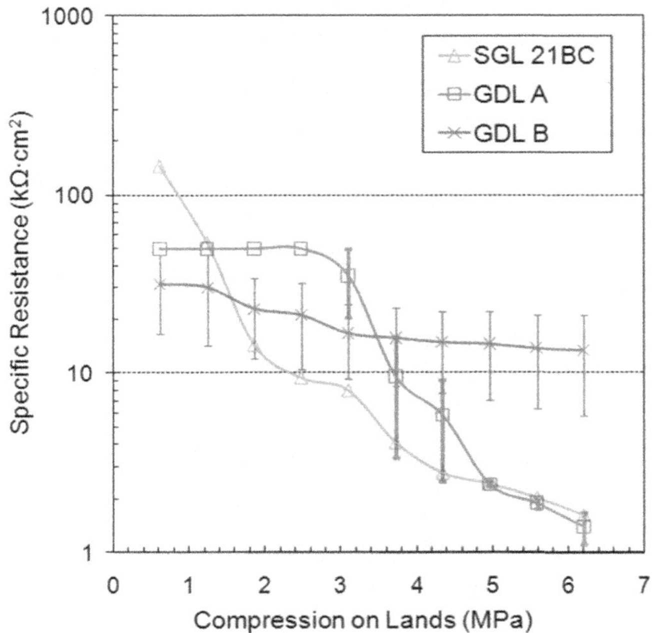

FIGURE 2.23 Shorting resistance of NRE-211 with several gas diffusion layers as a function of compression pressure. Land compression assumes that all pressure is in land regions and none in the channels. Error bars represent the min-max range divided equally on both sides.

resistance. For example, GDL type A can withstand compressions up to 3 MPa before suffering from resistance loss and GDL type B does not show significant loss of resistance at pressures up to 6 MPa. Both the carbon fiber substrate and the microporous layer can have a significant impact on shorting resistance dependence on compression. Note that in these tests using a constant imposed potential of 0.5 V no hard or severe shorts were created.

In a variation of the cell compression test, Mittelsteadt and Liu measured the membrane ohmic resistance during a slow (30 min wet followed by 30 min dry) humidity cycling protocol at 95°C under a constant land compression pressure of 2.75 MPa. They found that membrane electronic resistance decreases as the sample experiences more humidity cycles. They also observed that membrane shorting resistance closely follows the humidity profile, with a higher resistance during the wet step. This result supports the mechanism that shorting can be induced by progressive penetration of the carbon fibers into the membrane.

5.3. Voltage Induced Hard Shorts

The shorting test developed by Mittelsteadt and Liu provides a simple accelerated test to evaluate the fuel cell material set based on the accumulation and progression of soft shorts. The constant potential tests, however, do not enable the detection of very localized severe shorts, where pinholes are most likely to occur. The test also does not address the hard short failure mode that is observed in fuel cell stacks. To address both needs, a variation of Mittelsteadt and Liu's test was developed to study hard shorting.

The premise behind the hard shorting test is that, if the local shorting resistance from a soft short is reduced sufficiently by mechanical penetration of conductive objects under increasing cell compression, the soft short could turn into an easily detectable hard short under an elevated cell potential like that found during cell reversal caused by the high ohmic resistance during membrane dry-out. Thus, the compressive force measured upon detection of a voltage-induced hard short is a convenient metric to compare the soft short propensity of material sets or cell designs. Similar to the constant potential test protocol, a stepwise increasing compression scheme was used, where the cell was held at each compression level for one hour. Since the rate of PFSA membrane's response to mechanical stresses increases about tenfold for every 10°C temperature increase (Section 3.2), elevated temperatures (95°C) were used to accelerate penetration depth at each compression level. Thus, each one-hour compression step at 95°C is equivalent to testing for 300 hours at a typical stack operation temperature of 70°C in a fuel cell vehicle. Although testing could be even further accelerated by elevated humidity, in order to avoid the deleterious electrolysis reaction from the water in the MEAs, tests were done at a dry condition, where cell potential reversal is most likely in an operating fuel cell stack. Dry nitrogen is introduced to both sides of the cell and the cell

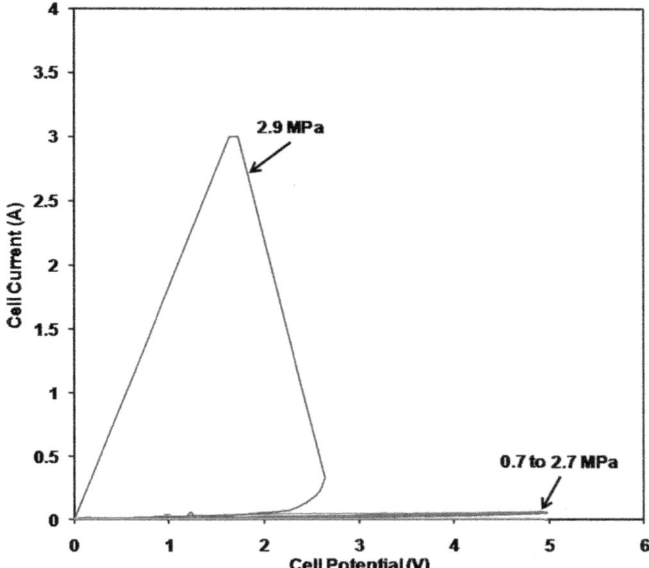

FIGURE 2.24 A typical cell current versus applied potential of a bare 50 μm Nafion® N112 membrane sandwiched between two sheets of Toray TGP-H060 GDL using the GM hard shorting test protocol. Hard short occurs at a cell compression of 2.9 MPa.

compression is stepwise increased in 200 kPa increments up to 3.5 MPa. At the end of each 60-minute compression step, the cell potential is cycled between 0 V and 5 V twice at a ramp rate of 50 mV/s by an external power supply. The maximum allowable current was set at 3 A to prevent hardware damage when a hard short occurs. If no hard short is detected at a given compression level, the cell compression is increased to the next level. The compression pressure and the cell voltage upon detection of a hard short, as indicated by rapid rise in cell current, is recorded.

Figure 2.24 illustrates the measured cell current vs. applied potential of a bare 50 μm Nafion® N112 membrane sandwiched between two sheets of Toray TGP-H060 GDL using the protocol described above. Voltage cycling at compression levels from 0.7 MPa to 2.7 MPa yielded a linear cell current/voltage relationship with a relatively high specific resistance of 3,960 $\Omega \cdot cm^2$. After one hour of conditioning at 2.9 MPa, the voltage cycle yielded a current/voltage curve that was linear up to about 2 V with a specific resistance of 2,600 $\Omega \cdot cm^2$ above which the current rose rapidly. At a cell potential of 2.6 V, the hard short occurred, reaching the maximum allowable current of 3 A. The subsequent voltage ramp followed a linear current/voltage profile with a 27 $\Omega \cdot cm^2$ specific resistance, indicating a severe short. Postmortem inspection revealed a 1.0 mm-diameter pinhole in the membrane, surrounded by a dark burn ring.

One of the key observations from Mittelsteadt and Liu's study was the relatively high standard deviation in the specific ohmic resistance, which suggests a strong statistical nature of membrane shorting [54]. Since the compression load at the detection of a hard short provides a clean and precise signal for shorting failure, the statistical nature of the shorting behavior can be easily studied and different material sets can be compared. Here, four material sets, all tested using a Nafion® N112 membrane, are studied to investigate the mechanical buffering effect of the electrode and microporous layers. The first set was the baseline in which the N112 is sandwiched between two layers of Toray TGP-H060 GDL. In the second set, 60 μm Gore Carbel® MP30Z micro-layers were included between the GDL and membrane. In the third set, the bare membrane in the baseline set is replaced with a N112 membrane coated with 12 μm electrodes (N112 MEA). The fourth set used the Carbel MP30Z micro-layers between the Toray TGP-H060 GDL and N112 MEA. At least 11 samples were tested for each material set. Two-parameter Weibull probability density functions and the associated parameters were determined. Figure 2.25 illustrates the shorting probability (Weibull probability density function) vs. the nominal cell compression when hard short occurs for the four material sets. The probability density function curves to the left indicate poorer shorting resistance, while the ones to the right show more resistance to shorting. The baseline material set (Toray 060/N112 PEM) has the poorest shorting resistance, while the other three sets perform similarly. All four have similar shape parameters with a range from 2.5 to 4.1. Note that a shape parameter of 3 approximates a normal curve, while a shape between 2 and 4 is still considered fairly normal. This result indicates that additional layers between the membrane and the GDL, either Carbel MP30Z micro-layers, electrode layers or the combination of both, provide significant mechanical buffering that reduces the probability of mechanical penetration of shorting objects. Notably, all 17 samples of the baseline material set failed at pressures below the maximum compression of 3.5 MPa, while many samples of the other three material sets did not fail, as shown under the 'C' column in the Table of Statistics inset in Fig. 2.25. Although extra layers between the GDL and membrane provide some protection against the mechanical penetration of carbon fibers from the GDL, there are still other objects that are able to penetrate into the membrane under increasing compression to create shorts. These objects include hard Pt/C aggregates or nuggets in the electrodes (Fig. 2.2(d)) and debris. The results suggest that a clean MEA fabrication process in addition to the mechanical buffering layers is needed to completely prevent membrane shorting.

An interesting observation that can be drawn from this study is that the shorting behavior of Nafion® does not follow the dielectric breakdown of polymeric materials where there is an increasing probability of breaking down when the electric field approaches the polymer's dielectric strength. For example, a polytetrafluoroethylene (PTFE) film has a dielectric strength of greater than 20 kV/mm [118]. Figure 2.26(a) illustrates the percentage

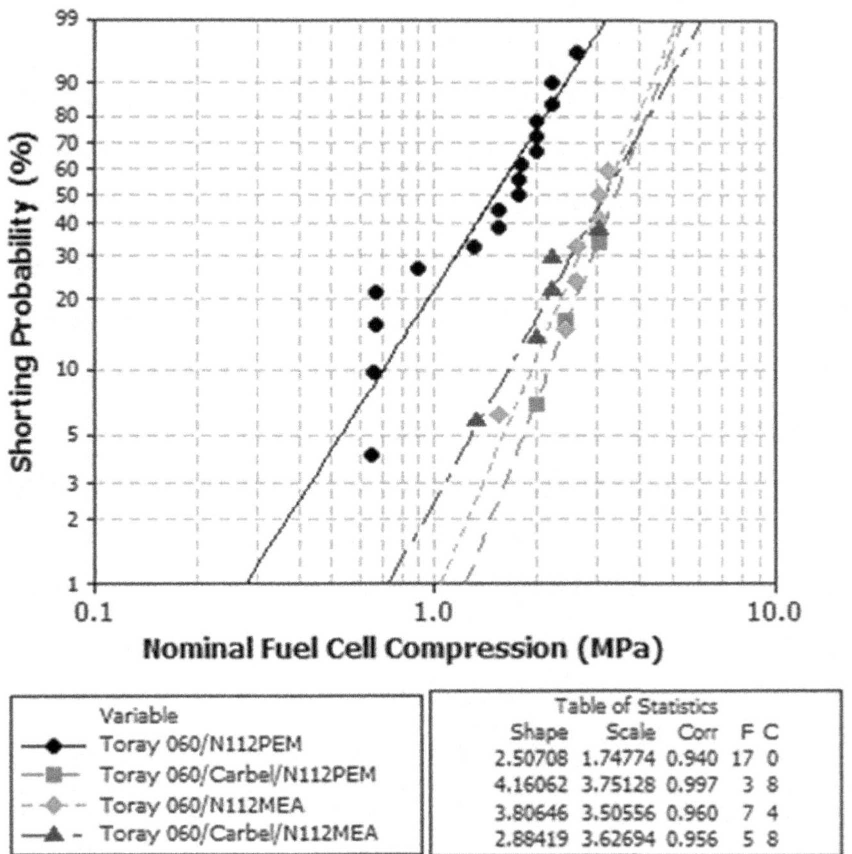

FIGURE 2.25 Weibull shorting probability density versus nominal cell compression for the four material combinations.

histogram of the shorting voltage for the 31 N112 samples that had hard shorts. The mean voltage at the onset of the hard shorts is 2.62 V with a standard deviation of 0.44 V. The maximum hard short voltage is 3.6 V and the minimum is 1.65 V. Although the maximum potential of 5 V was repeatedly applied during the tests, none of the 52 N112 samples shorted near this voltage. To date, General Motors has conducted several hundred shorting tests on various membranes, diffusion media, and electrode combinations. Figure 2.26(b) illustrates the histogram of the shorting voltage for the 217 PFSA membrane samples that have shorted. The mean shorting voltage is 2.76 V with a standard deviation of 0.80 V. Not a single sample developed a hard short below 1 V. Since 1 V represents the typical open circuit voltage in an automotive PEM fuel cell, the results suggest that it is unlikely that hard shorts will develop within a fuel cell stack under normal operation.

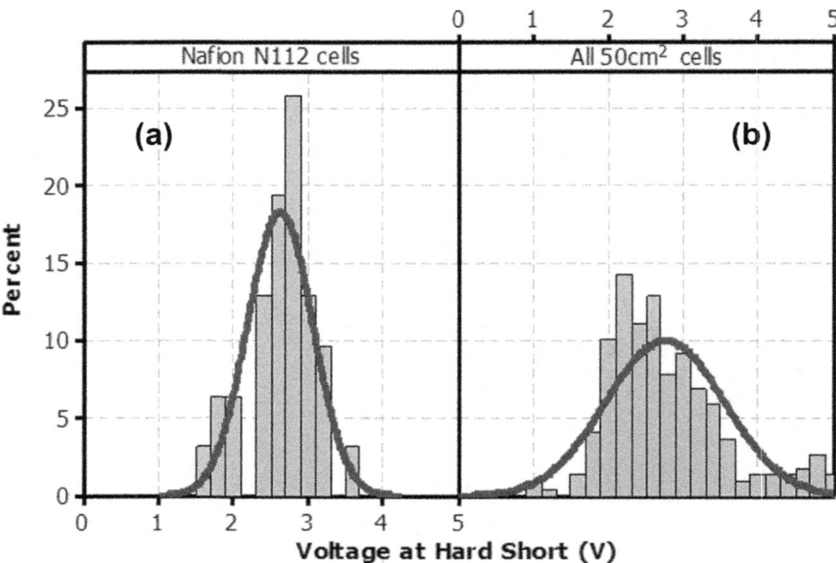

FIGURE 2.26 Percentage histogram of the shorting voltage for (a) the 31 N112 samples and (b) 217 samples of various membranes, diffusion media, and electrode types using GM hard shorting test protocol.

5.4. Thermal-Electrical Analysis

The hard shorting test has provided a versatile platform for evaluating various material systems. The test can be easily adapted to a variety of fuel cell hardware platforms. A full scale active area single fuel cell fixture equipped with a current distribution measurement device has been used to investigate the thermal-electrical membrane response during hard short events. A single fuel cell with a 360 cm^2 active area was built with a Gore Primea® MEA, Grafil U-105 GDL (Mitsubishi Rayon Co, LTD, Tokyo, Japan) coated with a proprietary microporous layer, and flowfield plates made of vinyl ester based composite material. The current distribution measurement device was incorporated into the anode flowfield plate to allow for local current and high frequency (1 kHz) resistance (HFR) measurements of 200 individual elements throughout the active area [116]. The cell was built at a nominal cell compression of 0.75 MPa. As in the 50 cm^2 cell hard short test, the cell was conditioned for one hour at 95°C and then subjected to two voltage cycles between 0 and 5 V. The step increment for cell compression was 0.25 MPa.

Figure 2.27 shows the cell current and potential (I/V) curves at 0.75, 1.25, and 1.5 MPa compression. Similar to the hard short test using a 50 cm^2 cell, the I/V curves at lower cell compression (0.75 and 1.25 MPa) remain linear over the whole range of potential from 0 to 5 V. Once the cell compression is raised

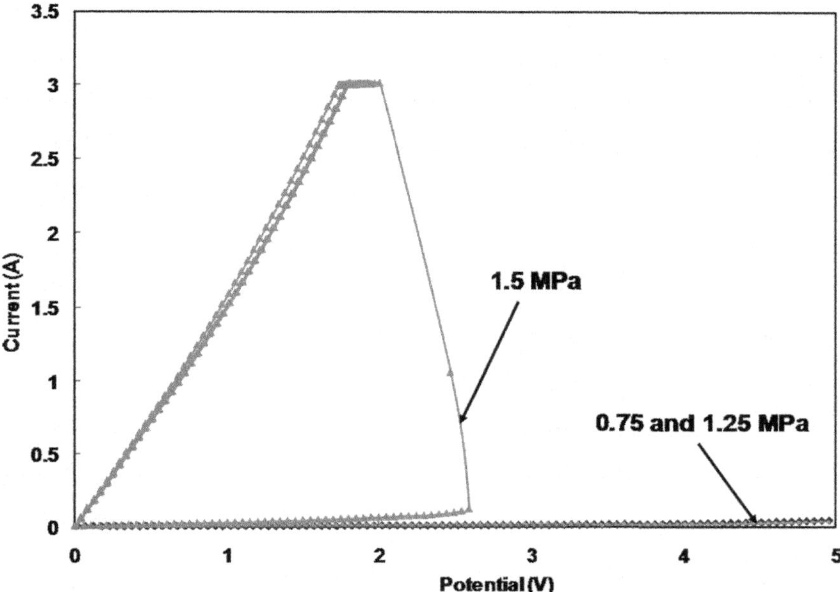

FIGURE 2.27 Cell current vs. applied potential of a 360 cm^2 large-active-area single fuel cell using the GM hard shorting test protocol.

to 1.5 MPa, the hard short occurs at 2.6 V and then the current jumps to 3 A with a corresponding reduction of cell voltage to 2 V. As the cell voltage was ramped down to zero and subsequently back up to 2 V, the I/V curves show a steeper slope with a cell resistance of 236 $\Omega \cdot cm^2$. It is interesting to note that the hard short voltage in this large active area test fell within the mean breakdown voltage range measured on 50 cm^2 cells (Fig 2.26).

Figure 2.28 shows current distribution maps captured during the first 0–5 V voltage cycle at 1.5 MPa. One of the current distribution segments, denoted as E5, stands out in this series of plots with a significantly higher current density than the rest of the 200 segments, as indicated by the darker color. Under each current distribution plot, the values of total cell voltage and the corresponding current density at segment E5 are shown. The current density in segment E5 steadily increases as the potential is ramped from 0.495 V to about 2.6 V. At 2.629 V, the hard short occurs and the current density jumps from 18.5 to 411 mA/cm^2 with the cell voltage decreasing to 2.014 V. The current distribution map at 2.014 V shows that the current in segment E5 spills over to the neighboring segments through the GDL and bipolar plates. Even though cell resistances (Fig. 2.27) do not show signs of global shorting (all well above 10,000 $\Omega \cdot cm^2$) before the hard short occurred, close examination of the local resistance of the E5 segment shows that there was progressive deterioration of shorting resistance as the segment resistance was 34,200, 1,180, and 149 $\Omega \cdot cm^2$

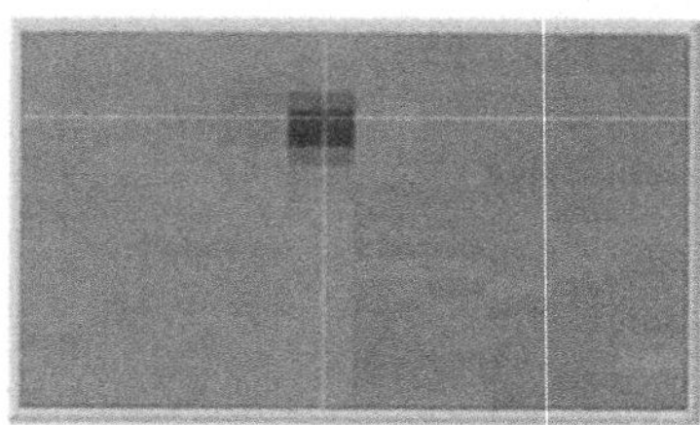

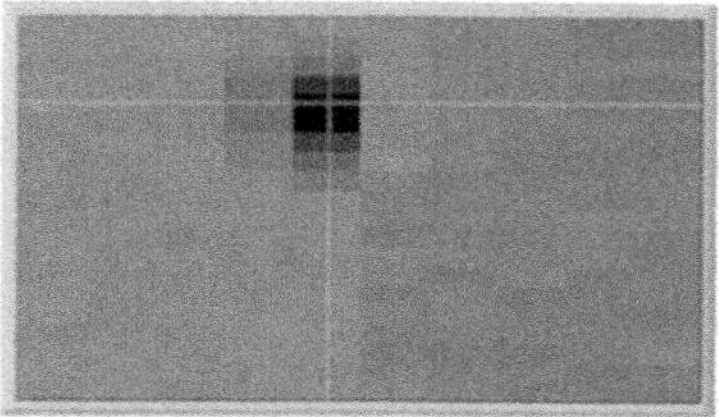

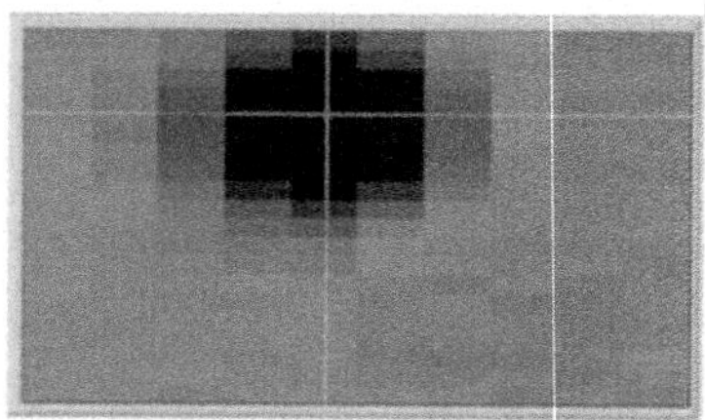

FIGURE 2.28 Current distribution contour maps captured from the GM hard shorting test for the 360 cm^2 large-active-area single fuel cell during 0-5 V voltage cycling at a cell compression of 1.5 MPa. Captions under the contour plots show the voltage and the current density of the E5 segment where the hard short occurs.

after compression conditioning at 0.75 MPa, 1.25 MPa, and 1.5 MPa, respectively.

To investigate the thermal and electrical response of the material set during the hard short event, and to elucidate the significance of the independence of hard short voltage on various material systems, a coupled thermal-electrical analysis was performed to examine the heat generation from the short and temperature distribution of the material surrounding the shorting location. Due to the large aspect ratio of the layered materials within a cell segment, modeling the entire segment is impractical. Therefore, a simplified axi-symmetrical finite element model using a commercial finite element program, ABAQUS, [119] was developed to represent half of the repeating unit thickness over the land region. Figure 2.29 shows the schematic of the finite element model. The radius of the entire model is 0.34 mm with the short epicenter located at the center of the model (r = 0 mm). A coupled thermal-electrical element type, DCAX4E, was used for all materials. Table 2.11 lists the material properties used in the analysis. The initial temperature for all the elements was set to the cell test temperature of 95°C. The boundary conditions were given as 95°C and 0 V at the outer surface of flowfield plate; zero heat and electrical flux at radii of $r = 0$ mm, 0.34 mm, and at the symmetric

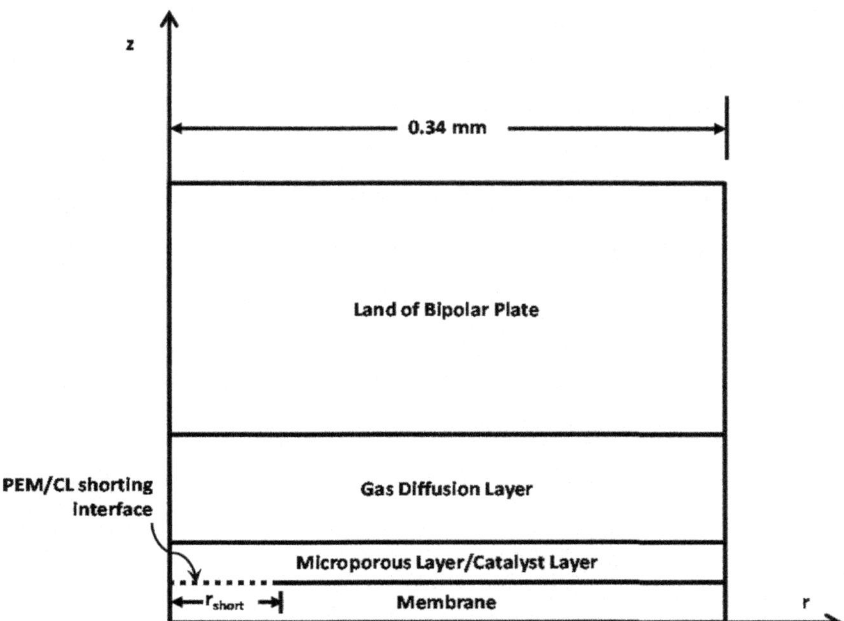

FIGURE 2.29 Schematic of the axi-symmetrical half-cell finite element model for the coupled thermal-electrical analysis. A PEM/CL shorting interface with a radius of r_{short} is used to simulate the shorting through the membrane.

TABLE 2.11 Material Property Data Used in the Thermal-Electrical Finite Element Model

	Bipolar Plate	GDL	MPL & CL	PEM
Thickness (mm)	0.78	0.18	0.062	0.018
Electrical Conductivity (S/cm)	8	40	40	1.00E−06
Mass Density (g/cm^3)	1.82	0.448	0.448	2
Thermal Conductivity (W/m·K)	19.2	0.96	0.96	0.16
Specific Heat (J/g·K)	0.841	0.8	0.8	1.1
Plate/GDL Interfacial Thermal Conductance (W/cm^2·K)			3.16	
Plate/GDL Interfacial Electrical Conductance (W/cm^2·K)			50	

plane (middle) of the membrane. Between the membrane and catalyst layer (CL), a PEM/CL shorting interface was installed, with a defined interfacial shorting resistance, R_{short}, to simulate the shorting through the membrane via the Gap Electrical Conductance function in ABAQUS [119]. By ramping up the potential at the PEM/CL shorting interface from 0 to 2.5 V in the half cell model, the full cell voltage ramp from 0 to 5 V was simulated.

There were two unknowns in this model that could not be provided by the experimental data and thus needed to be determined prior to the final coupled thermal-electrical analysis. One was the size of the PEM/CL short, represented by the shorting radius, r_{short}, and the other was the interfacial shorting resistance, R_{short}. To determine R_{short}, an electrical analysis was first performed assuming $R_{short} = 0$. By calculating (integrating through) the total current flowing through the model, the bulk electronic resistance of the half cell model minus the shorting resistance R_{short} was determined. The interfacial resistance, R_{short}, was assumed to be the difference between half of the experimentally measured resistance, R_{exp}, in the shorted segment and this calculated value, R_{model}, as shown in Eqn 2.12:

$$R_{short} = (R_{exp}/2) - R_{model} \tag{2.12}$$

Although efforts were made by postmortem examination to determine the size of short (r_{short}) before the hard short occurred, the damage in the MEA was too severe to permit a reasonable estimation. Instead, as GDL carbon fiber penetration was considered the most likely shorting initiator, the carbon fiber diameter (7 μm) was assumed to be a reasonable starting point for estimating the size of a soft short. Here, two shorting radii were selected for our analysis; $r_{short} = 0.006$ mm and 0.014 mm, and the sensitivity of the material's response to r_{short} was assessed. In both cases, nearly identical bulk resistances of $R_{model} = 14\ \Omega$ were obtained in the half-cell model. For this example, the instantaneous resistance upon hard short formation at the E5 segment was calculated to be 76 Ω, given by $R_{exp} = V/I$ at the shorting voltage of 2.6 V and the measured current of 0.033 A. Therefore, the interfacial shorting resistance was 24 Ω and the shorting resistance through the entire membrane was 48 Ω ($2 \cdot R_{short}$).

After determining the interfacial shorting resistance, R_{short}, the coupled thermal-electrical analysis was done to determine the temperature distribution within the model by ramping the voltage at the PEM/CL interface from 0 to 3 V. Figure 2.30 shows the temperature distributions along the mid-plane of membrane from the epicenter of the short for $r_{short} = 0.006$ mm and for cell potentials between 0 to 6 V. The highest temperature is, expectedly, found at the epicenter or $r = 0$ mm. In all cases, except 0 V, the membrane temperature decreases rapidly with increasing distance from the epicenter. Very little difference in temperature distribution is seen between the cases of $r_{short} = 0.006$ and 0.014 mm, suggesting that the results may be relatively insensitive to the actual size of the short.

Figure 2.31 shows results of a parametric study of the membrane temperature at the shorting epicenter as a function of cell potential for a wide range of

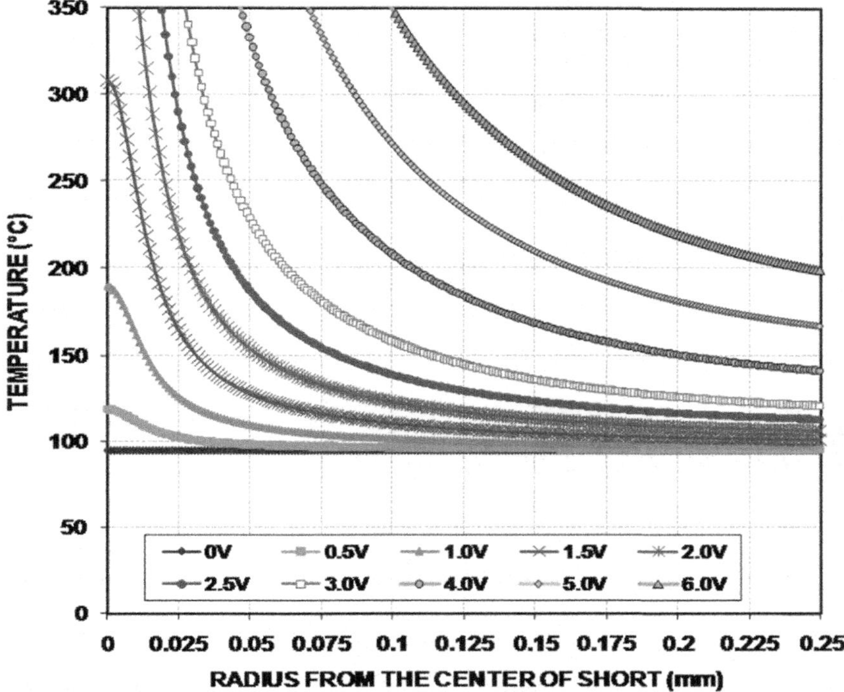

FIGURE 2.30 Temperature distribution, determined from the thermal-electrical finite element analysis, from the epicenter of the short at various cell potentials for a shorting radius, r_{short}, of 0.006 mm.

membrane shorting resistances. In the example discussed above, the membrane shorting resistance is 48 Ω. In that case, the temperature at the short epicenter increases rapidly from 95°C to more than 1,000°C as the cell potential is increased from 0 to 5 V. Samms *et al.* found that Nafion® has an initial decomposition temperature at 280°C and a peak decomposition temperature between 430°C and 470°C [120]. When the membrane decomposes and rapidly loses its mechanical integrity at these temperatures, the highly conductive catalyst and GDL materials from both sides of the membrane would come into contact under compression, allowing a large amount of current to pass through, and creating thermal runaway. The end result is a hard short. In the hard short study of the large active area cell, with a shorting resistance of 48 Ω, the 280°C to 470°C decomposition temperature range corresponds to a voltage range between 1.5 and 2 V, which is less than 1 V lower than the experimentally observed 2.6 V.

Temperature-cell potential curves are also shown in Fig. 2.31 for the membrane shorting resistances ranging from 6 to 8,000 Ω. Note that an 8,000 Ω short represents a healthy cell with virtually no short, as it is equivalent to

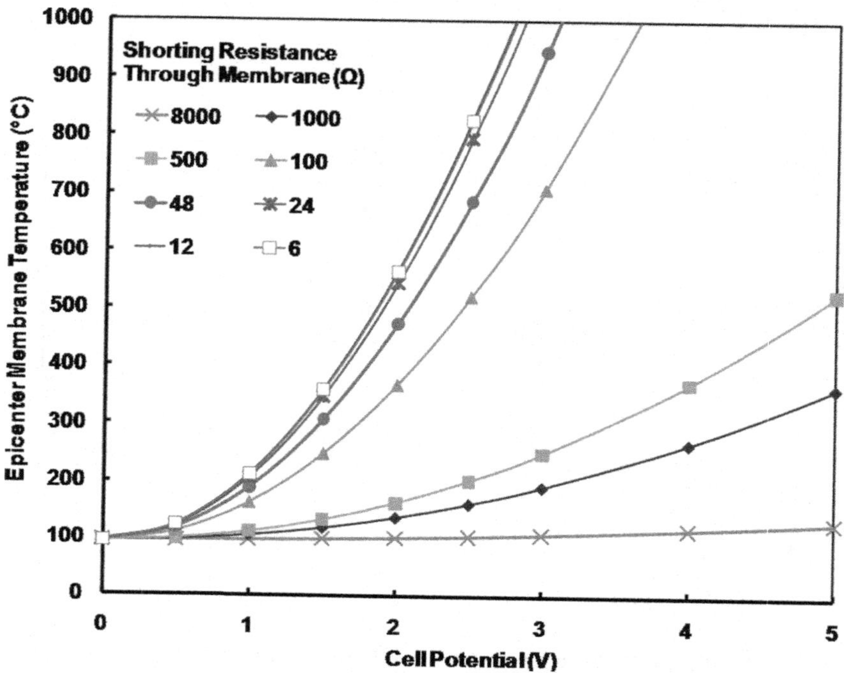

FIGURE 2.31 Membrane temperature at the epicenter of short as a function of cell potential for various shorting resistances through the membrane.

a shorting resistance of 14,400 $\Omega \cdot cm^2$ within a single 1.8 cm^2 current distribution segment. Although the 6 Ω resistance represents the most severe short in this study, it is still more than two orders of magnitude higher than that of a single 1 mm long carbon fiber poking completely through the membrane. There is very little difference in the temperature profiles for starting membrane shorting resistances ranging from 6 Ω to 24 Ω, suggesting that the membrane shorting resistance is sufficiently low so that the shorting current, and therefore the heat generation, is limited by the combined resistances in the bulk materials. On the other hand, as the membrane shorting resistance is sufficiently large, it can become the rate limiting factor. Therefore, the epicenter temperature strongly depends on the resistance of the soft short at the time of a voltage excursion. In the 8,000 Ω case, the epicenter only heats slightly as the potential is ramped from 0 to 5 V, and remains well below Nafion's decomposition temperature. This result suggests that thermal damage from a cell potential reversal should not occur in a healthy cell because thermal runaway cannot be triggered.

Figure 2.31 provides a convenient way to examine the conditions for the onset of hard shorts. Assuming that the PFSA membrane must reach Nafion's decomposition temperature (280°C to 470°C) to initiate a hard short, the critical cell shorting potential is between 1.4 V and 4.5 V for a membrane soft short less than 1,000 Ω. Equally important, when cell potential is below the OCV potential of 1 V, the temperature at the epicenter is below 200°C for even the lowest shorting resistance, suggesting that a hard short caused by thermal runaway is not likely to occur during normal fuel cell operation regardless of the magnitude of the soft short. This conclusion is in excellent agreement with the hard short voltage data compiled over several hundred tests (Fig. 2.26), as almost all of these cells developed hard shorts at cell potentials between 1.5 and 5 V.

It is interesting to note that during the course of this study, we also found that hard shorts developed over a similar voltage range when these shorting tests were run on a Kapton® polyimide film compressed between two sheets of GDL. As Kapton has a specific heat of 1.09 J/g·K and a thermal conductivity of 0.12 W/m·K, both very similar to that of Nafion®, the thermal-electrical finite element analysis and the temperature-cell potential results in Fig. 2.31 can be readily applied to a thin Kapton film. With a decomposition temperature between 500 and 600°C [121], hard shorts from thermal runaway in Kapton films are predicted to occur at a voltage in the range from 1.8 to about 5 V. This prediction agrees well with tests using a 12 μm Kapton film, which shorted at potentials ranging from 1.6 to 4.7 V, with a mean shorting voltage of 3.5 V. It should be noted that, since polyimide (e.g. Kapton) is one of the most thermally, dimensionally, mechanically, and dielectrically stable polymers, the fact that hard shorting occurred within the cell potential predicted by the finite element analysis indicates that soft shorts must have been induced in Kapton by mechanical penetration. This suggests that there is not likely to be a material solution to developing a PEM that is completely soft or hard short resistant.

5.5. Mitigation of Membrane Shorting

As mentioned above, it is not expected that the nature of the PEM material will enable membrane shorting to be prevented. All polymeric materials will decompose at the temperature ranges experienced when a potential excursion greater than about 1.5 V occurs at a soft short site of sufficiently low electrical resistance. The risk of shorting can be reduced by using thicker membranes, which are less susceptible to penetration by conductive objects, but thicker materials invariably result in increased proton transport resistance and a corresponding performance penalty. MPL and electrode materials have a far greater impact on membrane shorting because of their ability to buffer the membrane from GDL fibers. Additionally, use of defect-free materials, by eliminating catalyst agglomerates or nuggets (Fig. 2.2(d)) or other potentially damaging moieties, can reduce the risk of shorting. Similarly, clean manufacturing processes that minimize the potential for particulates or debris trapping between the material layers during assembly should be used. Cell design and operating strategy also offer opportunities to mitigate membrane shorting. The risk of shorting decreases with lower cell compression, but the compression must not be so low as to cause an increase in cell contact resistance, or increase the risk of mechanical membrane failure caused by humidity cycling (Section 3). Particularly, care must be taken during stack design to avoid local high compression regions. While a small degree of shorting may be beneficial in reducing the OCV and, thus, perhaps lowering the rate of chemical degradation or start-up/shut-down driven degradation [122], any 'intentional' shorting must be done extremely carefully so not to increase the risk of membrane failure.

The approaches discussed above all deal with reducing the risk of soft short formation. The best strategy to prevent soft shorts from turning into hard shorts is to prevent voltage excursions during fuel cell operation. This can either be accomplished by active cell voltage control, or by appropriate cell operating strategy and design. By keeping the membrane well hydrated, the risk of potential reversal caused by membrane dry-out and increased resistance is reduced. This can be achieved globally by controlling system temperature, pressure and humidification and locally by designing flow fields that enable uniform cell level humidification. Cell designs that enable uniform current density are preferable to reduce the risk of hot and dry spots that are most susceptible to local reversal.

6. SUMMARY AND FUTURE CHALLENGES

Durability of proton exchange membranes is a critical limitation to the commercialization of automotive fuel cell systems. Fuel cell stacks in which hundreds of cells operate in series will fail if even one cell develops a hole that enables direct combination of the H_2 and O_2 feed gases, or a short that causes current to flow through the membrane rather than through the external, power generating circuit. Three critical degradation mechanisms that can lead to membrane failure are chemical degradation, mechanical degradation, and membrane shorting. Effective study and mitigation of each membrane failure mode involves the following tasks:

- **i.** Fundamental, model-based, mechanistic understanding of the particular membrane failure mode under a given stressor.
- **ii.** Development of appropriate accelerated stress tests (ASTs) to enable rapid material and mitigation concept screening.
- **iii.** Incorporation of relevant *in-situ* diagnostics to track membrane health during fuel cell operation.
- **iv.** Thorough postmortem analysis to verify failure modes and identify root causes.
- **v.** Development of effective mitigation strategies to prevent or reduce membrane degradation.
- **vi.** Predictive models to predetermine membrane life in an operating fuel cell environment.

Both chemical and mechanical degradation of PFSA membranes have been extensively studied, and effective mitigation strategies have been developed that can significantly extend PFSA PEM lifetimes. ASTs have been developed that can be used to screen various membrane types and mitigation strategies, and effective *in-situ* diagnostics have been developed to track degradation of PFSA membranes during fuel cell operation. The failures observed in these accelerated tests, as seen by postmortem analysis, mimic the failures of PFSA membranes in field tests.

Future work is required to study the degradation of alternative PEM chemistries, such as hydrocarbon membranes. First of all, there is limited field data for these materials which can be used to determine the root causes of failure and validate that the ASTs are accelerating the relevant stressors. Secondly, new diagnostics may be required to monitor the health of alternative PEM chemistries. For example, monitoring fluoride release from cell effluents will not provide useful information about the chemical degradation of hydrocarbon-based PEMs. The approaches described in this chapter can be applied to such alternative PEMs, but new fundamental models for both chemical and mechanical degradation are likely required. Even for the case of PFSA membranes, further work is required to develop the predictive models required to enable accurate estimation of membrane life in a real system.

There have been limited studies to date on the impact of combined chemical and mechanical stresses that will be unavoidably experienced during automotive fuel cell operation. While evidence exists to support the idea that chemical degradation can be accelerated by mechanical stresses and vice versa, much more work is required to understand the fundamentals of combined chemical and mechanical stress and to develop predictive life models that are applicable when both stresses are present. However, one can effectively minimize the combined impact of chemical and mechanical stress by simultaneously incorporating both chemical and mechanical degradation mitigation strategies.

This chapter includes the first in-depth fundamental study of membrane shorting of PEMs in fuel cell systems. Shorting is a two step process during which soft shorts are created by penetration of conductive materials through the thickness of the membrane. Soft shorts become hard shorts when there is a voltage excursion that leads to high local temperatures and subsequent thermal decay of the PEM. These hard shorts are the severe events that lead to fuel cell stack failure. Results and analysis suggest that the PEM type is not the limiting factor in preventing membrane shorting, and that mitigation is best achieved by a combination of design and operating strategies.

ACKNOWLEDGMENTS

The authors would like to thank the following individuals for their contributions either by generating durability data, materials analysis, or method development: Michael Budinski, Wenbin Gu, Steven Falta, Ruichun Jiang, Yonqiang (Ron) Li, Daniel Miller and Kelly O'Leary from General Motors Electrochemical Energy Research Lab, Courtney Mittelsteadt and Han Liu from Giner Electrochemical Systems, and Scott Case, David Dillard, Michael Ellis and Kshitish Patankar from Virginia Polytechnic Institute and State University.

GLOSSARY

Acronyms

AST	accelerated stress tests
BPSH	sulfonated biphenyl sulfone
CL	catalysts layer
DENT	double-edge notch test
DMA	dynamic mechanical analysis
DOE	Department of Energy
ePTFE	expanded poly [tetrafluoroethylene]
EW	equivalent weight
FRR	fluoride release rate
GDL	gas diffusion layer
HPA	heteropolyacid
IR-corrected	resistance losses subtracted
MEA	membrane electrode assembly
MPL	microporous layer
OCV	open circuit voltage
PEFC	polymer electrolyte fuel cell
PEM	proton exchange membrane
PFSA	perfluorosulfonic acid
PMA	phosphomolybdic acid
POSS	polyhedral oligomeric silsesquioxane
PTFE	polytetrafluoroethylene
PVDF	polyvinylidene difluoride
PWA	phosphotungstic acid
RH	relative humidity
SEM	scanning electron micrographs
SHE	standard hydrogen electrode
SiWA	silicotungstic acid
SPEEK	sulfonated poly-ether-ether-ketone
USFCC	United States Fuel Cell Council

Nomenclature

a	blister radius
a_H	hygral shift factor
a_T	thermal shift factor
a_{TH}	hygrothermal shift factor
D	diffusion coefficient
E	uniaxial relaxation modulus
E_a	activation energy
h	membrane thickness
k_i	first order reaction rate constant
L	pulling force in each leg of knife slit specimen
r_{short}	soft short radius
p	blister pressure
P	applied cutting force
R_{exp}	measured short resistance in a given cell segment

R_{model}	resistance of entire half cell model
R_{short}	interfacial soft short resistance
T	temperature
t	time
t_f	failure time
W	knife slit tearing weight
β	linear coefficient of hygral expansion
δ_{ij}	Kronecker delta
ε_{ij}	strain tensor
θ	cutting angle
λ	number of water molecules per sulfonic acid
ξ	time
σ	applied stress
σ_{ij}	stress tensor

REFERENCES

[1] D.L. Feldheim, D.R. Lawson, C.R. Martin, Influence of the Sulfonate Countercation on the Thermal Stability of Nafion® Perfluorosulfonate Membranes, J Polymer Sci. Polymer Phys. 31 (1993) 953–957.

[2] S.H. de Almeida, Y. Kawano, Thermal stability and decomposition of Nafion® membranes with different cations, J. Therm. Anal. 58 (1999) 569–577.

[3] US DOE Multi-Year Research, Development and Demonstration Plan. Planned Program Activities for 2005–2015, http://www1.eere.energy.gov/hydrogenandfuelcells/mypp/, 2007.

[4] Y.H. Lai, C.K. Mittelstaedt, C.S. Gittleman, D.A. Dillard, Viscoelastic Stress Model and Mechanical Characterization of Perfluorosulfonic Acid Polymer Electrolyte Membranes, Proceedings of the 3rd International Conference Fuel Cell Science, Engineering, and Technology, Ypsilanti, MI, (2005) 161–167.

[5] E.F. Holby, W. Sheng, Y. Shao-Horn, D. Morgan, Pt nanoparticle stability in PEM fuel cells: influence of particle size distribution and crossover hydrogen, Energy Environ. Sci. 2 (2009) 865–871.

[6] V. Ramani, H.R. Kunz, J.M. Fenton, Effect of Nafion®/HPA Composite Membrane Composition on High Temperature/Low Relative Humidity PEMFC Performance, Journal of Membrane Science 232 (2004) 31–44.

[7] B.H. Pan, J.Y. Lee, Immobilisation of phosphomolybdic (PM) acid by Nafion and the electrochromism of the resulting PM-Nafion films, J. Mater. Chem. 7 (1997) 187–191.

[8] H. Tian, O. Savadogo, Silicotungstic Acid Nafion Composite Membrane for Proton-Exchange Membrane Fuel Cell Operation at High Temperature, Journal of New Materials for Electrochemical Systems 9 (2006) 61–71.

[9] C.Y. Yuh, L. Lipp, P. Patel, R. Kopp, Membrane-Electrode Assembly for High-Temperature PEMFC, Prepr. Pap.-Am. Chem. Soc, Div. Fuel Chem. 48 (2) (2003) 893–894.

[10] X. Zhang, S.W. Tay, Z. Liu, L. Hong, Restructure proton conducting channels by embedding starburst POSS-g-acrylonitrile oligomer in sulfonic perfluoro polymer matrix, J. Membr. Sci. 329 (2009) 228–235.

[11] S.H. Kwak, T.H. Yang, C.S. Kim, K.H. Yoon, Polymer composite membrane incorporated with a hygroscopic material for high-temperature PEMFC, Electrochimica Acta 50 (2004) 653–657.

[12] S. Cleghorn, J. Kolde, W. Liu, Catalyst coated composite membranes, in: W. Vielstich, H. Gasteiger, A. Lamm (Eds), Handbook of Fuel Cells Volume 3: Fundamentals, Technology and Applications, John Wiley & Sons, New York, 2003, pp. 566–575.

[13] F.Q. Liu, B.L. Yi, D.M. Xing, J.R. Yu, H.M. Zhang, Nafion/PTFE composite membranes for fuel cell applications, J. Membr. Sci. 212 (2003) 213–223.

[14] H. Liu, M. Chen, J. Willey, C. Mittelsteadt, A. LaConti, High Performance, Dimensionally Stable Membrane, 2008, Fuel Cell Seminar & Expo., Phoenix, AZ, 2008, paper 1384.

[15] F.A. Landis, R.B. Moore, Blends of a Perfluorosulfonate Ionomer with Poly(vinylidene fluoride): The Effect of Counterion Type on Phase Separation and Crystal Morphology, Macromolecules 33 (2000) 6031–6041.

[16] Y.Z. Fu, A. Manthiram, M.D. Guiver, Blend Membranes Based on Sulfonated Polyetheretherketone and Polysulfone Bearing Benzimidazole Side Groups for Fuel Cells, Electrochem, Commun 8 (2006) 1386–1390.

[17] C. Gibon, S. Norvez, S. Tencé-Girault, J.T. Goldbach, Control of Morphology and Crystallization in Polyelectrolyte/Polymer Blends, Macromolecules 41 (2008) 5744–5752.

[18] J.A. Kerres, Blended and Cross-Linked Ionomer Membranes for Application in Membrane Fuel Cells, Fuel Cells 5 (2005) 230–247.

[19] J. Choi, K.M. Lee, R. Wycisk, P.N. Pintauro, P.T. Mather, Nanofiber Network Ion-Exchange Membranes, Macromolecules 41 (2008) 4569–4572.

[20] D.E. Curtin, R.D. Lousenberg, T.J. Henry, P.C. Tangeman, M.E. Tisack, Advanced Materials for Improved PEMFC Performance and Life, Journal of Power Sources 131 (1–2) (2004) 41–48.

[21] E. Endoh, Development of Highly Durable PFSA Membrane and MEA for PEMFC Under High Temperature and Low Humidity Conditions, ECS Trans. 16 (2) (2008) 1229–1236.

[22] F.D. Coms, H. Liu, J.E. Owejan, Mitigation of Perfluorosulfonic Acid Membrane Chemical Degradation Using Cerium and Manganese Ions, ECS Trans. 16 (2) (2008) 1735–1747.

[23] P. Trogadas, J. Parrondo, V. Ramani, Degradation mitigation in polymer electrolyte membranes using cerium oxide as a regenerative free-radical scavenger, Electrochem. Solid State Lett. 11 (2008) B113–B116.

[24] C.S. Gittleman, Automotive Perspective on PEM Evaluation, DOE High Temperature Membrane Working Group, Washington, DC, 2009, http://www1.eere.energy.gov/hydrogenandfuelcells/pdfs/htmwg_may09_automotive_perspective.pdf.

[25] USFCC Protocol on Fuel Cell Component Testing.http://www.usfcc.com/resources/Trans-H2Quality-Primer-04-003.pdf, 2004.

[26] C.S. Gittleman, Y.H. Lai, D.P. Miller, Durability of Perfluorosulfonic Acid Membranes for PEM Fuel Cells. Extended Abstract, AIChE 2005 Annual Meeting, Cincinnati, OH, 2005.

[27] S.M. MacKinnon, T.J. Fuller, F.D. Coms, M.R. Schoeneweiss, C.S. Gittleman, Y.-H. Lai, R. Jiang, A. Brenner, Design and Characterization of Alternative Proton Exchange Membranes for Automotive Applications, in: J. Garche, C.K. Dyer, P.T. Moseley, Z. Ojumi, D.A.J. Rand, B. Scrosati (Eds), Encyclopedia of Electrochemical Power Sources, Elsevier, 2009, pp. 741–754.

[28] M.F. Mathias, R. Makharia, H.A. Gasteiger, J.J. Conley, T.J. Fuller, C.S. Gittleman, S.S. Kocha, D.P. Miller, C.K. Mittelsteadt, T. Xie, S.G. Yan, P.T. Yu, Two fuel cell cars in every garage, Interface 14 (2005) 24–35.

[29] A.B. LaConti, M. Hamdan, R.C. McDonald, Mechanism of Membrane Degradation for PEMFCs, in Mechanisms of Membrane Degradation for PEMFCs, in: W. Vielstich,

A. Lamn, H.A. Gasteiger (Eds), Handbook of Fuel Cells: Fundamentals, Technology and Applications, Vol. 3, John Wiley & Sons, New York, 2003, pp. 647–662.

[30] R. Borup, J. Meyers, B. Pivovar, Y. Kim, R. Mukundan, N. Garland, D. Meyers, M. Wilson, F. Garzon, D. Wood, P. Zelenay, K. More, K. Stroh, T. Zawodzinski, J. Boncella, J. McGrath, M. Inaba, K. Miyatake, M. Hori, K. Ota, Z. Ogumi, S. Miyata, A. Nishikata, Z. Siroma, Y. Uchimoto, K. Yasuda, K. Kimijima, N. Iwashita, Scientific Aspects of Polymer Electrolyte Fuel Cell Durability and Degradation, Chem. Rev. 107 (2007) 3904–3951.

[31] W. Liu, M. Crum, Effective Testing Matrix for Studying Membrane Durability in PEM Fuel Cells: Part I. Chemical Durability, ECS Trans. 3 (1) (2006) 531–540.

[32] H. Liu, J. Zhang, F. Coms, W. Gu, B. Litteer, H. Gasteiger, Impact of Gas Partial Pressure on PEMFC Chemical Degradation, ECS Trans. 3 (1) (2006) 493–505.

[33] T. Kinumoto, M. Inaba, Y. Nakayama, K. Ogata, R. Umebayashi, A. Tasaka, Ya. Iriyama, T. Abe, Z. Ogumi, Durability of Perfluorinated Ionomer Membrane Against Hydrogen Peroxide, J. Power Sources 158 (2) (2006) 1222–1228.

[34] M. Inaba, H. Yamada, R. Umebayashi, M. Sugishita, A. Tasaka, Membrane Degradation in Polymer Electrolyte Fuel Cells under Low Humidification Conditions, Electrochemisty 75 (2007) 207–212.

[35] F.D. Coms, The Chemistry of Fuel Cell Membrane Chemical Degradation, ECS Trans. 16 (2) (2008) 235–255.

[36] M. Danilczuk, F.D. Coms, S. Schlick, Visualizing Chemical Reactions and Crossover processes in a Fuel Cell Inserted in the ESR Resonator: Detection by Spin Trapping Oxygen Radicals, Nafion-Derived Fragments, and Hydrogen and Deuterium Adducts, J. Phys. Chem. B 113 (2009) 8031–8042.

[37] W. Liu, D. Zuckerbrod, *In-situ* Detection of Hydrogen Peroxide in PEM Fuel Cells, J. Electrochem. Soc. 152 (2005) A1165–A1170.

[38] S.W. Benson, Thermochemical Kinetics, Second ed., Wiley, New York, 1976.

[39] N. Cipollini, Chemical Aspects of Chemical Degradation, ECS Trans. 11 (1) (2007) 1071–1082.

[40] H. Liu, H. Gasteiger, A. Laconti, J. Zhang, Factors Impacting Chemical Degradation of Perfluorinated Sulfonic Acid Ionomers, ECS Trans. 1 (8) (2006) 283–293.

[41] V.O. Mittal, H.R. Kunz, J.M. Fenton, Membrane Degradation Mechanisms in PEMFCs, J. Electrochem. Soc. 154 (2007) B652–B656.

[42] T. Madden, D. Weiss, N. Cipollini, D. Condit, M. Gummalla, S. Burlatsky, V. Atrazhev, Degradation of Polymer-Electrolyte Membranes in Fuel Cells, J. Electrochem. Soc. 156 (2009) B657–B662.

[43] S. Burlatsky, V. Atrazhev, N. Cipollini, D. Condit, N. Erikhman, Aspects of PEMFC Degradation, ECS Trans. 1 (8) (2006) 239–246.

[44] J. Zhang, B. Litteer, W. Gu, H. Liu, H. Gasteiger, Effect of Hydrogen and Oxygen Partial Pressure on Pt Precipitation within the Membranes of PEMFCs, J. Electrochem. Soc. 154 (2007) B1006–B1011.

[45] H. Christensen, K. Sehested, Reaction of Hydroxyl Radicals with Hydrogen at Elevated Temperatures: Determination of the Activation Energy, J. Phys. Chem. 87 (1983) 118–120.

[46] H. Christensen, K. Sehested, H. Cortfitzen, Reaction of Hydroxyl Radicals with Hydrogen Peroxide at Ambient and Elevated Temperatures, J. Phys. Chem. 86 (1982) 1588–1590.

[47] H. Schwarz, Free Radicals Generated by Radiolysis of Aqueous Solutions, J. Chem. Ed. 58 (1981) 101–105.

[48] G.V. Buxton, C.L. Greenstock, W.P. Helman, A.B. Ross, Critial Review of Rate Constants for Reactions of Hydrated Electrons, Hydrogen Atoms and Hydroxyl Radicals (•OH/•O^-) in Aqueous Solution, J. Phys. Chem Ref. Data 17 (1988) 513–886.

[49] D.R. Burgess Jr., M.R. Zachariah, W. Tsang, P.R. Westmoreland, Thermochemical and Chemical Kinetic Data for Fluorinated Hydrocarbons, Prog. Energy Combust. Sci. 21 (1996) 453–529.

[50] S. Hommura, K. Kawahara, T. Shimohira, Y. Teraoka, Development of a Method for Clarifying the Perfluorosulfonated Membrane Degradation Mechanism in a Fuel Cell Environment, J. Electrochem. Soc. 155 (2008) A29–A33.

[51] T. Kinumoto, M. Inaba, Y. Nakayama, K. Ogata, R. Umebayashi, A. Tasaka, Y. Iriyama, K. Ogumi, Durability of Perfluorinated Ionomer Membrane Against Hydrogen Peroxide, J. Power Sources 158 (2006) 1222–1228.

[52] C. Chen, T. Fuller, H_2O_2 Formation under Fuel-Cell Conditions, ECS Trans. 11 (1) (2007) 1127–1137.

[53] K. Kreuer, On the Development of Proton Conducting Polymer Membranes for Hydrogen and Methanol Fuel Cells, J. Membr. Sci. 188 (2001) 29–39.

[54] C.K. Mittelsteadt, H. Liu, Permeability Conductivity, Ohmic Shorting of Ionomeric Membranes, in: W. Vielstich, H. Yokokawa, H.A. Gasteiger (Eds), Handbook of Fuel Cells: Advances in Electrocatalysis, Materials, Diagnostics and Durability, Vols 5 & 6, John Wiley & Sons, Chichester, 2009, pp. 345–358.

[55] S.J. Paddison, J.A. Elliott, Molecular Modeling of the Short-Side-Chain Perfluorosulfonic Acid Membrane, J. Phys Chem A 109 (2005) 7583–7593.

[56] S. Li, W. Qian, F. Tao, Ionic Dissociation of Methanesulfonic Acid in Small Water Clusters, Chem. Phys. Lett. 438 (2007) 190–195.

[57] R.E. Noftle, G.H. Cady, Preparation and Properties of Bis(trifluoromethylsufuryl) Peroxide and Trifluoromethyl Trifluoromethanesulfonate, Inorg. Chem. 4 (1965) 1010–1012.

[58] M.K. Kadirov, A. Bosniakovic, S. Schlick, Membrane-Derived Fluorinated Radicals Detected by Electron Spin Resonance in UV-Irradiated Nafion and Dow Ionomers: Effect of Counterions and H_2O_2, J. Phys. Chem. B 109 (2005) 7664–7670.

[59] C. Zhou, M.A. Guerra, Z. Qiu, T.A. Zawodzinski Jr., D.A. Schiraldi, Chemical Durability Studies of Perfluorinated Sulfonic Acid Polymers and Model Compounds under Mimic Fuel Cell Conditions, Macromolecules 40 (2007) 8695–8707.

[60] J. Healy, C. Hayden, T. Xie, K. Olson, R. Waldo, M. Brundage, H. Gasteiger, J. Abbott, Aspects of the Chemical Degradation of PFSA Ionomers used in PEM Fuel Cells, Fuel Cells 5 (2) (2005) 302–308.

[61] M. Aoki, H. Uchida, M. Watanabe, Decomposition Mechanism of Perfluorosulfonic Acid Electrolyte in Polymer Electrolyte Fuel Cells, Electrochem. Comm. 8 (2006) 1509–1513.

[62] E. Endoh, S. Hommura, S. Terazono, H. Widjaja, J. Anzai, Degradation Mechanism of the PFSA Membrane and Influence of Deposited Pt in the Membrane, ECS Trans. 11 (2007) 1083–1091.

[63] A. Ohma, S. Yamamoto, K. Shinohara, Analysis of Membrane Degradation Behavior During OCV Hold Test, ECS Trans. 11 (2007) 1181–1192.

[64] G. Davies, L.J. Kirschenbaum, K. Kustin, The Kinetics and Stoichiometry of the Reaction between Manganese (III) and Hydrogen Peroxide in Acid Perchlorate Solution, Inorg Chem. 7 (1968) 146–154.

[65] G. Czapski, B.H. Bielski, N. Sutin, The Kinetics of the Oxidation of Hydrogen Peroxide by Cerium(IV), J. Phys. Chem. 67 (1963) 201–203.

[66] H.A. Mahlman, R.W. Matthews, T.J. Sworski, Reduction of Cerium(IV) by Hydrogen Peroxide. Dependence of Reaction Rate on Hammett's Acidity Function, J. Phys. Chem. 75 (1971) 250–255.

[67] E. Endoh, Highly Durable MEA for PEMFC under High Temperature and Low Humidity Conditions, ECS Trans. 3 (1) (2006) 9–18.

[68] T.A. Greszler, T.E. Moylan, H.A. Gasteiger, Modeling the impact of cation contamination in a polymer electrolyte membrane fuel cell, in: W. Vielstich, H. Yokokawa, H.A. Gasteiger (Eds), Handbook of Fuel Cells: Advances in Electrocatalysis, Materials, Diagnostics and Durability, Vols 5 & 6, John Wiley & Sons, Chichester, 2009, pp. 728–748.

[69] V.A. Sethuraman, J.W. Weidner, A.T. Haug, L.V. Protsailo, Durability of Perfluorosufonic acid and hydrocarbon membranes: effect of humidity and temperature, J. Electrochem. Soc. 155 (2) (2008) B119–B124.

[70] L. Zhang, S. Mukerjee, Investigation of Durability Issues of Selected Nonfluorinated Proton Exchange Membranes for Fuel Cell Application, J. Electrochem. Soc. 153 (2006) A1062–A1072.

[71] M. Aoki, N. Asano, K. Miyatake, H. Uchida, M. Watanabe, Durability of Sulfonated Polyimide Membrane Evaluated by Long-Term Polymer Electrolyte Fuel Cell Operation, J. Electrochem. Soc. 153 (2006) A1154–A1158.

[72] M. Gross, G. Maier, T. Fuller, S. MacKinnon, C. Gittleman, Design rules for the improvement of the performance of hydrocarbon-based membranes for proton exchange membrane fuel cell, in: W. Vielstich, H. Yokokawa, H.A. Gasteiger (Eds), Handbook of Fuel Cells: Advances in Electrocatalysis, Materials, Diagnostics and Durability, Vols 5 & 6, John Wiley & Sons, Chichester, 2009, pp. 283–299.

[73] Y.H. Lai, D.A. Dillard, Mechanical durability characterization and modeling of ionomeric membranes, in: W. Vielstich, H. Yokokawa, H.A. Gasteiger (Eds), Handbook of Fuel Cells: Advances in Electrocatalysis, Materials, Diagnostics and Durability, Vols 5 & 6, John Wiley & Sons, Chichester, 2009, pp. 403–419.

[74] A. Reyna-Valencia, S. Kaliaguine, M. Bousmina, Tensile mechanical properties of sulfonated poly(ether ether ketone) (SPEEK) and BPO_4/SPEEK membranes, J. Appl Polymer Sci. 98 (2005) 2380–2393.

[75] K.A. Patankar, D.A. Dillard, S.W. Case, M.W. Ellis, Y.H. Lai, M.K. Budinski, C.S. Gittleman, Hygrothermal characterization of the viscoelastic properties of Gore-Select® 57 proton exchange membrane, Mechanics of Time–Dependent Materials 12 (2008) 221–236.

[76] Y. Li, J.K. Quincy, S.W. Case, M.W. Ellis, D.A. Dillard, Y.H. Lai, M.K. Budinski, Characterizing the fracture resistance of proton exchange membranes, J. Power Sources 185 (2008) 374–380.

[77] Y.H. Lai, C.K. Mittelsteadt, C.S. Gittleman, D.A. Dillard, Viscoelastic stress model and mechanical characterization of perfluorosulfonic acid (PFSA) polymer electrolyte membranes, J. Fuel Cell Sci. Technol. 6 (2009) 0210021–02100213.

[78] D.A. Dillard, Y. Li, J.R. Grohs, S.W. Case, M.W. Ellis, Y.H. Lai, M.K. Budinski, C.S. Gittleman, On the use of pressure-loaded blister tests to characterize the strength and durability of proton exchange membranes, J. Fuel Cell Sci. Technol. 6 (2009) 0310141–0310148.

[79] Y. Li, J.R. Grohs, M.T. Pestrak, D.A. Dillard, S.W. Case, M.W. Ellis, Y.H. Lai, C.S. Gittleman, D.P. Miller, Fatigue and creep to leak tests of proton exchange membranes using pressure-loaded blisters, J. Power Sources 194 (2009) 873–879.

[80] J.R. Grohs, Y. Li, D.A. Dillard, S.W. Case, M.W. Ellis, Y.H. Lai, M.K. Budinski, C.S. Gittleman, Evaluating the time and temperature dependent biaxial strength of Gore-Select® series 57 proton exchange membrane using a pressure loaded blister test, J. Power Sources 195 (2010) 527–531.

[81] M.T. Pestrak, Y. Li, S.W. Case, D.A. Dillard, M.W. Ellis, Y.H. Lai, and C.S. Gittleman, The effect of catalyst layer cracks on the fatigue lifetimes of MEA, J Fuel Cell Sci. Technol. 7 (2010) 0410091–04100910.

[82] K.A. Patankar, D.A. Dillard, S.W. Case, M.W. Ellis, Y.H. Lai, Y. Li, M.K. Budinski, C.S. Gittleman, Characterizing fracture energy of proton exchange membranes (PEM) using a knife slit test, J. Polym. Sci., Part B: Polym. Phys. 48 (3) (2010) 333–343.

[83] Y. Tang, M.H. Santare, A.M. Karlsson, S. Cleghorn, S.W.B. Johnson, Stresses in proton exchange membranes due to hygro-thermal loading, J. Fuel Cell Sci. Technol 3 (2006) 119–124.

[84] A. Kusoglu, A.M. Karlsson, M.H. Santare, S. Cleghorn, W.B. Johnson, Mechanical response of fuel cell membranes subjected to a hygro-thermal cycle, J. Power Sources 161 (2006) 987–996.

[85] A. Kusoglu, A.M. Karlsson, M.H. Santare, S. Cleghorn, W.B. Johnson, Mechanical behavior of fuel cell membranes under humidity cycles and effect of swelling anisotropy on the fatigue stresses, J. Power Sources 170 (2007) 345–358.

[86] Y. Tang, A. Kusoglu, A.M. Karlsson, M.H. Santare, M.H.S. Cleghorn, W.B. Johnson, Mechanical properties of a reinforced composite polymer electrolyte membrane and its simulated performance in PEM fuel cells, J. Power Sources 175 (2008) 817–825.

[87] X. Huang, R. Solasi, Y. Zou, M. Feshler, K. Reifsnider, D. Condit, S. Burlatsky, T. Madden, Mechanical endurance of polymer electrolyte membrane and PEM fuel cell durability, J. Polym. Sci., Part B: Polym. Phys. 44 (2006) 2346–2357.

[88] R. Solasi, Y. Zou, X. Huang, K. Reifsnider, D. Condit, On mechanical behavior and in-plane modeling of constrained PEM fuel cell membranes subjected to hydration and temperature cycles, J. Power Sources 167 (2007) 366–377.

[89] R. Solasi, Y. Zou, X. Huang, K. Reifsnider, A time and hydration dependent viscoplastic model for polyelectrolyte membranes in fuel cells, Mech. Time-Depend, Mater 12 (2008) 15–30.

[90] L. Yan, T.A. Gray, K.A. Patankar, S.W. Case, M.W. Ellis, R.B. Moore, D.A. Dillard, Y.H. Lai, Y. Li, and C.S. Gittleman, "The nonlinear viscoelastic properties of PFSA membranes in water-immersed and humid air conditions" Experimental Mechanics on Emerging Energy Systems and Materials, Vol. 5, Conference Proceedings of the Society for Experimental Mechanics Series, 16 (2011) 163–174.

[91] M.A.R. Sadiq, Al-Baghdadi, A CFD study of hygro-thermal stresses distribution in PEM fuel cell during regular cell operation, Renewable Energy 34 (2009) 674–682.

[92] M.A.R. Sadiq, Al-Baghdadi, H.A.K. Shahad, Al-Janabi, Effect of operating parameters on the hygro-thermal stresses in proton exchange membranes of fuel cells, Int. J. Hydrogen Energy 32 (2007) 4510–4522.

[93] S. Yesilyurt, Modeling and simulations of polymer electrolyte membrane fuel cells with poroelastic approach for coupled liquid water transport and deformation in the membrane, J. Fuel Cell Sci. Technol. 7 (2010) 0310081–0310089.

[94] J.D. Ferry, Viscoelastic Properties of Polymers, Third ed., Wiley, New York, 1980.

[95] K.A. Patankar, Linear and nonlinear viscoelastic characterization of proton exchange membranes and stress modeling for fuel cell application, PhD dissertation, Virginia Polytechnic Institute and State University 2009.

[96] Y. Tang, A.M. Karlsson, M.H. Santare, M. Gilbert, S. Cleghorn, W.B. Johnson, An experimental investigation of humidity and temperature effects on the mechanical properties of perfluorosulfonic acid membrane, Mater. Sci. Eng., A 425 (2006) 297–304.

[97] D.R. Morris, X. Sun, Water-sorption and transport properties of Nafion 117 H, J. Polym. Sci., Part B: Polym. Phys. 50 (1993) 1445–1452.

[98] F. Bauer, F. Denneler, S. Willert-Porada, Influence of temperature and humidity on the mechanical properties of Nafion® 117 polymer electrolyte membrane, J. Polym. Sci., Part B: Polym. Phys. 43 (2005) 786–795.

[99] D.L. Flaggs, F.W. Crossman, Analysis of the viscoelastic response of composite laminates during hygrothermal exposure, J. Composite Mater 15 (1981) 21–40.

[100] M. Budinski, C.S. Gittleman, Y.H. Lai, B. Liteer, D.P. Miller, Characterization of Perfluorosulfonic Acid Membranes for PEM Fuel Cell Mechanical Durability, Extended Abstract, the AIChE 2004 Annual Meeting, Cincinnati, OH, 2004.

[101] R. Jiang, H.R. Kunz, J.M. Fenton, Composite silica/Nafion® membranes prepared by tetraethylorthosilicate sol-gel reaction and solution casting for direct methanol fuel cells, J. Membrane Sci. 272 (2006) 116–124.

[102] A.S. Arico, V. Baglio, V. Antonucci, I. Nicotera, C. Oliviero, L. Coppola, P.L. Antonucci, An NMR and SAXS investigation of DMFC composite recast Nafion membranes containing ceramic fillers, J. Membrane Sci. 270 (2006) 221–227.

[103] A. Sacca, A. Carbone, E. Passalacqua, A. D'Epifanio, S. Licoccia, E. Traversa, E. Sala, F. Traini, R. Ornelas, Nafion-TiO_2 hybrid membranes for medium temperature polymer electrolyte fuel cells, J. Power Sources 152 (2005) 16–21.

[104] D.H. Jung, S.Y. Cho, D.H. Peck, D.R. Shin, J.S. Kim, Zeolite-based composite membranes for high temperature direct methanol fuel cells, J. Power Sources 118 (2003) 205–212.

[105] M.K. Song, S.B. Park, Y.T. Kim, K.H. Kim, S.K. Min, H.W. Rhee, Characterization of polymer-layered silicate nanocomposite membranes for direct methanol fuel cells, Electrochimica Acta 50 (2004) 639–643.

[106] A.L. Moster, B.S. Mitchell, Mechanical and Hydration Properties of Nafion®/Ceramic Nanocomposite Membranes Produced by Mechanical Attrition, J. Appl. Polymer Science 111 (2009) 1144–1150.

[107] J.E. McGrath, D.G. Baird, High-Temperature, Low Relative Humidity, Polymer-type Membranes Based on Disulfonated Poly(arylene ether) Block and Random Copolymers, DOE Hydrogen Program FY 2008 Progress Report, (2008) 982–986.

[108] X. Yu, A. Roy, S. Dunn, J. Yang, J.E. McGrath, Synthesis and Characterization of Sulfonated-Fluorinated, Hydrophilic-Hydrophobic Multiblock Copolymers for Proton Exchange Membranes, J. of Polymer Sci.: Part A: Polymer Chemistry 47 (2009) 1038–1051.

[109] H.B. Park, C.H. Lee, J.Y. Sohn, Y.M. Lee, B.D. Freeman, H.J. Kim, Effect of crosslinked length in sulfonated polyimide membranes on water uptake, proton conduction, and methanol permeation properties, J. Membr. Sci. 285 (2006) 432–443.

[110] S. Zhong, X. Cui, H. Cai, T. Fu, C. Zhao, H. Na, Crosslinked sulfonated poly(ether ether ketone) proton exchange membranes for direct methanol fuel cell applications, J. Power Sources 164 (2007) 65–72.

[111] Y.H. Lai, P. Rapaport, C. Ji, V. Kumar, Channel intrusion of gas diffusion media and the effect on fuel cell performance, J. Power Sources 184 (2008) 120–128.

[112] Y.H. Lai, D.P. Miller, C. Ji, T.A. Trabold, Stack compression of PEM fuel cells, Proceedings of FUELCELL2004, 2rd International ASME Conference on Fuel Cell Science, Engineering, and Technology, Rochester, NY, 2004.

[113] S. Kinoshita, H. Shimoda, Performances of Highly Durable PFSA Polymer based MEAs at High Temperature, Low RH & Dry-Wet Conditions, Progress MEA 2008, La Grande Motte, France, 2008.

[114] G. Escobedo, M. Gummalla, R.B. Moore, Enabling Commercial PEM Fuel Cells with Breakthrough Lifetime Improvements, DOE Hydrogen Program FY 2006 Progress Report, (2006) 706–712.

[115] D. Liu, S. Case, Durability study of proton exchange membrane fuel cells under dynamic testing conditions with cyclic current profile, J Power Sources 162 (1) (2006) 521–531.

[116] R.N. Carter, W. Gu, B. Brady, K. Subramanian, H.A. Gasteiger, MEA Degradation Mechanisms Studies by Current Distribution Measurements, in: W. Vielstich, H. Yokokawa, H.A. Gasteiger (Eds), Handbook of Fuel Cells: Advances in Electrocatalysis, Materials, Diagnostics and Durability, Vols 5 & 6, John Wiley & Sons, Chichester, 2009, pp. 829–843.

[117] M. Fowler, R.F. Mann, J.C. Amphlett, B.A. Peppley, P.R. Roberge, Conceptual reliability analysis of PEM fuel cells, in: W. Vielstich, A. Lamn, H.A. Gasteiger (Eds), Handbook of Fuel Cells: Fundamentals, Technology and Applications, Vol. 3, John Wiley & Sons, New York, 2003, pp. 663–677.

[118] DuPont™ Teflon® PTFE Product Information Sheets. http://www2.dupont.com/Teflon_Industrial/en_US/tech_info/prodinfo_ptfe.html, 2009.

[119] ABAQUS Analysis User's Manual, HKS Inc., 2007.

[120] S. Samms, S. Wasmus, R. Savinell, Thermal stability of Nafion in simulated fuel cell environments, J. Electrochem. Soc. 143 (1996) 1498–1504.

[121] A. Lua, J. Su, Isothermal and non-isothermal pyrolysis kinetics of Kapton® polyimide, Polymer Degradation and Stability 91 (2006) 144–153.

[122] J.P. Meyers, R.M. Darling, Model of Carbon Corrosion in PEM Fuel Cells, J. Electrochem. Soc. 153 (2006) A1432–A1442.

Chapter 3

Electrochemical Degradation: Electrocatalyst and Support Durability

Shyam S. Kocha*

Fuel Cell Laboratory, Nissan Technical Center North America, Farmington Hills, MI, USA,

**Currently at the National Renewable Energy Laboratory, Golden CO 80401, USA*

1. INTRODUCTION

Stack technology for direct hydrogen proton exchange membrane fuel cells (PEMFCs) for automotives has advanced to levels where commercialization is projected to commence around 2015. Currently (2010), there are an estimated 200 operational fuel cell vehicles (FCVs) and 20 fuel cell buses; these FCVs typically employ compressed hydrogen stored in tanks at 350–700 bar, can be refueled in about 3–4 min, and provide a driving range comparable to internal combustion engine (ICE) vehicles. About 10 million tons of hydrogen are produced in the US every year and currently ~60 hydrogen fueling stations supply fuel to the vehicles. However, several fundamental and practical engineering challenges need to be addressed in order to clear the pathway towards automotive PEMFC commercialization. The challenges lie in both the areas of durability/cost of the fuel cell components (catalyst, membrane, diffusion media, bipolar plates, etc.), and the balance of plant (thermal, fuel, air, and water management). In this chapter, we focus on the electrochemical degradation of the most critical fuel cell components – namely the catalyst and the support it is dispersed on.

Cost analyses (by TIAX, LLC and DTI) assuming 500,000 units/year indicate that improvements in technology have resulted in a lowering of the projected transportation fuel cell system cost from \$275/kW in 2002 to \$51/kW in 2010 – well on the way towards the target of \$30/kW in 2015 [1–3]. The cost of the stack alone is about half the cost of the entire fuel cell system including the balance of plant. The cost of the Pt-based catalyst in the stack has been estimated to account for ~50% of the stack cost which amounts to 22% of the total system cost [4]. Thus, the high catalyst material cost of ~\$50 g_{Pt} demands

Polymer Electrolyte Fuel Cell Degradation. DOI: 10.1016/B978-0-12-386936-4.10003-X

that we maximize the current drawn from every gram of Pt while simultaneously doubling its durability from current levels of 2,000–3,000 h to 5,000 h.

Although this chapter is devoted to electrochemical degradation, it is understood that performance, durability and cost are intertwined phenomena and cannot be treated as detached or isolated problems – for example, an increase in durability is easily obtainable by employing a higher Pt loading with concomitant higher costs. In our attempt to explore the intricacies of electrochemical degradation of catalysts and supports in PEMFCs, we briefly describe the close interaction between the catalyst, ionomer and carbon in conventional catalyst layers (Section 1.1) and subsequently present an overview of the practical automotive targets for performance and durability in Section 1.2.

1.1. The Catalyst Layer

In $H_2|$Air PEMFC stacks, the key electrochemical reactions that consume reactants and generate power, heat and water take place in each cell within a ~50 μm thickness. The hydrogen oxidation reaction (HOR), which is a fast $1e^-$ process, takes place at the anode, and the oxygen reduction reaction (ORR), which is a slow $4e^-$ process, takes place at the cathode. The ~50 μm thick region consists of the proton exchange membrane (~15–30 μm) that separates the anode (~10 μm) and cathode (~10 μm) catalyst layers (CL or electrodes). We refer to this 3-layer sandwich as the catalyst coated membrane or CCM. The 100–200 μm carbon-based diffusion media (DM), with or without a thin (10–20 μm) microporous carbon layer (MPL), on either side of the electrodes plays a central role in providing mechanical support, electronic contact and distribution/management of reactants, water and heat. The 5-layer structure, consisting of the 3-layer CCM and two DM+MPL, is referred to as the membrane electrode assembly or MEA [5].

The (conventional) catalyst layer is typically composed of Pt nanoparticles supported on a high surface area carbon black in close contact with a controlled amount of ionomer. The carbon black support allows the Pt nanoparticles to have a high dispersion (2–3 nm) and provides a porous, electronically conductive structure; this structure plays a critical role in providing transport of reactants and electrons to the Pt nanoparticles as well as removal of water and inert gases. The ionomer, which presumably coats the entire support surface, maintains discrete hydrophobic and hydrophilic domains for reactant and protonic access to the Pt nanoparticle active sites.

Not surprisingly, in addition to the HOR and ORR, undesired parasitic and corroding electrochemical reactions that degrade the MEA components also take place and limit the life of a fuel cell. The undesirable reactions include:

i. Pt dissolution and diffusion through the ionomer; re-deposition on another Pt particle forming a larger particle or diffusion into the membrane to form a band;

ii. carbon corrosion to carbon dioxide and the resulting disintegration of the CL leading to Pt agglomeration; formation of carbon surface oxides leading to hydrophilic or wettable pores that flood easily and impede the transport of oxygen to the Pt;

iii. generation of reactive species such as hydrogen peroxide (H_2O_2), hydroperoxyl radical (•OOH) and especially hydroxyl radical (•OH) that cause chemical degradation of the membrane/ionomer;

iv. reversible and irreversible adsorption of contaminants from air, fuel, MEA and bipolar plate degradation products on Pt nanoparticle sites.

Overall, these reactions result in a loss of electrochemically active catalyst sites that becomes manifest as a loss in mass activity of the catalyst, a rise in membrane/ionomer resistance and an increase in mass-transport resistance in the catalyst layer, all of which result in fuel cell performance degradation over time.

The conventional catalyst layer – using Pt dispersed on a carbon support described above – is currently incorporated in most PEMFC stacks used in FCVs. Advanced materials (possessing a higher activity and dissolution resistance) that are likely to be introduced in the near future include supported binary and ternary Pt-alloy nanoparticles, extended thin film electrocatalyst structures (ETFECS), supported 'core-shell' catalysts that employ monolayer Pt or Pt-alloy shells over a base metal core and catalysts dispersed on highly corrosion-resistant non-carbon conductive supports. Some of the advanced materials that are employed in extremely thin catalyst layers exacerbate mass-transport issues, so these need to be addressed.

The electrode degradation will also be affected by the composition and properties of the membrane and the ionomer used in the catalyst layer. In this chapter we focus on PEMFCs incorporating PFSA based membranes that can operate satisfactorily below 90°C in a reasonable RH range. Research on low humidity/high temperature membranes is currently a subject under intensive research since it promises more facile water and thermal management; low RH additionally suppresses Pt catalyst dissolution. These topics are discussed in the chapter on membrane durability (Chapter 2).

Thus, we can expect the overall electrode degradation rates and decay mechanisms to be convoluted by complex interdependent phenomena related to the choice of component materials and operating conditions. Targets for the performance and durability of the entire fuel cell stack can be easily set (based on ICE) but apportionments into electrocatalyst and support degradation rates that represent realistic or accelerated modes of operation are imperative. Attempts to set these targets are the subject of the following section.

1.2. Practical Targets for Electrocatalyst Activity, Cost and Durability

The targets for performance, durability and cost of H_2|Air PEMFC stacks are essentially based on today's ICEs used in automotives. Since expensive,

Pt-based catalysts are still the most active materials for cathodes and anodes of PEMFCs, it is imperative that we improve their activity and durability to accomplish the task of lowering costs through lowered Pt loadings. Obviously, performance and durability could be improved through the use of excess Pt, if it were not for cost and availability constraints. The target for allowable Pt content in fuel cell stacks is based on several factors such as available world Pt resources (~76 million kg), current usage in automotive catalytic convertors, and cost. At this time (Jan. 2010), the cost is ~$50/$g_{Pt}$ [6] although the historic mean over the last hundred years (1900–2000) is ~$25/$g_{Pt}$ [7]. One of the rationalizations for the target amount of Pt usage in PEMFCs is roughly based on the amount of Pt/precious metals currently being used in certain catalytic converters of ultra low emission ICE automobiles. Depending on the type of catalyst (Pt, Pd, Rh, etc.) used in ICE catalytic converters, the amount of Pt/precious metal may be as high as 5–10 g and constitutes ~50% of the total Pt usage today. This provides a justification for limiting the amount of Pt in a stack to ~10 g (in a 100 kW stack) or ~0.1 g_{Pt}/kW. (The target recommended by US DOE is ~0.2 g_{Pt}/kW [8] based on 2002 dollar value and $15/$g_{Pt}$.) 10 g of Pt per 100 kW stack at $50/$g_{Pt}$ translates to a cost of ~$500/stack or $5/kW. Approaching this goal will ensure that additional demand for Pt in fuel cells is small compared to that being used today and concerns about Pt resource availability can be set aside. Recycling of Pt has also been studied and the conclusion is that more that 95% of the Pt in fuel cell stacks can be recycled [9].

The reduction in the total amount of Pt catalyst used in stacks can be accomplished in three ways:

i. Developing new Pt-based catalysts that have a higher mass activity (mA/mg_{Pt}) – the pursuit of a higher activity catalyst is not intended for the direct purpose of raising the performance of the cell, but instead for lowering the catalyst loading (mg/cm^2) in stacks while simultaneously maintaining the same level of performance that is achievable today.

ii. Improving the catalyst layer structure/microporous layer/diffusion media/flow field design – the ensuing enhancement in mass-transport would result in higher limiting current densities (I_L) and therefore higher peak power densities (W/m^2); this would enable a reduction in stack size (higher kW/L) and the total amount (g_{Pt}/kW) of catalyst used. We note that although the fuel cell stack encounters peak or rated load for less than 5% of the total operation time, the size of the stack is determined by the peak power produced; furthermore, a stack efficiency of 55% is desirable at rated power.

iii. Improving the durability of the catalyst and support, i.e., lowering the degradation rate (μV/hr) at lower Pt loadings (mg_{Pt}/cm^2) – this would permit the use of a lower initial loading of the catalyst while maintaining acceptable performance ($< 10\%$ loss) at the end of life.

At this time, although not widely reported, anode Pt loadings have been reduced to ~0.05 mg/cm^2 while the cathode loadings are in the range of 0.2–0.35 mg/cm^2 with stack Pt content ~30 g.

The durability or degradation rate of the PEMFC stack is complex, as it is affected to different degrees by the modes of operation of the fuel cell, environmental conditions as well as choice of materials and cell design. The load or 'drive cycle' that a PEMFC stack is subject to is strongly dependent on the driving habits of the vehicle operator. Under normal operation of a vehicle, the stack experiences a variety of loads that may be constant or cyclic; a transient response time target of just 1 s from 10% to 90% of rated power has already been achieved. The automotive stack is also subject to a range of ambient environmental conditions of pressure, temperature and humidity and air quality.

The principal modes of operation of PEMFCs in automotives that can cause electrode degradation are summarized in Table 3.1 as:

- **i.** OCV/idling/low-load;
- **ii.** Acceleration/deceleration;
- **iii.** Start-up/shut-down cycles;
- **iv.** Ambient air quality; and
- **v.** Sub-zero temperatures [10,11].

In actual practice, automotive engineers have developed 'simulated drive cycles' that represent a properly weighted, complete range of driving conditions encountered by an automobile under city/highway driving. A PEMFC

TABLE 3.1 Impact of Automotive Modes of Operation and Environment on Electrode Degradation

Mode, Environment	Root Cause of Degradation	Degradation Pathway
OCV/idling/low-load	Formation of peroxy radicals	Membrane degradation, Pt ECA loss
Acceleration/ deceleration	Cathode potential cycling	Pt dissolution, particle growth, Pt ECA loss
Start-up/shut-down	Cathode potential spikes of 1.5 V	Support corrosion/Pt particle agglomeration
Ambient air quality	Adsorption of air contaminants	Poisoning of cathode catalyst sites
Sub-zero temperatures	Freezing water/fuel starvation	Support corrosion, electrode degradation

stack is subjected to such an integrated drive cycle in a test bench to verify its durability and fuel consumption. However, when carrying out research and development of new materials and clarifying the effect of different operating conditions, the main modes of operation are often applied independently to quantify the performance and degradation contribution specific to each mode. Within each particular mode of operation, for example load cycling, one typically studies the effect of several different cycle profiles in isolation or profiles that simulate a worst case scenario. At times, profiles or operating conditions are modified so that they serve as an 'accelerated test' that lowers the total test time while maintaining the degradation pathway. Thus a set of clear and simple standardized test protocols and unified targets have to be comprehensively defined for rapid progress in the field.

Durability targets have thus been addressed independently by organizations such as the US DOE Office of Energy Efficiency and Renewable Energy's Hydrogen Fuel Cells and Infrastructure (HFCIT) program in close association with national laboratories, universities and industry [12], the New Energy and Industrial Technology Development Organization (NEDO) in Japan and the European Hydrogen and Fuel Cell Technology Platform (HFP) and Implementation Panel (IP) in Europe. The Fuel Cell Commercialization Conference of Japan (FCCJ) has also published a booklet defining performance, durability and cost targets in collaboration with Nissan, Toyota and Honda [13]. Table 3.2 outlines some of the key targets for transportation fuel cell stacks operating on H_2|Air based on multiple sources [13–15]. Overall, the general 2015 target for fuel cell stack durability (in cars) is roughly defined to be 5,000 h/10 years (150,000 driven miles) at the end of which ~10% performance degradation is allowable. The automotive conditions for durability include ~30,000 cycles of start-up/shut-down, ~300,000 cycles of wide span load, a few hundred hours of idling.

So far 2,000–3,000 hours of stack life (validated in vehicles having 58% efficiency and 254 mile range) with low degradation rates has been demonstrated [2] and steady progress is being made towards the final goal. An additional durability enhancement by a factor of ~1.5–2 is needed for attaining the 2015 commercialization targets of 5,000 h/150,000 miles.

We continue this chapter by reviewing the literature (Section 2), outlining the experimental set-ups employed and electrochemical diagnostic techniques used to measure degradation rates (Section 3), discussing some of the fundamental and applied research carried out at Nissan on electrochemical degradation at the half-cell, subscale cell and stack platforms (Section 4), and lastly summarizing the future challenges that have to be met to achieve commercialization (Section 5).

2. SIGNIFICANT LITERATURE

In our attempt to methodically review the extensive literature on electrochemical degradation related to PEMFCs, we divide this section into four

TABLE 3.2 Selected Durability Targets for Automotive PEMFCs Based on Multiple Sources [8,13,14]

Durability Parameter	Units	Target
Stack life	h, years	5000, 10
Stack start-stop cycles	cycles	30,000
Stack major (0.6–0.95 V) load cycles	cycles	300,000
Stack performance degradation after 5000 h	%	10
Catalyst (MEA) ECA degradation after 5000 h	%	<40
Catalyst (MEA) mV degradation after 5000 h	mV	<30
Catalyst/support (MEA) start-stop degradation	mV/cycle	<1
Unassisted cold start-up temperature	°C	−40
Membrane H_2, O_2 crossover, 100 kPa, 80°C	mA/cm^2	<2

sub-sections; namely **2.1 Catalyst Durability**; **2.2 Support Durability**; **2.3 Contamination**; **2.4 Sub-zero Operation**. The two processes of catalyst degradation and support degradation are not always mutually exclusive, in that each can influence the degradation rate of the other. Pt-based catalysts are known to catalyze carbon corrosion and, conversely, a low surface area support or a support that does not permit robust anchoring of Pt on its surface will accelerate catalyst nanoparticle agglomeration. Consequently, some overlap is to be expected between the two sections. Contamination and subzero operation affect both catalyst and support degradation and hence are discussed in separate sections.

2.1. Catalyst Durability

In this section, we discuss the degradation of Pt and Pt-based catalysts in acidic liquid electrolytes as well as in PEMFCs. A great deal of the fundamental understanding of the observations and mechanisms of catalyst degradation springs from the vast amounts of work (since the 1960s) in the area of catalyst oxidation and dissolution in liquid electrolytes. Today, much of the screening for activity and durability (of novel catalysts available in small quantities of <1 g) is still being carried out in thin film rotating disk electrodes (TF-RDE) or other half-cell set-ups using liquid electrolytes such as dilute perchloric and sulfuric acid to mimic the PFSA ionomer/membrane of PEMFCs [16]. Moreover, in order to isolate the catalyst, and characterize its elementary behavior and identify and monitor the electro-oxidation and corrosion species, the use of liquid environments is indispensable. In fact, much of the research on the electrochemical degradation of Pt in PEMFCs being performed today is an evolution and transference of the understanding of the dissolution mechanisms that have been studied in liquid electrolytes. We also explore the significant literature related to the successful development of phosphoric acid fuel cells (PAFCs – conc. phosphoric acid, 160–205°C) for commercial use. Although typical PEMFC operating temperatures fall in the range 50–90°C and promise to be more benign, the findings from PAFC research are often directly applicable.

Section 2.1.1 is devoted to reviewing the electrochemical degradation of *Pt catalysts* in liquid electrolytes and PEMFC environments. In **Section 2.1.2** we examine the research conducted on the electrochemical degradation of *Pt-based alloys* in liquid electrolytes and PEMFC environments. **Section 2.1.3** reviews the most recent developments related to electrochemical degradation of *non–conventional catalysts* that offer pathways to achieve catalyst target requirements.

2.1.1. *Electrochemical Degradation of Pt*

2.1.1.1. Electrochemical Degradation of Pt in Acid Electrolytes

Thermodynamics Pourbaix diagrams or potential vs. pH plots reveal a narrow range close to 1.0 V where Pt is thermodynamically susceptible to dissolution [17] as shown in Fig. 3.1. Some of the Pt oxide species are also likely to dissolve chemically or electrochemically. To confirm and quantify this expected Pt dissolution from bare Pt as well as Pt/C, researchers have carried out experiments to measure the equilibrium concentrations of Pt ions in H_3PO_4, H_2SO_4 and $HClO_4$ using inductively coupled plasma-mass spec (ICP-MS) [16,18–20]. The thermodynamic equilibrium potential for the dissolution of bare metallic Pt is as follows [17]:

$$Pt = Pt^{2+} + 2e^- \quad E_0 = 1.188 + 0.0295 \log[Pt^{2+}] \text{ V} \tag{3.1}$$

Pt also forms oxides as follows:

$$Pt + H_2O = PtO + 2H^+ + 2e^- \quad E_0 = 0.98 - 0.0591 \text{ pH V} \tag{3.2}$$

$$PtO + H_2O = PtO_2 + 2H^+ + 2e^- \quad E_0 = 1.045 - 0.0591 \text{ pH V} \tag{3.3}$$

These oxides may dissolve chemically or electrochemically according to:

$$PtO + 2H^+ = Pt^{2+} + H_2O \quad \log[Pt^{2+}] = -7.06 - 2 \text{ pH} \tag{3.4}$$

$$PtO_2 + 4H^+ + 2e^- = Pt^{2+} + 2H_2O \quad E_0 = 0.837 - 0.118 \text{ pH} - 0.0295 \log[Pt^{2+}] \text{V} \tag{3.5}$$

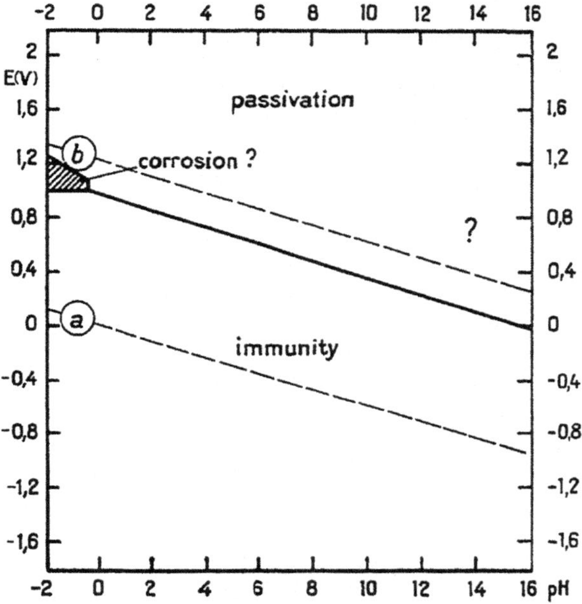

FIGURE 3.1 Potential-pH (Pourbaix diagram) of platinum depicting the domains corresponding to passivation, corrosion and immunity. [17]

For dispersed nanoparticulate Pt on carbon supports in PEMFCs, the Nernst potential for Pt dissolution is expected to decline from 1.188 V for bulk Pt to 1.160 V for 5 nm Pt particles, and to 1.088 V for 1 nm Pt particles as demanded by the Kelvin equation.

Bindra et al. [18] have reported a logarithmic (Nernstian) dependence (10^{-7}–10^{-4} M) of the equilibrium concentration of dissolved Pt (from 25 cm^2 poly-Pt foils) with potential in 96% phosphoric acid (at 176 and 196°C) in the range 0.8–1.0 V. Ferreria et al. [21] (General Motors, MIT) examined equilibrium Pt ion concentrations in 0.5 M H_2SO_4 using 2–3 nm Pt/C nanoparticles coated on carbon fiber diffusion media in the potential range 0.85–1.1 V at 80°C under N_2. The potential dependence of the concentration of dissolved Pt in the range 0.85–1.1 V was found to be lower than the predicted Nernstian behavior of ~29.5 mV/dec. Wang et al. [20,22] (ANL) carried out studies to determine the equilibrium concentration of dissolved Pt as a function of potential (0.65–1.5 V) for polycrystalline Pt wire and 10 wt% Pt/C in 0.57 M $HClO_4$ at 25°C under Ar; equilibrium Pt concentrations were recorded with time and steady-state was attained in 80 h. They found that the equilibrium concentration increased from 2.0×10^{-9} M to 2.0×10^{-6} M linearly in the range 0.65–1.1 V and then declined and leveled off between 1.3–1.5 V. Figure 3.2(a) illustrates their experimental results, oxide growth and hypothesis; Fig. 3.2(b) compares

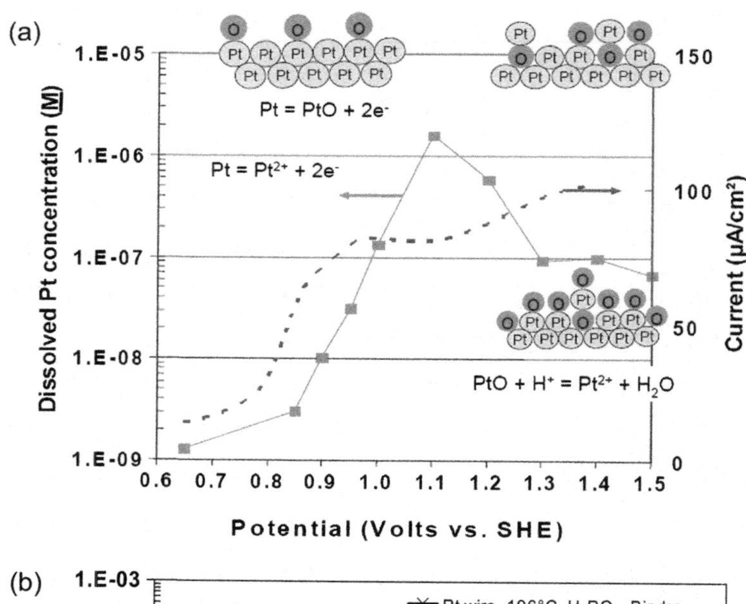

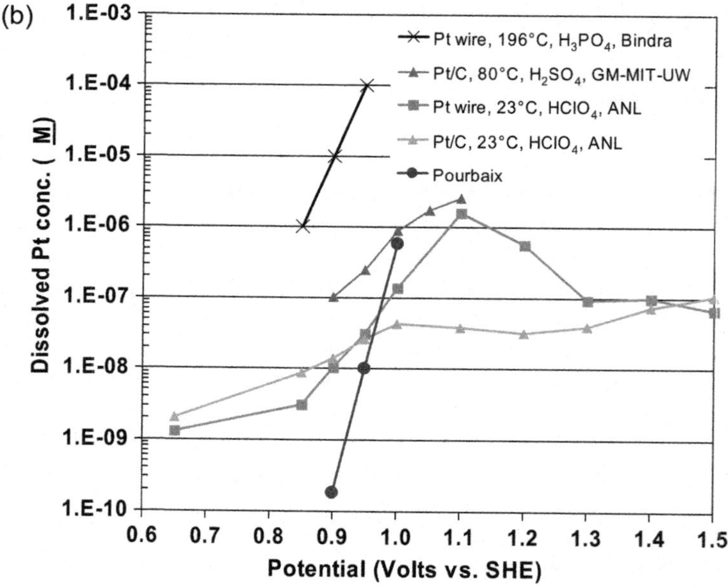

FIGURE 3.2 (a) Plot of the measured equilibrium concentration of Pt wire in 0.57 M perchloric acid at 23°C (red line with square markers) and portion of cyclic voltammogram (dashed blue line) showing oxide growth vs. potential. Schematic of oxide growth on Pt based on the well-known place exchange mechanism depicted over the appropriate potential regimes. (b) Summary of Pt dissolution data reported in the literature. *(Reprinted with permission, Wang et al. [20])*

their results to those from various research groups measured in different environments. Their hypothesis implies that:

i. Pt is not completely passivated with oxides in the 0.85–1.15 V range;
ii. in the 1.15–1.2 V range due to the place exchange mechanism, the surface is still not covered with a monolayer of oxide; and
iii. at high potentials of 1.3–1.5 V, the surface is passivated with PtO and continues to dissolve at a lower rate corresponding to $PtO + 2\ H^+ \rightarrow Pt^{2+} + H_2O$.

Interestingly, they reported no difference in the dissolution rates for Pt wire and Pt/C.

Mitsushima et al. [23] examined the dissolution of Pt black as a function of acid type, pH, temperature and atmosphere; they found that the Pt dissolution increased with temperature, $[H^+]$, and oxygen partial pressure. Figure 3.3 illustrates the logarithm of Pt/PtO solubility as a function of pH in three acids with the order of Pt solubility being $H_2SO_4 > HClO_4 > CF_3SO_3H$ at 25°C in air. Using the dithizone-benzene method and visible spectroscopy, they found that without the addition of 16% $SnCl_2$–HCl, no Pt ions could be detected in solution (Pt^{4+} conversion to Pt^{2+}). Based on this method of identification of dissolved Pt^{4+} species and the dependence of solubility on $[H^+]$ they postulated that Pt solubility in O_2-containing atmospheres might follow the reaction:

$$PtO_2 + H^+ + H_2O = Pt(OH)_3^+ \tag{3.6}$$

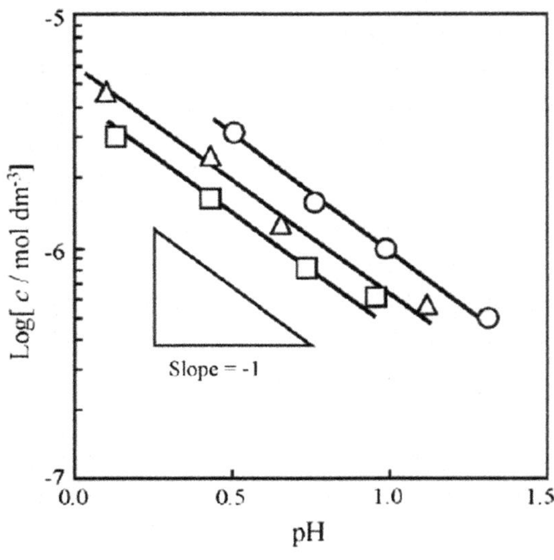

FIGURE 3.3 Solubility of Pt/PtO as a function of pH in sulfuric (circles), perchloric (triangles) and TMFSA (squares) measured at 25°C in air. *(Reprinted with permission, Mitsushima et al. [23])*

The dissolution of platinum low-index and nano-faceted surfaces has been investigated using atomic force microscopy (AFM) in combination with (ICP-MS) by Komanicky et al. [24]. They found that on low-index surfaces, platinum oxide passivates the surface, resulting in depressed dissolution rates at elevated potentials. On the other hand, they found that nano-faceted surfaces dissolved at an accelerated rate at higher potentials, indicating that edges and corner sites were the primary locations of dissolution.

Darling and Meyers [25,26] presented a model that described the oxidation and dissolution of Pt in PEMFCs taking into account Pt dissolution (to Pt^{2+}), PtO formation and PtO chemical dissolution; their model predicted Nernstian behavior for equilibrium Pt concentrations below 1.1 V, where a steep drop that is ascribed to oxide coverage occurs. They obtained good fits of their model to CV data for the oxide formation on commercial Pt/C catalysts.

We have discussed above the equilibrium concentrations of Pt reported by different groups that were acquired under various conditions of temperature, Pt material, acid type and concentration, blanketing atmosphere, detection methods, ICP-MS detection limits, etc.; we find that there is only rough agreement on the magnitude of equilibrium concentration values. At this time, there is no consensus on the mechanism of Pt dissolution, the identity of the dissolved Pt species (that originates from both bare Pt and Pt oxides in different electrolytes) and further extensive work is warranted.

Oxide Species of Pt XPS studies have revealed [27] that oxygen exists on the surface of Pt in the physisorbed state below 25 K, in the chemisorbed state below 135 K and in the atomic phase above 150 K; but it is not known if these data are relevant in the presence of water and electrochemical conditions. Automotive PEMFCs operate in a regime that spans 0.6–1.0 V, in which the Pt surface evolves from a relatively oxide-free surface (~0.6 V), to one where it is significantly covered with oxides species. Identifying the oxide species on the surface of Pt is important not only to facilitate the understanding of the kinetics of the ORR [28–33] but also the dissolution of Pt [34]. To attempt to understand Pt dissolution over the entire PEMFC operating potential range, it becomes necessary to characterize the surface oxide species which may contribute towards passivating the bare Pt, but may also themselves dissolve at a different rate from that of bare Pt.

The various stages of oxide formation, transition from sub-monolayer, to monolayer and multilayers, have been studied using cyclic voltammetry [35,36], optical techniques [37–39] and XPS [40–43]. An excellent review of research carried out up to 1968 on the oxygen species/oxide coverage on Pt can be found in the work of Hoare [44]. His review discusses the work of researchers involved in the preparation of various oxides of Pt, including anhydrous PtO_2 as well as those hydrated to various extents. Hoare summarizes that PtO_2 is the only stable oxide of Pt, while PtO or $Pt(OH)_2$ in the hydrated form are oxidized in air to PtO_2. These results may not necessarily be valid for surface oxide species of Pt, but nevertheless provide some useful insights.

There are conflicting reports in the literature on the mechanism of the electrochemical oxidation of Pt. For an in-depth review of the work carried out up to 1995, the reader is directed to the seminal review by Conway [45]. A two-step mechanism was advanced, in which the first step is a fast reversible process followed by a place-exchange mechanism that describes oxide growth at higher potentials. In contrast, the concept of a one-step oxidation of Pt to anhydrous oxide was initiated by Harrington [46]; his AC voltammetry studies did not show evidence of a fast time constant related to OH^- adsorption above 0.8 V. He proposed that the 2-electron oxidation of Pt to Pt^{2+} was the rate determining step (rds). The results of Alsabet et al. [47] and Jerkiewicz et al. [48] also generally subscribed to this hypothesis for potentials above 0.80 V. Jerkiewicz et al. [48] applied a combination of EQCN, CV and AES techniques for the first time to forward a hypothesis for the mechanism.

The two principal hypotheses discussed above can be expressed as:

i. a two-step oxidation via hydroxide [45]:

$$Pt + H_2O \rightarrow Pt\text{-}OH_{ads} + H^+ + e \qquad 0.85 - 1.10\ V \qquad (3.7)$$

$$\begin{aligned} &Pt\text{-}OH_{ads} \rightarrow OH_{ads} - Pt \\ &OH_{ads} - Pt \rightarrow Pt\text{-}O + H^+ + e \qquad 1.10 - 1.40\ V \end{aligned} \qquad (3.8)$$

and,

ii. a single-stage oxidation (strong physisorption, discharge of water and place exchange) directly to an anhydrous oxide [46]:

$$Pt + H_2O \rightarrow Pt^{\delta+} - - - - - - - - - - O^{\delta-}H_2 \qquad 0.27 - 0.85\ V \qquad (3.9)$$

$$(Pt - Pt) - H_2O \rightarrow (Pt - Pt)^{\delta+} - - - - O_{chem}{}^{\delta-} + 2H^+ + 2e \qquad 0.85 - 1.15V \qquad (3.10)$$

$$Pt - O_{chem} + H_2O - - - - - - - - - - \rightarrow (Pt^{2+} — O^{2-}) + 2H^+ + 2e \qquad 1.15 - 1.40\ V \qquad (3.11)$$

Kucernak and Offer [49] reviewed the merits and demerits of both place-exchange mechanisms presented in the literature. In particular, they question the assumptions often made that there are no hydroxides or oxides below 0.70 V. They reported experimental results on the potential of zero total charge (pztc) to be ~235 mV for poly-Pt in 0.5 M H_2SO_4. They proposed that this result, together with evidence for CO oxidation at low potentials, lends credence to the existence of OH^- species at low potentials and in the double layer regime and to an extent even the HUPD regime.

The more subtle effects of the surface orientation of Pt single crystals on oxide growth rate were reported by Conway and Jerkiewicz [50]; they showed that the normalized oxide growth rate proceeds with the trend Pt(100) >Pt(111)>Pt-poly and exhibited logarithmic behavior in the range 0.90–1.80 V. The fact that the oxide growth continues to follow a rate specific to the initial crystal orientation supports a place exchange mechanism that takes place at the inner interface of the oxide and metal surface. In other work, Nagy and You [51] have reported that PtO is a mobile species and is capable of moving to energetically preferred sites that may result in more bare Pt being exposed to the electrolyte on the surface leading to facile dissolution.

Growth of oxide films on Pt at very high anodic potentials relevant to oxygen and chlorine evolution and hydrocarbon formation are discussed in the work of Tremiliosi-Filho et al. [52]. Birss et al. [53] also studied growth and reduction of oxide films on Pt using EQCM. They concluded that the compact α-oxide (PtO or PtO_2) was anhydrous in nature. When thin, hydrous β-Pt oxide films grown by the potential cycling were partially reduced at constant potential (0.65–0.30 V), and the remaining oxide film was reduced by sweeping the potential to 0 V, 20% more charge was passed as compared to a film that was reduced using a potential sweep [54,55]. They speculated that potentiostatic film reduction probably occurs by a different mechanism, in which no oxide dissolution occurs, in contrast to sweep reduction.

Impedance spectroscopy has been employed in an attempt to deconvolute the oxide species on Pt as a first step to understanding the mechanism of Pt/PtO dissolution. AC impedance which employs a single perturbation frequency, electrochemical impedance spectroscopy (EIS) which employs a wide range of frequencies (to a pre-formed steady state oxide film) and recently potentiodynamic electrochemical impedance spectroscopy (PDEIS) which employs a set of frequencies applied at each potential step (step voltammetry) have all been studied. In particular, PDEIS may help to separate the faradic charging currents from the double layer charging currents as reported in preliminary work by Ragoisha et al. [56]. This would allow a more accurate estimate of the pseudo-capacitance related to oxide coverage from cyclic voltammograms.

A definitive fundamental understanding of the nature of the oxide species, the mechanism of their formation, the structure of the film, and the dissolution of oxide species remain areas for ongoing long-term research.

Pt Dissolution Under Cycling in Liquid Electrolytes Tafel and Emmert reported in 1905 [57] that although Pt is considered to be a noble metal, it can dissolve significantly when subjected to potential cycling. They pointed out with great foresight that, with the prevalent use of Pt as a counter electrode in electrochemistry, there were likely to be serious issues of contamination of the non-Pt working electrodes studied in the literature. Today, scientists are still grappling with the problem of Pt dissolution, its mechanism and with attempts to suppress it.

Johnson et al. [34] carried out a comprehensive study using a rotating ring-disk electrode (RRDE) to survey the oxide formation, the dissolved species in the anodic and cathodic sweep in combination with spectrophotometry to identify the Pt species. They found unaccounted excess charge of 50 μC/cm^2 in H_2SO_4, and 32 μC/cm^2 in $HClO_4$ for scans in the range 0.4–1.4 V. This indicated the generation of soluble Pt species which they presumably collected at the ring electrode set at different potentials (~0.35, 0.75, 1.05 and 1.65 V). They observed peaks which they attributed to the oxidation and reduction of Pt^{2+} species at the ring electrode. They also carried out independent dissolution experiments cycling a 60 cm^2 Pt gauze (in 100 ml of acid) 100 times between 0.20–1.2 V at ~10 mV/s. Pt^{2+} was identified as the primary dissolution product using absorbance measurements with less than 10% of Pt^{4+}. They found 20–30 μg of Pt in solution or 3.5–4.5 μC/cm^2 that closely matched the charge from RRDE experiments further verifying Pt^{2+} as the dissolved species. The authors inferred that Pt^{2+} detected during the cathodic reduction indicated that PtO_2 must have been the species on the oxidized electrode surface. (Reaction thermodynamics: $PtO_2 + 4H^+ + 2e \rightarrow Pt^{2+} + 2H_2O$ $E^0 = 0.837$ V vs. RHE).

Rand and Woods [58] investigated the dissolution of Pt, Pd, Rh and Au in 1 M H_2SO_4 while carrying out cyclic voltammetry. Their results indicated both Pt^{2+} and Pt^{4+} were identified after 200 cycles (0.41–1.46 V, 40 mV/s). They concluded that an anodic dissolution mechanism, in which the difference between the total anodic and cathodic charges per cycle (Q^a–Q^c) was >0 and matched the quantity of metal ions detected in the electrolyte, was in operation.

Recently, the effect of cycle profile on a Pt wire for very large voltage cycles (lower limit of 0.2–0.5 V and upper limit of 1.8–2.4 V) measured in H_2SO_4 was reported [19]. This study hypothesized that PtO_x species dissolve during the slow cathodic sweep of a cycle based on current detection on the ring at both oxidizing and reducing potentials. The authors also concluded that Pt was electrochemically oxidized to Pt^{4+} in the anodic sweep since currents were detected on the ring held at reducing potentials only.

Dam and de Bruijn [59] have investigated the dissolution of thin Pt films on Au substrates in 1 M $HClO_4$ between 40–80°C in the range 0.85–1.4 V using electrochemical quartz crystal microbalance (EQCM). By monitoring the time dependence of the electrode weight under potentiostatic conditions, they observed an initial mass increase ascribed to net oxide growth followed by a mass decrease due to net dissolution. They verified a log-log relationship for the dissolution rate up to 1.15 V, above which the oxide film passivated the surface and lowered the net dissolution rate.

Mechanisms for Crystallite Growth in Liquid Electrolytes There are many reports on the crystallite growth of metals on supports at high temperatures of 250–800°C and various physical interpretations have been proposed to describe the resultant surface area decrease with time. Two fundamentally different mechanisms have been proposed for metal transport on the surface of the supports:

i. metal clusters or particles migrate on the surface of the support in a manner similar to Brownian motion and on collision coalesce; the rate limiting step could be either the surface diffusion or coalescence;
ii. atoms from small crystallites or particles evaporate or diffuse as adatoms on the support surface and condense on larger ones similar to Ostwald ripening (dissolution–diffusion–re-deposition) in solutions.

Figure 3.4 illustrates both Ostwald ripening as well as agglomeration for Pt nanoparticles on a carbon substrate coated with a thin film of ionomer or electrolyte.

Bett et al. [60,61] studied the growth of Pt crystallites supported on graphitized carbon in both gaseous and liquid environments and discussed the mechanism in terms of:

i. the migration of Pt crystallites from trap sites on the graphitized carbon followed by coalescence on the carbon surface (Smoluchowski model); and
ii. the movement of Pt adatoms on the carbon surface (Ostwald ripening).

Treatment of the model resulted in a general form of the rate expression for the catalyst sintering that follows a power law expressed as:

$$S^{-n} = S_o^{-n} + kt \tag{3.12}$$

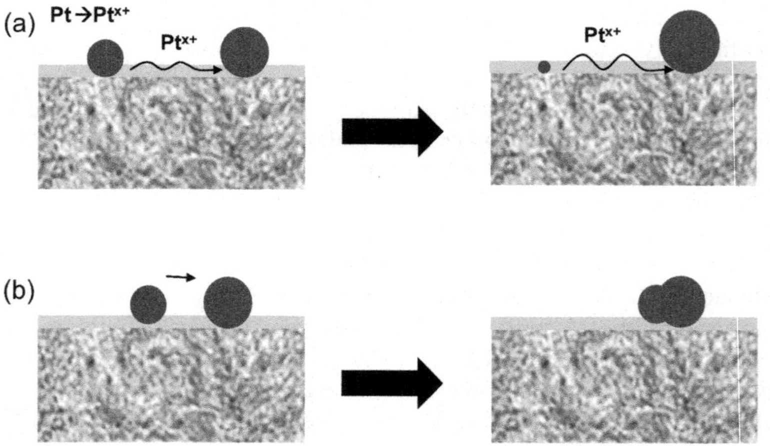

FIGURE 3.4 Schematic illustrating the basic principles involved in the conceptual dissolution mechanisms for Pt in acid media and classified into Ostwald ripening and particle agglomeration. The Pt particles are seated on a carbon support that is coated with a thin layer of ionomer. Note that although Pt ions are depicted, uncharged Pt adatoms have also been hypothesized to diffuse towards larger particles.

where S_o and S are the Pt surface areas per unit volume at time t = 0 and t = t, 'k' is a constant and the order 'n' is a function of the mechanism. (In empirical modeling of the degradation of Pt catalysts in PEMFCs, this equation has been used frequently in the fitting of ECA vs. time or number of cycles but without sufficient basis or justification for the values of 'k' and 'n').

The authors concluded that the Smoluchowski model was the dominant process for gas phase sintering of Pt on graphitized carbon, based on evidence that indicated a dependence on Pt loading or concentration. They also studied the effect of liquid environments such as ethylene glycol, bromobenzene, 96% phosphoric acid, water and toluene on the sintering process. In liquid electrolytes at 100–200°C, the loss in Pt surface area is accelerated and comparable to losses that occur in the gas phase at 600°C, i.e. the activation energies are in the same range. In addition, the area losses in liquid environments were found to be independent of the Pt loading (5, 20 wt%) and potential, leading them to propose that the sintering was the result of two-dimensional Ostwald ripening.

Blurton et al. demonstrated that when catalytically oxidized graphite supports were used instead of conventional graphite supports, a lower loss of Pt surface area that was independent of the operating potential resulted [62]. Blurton et al. [62] and Gruver [63] both suggested that this lack of potential dependence on the surface area loss demonstrated that crystallite migration and coalescence was the predominant pathway for degradation. Connolly et al. [64] reported large surface area losses on Pt/C in different electrolytes in the temperature range 100–200°C. For unsupported Pt black catalysts, Kinoshita et al. [65] and Stonehart and Zucks [66] also found significant loss of area and concluded that particle growth occurred due to surface diffusion of Pt atoms on crystallites from high to low surface energy sites.

Kinoshita et al. [67] reviewed the sintering of Pt catalysts and examined the effect of Pt loading, type of support, effect of potential and environment. They performed potential cycling experiments on Pt/C in 1 M H_2SO_4 at 1 cycle per min in the range 0.05–1.25 V and reported an ECA loss of 70% after 3,500 cycles and attributed it to Pt dissolution. They investigated the dissolution rate of platinum (low surface area Pt sheets, moderate surface area Pt blacks and high surface area Pt/graphitized carbon) having a wide range of initial surface areas by applying both triangular and square wave cycling. For the Pt sheets, the dissolution rate with slow triangular wave cycling resulted in ~4.5 μg Pt/cm^2_{real}/cycle; the Pt surface roughened and a significant increase of the (111) orientation was observed. On the other hand, for dispersed Pt nanoparticles, a lower dissolution rate was measured presumably as a result of the confinement/trapping of the soluble Pt complexes within the porous electrode structure; particle growth and surface area losses for the Pt catalysts were ascribed to Ostwald ripening.

Using the results of Yeager [68], Ross [69] estimated the Pt flux for two cases, one involving the migration of Pt from a cathode to anode and a second Pt migration from small particles to larger particles (analogous to Ostwald ripening). Assuming an OCV of 0.95 V, the dissolution rate was estimated to be 0.07 $mg_{Pt}/cm^2/h$ or 6×10^{13} atoms/cm^2/s, for a diffusion length of 1,000 μm, diffusion coefficient of D~10^{-5} cm^2/s and Pt ion concentration of 10^{-3} M. For the second case of Ostwald ripening, the dissolution rate amounted to ~6×10^{18} atoms/cm^2/s. In actual practice, a complex mechanism with both competing processes takes place simultaneously.

Aragane et al. [70] were among the first to report detailed findings on the change in Pt distribution within PAFC components (anode, cathode and matrix) after thousands of hours of operation; they applied electron probe microanalysis (EPMA), scanning electron microscopy (SEM), and transmission electron microscopy (TEM) to examine the morphology and size of Pt particles and X-ray diffraction (XRD) utilizing the Scherrer equation to calculate Pt crystallite size. At the beginning of life (BOL) the Pt distribution was uniform in the anode and cathode and none was detected in the matrix; after 1,500 h, 30% of the Pt was lost from the cathode and some Pt was detected in the matrix; after 4,500 h, 60% of the Pt was lost from the cathode and Pt was detected in the matrix as well as the matrix-anode interface. Evidence for dissolution of Pt in the cathode and its movement across the matrix and re-deposition at the anode was unambiguous.

Although these tests were conducted at constant current density of 200 mA/cm^2, it was interrupted to carry out polarization curves and OCV holds (hot idle at 190°C) accruing to 90 min over the lifetime and affecting the results. The authors, therefore, carried out another set of cleaner experiments to examine the potential dependence of Pt dissolution using fixed operating potentials. They found a clear dependence of potential on the extent of Pt dissolution with minimal Pt loss from the cathode below 700 mV while almost all the Pt dissolved after 1,500 h at OCV. They recommended additional studies to understand the complex effects of acid concentration in the matrix at different current densities on Pt distribution in the components.

In a later study, Aragane et al. [71] calculated the dissolution rate of Pt at different temperatures (110°C–240°C) and potentials (0.60–1.0 V) and quantified the effect of operational voltage on the performance decay rate of PAFCs. They found that higher potentials (especially OCV) resulted in higher Pt dissolution at the cathode, but higher current densities (lower potentials) accelerated the migration of Pt from the cathode to the anode. They attributed this phenomenon to acid concentration in the matrix and the associated hydrogen crossover that caused re-deposition of the Pt. They concluded that Pt dissolution rate for PAFCs is low for temperatures below 200°C, 840 mV leading to reasonable decay rates of ~1 mV/1,000 h. In general, commercial PAFCs are designed to studiously avoid OCV conditions and operating conditions have been optimized to obtain lifetimes exceeding 40,000 h.

Effect of Pt Particle Size on Activity and Durability An obvious route to improving the catalyst mass activity (mA/mg_{Pt}) and reduce the amount of Pt used was to obtain smaller particles by finding better ways to disperse them on the carbon support. But pioneering work by Bregoli [72] demonstrated that one cannot indefinitely reduce the particle size. Bregoli's work (see Fig. 3.5) shows that as the particle size is decreased and the corresponding surface area raised from 10 m^2/g to 70 m^2/g, the specific activity falls from 100 $\mu A/cm^2$ to 40 $\mu A/cm^2$ i.e. by more than a factor of 2.

There is much controversy over the existence of the 'particle size effect', but evidence from numerous groups demonstrates its validity. Kinoshita proposed that the ORR on dispersed Pt particles (cubo-octahedral) is a structure-sensitive reaction [73]; as a result of the changing distribution of surface atoms at the (100) and (111) crystal faces with an increase in particle size, an effect on specific activity was observable. More recently, researchers at GM [16], evaluated the activity of various Pt/C, Pt black and bulk Pt catalysts in $HClO_4$ using thin-film rotating disk electrodes (TF-RDE) and confirmed the particle size effect. The activity of bulk polycrystalline Pt was shown to be higher than commercial Pt/C by a factor of almost 10 as seen in Fig. 3.6. The effect of particle size is of considerable interest due to both its effect on catalyst activity and durability. Generally, researchers have found that larger particles, by virtue of their lower surface energy, are more resistant to dissolution. The explanation offered by those who observe this effect is that smaller Pt particles have more edge and corner sites that are not well-coordinated and are therefore more oxophyllic and susceptible to oxidation and dissolution. Pt particles having a larger diameter may be formed as a result of either poor/non-optimal

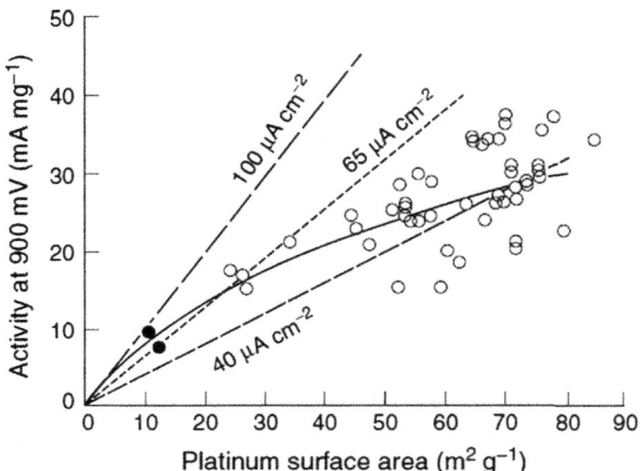

FIGURE 3.5 Illustration of the particle size effect for Pt black and Pt/Vulcan XC-72 catalysts in 99 wt% PA, 177°C. The mass activity is plotted vs. the surface area and lines of constant specific activity are overlaid as dashed lines. *(Reprinted with permission, Bregoli [72])*

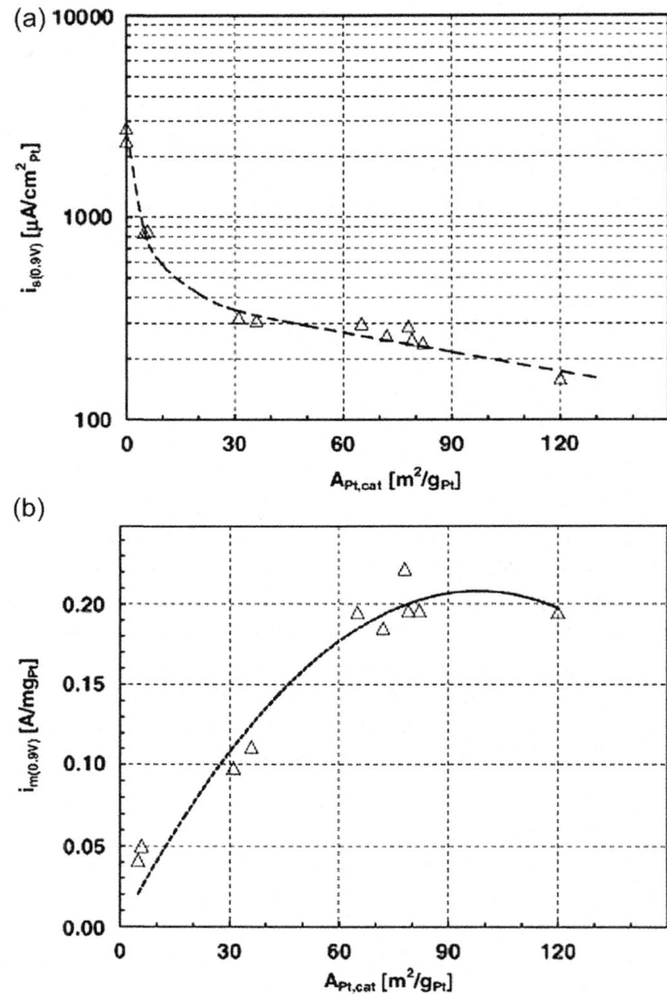

FIGURE 3.6 Specific and mass activities (at 0.90 V vs. RHE) vs. ECA for poly-Pt, Pt black and Pt/C measured in RDE set-ups using 0.1 M $HClO_4$ at 60°C. The trend in activity suggests a particle size effect with catalyst activity increasing significantly for larger particles and bulk films. *(Reprinted with permission, Gasteiger et al. [16])*

preparation procedures or intentionally prepared by subjecting the catalyst to controlled heat treatment as discussed in the next section.

Effect of Pt Heat Treatment on Activity and Durability Heat treatment of Pt is usually carried out at temperatures in the range 700–1200°C under N_2 or a reducing atmosphere. Particle growth of the catalyst during heat treatment steps as part of the preparation procedure of Pt and its alloys can result in both higher activity and durability. Jalan and Taylor [74] have reported that heat treatment of Pt catalyst lowers the Pt dissolution rate in addition to reducing the initial surface area. Despite the initial loss in surface area (140 m^2/g to 55 m^2/g), the performance of the heat treated Pt on Vulcan was found to be higher over a period of 10,000 h. Kocha et al. [75] have also proposed heat-treating Pt/C under controlled conditions to obtain 'voltage-cycling resistant' catalysts with superior activity to conventional Pt/C. We note here that Pt-alloy catalysts are typically prepared by a process that involves a heat-treatment step that facilitates good alloying and minimizes the leaching out of the base metal into the electrolyte; the separation of the individual contribution of particle size, heat treatment and alloying is discussed in Section 2.1.2.2 that deals with Pt-alloys.

2.1.1.2. Electrochemical Degradation of Pt in PEMFCs

Pioneering research at LANL [76] in the 1990s resulted in a reduction of the Pt loading on the cathode from ~4 mg/cm^2 (using Pt blacks) to ~0.40 mg/cm^2 (carbon supported Pt). The lower loaded 0.40 mg/cm^2 electrodes were initially PTFE bonded (as in PAFCs) and additionally impregnated with solubilized ionomer sprayed on the surface of the electrode in contact with the membrane. It is known that the permeability of oxygen through hydrated Nafion and water is much higher than in PTFE and the purpose of incorporating PTFE was mainly to provide hydrophobic pores that are not flooded with water and allow reactant access. Wilson et al. [76] from LANL took a major step away from this tradition and developed catalyst layers in which PTFE was completely eliminated. The solubilized ionomer and supported catalyst were cast together to obtain a thin film having low Pt loadings in which contact between ionomer and catalyst was maximized. These thin cast catalyst layers exhibited high catalyst utilizations and gains in performance.

Wilson and co-workers subsequently investigated the degradation of these thin film catalyst layers [77]. They carried out life tests by subjecting the cells constituted of catalyst layers having loading of 0.12–0.25 mg/cm^2 to a constant voltage of 0.50 V under H_2|Air, 300|500 kPa and 80°C. At that point in time, the harsh automotive conditions involving load cycling, OCV holds were not well-documented and 0.50 V hold represented a demanding peak power condition. They reported only 10% losses in cell performance (when either purified hydrogen or the anode cleaned of contaminants with air-bleed was used) after 4,000 h. They measured their surface area using XRD (rather than ECA) and found a loss of 60% of the initial area without corresponding losses in cell performance and attributed this discrepancy to an enhancement of specific activity of Pt with increase in particle size.

Our current understanding leads us to believe that the losses under a steady-state load of 0.5 V would be low since Pt is under reducing conditions. XRD peaks for the Pt (111) and (200) lattice planes (obtained after scraping off the catalyst from the membrane) showed particle size growth based on the narrowing of peaks after 2,200 h of operation. Based on distribution plots of mass fraction versus particle diameter, they found that 90% of the particles <3 nm had disappeared. On the other hand, a plot of the population distribution vs. particle diameter showed a high concentration of small particles and tailing population density for larger particles for both fresh and 2,200 h tested samples. Based on these results and reports of Blurton et al. [62] and Bett et al. [60], they found the evidence to be weighted towards a mechanism of crystallite migration and coalescence as opposed to Ostwald ripening. Around the same time, researchers had discovered a process to convert PFSA ionomers into a thermoplastic form by ion-exchanging with large hydrophobic counter ions such as tetrabutylammonium (TBA^+) that suppressed the ionic interactions between side chains. In a subsequent publication, Wilson et al. [76] prepared electrodes using TBA^+, and demonstrated that combined with appropriate heat-treatment of ~200°C (where TBA^+ becomes viscoelastic) the MEA becomes more durable.

Tada [78] of TKK (Tanaka Kikinzoku Kogyo) examined the particle size distribution in TEM for 30 wt% Pt supported on 300 m^2/g C that were operated at 1.5 A/cm^2 under $H_2|O_2$ (25 cm^2 cell) for a duration of 4,000 h. TEM images showed that the anode particle size grew from 3 nm to 4 nm after 300 h and thereafter remained constant; the cathode particle size increased to 5 nm after 4,000 h. He cited crystallite migration as the mechanism for Pt agglomeration in concurrence with the work of Wilson et al [77]. St-Pierre et al. [79] (Ballard) emphasized the detrimental effects of improper levels of humidification and their relationship to MEA contamination and durability. For their Mk513 stack, consisting of 8 cells, the degradation rate was 1 mV/h (over 5,000 h), whereas for their Mk5 single cell, the degradation rate was 60 mV/h (over 2,000 h). They attributed the higher losses for the Mk5 single cell to excess exposure to water that led to an increase in mass transport losses which they verified from polarization curves. Patterson (UTC) [80] reported that potential cycling (0.87–1.2 V) of Pt/C catalysts in MEAs of subscale fuel cells operated at 65°C resulted in the dissolution of Pt with a corresponding loss in ECA of as much as 50% after 6,500 cycles. This potential range includes potential regimes where both Pt and carbon simultaneously corrode but excludes potentials where Pt is reduced and free of oxide species.

Ferreira et al. [21] carried out an extensive study of the catalyst layer degradation based on the testing of one MEA sample (Pt/Vulcan, 50 cm^2 cell) for potential cycling (0.6–1.0 V, 80°C, H_2-N_2, 100 % RH, 10,000 cycles/100 h, triangular cycle profile). The measured ECA loss after cycling was determined to be ~63%, which is equivalent to a ~30 mV loss at the cathode and corresponds to a degradation rate of 300 μV/h over the 100 h test period. The degraded MEA after 10,000 cycles was diced into smaller pieces and evaluated by both TEM and XRD. What is particularly interesting is their use of low angle glancing X-ray diffraction technique which allows one to collect the diffracted intensities from only the cathode (without interference from the anode) while it is still attached to the MEA. They compared the X-ray diffraction pattern for the original Pt/Vulcan pristine powder to that of the anode and cathode of the cycled MEA and applied the Scherrer equation to estimate the average Pt crystallite size based on the (111), [200] and [220] full width half maxima. The diffraction peaks on the cycled MEA showed signs of a narrowing and analysis indicated that the Pt-volume averaged crystal size had increased from 2.3 nm to 10.5 nm. Fig. 3.7 shows the XRD peaks for fresh and cycled MEAs.

TEM analysis of unused Pt/Vulcan powder catalyst and used (cycled) MEAs (scraped off samples) were compared to reveal a larger mean particle size with a broader distribution for the used MEA, as expected from ECA results. In order to examine in more detail the uniformity of particle size growth over the entire thickness of the catalyst layer (from DM to membrane) they performed TEM studies systematically across the thickness of the cathode. Their analysis indicated that near the DM/cathode interface, the particles had grown in size and were larger but of uniform size; on the other hand, closer to the membrane/cathode interface, the particles tended to be non-spherical and were also found embedded in the ionomer uncontacted to the

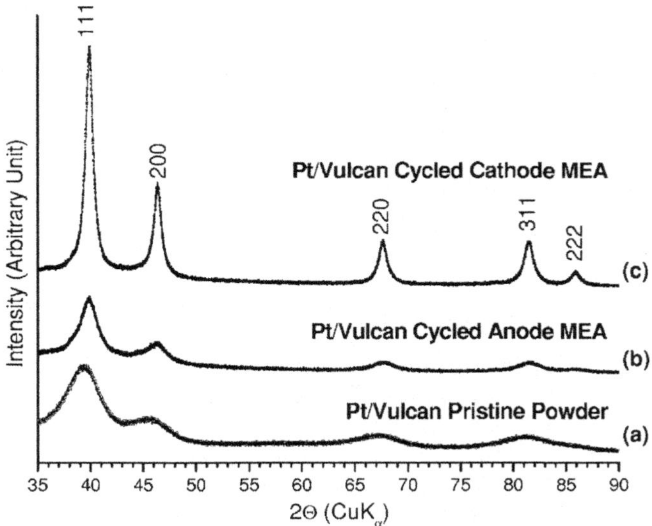

FIGURE 3.7 Baseline XRD patterns of unused Pt/Vulcan catalyst powder collected at normal incidence compared to 1° incidence patterns from the anode and cathode of a potential cycled (0.60–1.0 V, 10,000 cycles, 80°C) Pt/Vulcan MEA. Cycled, degraded cathodes are shown to exhibit a narrowing of peaks. *(Reprinted with permission from [21])*

carbon support. They concluded that the Pt particle growth occurs via two parallel mechanisms:

i. Ostwald ripening, i.e. small particles of Pt dissolve during high potentials and diffuse over nm length scales to re-deposit on larger particles (to lower their surface energy) contributing to 50 % of the ECA loss; and
ii. migration of Pt ions over micrometer length scales with the subsequent precipitation in the ionomer phase on contact with crossover hydrogen permeating from the anode.

The first mechanism was predominant in the region closer to the GDL and the second closer to the cathode membrane interface. It should be noted that the authors did not take into account the Pt ion transport from the cathode towards the membrane due to a concentration gradient or the effect of the potential distribution in the ionomer of the cathode, both of which are likely to be significant contributors to the Pt distribution profiles.

The use of HR-TEM (high-resolution TEM) as well as HAADF-STEM (high angle annular dark field-scanning TEM) in combination with partial embedding of electrode and diamond-knife ultra-microtome has been shown by others to be an excellent technique for imaging the porous catalyst layer after degradation tests [82–84].

Borup et al. [85] carried out durability studies using steady-state holds, simulated drive cycles (20 min voltage drive cycles derived from US06) and subscale single cell potential cycling experiments using linear potential sweeps (triangular potential profile, 0.1 V to 0.8–1.5 V, H_2-N_2) with varying conditions to determine the effect of individual conditions on Pt particle growth. Figure 3.8 depicts the increase in XRD Pt:C ratio as the upper potential is raised to values where carbon corrodes. It should be noted that the lower potential of 0.1 V is much lower than a fuel cell would experience during normal operation. They also carried out SEM/EDS, XRF, XRD and TEM to characterize changes in the electrodes before and after the durability tests. The XRD measurements after the testing were carried out by scraping off about 90% of the catalyst layer. They conclude from their work that the ECA decreased as the upper potential of the applied potential cycle was increased from 0.75 to 1.2 V. Based on the narrowing of XRD diffraction peaks, they inferred that particle growth was the main cause of surface area loss with cycling rather than Pt loss. Due to their use of a linear sweep rate, the time at elevated potentials was also high. By plotting the loss in ECA versus the time the cathode was subject to potentials >0.90 V, they found the losses for sweep rates of 50 mV/s were steeper than at 10 mV/s; they concluded from these results that the number of cycles was the dominant effect and the time duration at high potential was a secondary effect. They also reported that the Pt particles grew at an increasing rate at higher temperatures and humidity. They attributed the lower growth of Pt particles at low RH to the lower solubility and mobility of Pt in the ionomer of the catalyst layer.

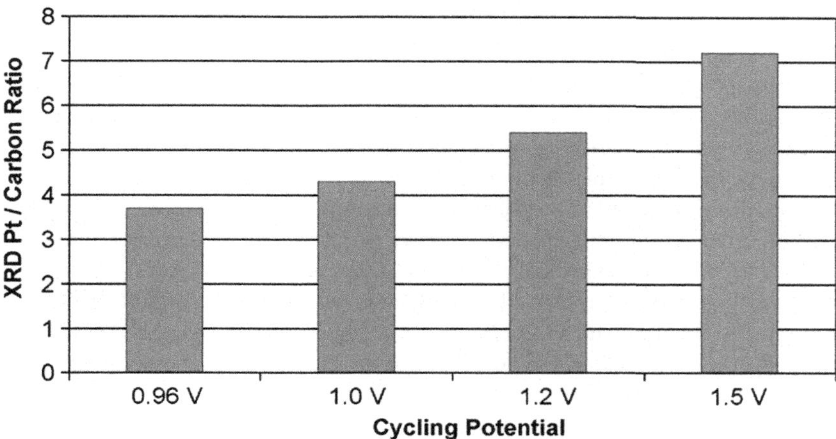

FIGURE 3.8 Ratio of XRD signals of Pt to carbon after executing potential cycling (1,500 cycles) at various potentials in the range where significant carbon corrosion takes place. *(Reprinted with permission, Borup et al. [85])*

We note that in all the cases of Pt dissolution within PEMFCs, the counter anion for the Pt^{x+} cations has not been clearly identified. Membrane degradation products of PFSA ionomers such as fluorides [86] and sulfates [87,88] have been measured under PEMFC operating conditions. Researchers have speculated that these anions could form complexes with Pt but it has not been verified experimentally.

Potential cycling in MEAs has been reported by Paik and co-workers [89] in several voltage ranges with the upper limit fixed at 1.3 V where significant carbon corrosion occurs in addition to Pt dissolution. They found higher losses for square wave profiles as compared to triangular profiles. Uchimura et al. [11] and Uchimura and Kocha [90] have also carried out an exhaustive investigation of the electrode degradation under cycling as a function of various parameters such as cycle profile potential range (0.60–0.95 V), anodic/cathodic sweep rates, temperature, RH and oxygen partial pressure on Pt/C and Pt-alloy catalysts in subscale fuel cells; they examined the catalyst (mass, specific activity) degradation rates along with TEM of MEA cross-sections of samples before and after testing. These recent results, as well as ongoing work, will be discussed in **Section 4**. In the following section (**Section 2.1.2**) we examine the electrochemical degradation of Pt-based alloys.

2.1.2. Electrochemical Degradation of Pt-based Alloys

This section examines the electrochemical degradation of Pt-based alloys in acid electrolytes (**Section 2.1.2.1**) and in PEMFCs (**Section 2.1.2.2**).

2.1.2.1. Electrochemical Degradation of Pt-based Alloys in Acid Electrolytes

While investigating the rate determining step for the ORR in acid medium, Vogel and Baris [91] proposed that the cleaving of the O-O bond in the peroxyl radical formed in a previous step was crucial. Thus, it appeared possible that by altering the Pt-Pt spacing, the dissociation of the O-O bond might be facilitated. A number of binary and ternary catalysts were subsequently prepared using the 'carbothermal reduction process' [92–94] to introduce base metals into the FCC lattice of Pt substitutionally causing lattice contraction. Jalan and Taylor [74] investigated the correlation between activity and nearest neighbor distance required for the dual site O-O cleaving mechanism and prepared a series of binary Pt-alloy electrocatalysts to modulate the lattice. They showed (Fig. 3.9) that the specific activity of binary Pt alloys increased with decreasing nearest neighbor distance (Pt = 2.775 Å, PtCr = 2.73 Å).

An obvious concern with the use of transition metals in alloys is the fact that Pourbaix diagrams indicate that Co, Fe, Cr, Mn, Ni, Cu, and V are susceptible to dissolution at pHs close to 0 in the range 0.3–1.0 V versus SHE. Figure 3.10 depicts the Pourbaix diagram for cobalt as an example [17].

Yet, some of the first reported studies (UTC) on binary and ternary Pt-transition metal alloys/C showed both enhanced electrocatalytic activity and

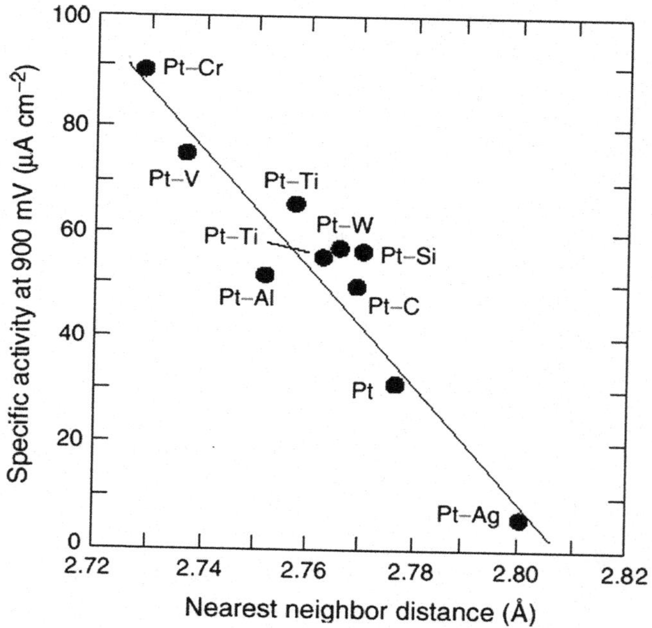

FIGURE 3.9 Plot of specific activity versus nearest neighbor distance for a number of Pt binaries. *(Reprinted with permission from Jalan and Taylor [74])*

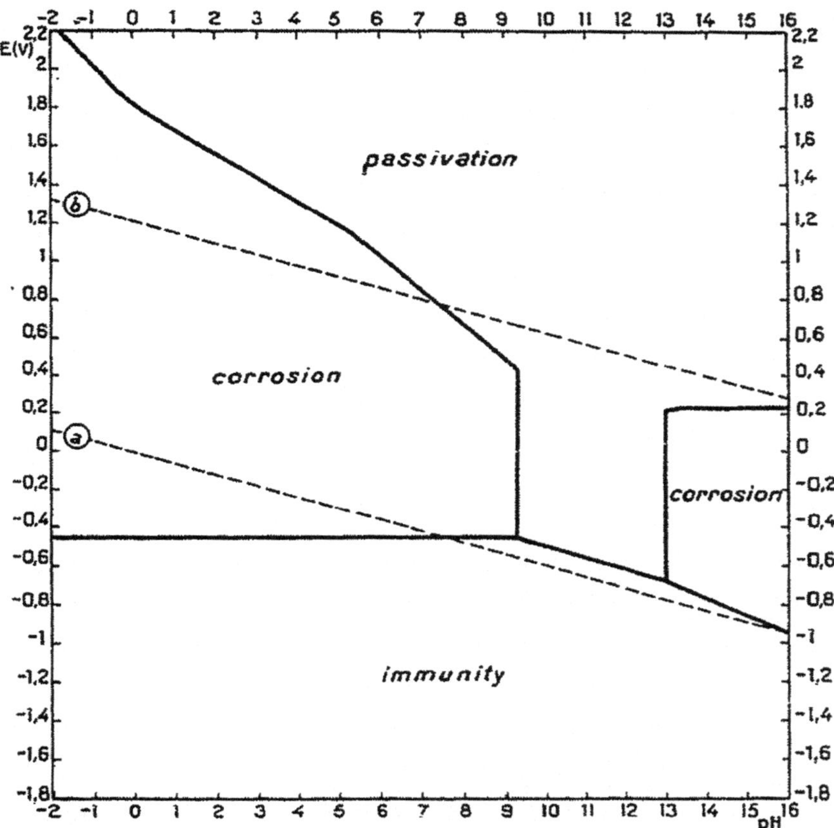

FIGURE 3.10 Pourbaix diagram of Co showing domains of passivation, corrosion and immunity. Co is expected to be fairly unstable in the range of PEMFC operating conditions. *(Reprinted with permission from Pourbaix [17])*

durability. Landsman and Luczak [95] showed that (Fig. 3.11) for a given binary catalyst, as the percentage of alloying element was varied, the activity passed through a maximum as the lattice parameter decreased. XAS analyses have indicated that the d-band center may also play a role in perturbing the activity [96]. In brief, various correlations between the improved activity of Pt-alloys over Pt may be described based on structural factors, suppression of anion adsorption, electronic factors and surface sensitive effects [97].

The earliest patents described the synthesis of Pt alloyed with about 20–40 atomic wt% of V, Mn, Mo, Ti and Al that showed higher activities than Pt, and later, PtCr and PtFe were reported with higher activities and improved durability. Despite the fact that the transition metal alloying components leached out into hot phosphoric acid, these catalysts maintained a higher activity. Finally, ternaries such as PtCrCo, PtCoNi, PtFeCu and quaternaries such as

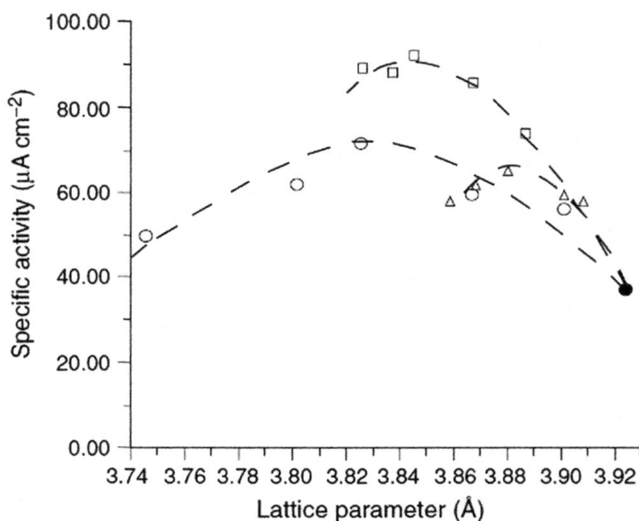

FIGURE 3.11 Plot of specific activity for several Pt-alloys/C vs. lattice parameter for the ORR in 99% H_3PO_4 at 177°C. Filled circles–Pt, hollow circles–PtCo, hollow squares–PtNi, and hollow triangles–PtCr. As the lattice parameter decreases with alloying, the specific activity increases to a maximum and then decreases. *(Reprinted with permission, Landsman and Luczak [95])*

PtFeCoCu were reported to possess high activity and high stability especially when the alloying phases were ordered. Luczak and Landsman demonstrated that PtCrCo/C exhibited a 2.5 times enhancement in mass activity compared to Pt/C [94]. Over long periods of time, the alloy catalysts in PAFCs showed loss in ECA and attempts were made to reduce the degradation rate by generating etch pits or having carbon deposit around the Pt and anchor it more firmly. Kocha et al. [98] have reported (Fig. 3.12) the loss in surface area of PtRhFe/C catalyst compared to standard ternary Pt-alloy/C catalysts in PAFCs over 20,000 h of accelerated tests (>40,000 h under normal operation). Both catalysts lose surface area over time, but the PtRhFe/C catalyst maintains a fairly high activity over its life. The reason for the preservation of high electrocatalytic activity in PAFCs despite the surface area loss over thousands of hours has been argued to be due to particle growth and associated higher specific activity but is still being debated.

Stamenkovic and co-workers [99] elucidated the fundamental relation between the experimentally determined d-band center and the ORR activity for Pt_3M (Pt-3d transition metal alloy) extended surfaces in perchloric acid. Their study details comprehensive sample preparation and characterization that includes Auger electron spectroscopy (AES) for surface cleanliness, low-energy ion scattering (LEIS) for determining the surface atomic layer composition and synchrotron based high-resolution ultraviolet photoemission spectroscopy (UPS) for electronic surface structure. Their AES/LEIS

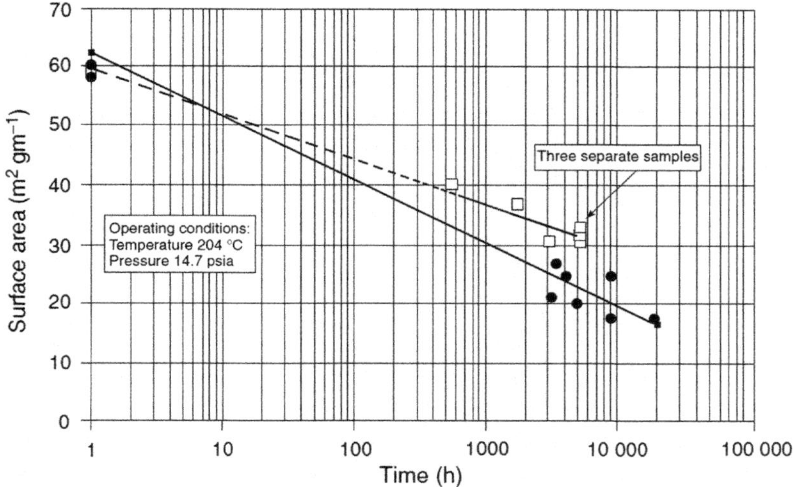

FIGURE 3.12 Plot of ECA vs. operating time for PtRhFe/C catalysts (squares) compared to typical standard Pt–alloys (filled circles) all having a loading of 0.5 mg/cm^2 and tested under H_2|Air at 204°C at a constant current density of 215 mA/cm^2 in phosphoric acid subscale cells. *(Reprinted with permission from Kocha et al. [98])*

spectra indicated that for sputtered samples, after immersion in acid electrolyte, the transition metal dissolves leaving a pure Pt 'skeleton' surface. On the other hand, samples that were annealed did not show any change after immersion in acid suggesting that a stable 'Pt-skin' had developed. These annealed surfaces with 'Pt-skin' also revealed features in the cyclic voltammogram that indicated a shift towards positive potentials in the onset of oxides. The relationship (volcano plot) between specific activity (0.90 V, 0.1 M $HClO_4$) and the d-band center values (established in UHV) is illustrated in Fig. 3.13.

2.1.2.2. Electrochemical Degradation of Pt-based Alloys in PEMFCs

The same binary and ternary Pt-alloy/C catalysts that exhibit enhanced activities in liquid acid electrolytes may also be expected to be good candidates in PEMFCs. This was confirmed by Mukerjee et al. in a series of papers that investigated binary Pt-alloys in subscale fuel cells for ORR kinetics complemented by X-ray techniques to study lattice parameter, stability and nature of surface species [100, 101]. Researchers at Johnson Matthey [102] also conducted extensive research on binary Pt-alloys such as PtFe, PtCr, PtNi, PtMn and PtTi (20 wt% Pt-alloy/Vulcan support, Pt:M 50:50) that were tested in 25 cm^2 subscale fuel cells; they reported ~25 mV activity gains, similar to that found in PAFCs. They also applied electron probe microanalysis (EPMA) on MEA cross-sections to look for evidence for base metal leaching from the catalyst. They discovered that Cr and Ti did not leach out

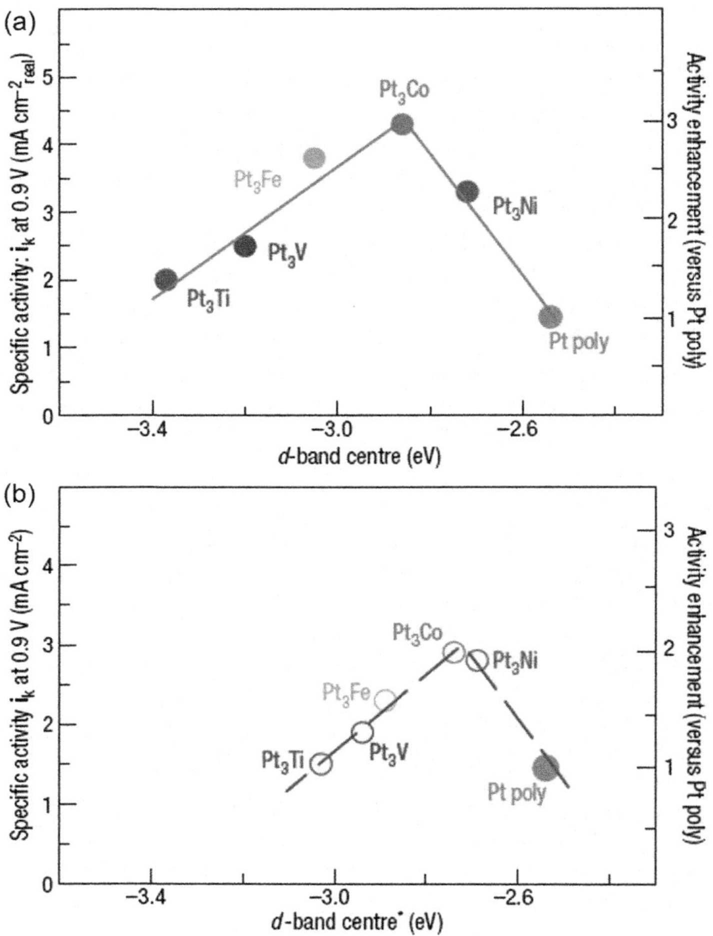

FIGURE 3.13 Specific activity and activity enhancement (in 0.1 M $HClO_4$) vs. the d–band center for Pt_3M alloys having: (a) Pt–skin and (b) Pt-skeleton surfaces illustrating volcano shaped trends. *(Reprinted with permission from Stamenkovic et al. [99])*

but Fe, Mn and Ni all leached out into the MEA, albeit with minimal performance loss over 200 h.

The concerns regarding the durability of Pt-alloys/C in PEMFCs are considerably greater than those in liquid electrolyte fuel cells. Unlike the case of an acid liquid electrolyte, there exist extremely limited numbers of protonic sites in the ionomer (catalyst layer and membrane); these sites are susceptible to being ion-exchanged by dissolved metal ions. When ion-exchanged, the ionomer/membrane will lose protonic conductivity and also tend to dehydrate. This phenomenon has been demonstrated by several groups by immersing of a sample of PFSA membrane in an aqueous solution of a given cation for a period of time. A 30 μm thick PFSA membrane contains ~5 μmole/cm^2 of protonic sites; the ionomer in a 10 μm catalyst layer contains only about 0.5 μmole/cm^2. Consider the case of the popular binary catalyst PtCo/C; if we were to assume that all the Co in PtCo or Pt_3Co catalyst dissolves, they would exchange with and consume about 0.6 μmoles/cm^2 of available protonic sites in the MEA. If we also allow for diffusion of the Co ions to the membrane, in the limit, they would occupy about 10% of the total sites. To experimentally verify this analysis, researchers at GM [103] evaluated the solubility of Co from PtCo/C by carrying out *ex-situ* tests in which various catalysts were immersed in 1 M H_2SO_4 at 95°C for several hours and the solution analyzed by ICP analysis. 35% of Co ions were found in the solution for unleached catalysts (Fig. 3.14(a)). Leaching tests in sulfuric acid are a good screening test for conventional Pt-alloys as they give some indication of the stability of the catalyst before MEA preparation. Eliminating excess or easily soluble Co from the surface of the PtCo/C catalyst before preparing and testing an MEA might enhance the activity while simultaneously reducing contamination of the ionomer. This premise was confirmed when it was found that if a 'pre-leached catalyst' was used to manufacture an MEA, the cell exhibited higher performance than cells using MEAs with unleached PtCo/C. Figure 3.14(b) illustrates the enhancement of activity of pre-leached Pt-Co/C MEAs [16]. Figure 3.14(c) shows EPMA micrographs that exhibit a band of Co in the membrane of the MEAs analyzed after testing in fuel cells.

Yu and co-workers [104] were among the first to evaluate the degradation of PtCo/C catalysts in comparison to Pt/C catalysts under potential cycling in subscale fuel cells (25 cm^2). They carried out experiments in which the PtCo/C and Pt/C catalysts (0.4 mg/cm^2, cathodes) were subject to potential cycles in the range 0.87–1.2 V (30 s, 30 s, square, or 1 cycle per min), under $H_2|N_2$, 65°C and 100% humidified gases. The choice of 0.87 V was justified as representing idling/low current density and 1.2 V as the 'air-air' OCV potential. $H_2|$Air I-V polarization curves used as diagnostics every 400 h indicated minimal performance loss for the PtCo/C compared to Pt/C over 2,400 cycles as seen in Fig. 3.15. EPMA images showed no deposition of Co in the membrane whereas a Pt band was formed in both cases at the end of cycling tests. The three outcomes of minimal change in resistance of the membrane, no Co band in EPMA, and low performance loss led them to conclude that the dissolution of Co from PtCo/C cathodes into the membrane was not measurably detrimental.

Kocha and Gasteiger [103] examined the degradation rates of Pt/C and PtCo/C catalysts in large area platform (800 cm^2) PEMFC short stacks over a period of ~1,000 h with periodic interruptions for the diagnostic measurements of the *in-situ* ECA and ORR kinetics. The stack was built with four cells each of baseline Pt/C and PtCo/C, so that reproducibility and benchmarking under identical conditions was inherently simplified. Their results confirmed that PtCo/C had a higher ORR activity and exhibited a corresponding higher cell performance compared to the

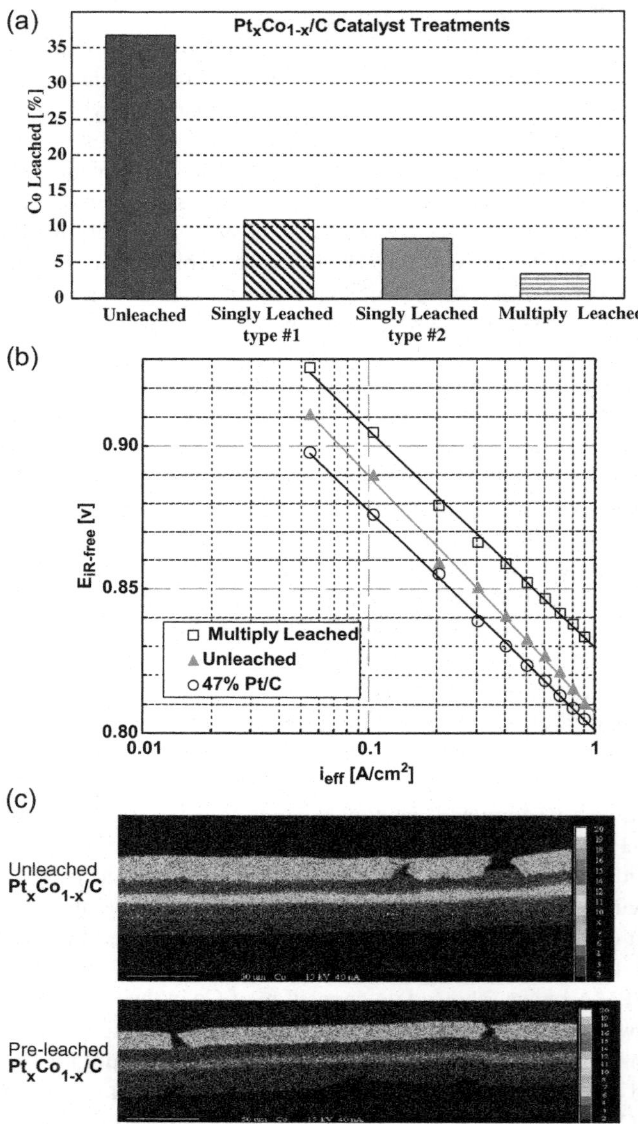

FIGURE 3.14 (a) Leaching of Co from several pre-treated PtCo/C catalysts. (b) Tafel plots under H_2-O_2 illustrating the enhanced ORR activity of pre-leached PtCo/C catalysts in comparison to unleached PtCo/C and baseline Pt/C measured in subscale PEMFCs. (c) EPMA micrographs for both leached and unleached catalysts showing Co re-deposition as a band in the membrane. *(Reprinted with permission from [16,103])*

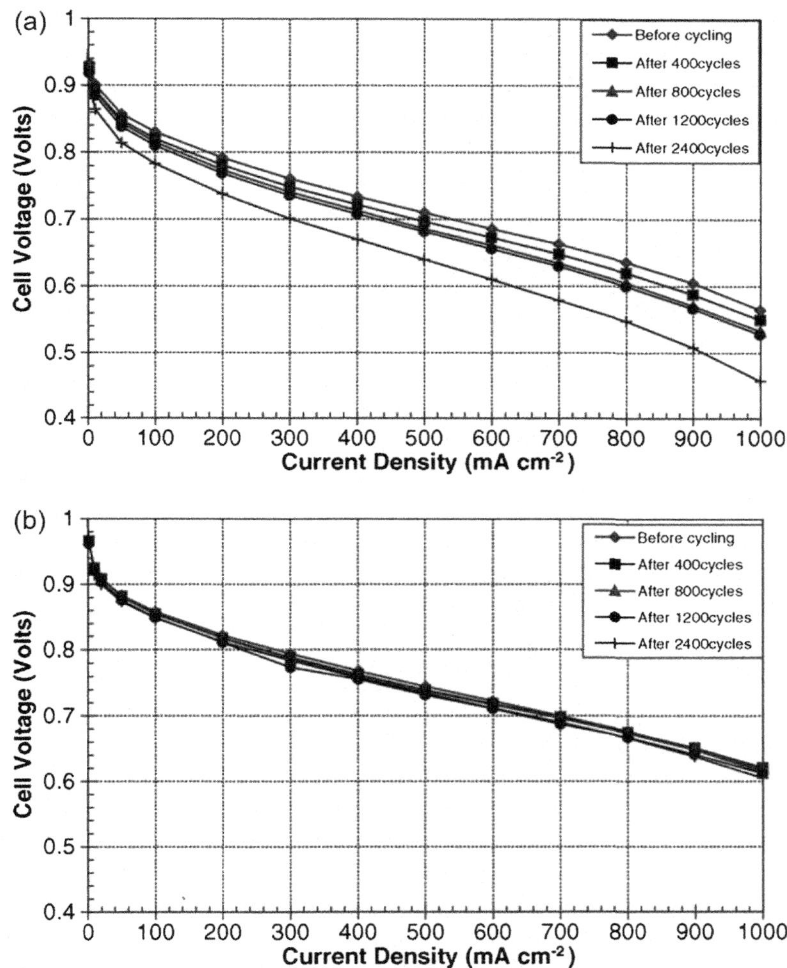

FIGURE 3.15 H_2|Air I-V polarization curves for: (a) Pt/C MEA and (b) PtCo/C MEA before and after potential cycling in the potential range 0.87–1.2 V at 65°C in subscale PEMFCs. *(Reprinted with permission from Yu et al. [104])*

baseline Pt/C. Figure 3.16 illustrates the loss in mass activity for Pt/C and PtCo/C in the short stack over 1,000 h; the mass activity of PtCo/C continued to be higher than that of Pt/C at 1,000 h when testing was stopped due to other stack issues. They debated whether the improved performance and initial durability was a function of alloying, heat treatment, particle size or a combination of all three parameters and resolved the issue in later work [105].

Hidai and co-workers have studied the degradation of PtCo/C (powder, decal, and MEA (before and after 10,000 cycles between 0.60–1.0 V)) using

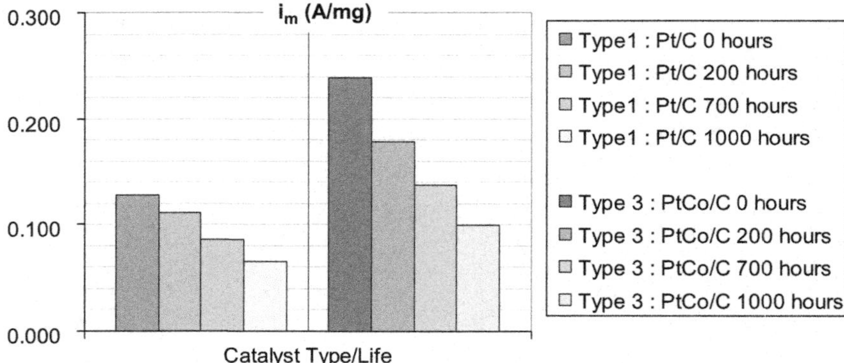

FIGURE 3.16 Mass Activity of the Pt baseline as well as Pt_xCo_{1-x}/C over 1,000 hours of testing in a full-scale PEMFC short stack. The ratio of activities for the Pt_xCo_{1-x}/C to Pt/C baseline decreases from 1.85 to 1.55 over the period of operation. *(Reprinted with permission from Kocha et al. [103])*

X-ray photo-emission spectroscopy (PES) from SPring-8 at the Japan synchrotron radiation research institute (JASRI) [106, 107]. The Pt 4f and Co 2p electronic states were probed using soft X-rays (1 keV, escape depth 1 nm) and hard X-rays (8 keV, escape depth 10 nm). The results from the 'hard X-ray' PES binding energy spectra indicated that Co existed mainly in the metallic form, and those from the 'soft X-rays' revealed an absence of Co signal peaks suggesting a Pt-skin on the surface. The authors postulated that any electrochemical degradation that occurs due to potential cycling takes place at the surface skin layer. Thus, we find a consensus from studies in liquid electrolytes and MEAs of PEMFCs [99,103,106,108,109] that leaching of non-noble elements from the surface of the Pt–based catalyst takes place to expose a Pt skin that ultimately takes part in the ORR.

Neyerlin et al. have reported on electrochemically and otherwise de-alloyed PtCu/C catalysts (that have Pt-rich surfaces and alloy-rich cores) and have reported activities higher than that of PtCo/C in RDE measurements and MEAs. The same group [110] further evaluated the durability (30,000 cycles, 0.5–1.0 V) of $Pt_{25}Cu_{75}$ and $Pt_{20}Cu_{20}Co_{60}$ (prepared by an impregnation/freeze-drying/annealing technique) in MEAs compared to baseline Pt/C. They concluded that higher annealing temperatures, larger particles sizes, and the alloying elements all contributed to higher durability of the alloy catalysts over Pt/C. Recently, they have also reported on the use of reciprocal space X-ray scattering techniques wherein *in-situ* SAXS and *in-situ* XRD were applied together to follow the size and compositional changes of Pt/C, heat-treated Pt/C and PtCu/C [111] as well as kinetics of atomic scale dissolution. Their results confirmed that Cu initially dissolved from the surface resulting in the formation of an enriched Pt skin that continued to degrade through the mechanisms outlined earlier in the literature [99,103,106,108,109].

Role of Particle Size, Heat Treatment and Alloying Effects on Durability Pt-alloys typically show lower dissolution rates compared to Pt/C but it was not apparent if the improved durability was dictated by the larger particle size, the heat treatment or the alloying element or perhaps a synergistic effect. To address the issue Makharia et al. [105] conducted a study to de-convolute the above-mentioned effects; they evaluated four catalyst samples namely: Pt/C (2–3 nm), Pt/C (4–5 nm), Pt/C-HT (4–5 nm) and PtCo/C (4–5 nm) under potential cycling between 0.6–1.0 V at 80°C and 100% RH. Their findings (Fig. 3.17) indicated that, although the effect of particle size alone on durability was a significant contribution, heat treatment had the most pronounced effect on cycling durability, confirming previous reports by Jalan and Taylor [74] for Pt/C in PAFC environments. A dissolution-resistant annealed 'Pt skin' that covers the catalyst nanoparticles is thus the predominant surface where ORR takes place in these materials.

2.1.3. Electrochemical Degradation of Non-conventional Catalysts

In this section, we briefly discuss some non-conventional catalysts that have been under development over the last few years. There are two major development pathways for novel advanced HOR and ORR catalysts: high specific

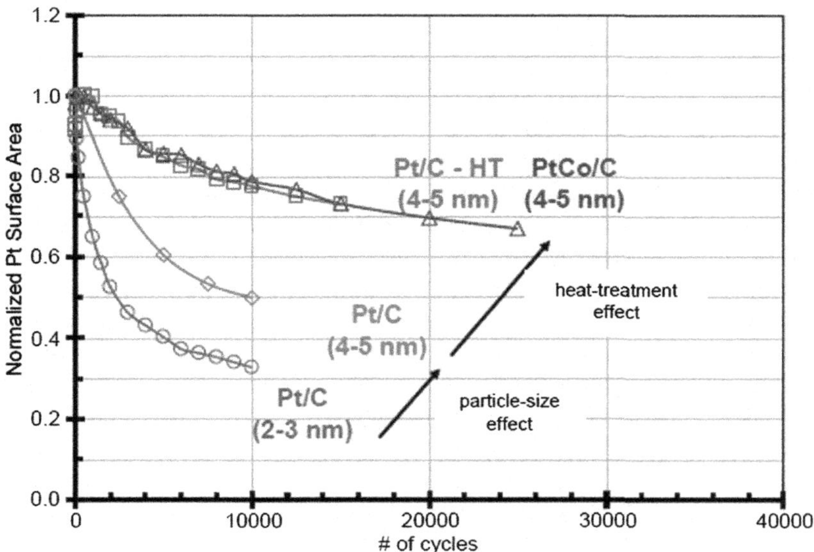

FIGURE 3.17 Particle size, heat treatment and alloying effects on the loss in Pt surface area under potential cycling under $H_2|N_2$ (0.60–1.0 V, 20 mV/s, 80°C, 100% RH). Heat treatment of catalysts is shown to result in improved voltage-cycling resistant catalysts. *(Reprinted with permission from Makharia et al. [105])*

activity ($\mu A/cm^2$) materials that have extended thin film electrode structures (ETFECS) (properties comparable to bulk poly-Pt) and 'core-shell catalysts' that have a non-precious metal core covered by a thin Pt shell and therefore high mass activities (mA/mg_{Pt}).

Debe and co-workers have been working over the last decade on the development of 'nano-structured thin film' catalysts (NSTF) [112–116]; these are Pt-based catalysts deposited on micron thick, highly oriented, densely packed (3–5 billion/cm^2), crystalline, organic whiskers (~50 nm cross-sections) as depicted in Fig. 3.18. The organic pigment N,N-di(3,5-xylyl)perylene-3,4:9,10 bis(dicarboximide) or perylene red is vacuum deposited and thermally annealed to produce these whiskers. These NSTF Pt catalysts form CCMs with catalyst layer thicknesses of <0.5 μm (compared to 10 μm for conventional Pt/C). The NSTF catalysts have low electrochemical areas of ~8–12 m^2/g – comparable to unsupported Pt-black catalysts – but $1/10^{th}$ that of conventional nanoparticle catalysts (70–100 m^2/g). They exhibit extremely high specific activities that are 5–10 times that of conventional nanoparticle Pt/C and comparable to bulk poly-Pt, with resultant mass activities comparable to Pt/C. Recent reports show that these catalysts are resistant to loss in surface area and Pt dissolution under constant potential holds, potential cycling, and simulated start-up/shut-down cycles. The Pt-alloy NSTF catalysts have been recently reported to suffer low degradation under OCV conditions (0.2 mg-Pt/cm^2 PtCoMn; 3M ionomer; mechanical stabilized support) over a period of ~7,000 h. Furthermore, these catalysts, by virtue of their non-conductive, corrosion-resistant support, exhibit high endurance under start-up/shut-down conditions where cathode potential spikes exceed 1.2 V. Ongoing work in this field includes the study of various alloys of Pt as well as attempts to improve the mass-transport within the very thin (1 μm) catalyst layers.

To address issues found with the NSTF catalysts, researchers are currently studying the general class of extended thin-film electrocatalyst structures or ETFECS [117]. This includes deposition of Pt on conductive high surface area corrosion resistant materials such as carbon nanotubes as well as spontaneous deposition (SGD) to form nanotubes of Pt by starting from nanotubes and naoplatelets of Cu, Ag and other materials. The SGD material may be used as unsupported materials or mixed with a support to help create an electrode structure conductive to mass-transport in fuel cell electrodes. Recent results indicate that the SGD catalysts can be prepared with ECAs greater than 30 m^2/g and show high mass activities when measured in RDE set-ups.

Adzic and co-workers [118–120] have been working for several years on a new class of catalysts known as 'core-shell' catalysts that consist of a monolayer (ML) of Pt deposited on a metal core. Pt ML on Pd nanoparticles, mixed metal-Pt on Pd nanoparticles, Pt ML on $AuNi_5/C$, Pt ML on alloy nanoparticles of PdFe/C, Pt ML on PdCo alloy surfaces, and Pt ML on oxide surfaces have all been investigated by the Adzic group at Brookhaven National Laboratory (BNL) [119]. Enhancement of activity with monolayer coverage of

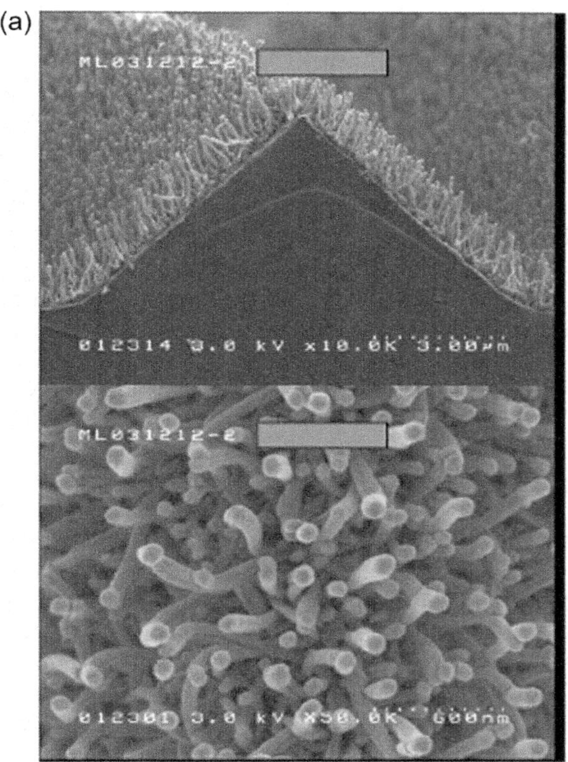

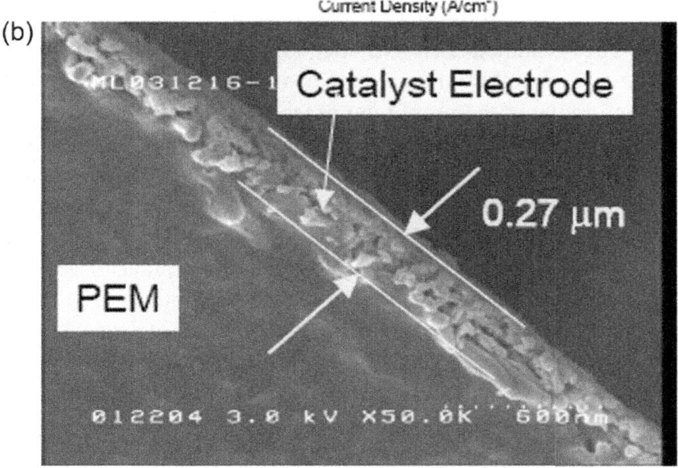

FIGURE 3.18 Structure of 3M NSTF catalysts (a) and catalyst layer on membrane (b). *(Reprinted with permission from Debe et al. [113])*

Pt on the surface has been ascribed to strain-induced d-band shifts, electronic ligand effects and ensemble effects. Mass activity improvements of a factor of five with high stability have been reported, based on total precious metal content. Interestingly, despite their monolayer coverage, these core-shell catalysts exhibit reasonable durability under standard potential cycling conditions in liquid electrolytes.

In one of the papers written by the group, Sasaki et al. [121] report on the synthesis of an HOR electrocatalyst by placing Pt atoms ($1/8^{th}$ monolayer) on Ru nanoparticles supported on carbon using an electrodeless deposition method. In 50 cm^2 fuel cells, these $PtRu_{20}$ catalysts showed high resistance to CO poisoning and good long term durability (~1,000 h) for extremely low loadings (17 $\mu g/cm^2$ Pt, 180 $\mu g/cm^2$ Ru).

Zhang et al. [120] have recently modified conventional Pt/C with Au clusters (30–40% coverage) through the displacement of a Cu monolayer. Thin films of 'Au cluster-modified Pt/C' as well as baseline Pt/C were cycled for 30,000 cycles, in the voltage range 0.6–1.1 V, at room temperature in oxygen saturated 0.1 M $HClO_4$ using an RDE set-up. The ECA and ORR activity losses for 'Au cluster-modified Pt/C' were negligible, whereas baseline Pt/C lost 45% of its initial area and 39 mV at the half wave potential (Fig. 3.19). Based on CVs as well as XANES spectra, the authors attribute the dissolution resistance of the Au modified-catalyst to the raised potential (Fig. 3.19(b)) at which Pt gets oxidized.

Ball et al. have described their preliminary results for the preparation of core-shell catalysts consisting of Pt ML on Ir, Pd_3Co, and Pd_3Fe cores using 'scalable chemistry' that can be extended to MEAs [122]. XPS and LEIS were used to confirm the coating of the cores with Pt and corresponding CVs also showed features indicative of Pt. They assessed the cyclic durability of the 'cores' as well as 'core-shells' independently in 1 M H_2SO_4 at 80°C between 0.60–1.0 V for 1,000 cycles. They obtained mass activities for Pt-shell on Ir-cores to be higher than that for conventional Pt-alloy/C and also found that the addition of the Pt monolayer shell to the core lowered the degradation of the catalysts.

The Norskov group [123] had predicted that Cu, Co, Ni and Fe might lower the d-band of Pd towards that of Pt. Wang et al. [124] selected base and noble metal combinations with the intent of modifying the electronic properties of the noble metal to be more 'Pt-like'. Pd-Cu was prepared and characterized by them for various molar ratios of Pd and Cu; they found that the Pd/Cu alloys having molar ratio of 1:1 (and treated at 600°C) exhibited 4–5 times greater activity (at 0.85 V in RDE experiments) compared to Pd, and also demonstrated reasonable durability in acid over the period of a few days. Although these Pd-Cu catalysts possess only ¼ the activity of conventional Pt/C, keeping in mind the cost of Pd is ¼ that of Pt, they warrant recognition as an alternative candidate to Pt/C.

Several research groups have been addressing the issue of Pt cost and durability by studying alternative non-Pt catalysts such as non-Pt alloys, transition metal chacogenides [125,126], macrocycles containing nitrogen [127–129] as

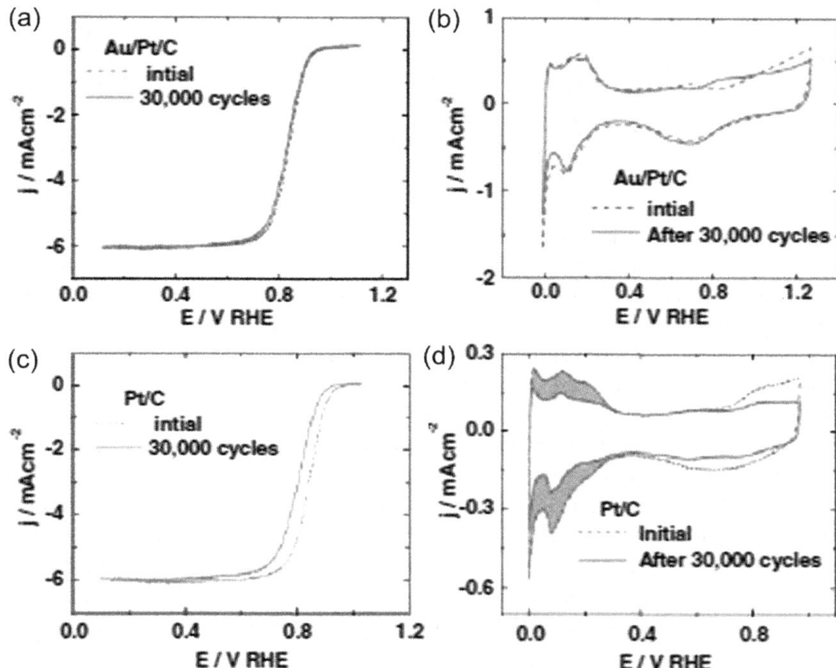

FIGURE 3.19 (a) and (c): ORR I-V curves measured in RDE at 10 mV/s and 1600 rpm in 0.1 M $HClO_4$ before and after potential cycling. (b) and (d): CVs measured before and after potential cycling. Potential cycling of AuPt/C and Pt/C carried out for 30,000 cycles between 0.6–1.1 V in O_2 saturated 0.1 M $HClO_4$. *(Reprinted with permission from Zhang et al. [120])*

well as certain carbides and nitrides [130]. Some of the initial research on macrocycle-based non-precious metal catalysts (NPMCs) was carried out by Jasinski [131], Jahnke et al. [132], Bagotsky et al. [133], and Gupta et al [134]. In a series of papers, Dodelet and co-workers, over a period of 10 years, have studied catalysts composed of Fe, C and N [135–137]. They have recently reported fairly high catalytic activities (evaluated at 0.90 V in a PEMFC) by stuffing the micropores of a ball milled mixture of carbon (Black Pearls) with phenanthroline and 1 wt% Fe (iron acetate) for a total loading of 5.3 mg/cm^2 [138]; their durability under automotive fuel cell conditions has not yet been demonstrated.

The minimum activity required for a 'costless' catalyst to be cost-effective in a PEMFC stack has been extensively examined [16] by taking into consideration the additional thickness of the catalyst layer associated with a low activity catalyst. Because the thickness of the catalyst layer and the concomitant mass-transport become limiting, it is recommended that the activity for NPMCs be reported in units of A/cm^3. The DOE target for the activity of a non-Pt catalyst measured at 0.80 V is 300 A/cm^3. Most of the effort in the area of NPMC research is currently being devoted to enhancing the catalytic activity and measuring it accurately. Identifying the catalytically active sites and quantifying the electrocatalytic activity in terms of the product of the 'number of active sites' and 'turnover frequency' is not a straightforward task. NPMCs, at this point in time, are not ready to be evaluated for long term electrochemical degradation until they first achieve activities approaching about ¼ to ½ that of Pt/C.

2.2. Support Durability

Prior to the advent of research into PAFCs in the 1960s, the majority of the work on carbon corrosion in aqueous media was directed towards the development of graphite anodes for the chlor-alkali industry. In the very early stages of fuel cell development, Pt black was used as the catalyst of choice without being dispersed on any support. The electrodes (gas diffusion electrodes or GDEs) were formed typically by incorporating PTFE as a binder to produce hydrophobic gas pores that enhanced mass-transport but also resulted in a lowered Pt utilization due to partial masking of catalyst sites responsible for the reaction. The Pt blacks had low surface areas in the range of 15–25 m^2/g corresponding to an average particle size of ~8–10 nm. These blacks were fairly durable since they were not dispersed on a high surface area support that corroded and the larger Pt particles were inherently more stable. But the low surface area and low utilizations resulted in very high loadings of catalyst (>5 mg/cm^2) with costs acceptable only to the space industry; Pt and precious metal blacks are still being used in alkaline fuel cells used in the space shuttle. Larger particles are more stable; however, due to their inherently poor dispersion (in the absence of a support), the particles have a tendency to coalesce and agglomerate with a consequent degradation in performance over time.

In homogeneous catalysis, the use of catalysts deposited on a support was well-established and this idea was transferred to fuel cells by supporting Pt on various carbon blacks, especially Vulcan. Carbon blacks have been the material of choice for the support on which Pt is dispersed since its first use in PAFCs by United Technologies for several reasons. Carbon blacks:

- **i.** have a high electronic conductivity,
- **ii.** have a high surface area (~80–2000 m^2/g),
- **iii.** form a porous structure,
- **iv.** are fairly free of contaminants, and
- **v.** are easily available at a reasonable cost.

Of the drawbacks, carbon is thermodynamically susceptible to corrosion in the entire fuel cell operating regime. From the Pourbaix diagrams [17], one can observe that the oxidation of carbon to gaseous CO_2 has a standard potential of 0.207 V and that for carbon to CO is 0.518 V. CO is unstable and the CO-CO_2 reaction has a standard electrochemical potential of −0.103 V. Fortunately, the kinetics of carbon corrosion are slow and substantial measurable corrosion (as measured from carbon corrosion currents or CO_2 evolution) takes place only above ~1–1.2 V in acidic media.

Catalyst support degradation emerged as an important issue during the development of commercial PAFCs operated at ~205°C at high potentials in hot concentrated phosphoric acid. Typical Vulcan blacks (240 m^2/g) corroded easily in PAFCs, leading to intense research on the corrosion rates and fundamental governing mechanisms for many available carbons; it culminated with the use of a graphitized corrosion-resistant Vulcan carbon in commercial PAFCs. These carbons had to undergo heat treatment at ~2,200°C for several hours in order to be graphitized with the concomitant drawback of lower surface area (~80 m^2/g) and higher material processing cost. Landsman and Luczak [95] have reported the effect of heat treatment (in the temperature range 0–3,000°C) of Vulcan XC-72 in hot phosphoric acid on corrosion currents for samples held at 1 V. Figure 3.20 illustrates how the corrosion rate of Vulcan XC-72 decreases from ~0.14 mA/mg_C to ~0 after heat treatment at 3,000°C.

Landsman and Luczak also provide details of a practical support evaluation method (for PAFCs) known as the '100 min corrosion current test'; carbon support is contained in a gold foil, immersed in 99% phosphoric acid at 175°C under N_2, held at a given potential for 100 min, and the ensuing corrosion current recorded. Kinoshita [142] also studied specific carbon corrosion currents ($mA/cm^2{}_C$) of Vulcan and graphitized Vulcan-2700 in phosphoric acid (135–160°C) and plotted them vs. potential to obtain various kinetic parameters as shown in Fig. 3.21.

Although the operating temperatures of PEMFCs are more benign than PAFCs, in automotive environments, PEMFC cathodes undergo short bursts of

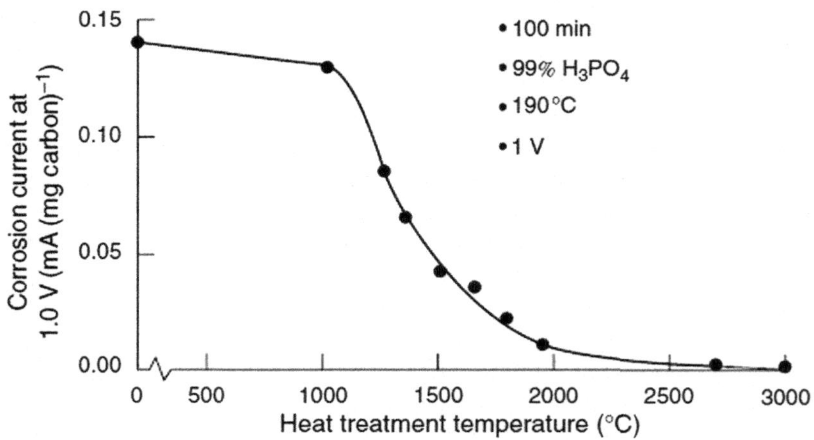

FIGURE 3.20 Effect of heat treatment on the corrosion currents for Vulcan XC-72 measured in hot concentrated H_3PO_4 for conditions shown in the figure. *(Reprinted with permission from Landsman and Luczak, [95])*

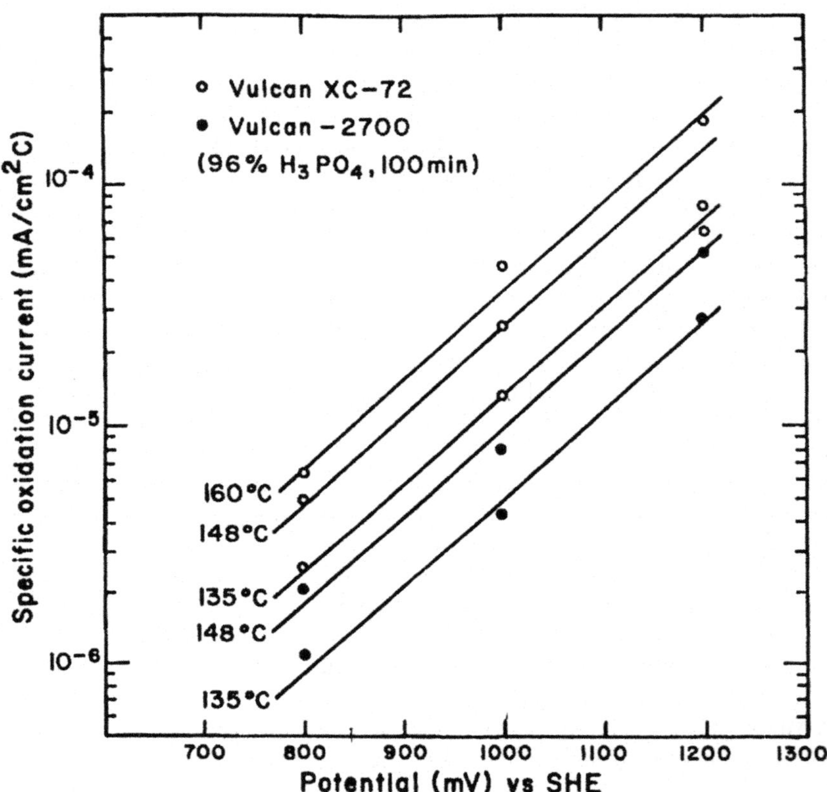

FIGURE 3.21 Log corrosion current vs. potential for Vulcan and graphitized Vulcan measured in phosphoric aid. Extracted parameter: $E_a = 55\,kJ/mol$; Tafel slopes ~300 mV/dec. Corrosion currents for Vulcan carbon are 5 times higher than for Vulcan 2700 graphitized carbon. *(Reprinted with permission from Kinoshita [142])*

potential transients that reach 1.5 V. This obviously leads to severe carbon corrosion and electrode degradation even with the use of graphitized carbon supports. This degradation only occurs during start-up/shut-down, and can be managed and limited through system controls, but these lead to higher costs. Research is therefore needed to produce an alternative support that has all of the positive attributes of carbon plus a high corrosion resistance at high potentials.

This section, covering support durability, has been divided into electrochemical degradation of carbon support in liquid electrolytes (**Section 2.2.1**), electrochemical degradation of carbon support in PEMFCs (**Section 2.2.2**), effect of Pt on the corrosion of carbon supports (**Section 2.2.3**), thermal degradation as a screening tool to measure electrochemical degradation (**Section 2.2.4**), and lastly, electrochemical degradation of alternative supports (**Section 2.2.5**).

2.2.1. Electrochemical Degradation of Carbon in Liquid Electrolytes

Kinoshita and Bett [139] studied the electrochemical oxidation of a high surface area carbon black (Neo Spectra-Columbian Carbon Co.) in 96% H_3PO_4 at 135°C, in which they measured the oxygen content of the carbon, as well as the CO_2 evolved as a function of potential and time. In their experiments, they formed carbon-10% PTFE-bonded electrodes supported by gold-coated tantalum screens and ensured complete wetting by vacuum filling. Since the CO_2 volume generated was small, and would dissolve into the electrolyte, it was saturated with CO_2. Their results on the oxidation of the carbon at different potentials indicated a current-time relationship of the form:

$$i_t\, t^n = k_t \qquad (3.13)$$

where:

'i_t' is the total anodic current
'k_t' is a constant
'n' is the slope of the log-log plot.

The value of 'n' varied from 0.42–0.76 and the corrosion currents ranged from 0.50–0.05 mA/mg_C, for potentials in the range of 1.0–0.70 V. Their experiments were carried out for 445–4750 min and the measured slopes were unchanged for times greater than 10 min.

They proposed that the total anodic current was the sum of the CO_2 evolution currents and the oxide formation current. By subtracting the total oxidation current from the current calculated from the evolved CO_2, they estimated the oxide growth. They found that the oxide growth followed the equation:

$$Q_{ox} = k_{ox}\, t^m \qquad (3.14)$$

where 'Q_{ox}' is the charge, and 'm' and 'k_{ox}' are constants. As expected, the amount of charge increased with increasing potential and time. Their work identified two simultaneous processes of oxide formation and CO_2 evolution; these processes were independent and the surface oxide continued to grow even after CO_2 became the major product. Of further interest in this seminal work on PEMFCs is that the water activity and dilution of the acid does have an impact on the corrosion rate of carbon.

In a subsequent publication [140], they investigated the surface oxides on carbon blacks using a potentiodynamic method in phosphoric acid. They obtained the surface concentrations of the quinine-hydroquinone redox couple that was observed in the range 0.05–1.28 V versus NHE. A calculation of the surface concentration of species on carbon was calculated from CVs to obtain 1.1×10^{-10} moles/cm^2 for Vulcan XC-72 and only 1.3×10^{-11} moles/cm^2 for Vulcan-2700 (Vulcan graphitized at 2,700°C). They inferred from these results that the application of heat treatment reduced the amount of oxide species on the surface.

Antonucci et al. [141] also evaluated the corrosion of several carbons under potentiostatic conditions in phosphoric acid at 170°C. They confirmed that the corrosion of all the carbons that they evaluated obeyed a log-log law as reported in prior literature [139,140]. Moreover, they found that the corrosion currents varied linearly with surface area implying a linear relationship to the density of sites. They also found that the specific corrosion current ($\mu A/cm^2_{real}$) plotted vs. pH revealed a volcano profile with a maximum at pH 8 suggesting that carbon slurries of different pH exhibit different behavior.

The typical log-log relationship of carbon corrosion with time observed by most researchers does not have a clear explanation but has been implicated in the physical mechanism of carbon corrosion. Most carbons have primary particles that exhibit some degree of graphitic order on their surface; Kinoshita found [97,142] that the carbon in graphitic planes has an order of magnitude lower corrosion rate than the edge atoms. Wissler has reported carbon black primary particles that have concentric layers that show a gradual decrease in ordered layering as one proceeds towards the center of the particle [143]. These observations, along with TEMs (Fig. 3.22) revealing a 'shell-like structure' for corroded Vulcan [63], indicate that the core of the particle suffers corrosion at the outset and corrosion subsequently proceeds outwards leaving a graphitic shell.

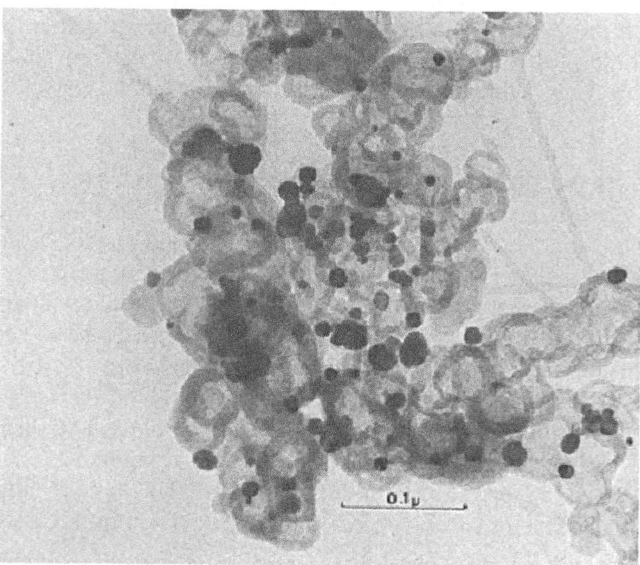

FIGURE 3.22 TEM image of a Pt/Vulcan XC-72C catalyst that was held and corroded for 1,000 h at 191°C in 95% H_3PO_4 at a potential of 835 mV. *(Reproduced with permission of Gruver [63])*

2.2.2. Electrochemical Degradation of Carbon Supports in PEMFCs

PEMFCs are operated at much lower operating temperatures of 60–90°C mainly due to the constraints imposed by the material properties of PFSA based membranes. At these temperatures one might not anticipate the carbon corrosion to be an important factor in the overall degradation of the electrodes. Unfortunately, PEMFCs for automotives are subject to potential cycling as part of normal vehicle drive cycles (0.6–1.0 V) as well as start-up/shut-down operations that, if left uncontrolled, can result in cathodic potential spikes reaching 1.5 V.

The measurement of carbon corrosion kinetic currents via *in-situ* measurements in PEMFCs presents us with a new set of challenges. The currents associated with carbon corrosion are ~0.01–0.1 mA/mg_C (~0.01–0.1 mA/cm^2) in PEMFCs operated at 80°C and ~1.0 V. The H_2 X-over currents (membrane permeability) for PFSA membranes are ~0.5–3.0 mA/cm^2 depending on the temperature, hydrogen partial pressure and thickness of the membrane [5,144]. Electronic shorting currents of similar magnitude may also be encountered [5]. Thus, even after many hours of testing, the carbon corrosion currents will be masked and overwhelmed by these larger magnitude parasitic currents . Direct measurement of the CO_2 evolution rate (from the fuel cell cathode exhaust) is thus warranted and has been reported by Roen et al. [145] and Yu et al. [146] in PEMFCs.

Yu et al. [146] used an experimental set-up that consisted of an on-line gas chromatograph (GC) and a methanizer in series with an ionization detector to obtain CO_2 evolution rates that were convertible to a current assuming a $4e^-$ step. The cell set-up included a subscale fuel cell, with H_2 flow maintained on the 'anode' (reference + counter) and N_2 on the working electrode (WE); the WE was held at various defined potentials. Figure 3.23 corroborates that the log-log relationship (log i_{CO2} vs. log t) observed in previous work on acid electrolytes also holds good for PEMFCs. Figure 3.23 is labeled to present their estimation (~1.05 V) of what is often referred to as the 'air-air potential' or the rest potential of the anode and cathode when both compartments are filled with air at 80°C.

Shao et al. [147] have recently investigated the corrosion of Vulcan carbon blacks using potential step cycling (upper potential = 1.4 V) to simulate automotive drive cycle conditions. Decreasing the lower potential limit (to 0.1 V) as well as increasing the frequency of potential cycling accelerated the corrosion rate; the higher corrosion rates were attributed to the consumption/regeneration of oxidized surface species on the carbon surface.

2.2.2.1. Electrochemical Degradation of Carbon Supports Under Automotive Operation Modes

In this section, we focus on the stresses encountered by the carbon support due to the peculiarities of automotive operating conditions; including frequent

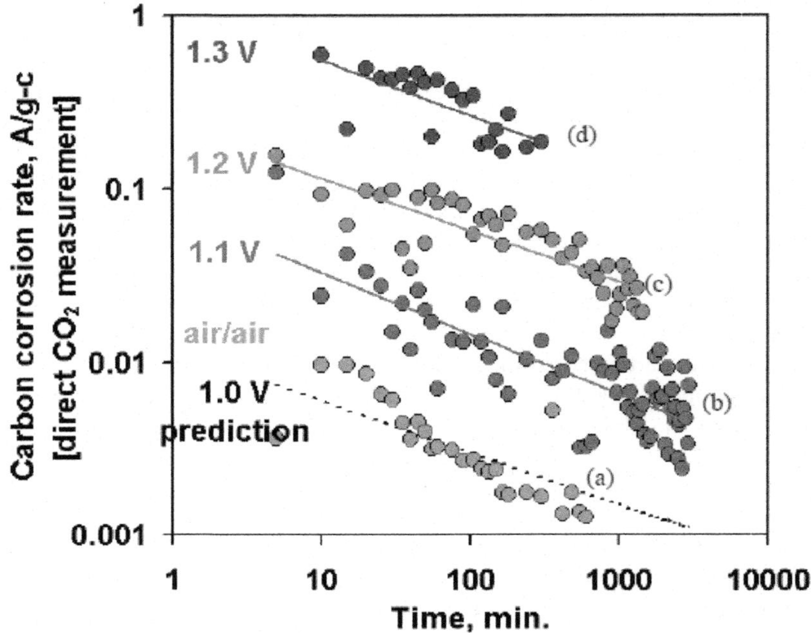

FIGURE 3.23 Log-log relationship of carbon corrosion rates with time (markers) estimated from direct CO_2 evolution measurement at 80°C in subscale PEMFCs. The lines represent prediction of kinetics for potentials in the range 1.0–1.3 V. Magnitude of 'air-air' mixed potential is marked slightly above 1.0 V. *(Reprinted with permission from Yu et al. [146])*

uncontrolled start-up/shut-down, mitigated start-up/shut-down, fuel starvation on the anode during sub-zero start-up and flooding and under steady state operation.

Carbon Corrosion Under Automotive Start-up/Shut-down Modes Initially a corrosion-resistant graphitized carbon was considered not only unnecessary under nominal PEMFC operating temperatures and potentials, but even detrimental to the cell performance. The severe degradation associated with the phenomenon of start-up/shut-down of fuel cell stacks has led to a re-evaluation of that perception. Research into more robust carbons and alternative supports for Pt has been reinvigorated by the findings on the start-up/shut-down phenomenon that involves potential excursions as high as ~1.5 V. It is estimated that a fuel cell stack in a vehicle will encounter about ~30,000 such start-up/shut-down cycles over its lifetime. The allowable degradation under these conditions has been targeted at 1 μV/cycle (~30 mV @ peak current density) [14]. This is in stark contrast to stationary fuel cell applications where the number of start-up/shut-downs is much lower and the luxury of applying carefully controlled conditions is available.

When a vehicle is shut down for an undefined rest period (or standing-time) the anode compartment eventually gets filled with air (O_2 and N_2) that has 'crossed over' through the membrane from the cathode as well as from leaks to the ambient. When the stack (now filled with air on both the anode and cathode) is re-started and the hydrogen flow turned on, a 'H_2-Air front' passes through the anode chamber flow fields from inlet to outlet. During such uncontrolled start-up of a PEMFC stack, unprecedented degradation and thinning of the catalyst layer occurs. The cathode potential rises to values as high as 1.5 V during the passage of the H_2-Air front; at these potentials, the carbon support on the cathode catalyst layer corrodes and is released as CO_2; as the carbon corrodes, the electrode structure disintegrates, loses thickness and Pt nanoparticles simply agglomerate into much larger particles of lower surface area. Through direct experimental observations and later through modeling to confirm their theory, researchers at UTCFC [148–151] elucidated the nature of the phenomena; they also provided preliminary mitigating solutions including high flow (N_2 or air) purges and the use of a voltage limiting device (or shorting resistor) to limit the potential spike. They elaborated on the intricacies of the degradation mechanism and validated their hypothesis using two different experimental set-ups [152]. In addition, Reiser et al. also reported that a similar phenomenon (of cathode degradation) occurs on a localized scale when the anode side is locally starved of hydrogen due to water droplets or blocked flow field channels [152].

One of the most basic solutions for mitigating the start-up/shut-down degradation is to limit the duration of the H_2-Air front on the anode by a fast purge [150]. Yu et al. [153] systematically investigated the losses incurred while purging the (H_2 filled) anode either with air or N_2 during shut-down in subscale cells. Over 200 purge cycles (Fig. 3.24), the average degradation rate is ~1000 μV/cycle for an air purge, and <100 μV/cycle for an N_2 purge, emphasizing that the losses are dictated by the oxygen concentration in the front. Additionally, they determined (Fig. 3.25) that the degradation rate was linearly related to front residence time in the range 0–1.3 s. A loss of 100 μV/cycle was calculated for a front passage time of 0.05 s (~peak fuel cell power flow rate); this is unacceptably large to be considered as a mitigating solution for automotive PEMFCs. Kocha [154] has also proposed mitigation techniques that involve drying out the membrane and ionomer in the catalyst layer (to λ ~3) by flowing dry air over the cathode before shut-down, so that during the subsequent start-up, the losses are minimized. This partial mitigation is due to the fact that both carbon and Pt corrode to lower extents under low RH conditions.

Extending the work reported in the literature, Takeuchi and Fuller [155–157] systematically (experiments and model) carried out studies to elucidate the effect of cell potential, ORR activity, humidity and water vapor pressure on carbon corrosion during start-up. For a subscale cell tested under the conditions of $H_2|N_2$, 80°C, 50% RH, 1.2–1.5 V, they found the exchange

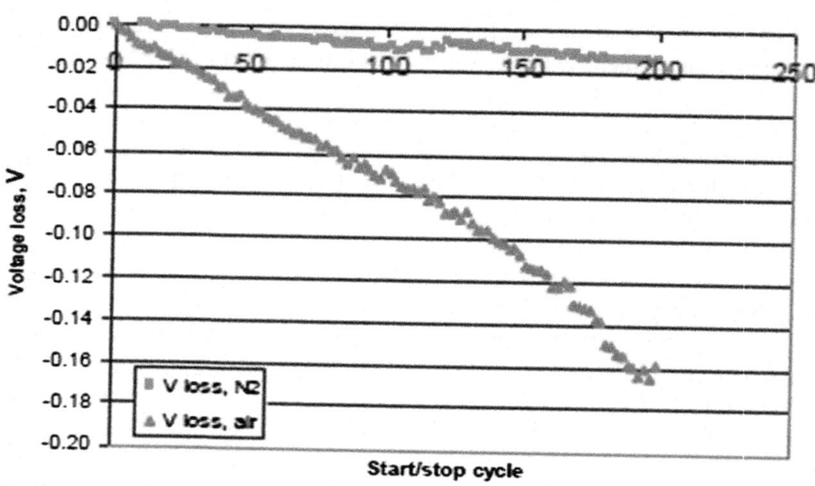

FIGURE 3.24 Degradation rate as a function of the composition of the purge gas that generates a front that passes through the anode flow fields during start-up and shut-down. *(Reprinted with permission from Yu et al. [153])*

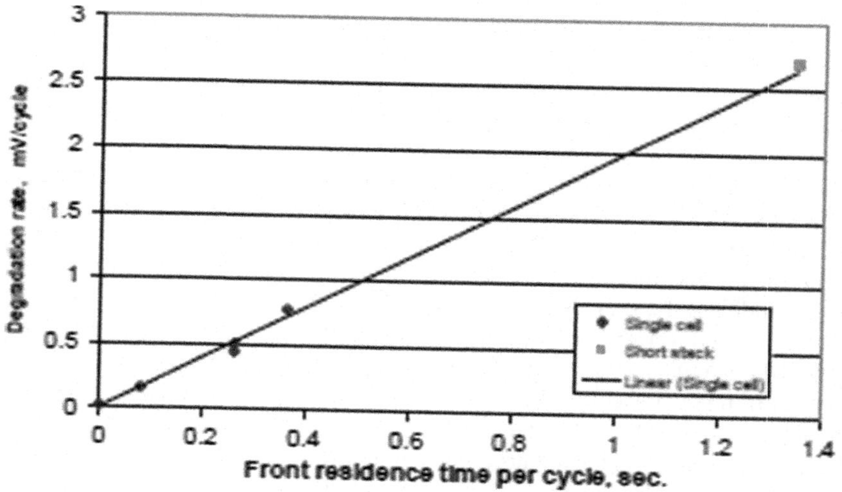

FIGURE 3.25 Degradation rate per cycle as a function of the H_2-Air front residence time at 80°C. *(Reprinted with permission from Yu et al. [153])*

current density for carbon corrosion was ~10^{-9} A/cm^2. Under H_2|Air operating conditions, they found that the carbon corrosion steeply decreased by an order of magnitude to fairly low values below 0.80 V; in addition, since the solution potential is lower under H_2|Air operation, for a given overall cell potential, the losses are higher than under H_2|N_2. Although the effect of humidity on carbon oxidation is complex since RH affects membrane conductivity, reactant crossover as well as ORR kinetics, they verified that carbon oxidation increased with humidity as described by the fundamental carbon oxidation equation.

Residual Carbon Corrosion (Mitigated Start-up/Shut-down) Even when the start-up/shut-down procedure includes mitigation by shorting the stack and suppressing the extent of the voltage spike, the stack may still experience potentials close to the 'air-air potential' for short periods of time. Researchers at GM [14] reported that the 'cumulative time' for which the cathode catalyst in an automotive stack might experience ~1.2 V at 80°C is about 100 hours (assumes thermodynamic air potential, 10 s × 30,000 start-up/shut-down cycles); in addition, the time under idle conditions or 0.90 V over the lifetime of the automotive stack may be in the thousands of hours. Based on these concerns and in order to verify the necessity of developing new corrosion-resistant supports for Pt, they undertook a study of carbon corrosion in PEMFCs. They derived an empirical kinetic equation relating the carbon corrosion current to temperature, potential and time and obtained fit parameters for the Tafel slope (150 mV/dec), activation energy (67 kJ/mol) and an exponent (0.30) for decay with time.

Based on this equation, one can estimate the carbon weight loss for Pt/C to vary from 5–10% corresponding to 3,000–10,000 hours at 0.90 V, 80°C, 100% RH. Figure 3.26 shows the increase in carbon loss wt% over 10,000 h at 0.90 V and 1.2 V for a 50 wt% baseline Pt/C as well as a 30% Pt-alloy/corrosion-resistant carbon catalyst.

They additionally carried out practical tests where they subjected two cathode catalysts in subscale fuel cells to 1.2 V at 80°C and 100% RH for different periods of time; they found that the mass transport losses as measured from I-V curves under H_2|Air were significantly increased with unacceptable losses after just 50 hours for Pt/C; on the other hand, the Pt-alloy catalyst on graphitized support showed negligible losses. In such practical tests, it is difficult to separate the kinetics from the contribution to mass transport losses by corrosion of the catalyst support and microporous carbon layer (MPL) at the interface between the DM and CL. We should also note that most of these carbon corrosion experiments were conducted at 100% RH which is the highest RH that the cathode would see during operation; at lower RH, the losses would be lower since both Pt dissolution as well as carbon corrosion are suppressed in the absence of water. The potential of 1.2 V and the temperature of 80°C should also be considered as an accelerated condition, since the 'air-air' potential is actually expected to be ~1.05 V and the stack temperature also is expected to be lower under idling conditions. Finally, the rest period or standing time after vehicle shut-down is variable in practice, and start-ups that occur within a short interval after shut-down will not result in degradation as long as hydrogen is present in the anode.

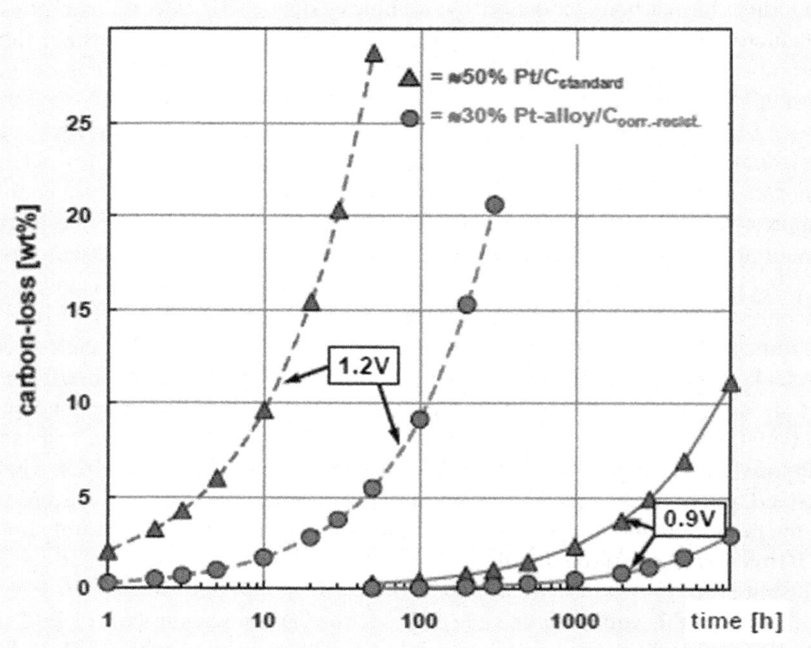

FIGURE 3.26 Carbon weight loss vs. time for a baseline 50 wt% Pt/C catalyst as well as a 30 wt% Pt-alloy/corrosion-resistant support. *(Reprinted with permission from Mathias et al. [14])*

Carbon Corrosion Under Anode Fuel Starvation The hydrogen flow rate in the anode compartments of automotive PEMFCs is often maintained at low stoichiometry/high utilization in order to attain high fuel efficiency and minimal exhaust of hydrogen to the atmosphere. Blockages can occur for several reasons in flow channels or portions of channels: due to localized flooding with water, formation of water droplets, and ice formation during start-up of a stack that been resting under sub-zero conditions. When hydrogen is unavailable in a localized area of a cell anode while the stack is still connected to a load, its potential will rise to levels above 1.2 V. In the absence of hydrogen, at these potentials, either the oxidation of water and/or carbon support takes place. The overpotentials for oxidation of carbon are lower and hence carbon corrosion is favored. Knights et al. [158] have reported lifetimes of 1,000–13,000 h for PEMFCs (49–1,280 cm^2, 4–20 cells) and discussed aging mechanisms based on operating conditions such as reactant flows, humidity, and temperature. Several anode structures in four-cell stacks were intentionally starved of hydrogen and allowed to go into 'voltage reversal' averaging −2 V. The time period for which the cell continued to operate before crossing −2 V was used as a measure of its durability. They propose the use of more robust carbons, anode alloy catalysts (Ru addition) that are selective to the oxidation of water, a higher wt% Pt/C, and enhanced water retention as possible solutions to alleviate the problem. Engineering solutions to mitigate this phenomenon include the recycling of anode reactants so that the total flow rates are high enough to prevent water from blocking the local areas or channels especially in the vicinity of the exit.

Carbon Corrosion Under Steady State Operation Classic works on the morphology and surface properties of carbon can be found in the works of Rivin and co-workers of Cabot Corporation [159, 160]. It is important to study and analyze the carbon blacks in terms of such properties as particle size, primary and secondary aggregates, and structure because they eventually determine the porous structure of the electrodes of fuel cells that may degrade over time. Operation of PEMFCs over thousands of hours can result in the formation of surface oxides, carbon corrosion and a concomitant change in wettability and pore size distribution that will affect the general catalyst layer structure, and result in an increase in mass-transport resistance and slow degradation of performance. A few researchers have reported that over long periods of time, some carbon corrosion and increase in wettability of the carbon surface can occur during normal operation of fuel cells under automotive conditions. Kangasniemi et al. [161] carried out a study of the formation of surface oxides on carbon in sulfuric acid using TG-MS, XPS as well as CVs. Maintaining potentials of 0.8 and 1.0 V appeared to increase the amount of lactones and carboxylic acids and could affect the hydrophobicity of the surface and possibly the mass-transport. Since the losses that occur during normal operation are quite low, at this time they have not been carefully separated and quantified from other losses such as catalyst activity and resistance.

2.2.3. Effect of Pt on Electrochemical Degradation of Carbon Supports

Pt has been reported in the literature to catalyze carbon support corrosion and accelerate the overall electrode degradation. In preliminary work Kinoshita and Bett [139] had reported that Pt did not catalyze the carbon corrosion reaction in hot conc. phosphoric acid as did Stonehart [162]. They argued that Pt dissolved at high potentials (as reported by Bindra et al. [18]) and therefore there was not enough Pt remaining to catalyze the electro-oxidation of carbon at high potentials. Passalacqua et al. [163] showed that Pt/Ketjen black clearly catalyzed the oxidation of carbon at potentials of 0.6, 0.8, and 1.0 V with the corrosion currents (mA/mg_C) being lower at 1.0 V due to the dissolution of Pt. They proposed a mechanism where the surface oxides of carbon undergo further oxidation to CO_2 in the presence of Pt or PtO as an intermediary, thus causing catalyst degradation. Differential electrochemical mass spectroscopy (DEMS) has also been applied to evaluate carbon corrosion (in sulfuric acid) and confirmed that the anodic peaks related to corrosion were indeed amplified in the presence of Pt catalyst [164,165].

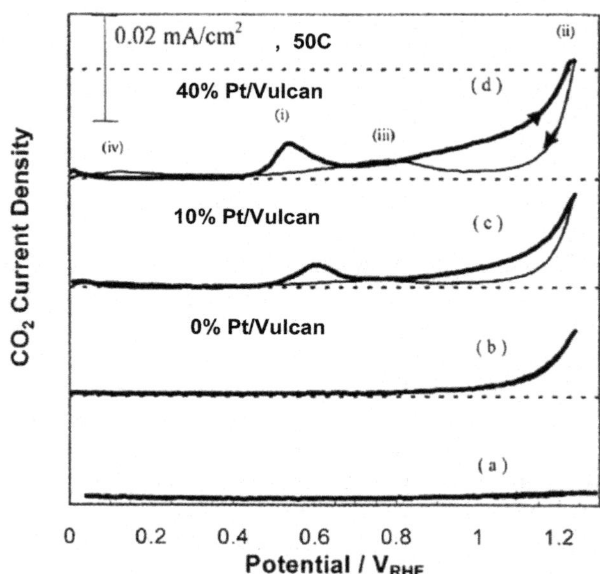

FIGURE 3.27 Plot illustrates the influence of Pt concentration or wt% in the range 0–40% on carbon corrosion currents at 50°C. *(Reprinted with permission from Roen et al. [145])*

Roen et al. [145] investigated the role of Pt on the electrochemical oxidation of carbon under conditions relevant to PEMFCs. They systematically studied the CO_2 mass-spec profiles (converted into faradaic current values vs. potential, in the range 0–1.3 V, 2 mV/s, 50°C) for MEAs prepared using un-catalyzed Vulcan, 10 wt% Pt/Vulcan, and 40 wt% Pt/Vulcan. Their results showed (Fig. 3.27) that corrosion currents increased with Pt loading in the range 0–40 wt% and the onset of CO_2 currents at potentials as low as 0.70 V was observable at 50°C. Their results revealed the existence of a second low potential peak in the potential range ~0.4–0.6 V which they attributed to the oxidation of CO. It appears that Pt nanoparticles indeed accelerate the corrosion of carbon supports, but a systematic study to quantify the acceleration of degradation measured in fuel cells has not yet been reported.

2.2.4. *Electrochemical vs. Thermal Degradation*

Although the durability of carbon must rightly be evaluated under fuel cell or electrochemically relevant conditions, it may also be informative to take a cursory look at its *ex-situ* degradation at high temperatures in the presence of various gases. In particular, it would be useful to know whether the electrochemical degradation of carbon supports can be simulated by a thermal degradation process and applied as a screening technique to evaluate new support materials.

Kinoshita studied the differential thermal analysis of PtO_2 on carbon in helium and air. One of the interesting outcomes of the study was that in the case of Pt/C, Pt appears to lower the oxidation temperature of carbon due to the formation of atomic oxygen generated on the surface [166]. More recently, Cai et al. [167] have reported an attempt to experimentally simulate the electrochemical degradation of carbon and Pt/C in PEMFCs using an 'accelerated thermal sintering' protocol. Their studies on the effect of oxygen and water concentrations on two Pt catalysts supported on carbon revealed that there is an 'oxygen pathway' (mass loss) as well as a 'water pathway' (mass gain) for carbon oxidation. They conclude that Pt surface area loss over 10 h under thermal degradation was comparable to the electrochemical degradation after 500 cycles in the potential range 0–1.2 V. They also observed that 5 h of thermal degradation produced a similar loss in carbon mass as a potential hold for 86 h at 1.2 V and 95°C. A simplified thermal test method might possibly provide a different route to screening alternative supports for corrosion resistance, although so far there is no consensus on whether fundamentally electrochemical degradation can truly be related to thermal degradation.

2.2.5. Electrochemical Degradation of Alternative Supports

Although the deleterious effect of start-up/shut-down on the carbon support can be ameliorated by the use of appropriate control system procedures, one would ideally like to minimize the attendant system complexity and associated costs. In addition, potential cycling in the range 0.6–1.0 V is unavoidable under automotive operation and new supports that anchor the Pt and render it immobile are highly desirable. Therefore, the search for a catalyst support more durable than the typically used carbon blacks for Pt catalysts has recently intensified to meet automotive durability targets.

Non-Conventional/Modified Carbon Supports Before discussing non-carbon supports, mention must be made of the work done on non-conventional carbon materials such as single walled and multi-walled carbon nanotubes (SWCNTs, MWCNTs) [168,169], carbon nanofibers, nanocoils, nanohorns, ordered mesoporous carbons (OMCs, 2–50 nm pores), carbon aerogels, carbon xerogels, carbon cryogels, etc. [170,171]. CNTs have been reported to have a dearth of sites for anchoring metal catalyst ions/nanoparticles resulting in poor catalyst dispersion for >30 wt% Pt loading. Some of the reports claim small improvements in performance and a few that the durability is higher at 0.9 V and 1.2 V compared to carbon blacks [147,172]. Girishkumar et al. [173] carried out potential cycling in $HClO_4$ and found that Pt/SWCNT had lower ECAs compared to Pt/C but lost only 16% of the initial area after 36 h while Pt/C lost 50% of its initial area. In the light of their high cost, evidence of a substantial improvement in oxidation resistance will be necessary for use in commercial fuel cells. Other highly corrosion-resistant materials are boron doped diamond and porous honeycomb diamond [174], but special methods to anchor the Pt may be needed to prevent Pt mobility and maintain Pt utilization.

Modification of supports to improve the durability of Pt catalysts has been suggested in many reports. Tseung and Chen [175] and Tseung and Dhara [176] suggested the use of SnO_2 (which has a higher surface energy than carbon) so that Pt ions might re-deposit on the high energy support preserving the surface area. Blurton et al. [62] investigated the oxidation of supports in oxygen at 400°C using iron or copper salts as catalyzers. They found the sintering rates of Pt supported on these modified materials were inhibited possibly due to better anchoring of Pt. Thus, the support on which Pt based catalysts are deposited can have an effect on the sintering rates in electrodes.

Non-Carbon Supports A successful replacement for carbon has to have almost all the properties of carbon blacks such as high electronic conductivity, high surface area, the ability to form a porous catalyst layer and chemical stability. Interestingly, it may be possible to lower some of these requirements and demand certain other ones that carbon does not possess. Since the ionic conductivity of the ionomer in the catalyst layer determines the resistance of the catalyst layer, the new support is not required to have conductivities as high as carbon blacks. If the new support can be tailored to have the optimal hydrophilicity/hydrophobicity and if it can somehow anchor the Pt firmly, then it does not need to have an ultra-high surface area. Furthermore, if the support possesses a degree of ionic conductivity in addition to electronic conductivity, it would be easier to form the catalyst layer with minimal addition of ionomer. Finally, a desirable property in our wish list would be selectivity for desired electrocatalytic reactions such as ORR on the cathode by modification of the electronic behavior of the catalyst through strong metal support interactions (SMSI) or through spillover effects. For example, although gold is known not to be suitable for most heterogeneous catalytic reactions, when small Au clusters are supported on transition metal oxides, they can catalyze CO oxidation at 200 K [177].

A variety of supports such as oxides, nitrides and carbides have been considered for use as a support for Pt in PEMFCs. Some of the more robust supports are non-conductive and need to be doped to provide reasonable conductivity. Metal oxides have been considered for use as an intrinsic catalyst and as a support for metal catalysts. Most oxides are susceptible to dissolution except in narrow ranges of pH. Oxides are often semiconductors and need to be suitably doped or modified to contain sub-stoichiometric quantities of oxygen; despite these modifications they usually have conductivities that are less than 100 S/cm. Oxides have high energy surfaces that give them a hydrophilic tendency; when oxides are immersed in water, the solvent molecules bond to the metal cations and transfer protons to the oxygen sites nearby, thus leading to a high surface coverage with OH groups. There are several reported studies on TiO_2 as a catalyst and as a possible support. TiO_2 is a poor catalyst for ORR with overpotentials of 1 V where TiO_2 can be reduced to TiO_{2-x}. TiO_2 can be substitution doped with a pentavalent cation such as niobium, or electrochemically, by biasing the electrode negative of the flat-band potential. Ebonex is a ceramic of mainly Ti_4O_7 and Ti_5O_9 that are part of a series of compounds known as the Magneli phases and show good conductivity and corrosion resistance. Ioroi et al. [178] evaluated Pt/Ti_4O_7 and compared its performance and durability to Pt/Vulcan; Ti_4O_7 supports showed higher resistance to corrosion for potentials >0.90 V but the performance was lower due to larger Pt particles. It appears that the performance of Pt/Ti_4O_7 is intrinsically lower due to poor dispersion of Pt and/or catalyst-support interactions, but further studies are warranted. WO_3 and SnO_2 also do not have much activity for ORR.

WOx prepared by hot wire deposition (HWD) as well as wet chemistry is being studied in greater depth at this time [179]. The blue colored sub-stoichiometric forms are more conductive than WO_3. Preliminary results indicate that these materials have a higher corrosion resistance than convention carbon blacks. Difficulties in the measurement of the ECA from HUPD have been reported due to the interaction of protons with WOx and its contribution to the total charge. Addition of 10% carbon to the inks have also been shown to improve the measured activity.

Transition metal (W, Zr, Ta, Mo) carbides have been studied as supports and reported to be generally unstable in acid solutions. The Ota group [180–182] have conducted extensive studies on valve metals compounds such as oxides, oxynitrides and carbonitrides that were prepared by RF magnetron sputtering. These materials exhibited measurable activity for the ORR; solubilities for ZrO_2, ZrON, TaON, TiO_2 were determined to be lower than that for Pt in 0.1 M H_2SO_4 at room temperature.

Halalay and co-workers [183] discuss in detail the preparation of Pt catalysts on TiC and TiN nanoparticle (20 m^2/g) supports as well as the measurement of corrosion rates in acid and cyclic durability under potential cycling. They recommend the addition of Vulcan carbon (weight ratio of Pt/TiN to carbon of 70:30) for improved performance. Chemical and electrochemical corrosion tests in a simulated fuel cell environment (0.5 M H_2SO_4 aqueous solution at 95° C.) were performed on several of these materials. In their measurements they determined that TiN and TiC had corrosion rates of 3.9 and 5.6 μmoles/m^2/week at room temperature. They also report the corrosion rates of commercial Pt/C at constant potential holds of 1.2V and 95°C to be −1.8 μmoles/m^2/hr while that for Pt/TiC was +20 μmoles/m^2/hr. They ascribed the increase in weight of the TiC supports to the formation of TiO_2 that might actually stabilize the Pt nanoparticles. In the potential cycling tests between 0 and 1.2 V, at 10 mV/s, they found that Pt/TiC and Pt/TiN + carbon did not lose their initial ECA after 500–1,000 cycles. In related work, Merzougui et al. report deposition of Pt on metal oxides along with a conductive matrix of particles using compounds such as borides, carbides, nitrides, silicides, carbonitrides, oxyborides, oxycarbides, or oxynitrides of metals such as cobalt, chromium, nickel, molybdenum, neodymium, niobium, tantalum, titanium, tungsten, vanadium, and zirconium that provide good ORR as well as corrosion resistance [184].

In summary, although the phenomena of carbon corrosion start-up/shut-down can be partially mitigated through controlled operating conditions, the development of new highly corrosion resistant, high surface area materials is still warranted in order to simplify system operation, cost and durability of the PEMFC stack. New materials that provide additional corrosion resistance are being studied and modified while still maintaining many of the properties of carbon supports.

2.3. Contamination

Electrode degradation of PEMFCs can occur as a result of various impurities found in the fuel feed, air stream, as well as corrosion by-products originating from cell components such as the bipolar plate, catalyst or membrane. All the components of the MEA including the membrane, ionomer, catalyst and carbon are susceptible to contamination; however, the focus of this section is catalyst contamination.

Hydrogen for PEMFCs can come from various sources including re-formed fossil fuels. Therefore, depending on the re-forming technique and degree of post-treatment, small quantities of impurities such as CO, CO_2, NH_3, H_2S, etc. are expected to be present. Trade-offs in the level of impurities is unavoidable since ultra-high purification would lead to higher hydrogen costs. The US FreedomCAR technology team has arrived at preliminary fuel mixture specifications that include: >99.9% H_2, 10 ppb H_2S, 0.1 ppm CO, 5 ppm CO_2, and 1 ppm NH_3. Hydrocarbons such as methane, benzene and toluene are other common impurity by-products from re-forming processes. Figure 3.28 illustrates the losses (~100 mV) incurred using FreedomCAR fuel mixtures over 1,000 h while operating at 0.8 A/cm^2 as reported by Garzon et al. [185].

A large amount of work has been conducted on impurities in hydrogen-rich fuels exiting a steam re-former or partial oxidizer where the levels of CO and CO_2 are quite high; only recently has there been work focused on extremely low levels of contaminants. Although CO poisoning is completely reversible, 10 ppm of CO is sufficient to poison the anode catalyst and reduce the performance precipitously. For such high levels of CO a practical solution is the use of PtRu/C as the catalyst, or carrying out air injection into the anode to oxidize the CO and mitigate the losses. Since anode Pt loadings have been reduced to 0.05 mg/cm^2 over the last few years, research on the effect of trace amounts of CO on performance has become relevant again. Studies in the range of <1 ppm CO are difficult to conduct since the degradation rates are low enough to make it difficult to separate the normal degradation of fuel cell cathodes from the degradation due to the anodic poisoning.

Atmospheric air contains ~78% nitrogen and ~20% oxygen with trace gases and particulates (depending on the local air quality) making up the remainder.

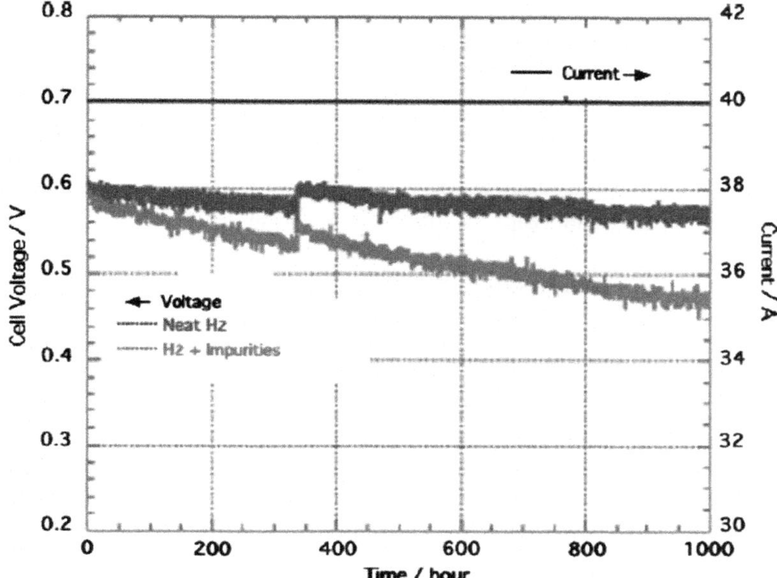

FIGURE 3.28 Cell voltage losses in 50 cm^2 subscale cells operated at 0.80 A/cm^2 for 100 h on pure H_2 as well as a mixture of hydrogen (99.9%), H_2S (10 ppb), CO (0.1 ppm), CO_2 (5 ppm) and NH_3 (1 ppm) as per FreedomCAR fuel specifications. *(Reprinted with permission from Borup et al. [171])*

Pollutants found in the atmosphere include nitrogen oxides (NO and NO_x), hydrocarbons, carbon monoxide (CO), ozone, sulfur dioxide (SO_2), fine primary and secondary particulate matter, and chloride salts from the ocean and de-icers. It should be noted that trace amounts of CO and other contaminants can be easily oxidized on the cathode catalyst surface in the presence of copious amounts of oxygen in air and high potentials.

The degradation of performance due to 2.5 ppm and 5 ppm SO_2 in the air stream was reported to be about 50 and 80% respectively by Mohtadi et al. [186]. The degradation is due to chemisorption of sulfur species on the Pt catalyst, and oxidation by application of high potentials (CV) reversed the process [185,186], but operation under normal potentials did not. The poisoning mechanism by NO_2 was reported [185,186] to be dependent on the time of exposure rather than bulk concentration and could be reversed by operation under clean air for 24 h. Although the exact mechanism is not clear due to the absence of surface species detected in CV profiles, it seems unlikely to be related to adsorption on Pt. The effect of battlefield contaminants such as sarin and mustard gas has been investigated and they were found to adsorb on the Pt surface with a gradual loss in performance that was found to be irreversible [187].

One must also keep in mind the impurities that may be present on the support which could poison and degrade Pt catalysts. Vulcan is a commonly used carbon support in PEMFCs, and is known to contain levels of organo-sulfur that can contaminate Pt. Swider and Rolison have examined the chemical state of sulfur in Pt/Vulcan catalysts (using XPS) and monitored the sulfates generated from it [188]. They found that Pt in intimate contact with carbon catalytically oxidizes sulfur in the presence of heat and water which are normally fortuitously present during hot-pressing of MEAs. In a follow-up paper [189], they demonstrate that the effect of sulfur poisoning can be reduced by de-sulfurizing the carbon surface, allowing one to employ MEA preparation methods devoid of heat and water if necessary.

Trace chloride impurities from the atmosphere (de-icers, sea mists) as well as from certain catalyst preparation precursors such as H_2PtCl_6 are known to drastically alter the CVs and lower the ORR activity [185] but these can be flushed out by fuel cell operation at high current densities with the generation of water. The effect of membrane degradation products on PEMFCs durability has been reported by Sugawara et al. [87] and will be discussed in detail in **Section 4**. Modeling of contamination is not an easy task but recently St. Pierre has generated a transient contamination kinetic model for anode and cathode contaminants and recommended approaches to validate it using fuel cell data [190].

2.4. Sub-zero Operation

All automotive systems are affected by sub-zero environmental conditions, but PEMFCs are especially vulnerable due to the ubiquitous presence of water in the stack. The principal targets that automotive PEMFC stacks are required to meet in 2015 are the ability to start unassisted from −40°C and a cold start-up time to 50% of rated power in 30 s (from −20°C) with a start-up energy consumption of < 5 MJ [8,191,192].

PEMFCs may suffer reversible or irreversible degradation of stack components, including membrane, catalyst support and electrode structure, during its lifetime, when subject to freezing conditions or freeze/thaw thermal cycles. The problem mainly arises when a PEMFC stack has been idle for a long period of time under sub-zero conditions and is required to start up. The use of insulation, heaters and antifreeze coolants are helpful for short periods of time, but the residual water within the fuel cell is liable to freeze eventually forming ice. The catalyst layer and the diffusion media are especially susceptible to repeated freeze-thaw cycles due to their porous nature and the inherent presence of water (loosely bound and free) that can expand and contract leading to a disruption of the porous structure that is so important for facilitating mass-transport. Significant disconnection of the catalyst particles from the ionomer and carbon support could result in degradation in performance due to the loss in protonic and electronic interfacial contact.

Elimination or minimization of the water in the cells by purging with gases or vacuum drying prior to shutting down the stack is one general approach that has been found to be effective. In short, 'keeping the stack warm' and 'removal of water' are essentially the two main strategies to counter the cold start issue. The issue of PEMFCs encountering sub-zero temperatures can thus, in a sense, be considered as a special case of start-up/shut-down, since no additional degradation actually occurs while the fuel cell is operating. Under sub-zero conditions, an operating fuel cell will generate more than sufficient heat to prevent water from freezing and is unlikely to exhibit an increased degradation rate.

The performance of a PEMFC is expected to be inherently lower at sub-zero temperatures, as the proton conductivity (and hence ohmic losses) of the membrane/ionomer as well as the catalytic activity of Pt for the ORR will be affected by sub-zero temperatures affecting the power producing capability of the stack, delaying cold starts. The catalyst itself is not expected to undergo any degradation. Thompson et al. [193] evaluated the ORR kinetics (Tafel slopes, exchange current densities, activation energy) at subfreezing temperatures (-40 to $+55°C$) using an experimental procedure where mass-transport due to ice formation was suppressed. They estimated the activation energy from Arrhenius plots to be in the range 25–35 kJ/mol in the temperature range -40 to 55°C. Their calculated exchange current density and specific activity were found to be off by a factor of 2–4 compared to their previous results and the literature. They observed that a lower water content and lower temperature resulted in both a lower catalytic activity and a positive shift (inhibition) of the onset of oxide formation in CV measurements. Based on these results, they argue that reduced proton activity (incomplete dissociation of protons) at low RH is a more suitable explanation for the loss in catalytic activity rather than a change in the rate determining step (rds) or changes in Pt oxide species coverage. We note that one of the interesting aspects of a cold start is that the severe cathode support degradation during start-up/shut-down is suppressed by the low ORR and COR reaction kinetics.

Additional reports on the effect of freeze/thaw cycles on MEAs of PEMFCs can be found in a number of papers from the Mench group (Chacko et al., Kim et al.) [194,195]. In short, the electrode structure, its pore distribution as well as hydrophobicity and hydrophilicity, extent of water removal at shutdown, flow field design, etc. can have a significant effect on the catalyst layer degradation under freeze/thaw thermal cycling conditions. FC vehicles from major auto manufacturers have been reported to be able to start from $-20°C$ without noticeable degradation, although ongoing research to understand the detailed mechanism is still underway.

3. EXPERIMENTAL SET-UP AND DIAGNOSTIC TECHNIQUES

3.1. Experimental Set-up

In this section, we briefly describe the cell configurations and experimental apparatus used in this work to obtain the results discussed in Section 4. The research was performed in the Advanced Electrocatalysis Group and MEA groups at Nissan. Experiments were primarily carried out in electrochemical half-cells as well as subscale fuel cells, with the exception of some short-stack and full size stack results reported in subsection 4.3 on durability of automotive PEMFC stacks in vehicles.

3.1.1. Electrochemical Half Cells

Rotating Disk Electrode (RDE) Set-up Three-electrode systems, consisting of a working electrode (WE), counter electrode (CE) and reference electrode (RE), were routinely employed. A photograph of the cell glassware connected to the rotator, pressure gauges, rotameters and located in a chemical hood is shown on the left side of Fig. 3.29. The right side of Fig. 3.29 is a close-up of the glass cell. The CE was a Pt gauze or Pt wire coated with Pt black to provide a surface area greater than 20 times the area of the WE. The RE used for all cases was a reversible hydrogen electrode (RHE) introduced into the main cell compartment using the identical electrolyte and avoiding complications of a salt bridge and liquid junction potentials. The RE consisted of a glass tube with a capillary tip tilted upwards and positioned close to the WE. Pt wire was inserted into this glass tube and sealed from atmospheric air intrusion. Another glass tube of smaller diameter was inserted into the main RHE chamber to bubble hydrogen through the electrolyte and over the Pt wire; the hydrogen was exhausted out of the RHE by allowing it to bubble through a small glass chamber filled with electrolyte, thus maintaining a seal from atmosphere.

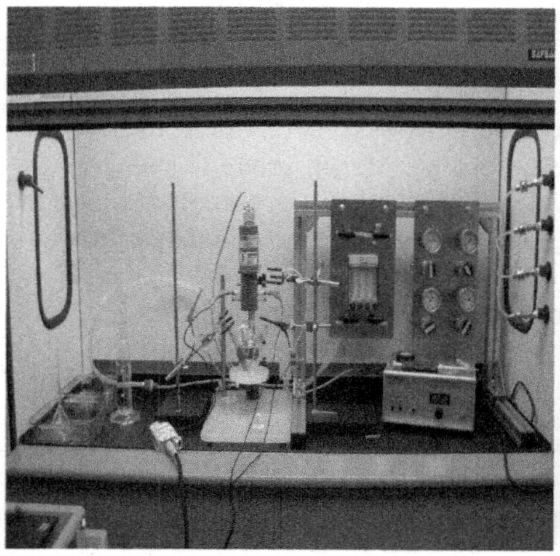
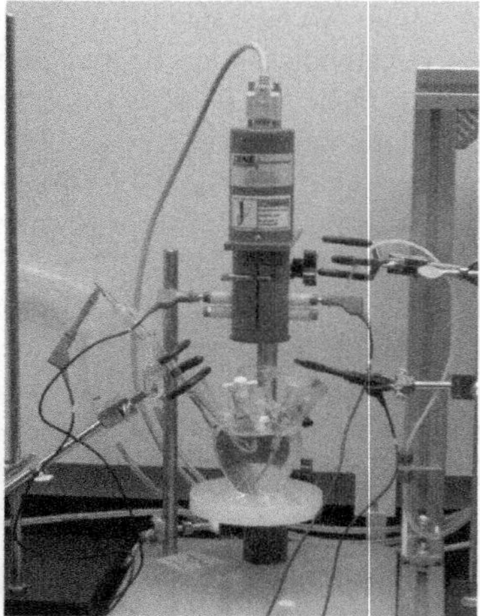

FIGURE 3.29 Photograph on the left depicts a fume hood with the complete electrochemical set-up for RDE experiments including gas flow meters, water traps and electrometer for connection to a potentiostat. Picture on the right is an expanded view of the rotator and cell glassware and connections.

The RDE-WE was a glassy carbon (Φ5 mm) substrate on which various catalyst ink dispersions were deposited. Cyclic voltammetry and hydrodynamic voltammetry were carried out using a potentiostat (HZ-3000, Hokuto Denko or Autolab) and rotating disk electrode (RDE) (HZ-301, Hokuto Denko or Pine Instruments). Cable lengths were restricted in length and covered with aluminum foil to reduce electrical noise. When using the Autolab potentiostat, linear sweep voltammetry technique or LSV (as opposed to step voltammetry) was selected to measure CVs and I-Vs.

Cleaning of Cell Glassware RDE experiments are extremely susceptible to contamination, especially in the case of bulk/thin film Pt-based catalysts, where the surface area of the WE is close to the geometric electrode area. This is especially true when measuring catalyst activity using electrolytes with low anion adsorption such as perchloric acid. Obtaining high values of electrocatalytic activity with bulk/disk electrodes allows us to benchmark the cleanliness of the system and to obtain reproducible values of activity for the more practical high surface area nanoparticle Pt-based catalysts supported on carbon. Different laboratories utilize varying techniques to clean glassware in order to eliminate trace amounts of cationic, anionic and organic impurities. In our studies the glassware is soaked in a 1:1 mixture of concentrated sulfuric acid/nitric acid mixture for 2–4 hours, followed by rinsing in DI water. Nitric acid, which acts as a strong oxidizing agent, may be substituted by chromic acid or preferably Nochromix (GODAX laboratories) which is a metal-free alternative prepared by using ammonium persulfate ($(NH_4)_2S_2O_8$). The glassware is then boiled in DI water for ~2 h followed by rinsing in DI water at least 2 times. Prior to any experiment, the cell is rinsed in the electrolyte of the experiment such as 0.1 M $HClO_4$. Obtaining a specific activity of about 2,000–2,500 μA/cm^2 for poly-Pt indicates a cell cleanliness adequate for evaluating high surface area nanoparticle electrocatalysts.

Electrolytes Sulfuric, perchloric and phosphoric acids were used in various experiments. Double distilled ultra-pure perchloric acid from GFS Chemicals was used in all RDE studies. 0.5 M H_2SO_4 was obtained from Wako Pure Chemical Industries. In order to obtain results in half-cells that somewhat mimic the PFSA or hydrocarbon polymer membranes used in real fuel cells, an appropriate electrolyte must be selected. Perchloric acid is often used in the literature since it exhibits anion adsorption on Pt similar to Nafion. A low concentration of perchloric acid such as 0.1 M is typically employed although it corresponds to a higher pH than applicable values for PFSA membranes. The lower concentration aids in minimizing the amount of Cl^- impurities in the electrolyte that may be present in the original acid or might be generated during the experiment due to decomposition of the acid over time. It is known that the dissolution of Pt is accelerated by the degradation products (SO_4^{2-}) of the PFSA membranes and so sulfuric acid is often used to carry out short-term durability tests at an accelerated rate; 0.5 M sulfuric acid was typically used for short-term degradation studies of Pt-based electrocatalysts.

Catalysts, Inks and Electrodes Catalyst powder samples were purchased from catalyst vendors including Tanaka Kikinzoku Kyogyo (TKK). Several commercially available TKK catalysts were evaluated including TEC10E50E (50 wt% Pt/C), for simplicity referred to in this manuscript as 'baseline Pt/C', TEC10E50-HT (50 wt% Pt/C-Heat treated (HT)), TEC10E50EA (50 wt% Pt/Gr C) as well as alloy catalysts such as 50 wt% PtCo/C and PtNi/C. The heat treated catalyst HT was especially chosen as a candidate for longer duration tests due to its stability under testing. Analysis was carried out using a transmission electron microscope (TEM) (H9000UHR, HITACHI) operating at 300 kV and having a resolution of 0.1 nm. Histograms of particle size were constructed from micrographs by measuring particle size for about 300 particles. X-ray diffraction (XRD) analysis was performed using an X-ray diffractometer (MXP18VAHF, Mac Science) operating with Cu Kα radiation generated at 40 kV and 300 mA. Scans were carried out for 2θ values lying between 5° and 90°. Divergence slit was 1.0° and scattering slit was 1.0°. Crystallite sizes were estimated from the (111) peak using the Scherrer equation.

The glass carbon (GC) electrode was polished on a polishing pad using alumina abrasive (0.05 μm). The electrode was pressed in a direction perpendicular to the pad, and polished by simulating the shape of the character 8. This was followed by ultrasonic cleaning and rinsing in distilled water.

Catalyst powders, ultra pure water (Milli-Q water, Milli-Q Co.Ltd) and isopropanol (Wako Pure Chemical Industries) as well as 5 wt% Nafion dispersion solution (Wako Pure Chemical Industries) were mixed using ultrasonic agitation for 30 min. The ratio of water and isopropanol added to catalyst ink was varied so as to obtain the best dispersion and the highest ECA value. Carbon support in the catalyst ink was kept constant and Nafion ionomer loading was kept extremely low < 0.36 mg/ml. For the baseline TKK catalyst TEC10E50E, the following ink recipe was obtained after considerable optimization. 18.5 mg of TEC10E50E was added to a mixture of 19 ml of ultra pure DI water and 6 ml of 2-propanol. 100 μL of 5% Nafion ionomer solution was added to this solution. This mixture (in a 50 ml sample bottle) was placed in an ice bath and ultrasonicated for 30 min. 10 μL of the so-prepared catalyst ink was immediately pipetted onto the previously cleaned GC electrode. The electrodes were dried in an oven under air at 40–60°C for about 5 min before being inserted in the electrochemical cell. The effective loading of the catalyst was typically ~17–18μg-Pt/cm^2_{geo}. For catalysts with different wt%, the loading was adjusted so that the thickness of the dispersed film was maintained invariant.

Measurement of CVs and I-Vs Cyclic voltammetry (CV) was carried out under nitrogen atmosphere in the cell glassware described above. Electrodes were conditioned to generate a reproducible surface for carrying out CV and I-V measurements. Depending on the electrocatalyst, about 20–100 potential cycles in the potential range 0.04–1.0 V are executed to condition the electrode. In order to carry out CVs, the cell was purged with a H_2-N_2 mixture for 30 min following which CVs were measured typically at 10 mV/s. Three sweeps were consecutively carried out with the third one being recorded. The entire process is schematically illustrated in Fig. 3.30.

Hydrodynamic voltammetry for the oxygen reduction reaction (ORR) evaluation employed linear sweep voltammetry (LSV) under oxygen atmosphere. LSV typically was carried out with voltage scans from 0.2 V–1.2 V, and scan rates employed were 10 mV/s. Rotation speeds of 3,600, 2,500, 1,600, 900 and 400 rpm were sequentially applied. Prior to the LSV at each rotation speed, the electrode was polarized at 0.05 and 1.1 V alternately to construct a reproducible Pt surface. The background current was also measured under identical conditions as the LSV for ORR under N_2 atmosphere without rotation. The details of the operating conditions for LSV measurements in RDE set-ups are delineated in Table 3.3.

For typical durability tests, initial or beginning of life (BOL) diagnostics were conducted as explained above after which the atmosphere was switched from oxygen to nitrogen. Appropriate protocols were applied for load cycling tests (0.60–1.0 V) and support corrosion tests (1.0–1.5 V; 2 s triangular cycles) for a given number of cycles. The gases were switched back from nitrogen to oxygen and end of life (EOL) ORR diagnostics were conducted using the same method as initial diagnostics. The BOL and EOL ECA values as well

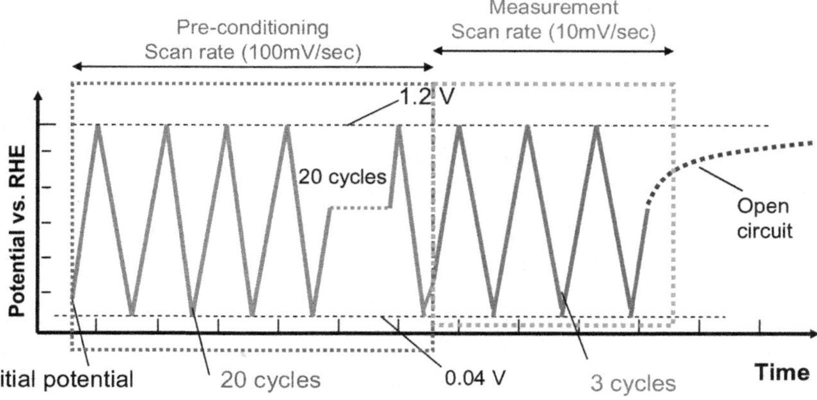

FIGURE 3.30 Schematic plot of the voltage profiles invoked to pre-condition the RDE catalyst thin-film and subsequently obtain a clean measurement of the CV. The pre-conditioning step may require from 20–100 cycles depending on the type of catalyst being evaluated.

TABLE 3.3 Operating Conditions for LSV/I-V ORR Polarization Curves in RDE Set-up. Saturation with N_2 Applied to Obtain a Background Measurement

Temperature/°C		25
Solution	**Electrolyte**	0.1 M $HClO_4$
	Dissolved gas	~30 mins saturation of electrolyte with N_2 ~30 mins saturation of electrolyte with O_2
Voltage Range/V		0.2—1.2
Scan Rate/mV/s		10
Rotating speed/rpm		(400, 900, 1,600, 2,500, 3,600)

as catalyst activity were compared to estimate the losses incurred under cycling. In some cases, at the end of experimentation, the electrolyte was analyzed for dissolved material.

H-type Cell Set-up Pt solubility under different holding potentials was examined with an H-type cell with the working electrode chamber and the counter electrode chamber separated by Nafion membrane. Pt black deposited on a Pt plate was used as the working electrode with a large area gold mesh as the counter electrode. The platinum substrate plate had a geometric area of 8 cm^2 while the HUPD area of the platinized plate was about 1,200 cm^2. The working compartment contained ~240 ml of 0.5 M sulfuric acid. Equilibrium concentration studies on Pt, Pt/C and other relevant catalysts in several acids of different concentrations and under different atmospheres were conducted. The Pt ion concentration was measured with ICP-MS instrumentation having a detection limit of 1 ppb.

EQCN Cell Set-up For EQCN (electrochemical quartz crystal nanobalance) studies, a quartz crystal (9 MHz, 5 mm ϕ) coated with Pt was used as the working electrode using Seiko-EG&G QCM934 instrumentation. Experiments were conducted on smooth surfaces using several different electrolytes.

3.1.2. Subscale Fuel Cells

Figure 3.31 depicts a subscale cell being evaluated in a test stand as well as a schematic of the cell and dimensions of the MEA. Commercial platinum catalyst (46 wt %, Pt/C, (TKK)) and a commercial Nafion membrane (NR-211-cs, thickness 25 μm) were used in Nissan in-house 25 cm^2 cell hardware. Anode and cathode Pt loadings were ~0.35 mg_{Pt}/cm^2 unless specified otherwise. Commercial platinum catalyst (46 wt %, Pt/C (TKK)) and an

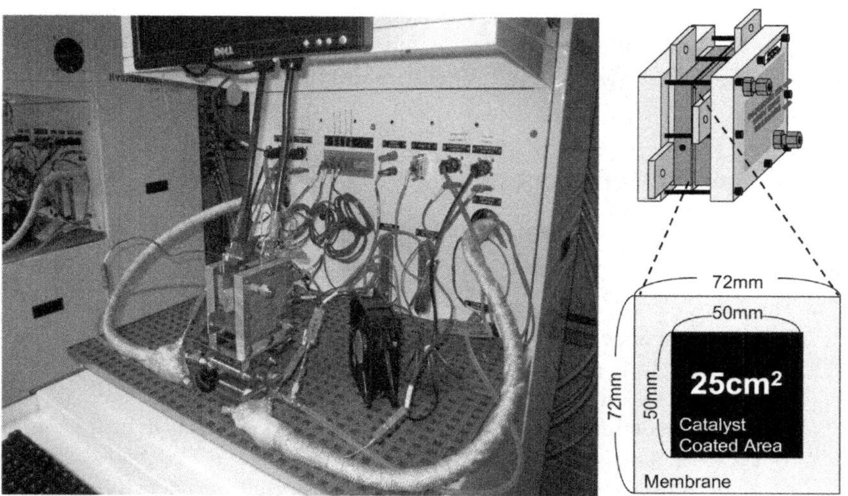

FIGURE 3.31 Photograph on the left shows a subscale fuel cell connected for testing in a test stand and schematic on right illustrates the subscale cell configuration and MEA dimensions.

ionomer (Nafion 2020-cs) were used for ink preparation employing an ionomer to carbon ratio of ~0.9 in this study. Catalyst layers were coated using the ink and hot pressed to a commercial Nafion membrane (NR-211, thickness 25 μm) between 25 cm^2 electrode decals at 150°C, 0.80 MPa for 10 minutes. Gas diffusion media (DM), CFP200 from Gore were employed. Catalyst coated membrane (CCM) and DM were sandwiched between graphite double serpentine flow fields. Pressure sensitive paper (Fuji Film, super low pressure type) was used prior to cell assembly to verify uniformity of contact pressure over the CCM active area. The fuel cells were typically operated at 80°C unless otherwise specified. Reactant back pressures were maintained at atmospheric pressure and inlet gas relative humidity at 100% unless otherwise specified.

Measurement of CVs and I-Vs The measurement of CVs in MEAs of subscale cells is more difficult than in liquid electrolytes and the details of the issues related to diagnostics and analysis are discussed in Section 3.2. CVs were measured by flowing hydrogen on the RE/CE electrode and an inert gas such as N_2 on the WE. 0 SLPM of N_2 (preceded by a N_2 purge) was maintained on the WE while 0.5 SLPM of hydrogen was maintained on the RE. The cell was brought to room temperature prior to every CV measurement for best results. I-V measurements of catalyst activity were carried out at 80°C and 100% RH reactants having $H_2|O_2$ stoics of 2:8; the protocol involved holding the current density constant for 15 min/point and a direction of scan that proceeded from high current densities to low current densities. H_2|Air performance curves were also measured to monitor mass-transport changes with time. The CVs and I-Vs were typically performed at BOL, EOL, and at various intermittent interruptions over the life for long term tests.

3.2. Diagnostic Techniques

The measurement of absolute degradation rates of PEMFCs is beset with complexities arising from:

i. The parameter used to quantify the degradation rate – starting from a given BOL performance, the parameters commonly used to report degradation rates in fuel cells are:

  a. absolute loss in m^2/g/h or m^2/g/cycle or % loss in surface area;

  b. loss in electrocatalytic activity of the catalyst in μA/cm^2/h or μA/cm^2/cycle or % loss in activity;

  c. % performance loss (μV/h) at rated power or current density.

  In all these expressions, the EOL is an arbitrary variable that depends on the duration for which data was collected or on the time to catastrophic failure. The instantaneous degradation rate is usually non-linear and changes (decreases) with elapsed time/number of cycles; therefore the degradation rates, either instantaneous or averaged, reported by different groups are not easily comparable.

ii. Diagnostic method used – for example, is the surface area measurement based on ECA, XRD, TEM or CO chemisorption?

iii. Test protocol used – for example, the pre-conditioning and hold time at each potential when measuring an ORR I-V polarization curve; the use of realistic conditions or accelerated protocols, etc.

iv. Influence of the diagnostic on the degradation rate – for example, interruption of long-term durability tests with frequent diagnostics is known to accelerate the measured degradation rates.

Despite these complications, some progress has been made in standardizing protocols by the US DOE to permit reasonable benchmarking of materials and systems. In this section, we review the main electrochemical diagnostic techniques available to us to enable quantification of the degradation rates of catalysts and supports in fuel cells.

3.2.1. Electrochemical Area (ECA)

Prior to the measurement of the electrochemical area, the original dry catalyst powder is usually characterized using BET, XRD, TEM and CO chemisorption. These measurements allow one to obtain an estimate of the upper limit for the catalyst surface area. Once a catalyst is formulated into an ink/slurry it can be evaluated as TF-RDE in liquid electrolytes or as catalyst layers in subscale fuel cells.

Since all electrochemical reactions occur at the surface of the catalyst, accurate and reproducible measurements of the surface area are critical, both as a measure of the initial condition, and also as a diagnostic to monitor the loss in area due to degradation. The ECA can be measured both *ex-situ* in acid electrolytes as well as *in-situ* in subscale fuel cells or large-platform short stacks. A cyclic voltammogram (CV) is performed on the working electrode under inert gases and the area under the peaks corresponding to hydrogen under-potential deposition (HUPD) on the Pt surface atoms in the potential range ~0.05–0.40 V is integrated to obtain the ECA (m^2/g). A pure double layer capacitance in the potential range of ~0.4–0.6 V and adsorption of one proton per Pt atom (210 $\mu C/cm^2$) is assumed in the ensuing calculations of surface area. The ECA is thus a measure of the number of active sites on the catalyst surface that are in contact with both protons and electrons for small currents pertaining to HUPD. Both Pt and the carbon support provide electronic conductivity.

When the ECA is measured in liquid electrolytes (*ex-situ* ECA), there is usually an excess of protons in the conductive electrolyte ensuring complete contact with Pt. On the other hand, in the catalyst layer of the MEA, the amount of ionomer used is intentionally limited since excess ionomer would result in plugging of the pores necessary for transport of gases and water. Pt particles that are not protonically and electronically connected to the catalyst layer would thus not participate in the oxygen reduction reaction (ORR) on the cathode or the hydrogen oxidation reaction (HOR) on the anode. Therefore, the ratio of *in-situ* ECA/*ex-situ* ECA is often looked upon as a first, simple measure of the utilization of the catalyst surface in the catalyst layer. Every attempt is made to optimize the ionomer and ionomer distribution in the catalyst layer to maximize performance at all current densities.

It is fairly easy to measure the *ex-situ* ECA in liquid electrolytes and obtain clear, well-resolved, distinguishable peaks for the hydrogen under-potential deposition or hydrogen adsorption peaks (HUPD or HAD) but this is not the case for the *in-situ* measurements in PEMFCs. *In-situ* ECA measurements are carried out in subscale fuel cells or stacks with H_2 on the RE+CE and N_2 on the WE. It is often found that the onset of hydrogen evolution occurs early (at positive potentials) and overlaps and smears the HUPD peaks making it difficult to measure the area under the peaks accurately. We observe (Fig. 3.32) that by using a low nitrogen flow and room temperature, the situation can be alleviated to obtain reasonably sharp, well-resolved peaks and reproducible ECAs. At low or zero N_2 flows, the hydrogen crossover from the RE+CE through the membrane results in a higher H_2 partial pressure on the WE and causes the H_2 onset to shift back towards 0 V.

As we improve the catalyst activity and use lower and lower loadings of Pt in the cathode as well as anode catalyst layers, the HUPD peaks contract; also, as Ostwald ripening and particle size growth occurs over time, the

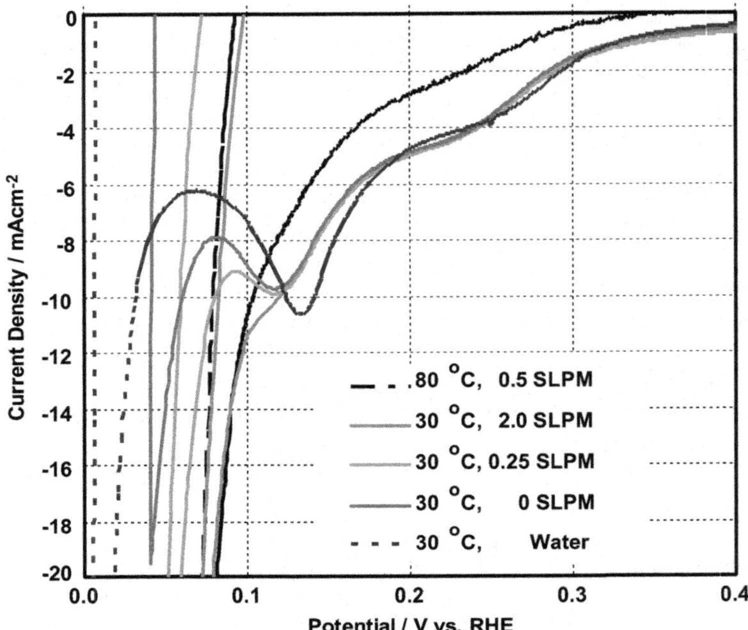

FIGURE 3.32 Cyclic voltammograms carried out *in-situ* in subscale fuel cells with hydrogen on the RE/CE side and various N_2 flows (or water) on the cathode/WE. Scan rates for the CVs were 20 mV/s and temperatures are as marked in the plot. *(Reprinted with permission from Uchimura and Kocha [90])*

peaks correspondingly shrink due to the lower residual area. Both these situations further warrant that we use the best measurement possible to be able to accurately measure the change in ECA over time while monitoring degradation. Figure 3.33 demonstrates the systematic trend of CVs with increasing number of potential cycles (0–15,000 cycles) as the catalyst gradually degrades (particle growth/surface area loss) with shrinking HUPD peaks.

3.2.2. Specific (i_s) and Mass Activity (i_m)

I-V curves carried out under $H_2|O_2$ are typically used to determine the activity of the catalyst. Extremely high flows of oxygen are maintained to ensure a uniform O_2 concentration from cell inlet to outlet. The cell approximates a differential cell and eliminates artifacts of utilization (or stoichiometry) on the measured currents. Saturated gases are required to ensure uniformly high conductivity of the membrane and ionomer in the catalyst layer over the entire cell area. At 0.90 V, where the currents are low, the overpotential losses (η) due to resistance, mass transport, current distribution and effects of oxygen utilization are all expected to be inconsequentially small. Hence, although the

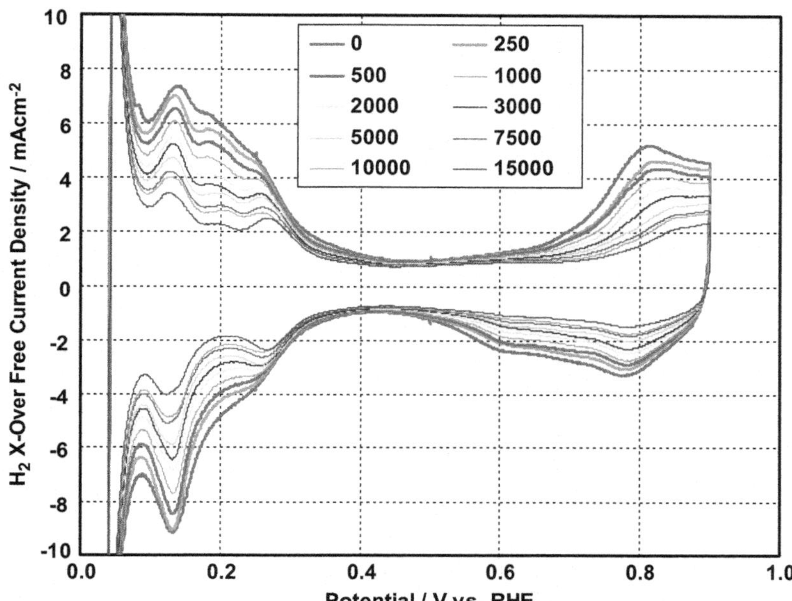

FIGURE 3.33 H_2-Xover corrected cyclic voltammograms executed on the cathode of a subscale 25 cm^2 PEMFC at intermittent intervals during cyclic durability testing over 15,000 cycles. Square wave profiles with 10 s hold time at $E_H = 0.95$ V and 2.5 s hold time at $E_L = 0.60$ V. Loss in HUPD charge/ECA as well as oxidation/reduction charge in the CV is due to Pt dissolution and particle growth.

entire I-V curve is usually measured, the current at 0.90 V has been traditionally used to obtain and compare the activity of catalysts.

In PEMFCs with thin membranes, however, the H_2 X-over current (permeation of hydrogen through the membrane) is not negligible and a simple correction (addition of $i_{X\text{-over}}$ to the measured current) must be made to account for it. The cell active area is usually kept in the range 25–50 cm^2 in order to facilitate uniform conditions over the cell area; smaller active areas may result in edge effects due to the larger ratio of perimeter to area. In short, all precautions are taken to ensure that we approximate a 'differential cell' that would allow us to easily measure a purely kinetic current due to the ORR. At this time, small currents due to carbon corrosion, any poisons, and Pt dissolution are not typically taken into account.

In spite of all these precautions, it is fairly difficult to benchmark the activity of any given catalyst especially between different laboratories. The main reason is the protocol used in testing the MEA in the fuel cell. Even though psuedo steady state data is taken at each point of the I-V curve (15 min/point), there is noticeable hysteresis in the forward and backward direction data of the I-V curve. This is due to the fact that different oxide species with different coverage and different structures grow on the surface of Pt over time; thus the pre-conditioning which determines the oxidation state of the Pt catalyst prior to each data point influences the measured ORR activity. The influence of oxide species reaction intermediates on the ORR, Tafel slope, and rate determining step have all been discussed at length in the literature. [28–31,33,196]. It has been postulated that the lower Tafel slope of ~60–70 mV/dec is due to oxygen adsorption under Temkin conditions while the higher ~120 mV/dec slope is ascribed to Langmuirian kinetics on oxide-free Pt surfaces [29,197].

Figure 3.34 illustrates the effect of direction of measurement; a change in measured activity by a factor of 2 is observable (at 0.90 V) despite pseudo steady state data acquisition at 15 min/point measured at each voltage [10,11]. One can imagine that extrapolating over several orders of magnitude to obtain the exchange current density 'i_0' would significantly amplify the error. Typical value of the exchange current density for the ORR [5] is ~1 × 10^{-8} A/cm^2_{Pt}.

It has been demonstrated that the ORR polarization curve over the entire range of potentials can be fit (model fits in Fig. 3.34) using a single value of i_0 that corresponds to the oxide-free low potential regime [198]. Markovic et al.

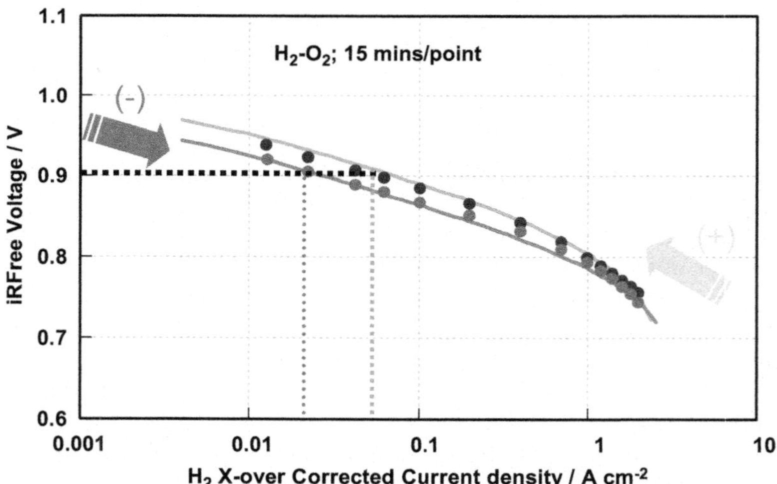

FIGURE 3.34 Tafel plots showing the effect of pre-conditioning of CCM and measurement protocol on the ORR activity. Both Tafel plots are based on protocols in which the data-acquisition was carried out at constant current density for 15 min per point, at 80°C and 100% RH. The higher curve (blue square markers) was obtained by beginning at the highest current densities (low potentials) and ending at low current densities (high potentials) and referred to as positive direction. The lower curve (red triangular markers) was begun at high potentials and ended at low potentials and is referred to as negative direction. *(Reprinted with permission from Uchimura and Kocha [10])*

[199] have invoked an extended Butler-Volmer equation incorporating oxygen species coverage θ as follows:

$$i = nFkc_{O_2}(1-\theta)^x \exp\left(-\frac{\beta FE}{RT}\right)\exp\left(-\frac{\gamma r\theta}{RT}\right) \qquad (3.15)$$

where 'θ' depends on potential E of the electrode. We have applied this expression using values of 'θ' obtained from integrating under the oxide reduction peaks from cyclic voltammograms corresponding to the data points on the I-V polarization curves. Thus, in an ideal scenario, one would like to report the ORR activity for a 'defined surface oxide coverage θ'. We have also reported that the oxide species on the surface of Pt affects the ORR reaction order [200]; the magnitude of the reaction order is important not only to decipher the mechanism but also to predict the effect of operating pressure (oxygen partial pressure) on practical fuel cell performance.

Often only small quantities of novel catalysts are available for experimentation. In such cases, they are tested in half-cells, typically as a TF-RDE using aqueous acid electrolytes such as perchloric acid or H_2SO_4. As mentioned previously, ECA measurements in acid electrolytes are useful to estimate the Pt utilization in MEAs, but what about the absolute value of Pt activity in liquid acid electrolytes vs. MEAs in PEMFCs? Firstly, the choice of acid is important since anion adsorption lowers the activity of Pt significantly. $HClO_4$ is a fairly reasonable choice of acid to mimic the low anion adsorption of PFSA ionomers and hence provides a closer representation of the activity of Pt covered with Nafion in the catalyst layer. On the other hand, Pt dissolves more readily in H_2SO_4 and is therefore more appropriate for potential cycling durability tests in TF-RDEs. TFMSA is perhaps the best choice for an electrolyte in terms of anion adsorption [201], but difficulties in purifying it has limited its use to few laboratories.

Finally, an issue of contention is whether the value of Pt activity measured as a flooded TF-RDE in 0.1 M $HClO_4$ (at a given sweep rate of 1–20 mV/s) is a direct measure of the absolute value of Pt activity measured *in-situ* in a PEMFC under pseudo steady state conditions of 15 min/point. Results tabulated in [16] seem to indicate that it is so, but this is a coincidence considering that the test conditions in the two platforms are vastly different. Nonetheless, testing in RDE is a useful tool for screening new catalyst materials; the activity and durability values, though relative, are extremely useful and time-saving.

3.2.3. H_2 X-over and Shorting

H_2 X-over and shorting are leakage currents and always present in practical PEMFCs. Hydrogen crosses over from the anode to the cathode by dissolving and diffusing through the PEM, reacts with the oxygen to produce water, and leads to a drop in cathode potential. The equivalent magnitude of the crossover in terms of a constant current is of the order of a few mA/cm^2

(for thin PFSA membranes) and can affect the measurement of the low ORR kinetic currents at 0.90 V. Electronic shorting of the membrane can occur due to various reasons and is a low resistance pathway for a leakage current. Both the H_2 X-over and shorting can be easily diagnosed from a CV carried out at slow sweep rates or a simplified one-point measurement (at ~0.5 V) [5]. A study of the permeability of typical PFSA membranes as a function of partial pressure of O_2, temperature and membrane thickness has been reported by Kocha et al. [144]. Figure 3.35 illustrates the results of a CV (and simulations) in a subscale cell under $H_2|N_2$ exhibiting limiting currents at low scan rates corresponding to H_2 X-over and a constant slope due to ohmic shorting. Although H_2 X-over and shorting are not related to catalyst or support degradation, these parameters have to be monitored closely as a significant increase in either of these quantities is an indication that catastrophic failure of the cell is imminent. Moreover, the quantitative value of the

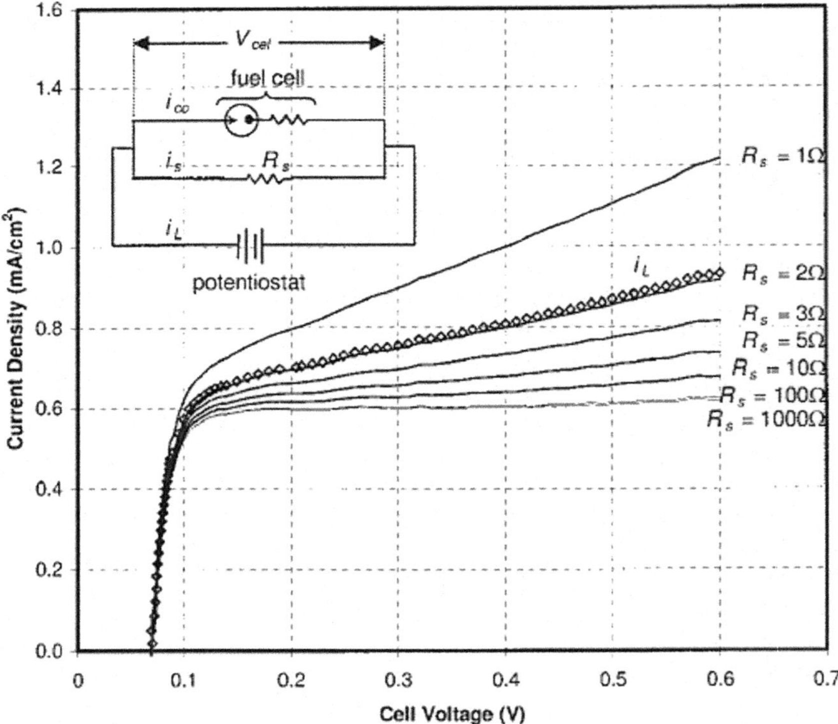

FIGURE 3.35 H_2 X-over currents measured in subscale PEMFCs using Nafion 112 membranes along with simulation for various degrees of electrical shorting. Diamond markers are actual data and correspond to a shorting resistance of 2 ohms; other curves are simulated for various shorting resistances as labeled. *(Reprinted with permission from Kocha et al. [144])*

hydrogen X-over and shorting will permit us to account for and distinguish other cathode side losses due to surface area, activity loss or mass transport resistance.

3.2.4. Electrochemical Impedance Spectroscopy (EIS)

Various adaptations of EIS are used as diagnostics of catalyst behavior as well as the membrane, ionomer and other resistances of the MEA. The single high frequency version of the technique known as high frequency resistance (HFR) involves the application of a single small amplitude AC frequency signal ~10 kHz and is a standard *in-situ* measurement applied to PEMFCs. HFR includes the membrane resistance, contact resistance and any other series electronic resistance that is encountered between the terminals of the fuel cell. It does not include the catalyst layer resistance. The HFR values are determined at each steady-state point of the I-V curve; subtraction of the HFR from the cell voltage V_{cell} gives us an 'iRFree Voltage' that is critical for isolating the catalyst activity. A typical value for the HFR for a 25 μm PFSA membrane is ~40–50 $mohm \cdot cm^2$.

In order to determine the 'catalyst layer resistance' (represented by a transmission line array of ionomer resistances and capacitances) a wider range of frequencies have to be applied to a fuel cell fed with hydrogen on the anode and nitrogen on the cathode/working electrode [5,202]. The values reported for the catalyst layer resistance (dominated by the ionomer resistance) for 100% RH in the literature are in the range of 50–150 $mohm \cdot cm^2$. Figure 3.36(a) illustrates the measurement of HFR and catalyst layer resistance R_{CL} from Nyquist plots along with the effect of RH; the value of the catalyst layer resistance increases as the RH is lowered due to the decrease in ionomer conductivity with water content. It should be noted that this value should be considered as an upper limit of R_{CL}; R_{CL} is actually lower under actual fuel cell operation wherein the current density across the thickness of the CL is non-uniform. There are several interfacial resistances encountered and included in the measured resistance; in recent work, Pivovar and Kim have discussed a method to quantify the membrane-electrode interface resistance [203].

EIS can also be applied to an operating fuel cell under H_2|Air to study the ORR, resistance and mass-transport [204,205] for cells with different DM components including before and after durability tests. Figure 3.36(b) depicts the Nyquist plots for the impedance spectra as a function of reactant gases used. The low frequency time constant observed under air is attributed to mass-transport and the intermediate frequency time constant to charge-transfer. Changes in these time constants provide insights into degradation mechanisms in the cathode catalyst layer.

For the purposes of carrying out fundamental studies on the oxide species on the surface of Pt-based catalysts and their mechanism of growth AC impedance [46], EIS [45] and PDEIS [56] are often employed in

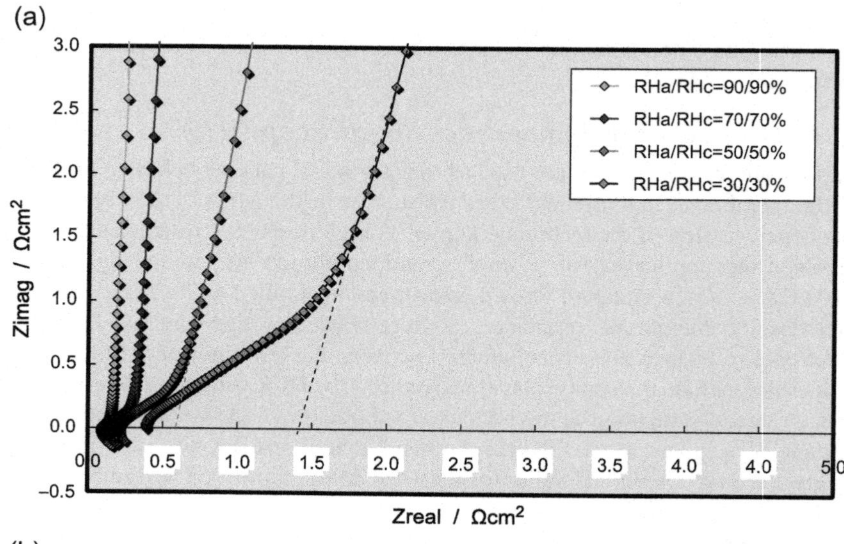

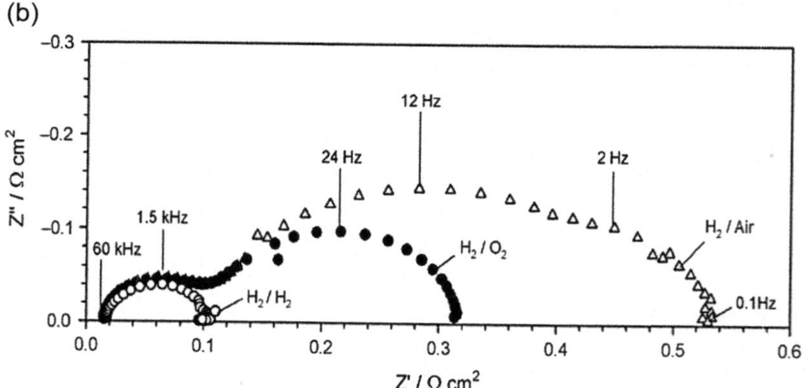

FIGURE 3.36 (a) EIS spectrum carried out in subscale fuel cells under H_2-N_2 to determine the catalyst layer resistance under several RH conditions. Dotted lines extrapolated to the real impedance axis can be analyzed to obtain the catalyst layer resistance R_{CL} [214]. (b) Effect of reactant gases at the cathode for a PEMFC operated at 210 mA/cm^2, 60°C and 100% RH on the EIS spectra. Time constants related to H_2|Air, H_2|O_2 and H_2|H_2 are delineated adjacent to the spectra. *(Reprinted with permission from Malevich et al. [205])*

combination with complementary techniques [48] such as electrochemical quartz crystal nanobalance (EQCN) and spectroscopies such as auger electron spectroscopy (AES).

3.2.5. H_2|Air, Oxygen Gain and Limiting Currents

H_2|Air I-V curves are often measured under conditions that include realistic RH, reactant pressure, and stoichiometry. Limiting currents due to mass-transport under H_2|Air are observed in the range 1–3 A/cm^2 depending on the quality and design of the MEA and fuel cell. These 'H_2|Air performance curves' include losses due to activity, resistance as well as mass-transport. Therefore, degradation rates measured under H_2|Air, usually at a fixed current density or under cycling, provide us with an 'overall degradation rate' that includes catalyst activity loss convoluted with mass transport losses. H_2|O_2 I-V curves described in the previous section are needed to obtain pure kinetics and diluted O_2 curves are sometimes invoked to further refine mass transport losses. The use of dilute O_2-N_2 or O_2-He allows us to obtain additional details on the initial value as well as changes that occur in mass transport resistance caused by carbon support corrosion, change in hydrophilicity of the carbon support, and also changes in the electrode structure such as porosity and pore distribution. Using dilute O_2, limiting currents can be reached at lower current densities and the shift in these limiting currents provides us with a measure of the 'mass-transport degradation' of the catalyst layer and diffusion media or MEA. Such measurements at varying oxygen partial pressure as well as the use of a diluent other than nitrogen are difficult and tedious experiments and so far few reports have been made detailing these studies [5]. Figure 3.37 illustrates the effect of dilution of O_2 to the limiting current measured in subscale fuel cells.

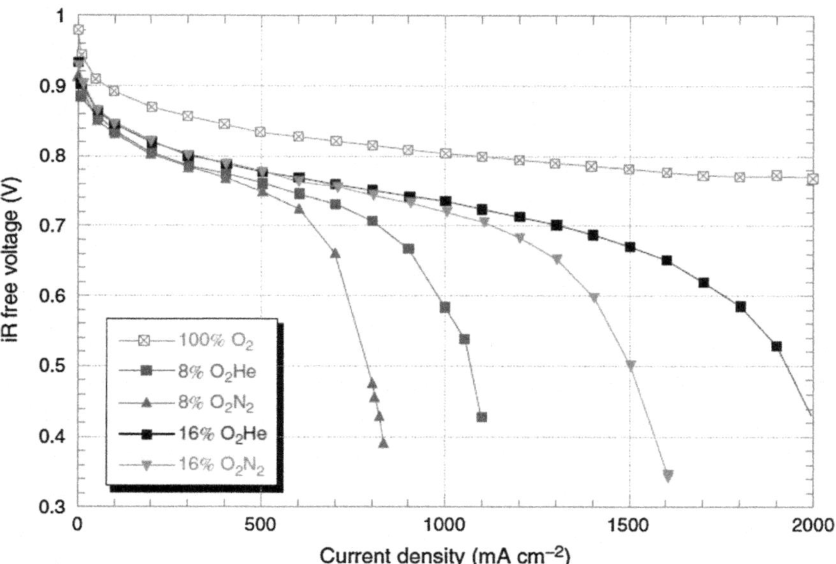

FIGURE 3.37 Diagnostic technique used for evaluation of the change in mass transport and pore structure of catalyst layers between BOL and EOL. Plot illustrates the change in limiting current caused by switching of inert gases for several concentrations of O_2. Data collected at 65°C, 100 kPa with pure H_2 on the anode flow fields. Catalyst loadings are 0.40 mg/cm^2. *(Reprinted with permission from Kocha [5])*

4. RESULTS AND DISCUSSION

In this section, we discuss recent results (carried out at the Advanced Electrocatalysis Group and MEA groups, Fuel Cell Laboratory, Nissan) regarding the electrochemical degradation of the catalyst and support, studied either in *ex-situ* tests or *in-situ* in MEAs of subscale or full scale cell platforms of PEMFCs. This section is subdivided into: **4.1 *Ex-situ* Catalyst Dissolution Measurements**, **4.2 *In-situ* Catalyst Degradation Under Automotive Operation**, and, **4.3 Durability of PEMFC Stacks in Vehicles**. Section 4.2 has been further subdivided into: **4.2.1 OCV and Idling**; **4.2.2 Load Cycling**; **4.2.3 Contamination**, and **4.2.4 Start-up Shut-down**.

4.1. *Ex-situ* Pt Dissolution Measurements

4.1.1. Equilibrium Concentration in Acid Electrolytes

As mentioned in the literature review the Pourbaix diagram indicates that Pt dissolves in a narrow potential-pH window near 1 V [17]. A handful of studies [16,18,20,22,23], have attempted to verify the potential-pH or Pourbaix diagrams by carrying out equilibrium dissolution measurements under constant potential in acid solutions. Although these experiments appear simple in principle, they are time consuming and difficult to conduct, and literature results are difficult to verify since the Pt/C and Pt surface area, its pretreatment, acid concentration and temperatures vary widely. In order to verify the magnitude of dissolution and the general trends, we conducted a series of equilibrium concentration studies on Pt, Pt/C and other relevant catalysts employing a standard H-type cell. Experiments were conducted in sulfuric and perchloric acids of different concentrations, under air and N_2 and at different temperatures over the potential span 0.9–1.4 V. The Pt ion concentration was measured with ICP-MS that had a detection limit of 1 ppb. It typically took ~24 h to attain steady state; during this period concentration measurements were systematically acquired and repeated. The detailed results of the investigation will be reported elsewhere [206]. Table 3.4 presents only selected results in comparison to the literature data corresponding to the peak dissolution rates at 1.1 V; the trends were found to be reasonable. Some of the general trends of relevance to PEMFCs are as follows:

- **i.** Pt dissolution is higher in H_2SO_4 than $HClO_4$ probably due to the easier formation of complexes;
- **ii.** acid concentration has a strong effect on dissolution; and
- **iii.** poly-Pt as well as Pt nanoparticles with larger diameters dissolve at lower rates than nanoparticle Pt/C most likely due to the lower surface energy of larger particles with better-co-ordinated surface atoms.

TABLE 3.4 Comparison of Pt Dissolution at 1.1 V in $HClO_4$ and H_2SO_4 Measured in This Work and Extracted from the Literature. Acid Composition, Concentration and Temperature of Measurement are as Indicated

	Reference	p-Pt; $HClO_4$; (M)	Pt/C; $HClO_4$; (M)	Pt/C; H_2SO_4; (M)
0.5 M, 80°C	[21] GM	–	–	2×10^{-6}
0.1 M, 80°C	Nissan	10^{-7}	5×10^{-7}	5×10^{-6}
0.57 M, 23°C	[22] ANL	10^{-6}	6×10^{-8}	–
1.0 M, 23°C	[19] YNU	–	–	5×10^{-6}

4.1.2. Potential Cycling in Liquid Electrolytes

In potential cycling experiments conducted in MEAs of PEMFCs (or gas diffusion electrodes [67]), it is difficult to estimate the amount of dissolved Pt in MEAs, as it can get trapped in the membrane as well as in the ionomer of the catalyst layer. Hence results obtained from liquid electrolyte studies are needed to corroborate dissolution mechanisms. In experiments similar to the determination of the equilibrium concentrations described above, we confirmed that dissolution of Pt under cycling is much higher than that at constant potential. We found that the Pt ion concentration increased drastically (by two orders of magnitude) from steady-state values after being subject to potential cycling between 0–1.4 V at a cycle scan rate of 100 mV/s over a few hours.

In the light of the fact that the durability/dissolution of Pt appears to be significantly influenced by the coverage of the surface with oxide species during cycling, we have performed a wide range of complementary electrochemical experiments in liquid electrolytes (chrono-potentiometery, chrono-amperometery, EQCN, RRDE, EIS, PDEIS etc.) to gain a deeper insight into the oxide species growth, coverage, and formation mechanism. We briefly examine the difficulties involved in arriving at clear conclusions based on selected results from EQCN and Pt dissolution experiments. EQCN was studied in combination with CVs in an attempt to determine the likely species of oxides covering Pt as well as to estimate the degradation rate during symmetric and asymmetric potential cycles. Figure 3.38 depicts the mass change (Δm – corresponding to a frequency change Δf) from EQCN and charge (Δn) from CVs overlayed on the same plot. Based on the change in charge and change in mass in the oxide potential regime, our measurements (not shown) typically resulted in an average value of $(\Delta m/\Delta n) \sim 8$. Since PtO, PtO_2 and Pt_2O would all produce the exact same ratio, it is not possible to distinguish between them with these experiments alone. The observed increase in mass in the

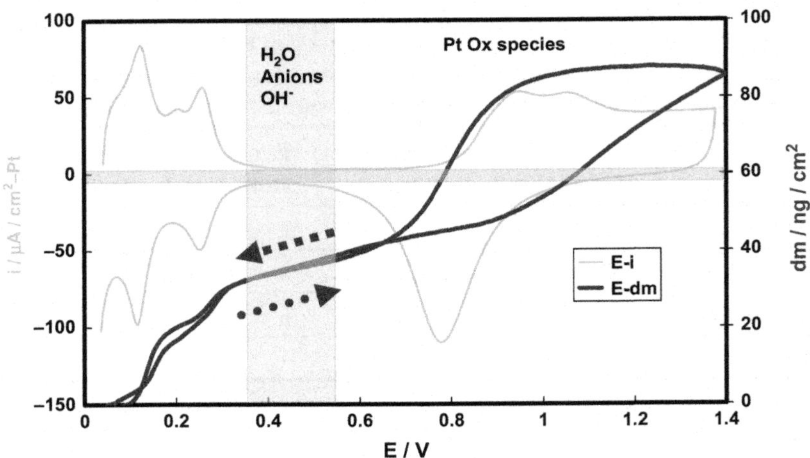

FIGURE 3.38 Plot of CV and EQCN profiles on a Pt film sample depicting the HUPD (0.05–0.35 V), double layer (0.35–0.55 V) and oxide regimes (0.55–1.4 V) [206].

double layer regime while the current in the CV is constant reveals that anion adsorption including OH^- contributions may play a role. Thus, there are obvious difficulties in applying EQCN to identify the adsorbed species.

EQCN was also applied to measure the loss in mass during dissolution caused by potential cycling. A Pt film electrode immersed in 0.5 M H_2SO_4 was subjected to a series of five asymmetric cyclic potential profiles with the upper and lower potentials of 1.4 and 0.4 V, respectively (Fig. 3.39(a), (b), (c)). The cycle period was 21 s. The pairs of scan rates (anodic, cathodic) for the five cases in mV/s were (1,000,50), (167,67), (95,95), (67,167), (50,1,000), respectively. Both the CVs and EQCN results for the five cases are separately shown in Fig. 3.39(b) and (c). Figure 3.40 illustrates the change in frequency and hence mass for two extreme profiles, one with a fast-anodic/slow-cathodic (1,000 mV/s, 50 mV/s), and the other with a slow-anodic/fast-cathodic sweep (50 mV/s, 1,000 mV/s). Over a large number of cycles, we find that the fast-anodic/slow-cathodic sweeps result in higher dissolution rates. These results corroborate similar outcomes on the impact of profile on degradation observed in MEAs of PEMFCs discussed later.

Kinetic models for the dissolution of Pt taking into account the electrochemical dissolution of Pt ($Pt = Pt^{2+} + 2e$), formation of PtO as well as the chemical dissolution of PtO ($PtO + 2H^+ = Pt^{2+} + H_2O$) have been reported by Darling and Meyers [25,26]. In our simulation results, shown in Fig. 3.41, the model and parameters used in the work of Darling and Meyers were employed with one difference; instead of simulating the oxide coverage from the model, the coverage was experimentally estimated directly from the measured current, assuming PtO as the oxide species. The electrochemical dissolution from

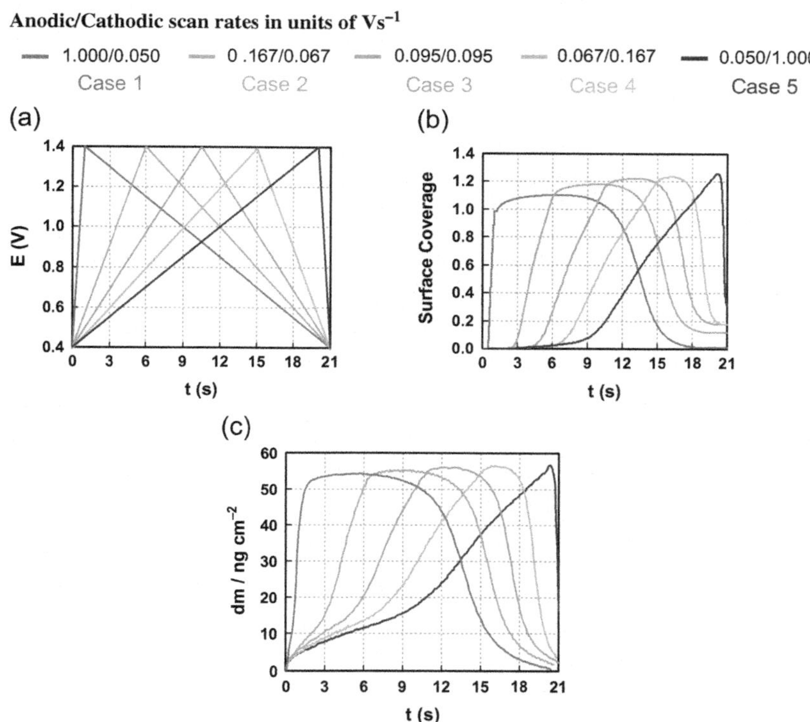

FIGURE 3.39 (a) Potential profiles applied to the working electrode for combined CV and EQCN studies; (b) surface coverage with oxides measured using oxide reduction charge from CVs; (c) change of mass measured using EQCN. Scan rates and case number for the anodic/cathodic sweeps are marked at the top. *(Uchimura et al. [11])*

exposed Pt and the chemical dissolution from Pt oxide were calculated for each point in time and integrated for one cycle. The simulation results indicated that the dissolution from the Pt oxide was negligibly small (10^{-29} moles/cm^2) compared with that dissolved from exposed Pt. Figure 3.41 shows that the estimated Pt dissolution rate for case 1 (highest anodic scan rate) was about six times higher than that for the other cases. Case 1 had the least amount of passivating oxide during excursions to high potentials facilitating dissolution of bare Pt.

4.2. *In-situ* Catalyst Degradation Under Automotive Operation

4.2.1. OCV and Idling

When a vehicle in operation comes to a rest, for example, at a traffic light, the PEMFC stack is required to produce very minimal power to support the automobile auxiliaries. Under these conditions, the current drawn is very low

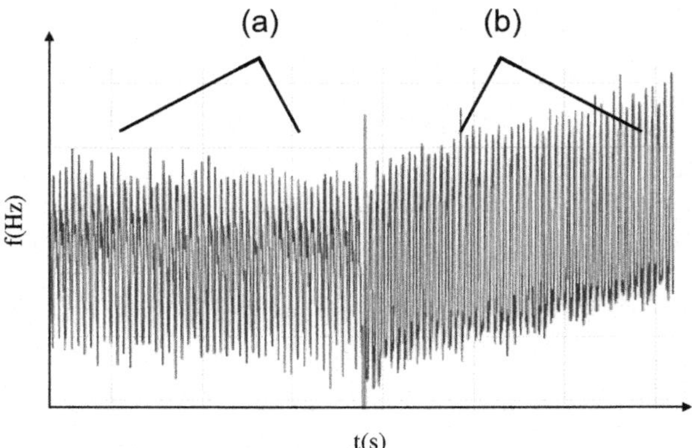

FIGURE 3.40 (a) Plot of frequency change vs. time obtained from EQCN experiments conducted to measure the degradation rate under different cycle profiles. The right half of the plot (b) with higher frequency (lower mass) corresponds to higher dissolution rates 0.6 ng/cm^2 under the fast anodic/slow cathodic scan rate cycles [206].

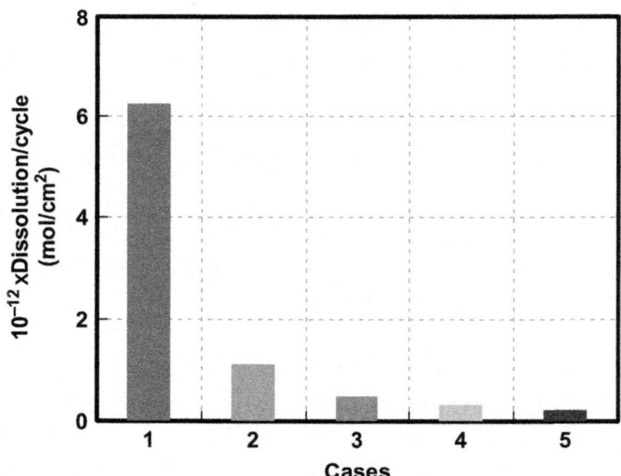

FIGURE 3.41 Magnitudes for Pt film dissolution subject to the potential profiles from Fig. 3.39 (a) and based on simulations using the oxide charge values shown in Fig. 3.39 (b). *(Uchimura et al. [11])*

and the corresponding potential of each individual cell in the stack is quite high and typically falls in the range 0.90–0.95 V.

The theoretical equilibrium potential of Pt under standard conditions of 25°C and 100 kPa of O_2 is about 1.229 V; the numerical value will differ under actual fuel cell operating conditions due to differences in partial pressure of reactants and temperature. The standard potential for the Pt/Pt^{2+} redox couple is ~1.18 V and therefore Pt electrodes are not thermodynamically stable at the reversible potential. Therefore, the cathode is at a mixed potential arising from the cathodic ORR and anodic reactions related to the formation of oxide species on Pt, Pt dissolution, carbon corrosion and, to some extent, the oxidation and reduction of impurity species. The following are some of the main reactions at the cathode leading to the mixed potential:

$$2H^+ + 2e^- \rightarrow H_2 \tag{3.16}$$

$$O_2 + 4H^+ + 4e^- \rightarrow 2H_2O \tag{3.17}$$

$$Pt + H_2O \rightarrow PtO + 2H^+ + 2e^- \tag{3.18}$$

$$Pt \rightarrow Pt^{2+} + 2e^- \tag{3.19}$$

$$C + 2H_2O \rightarrow CO_2 + 4H^+ + 4e^- \tag{3.20}$$

Based on the above, we can surmise that the loss or decay in voltage over time under OCV (in addition to H_2 X-over) can be due to a combination of:

i. formation of oxide species with the consequent lowering of the ORR kinetics;
ii. change in the rates of oxide formation/reduction; and
iii. change in the rates of Pt dissolution/re-deposition.

When held continuously for long periods of time at OCV or idling potentials, both the membrane and catalyst are known to undergo degradation. Although this chapter is devoted to catalyst and support degradation, in the case of degradation under OCV or idling operation it is not possible to exclude one phenomenon over the other. Recent work carried out by several groups at Nissan on OCV durability is discussed below.

Sugawara et al. [87] have investigated the underlying mechanism for the large voltage drop during OCV holds as well as provided partial *in-situ* mitigation techniques to counter the losses. In their study, they subjected a cell to OCV conditions for 48 hours at 80°C under 30% RH. Figure 3.42 depicts the decay in open circuit voltage as a function of time. CVs and I-V polarization curves were measured before and after the OCV hold test, along with fluoride and sulfate emission rates from collected exhaust water. In addition, a 'recovery technique' was executed after the hold that involved operating the cell at a low temperature and high humidity with the cathode under N_2 in order to condense water and flush the cathode. Fluoride and sulfate emission measurements from ion chromatography studies of the effluent water were obtained during the OCV test and also during the 'recovery technique' as illustrated in Fig. 3.43. During the 48 h OCV test the fluoride emission was found to be significant, at ~260 μmol (6% of the fluorine in the membrane) on both the anode and cathode sides, whereas the sulfate emission was found to be quite low at < 0.05 μmol. Whereas, after the recovery procedure was executed, more sulfate at 0.5 μmol was found in the effluent water.

Figure 3.44 shows the CVs before and after the OCV hold as well as after recovery technique. A shift in the onset of oxide formation as well as the formation of sharper peaks overlapping the HUPD peaks is observed. The

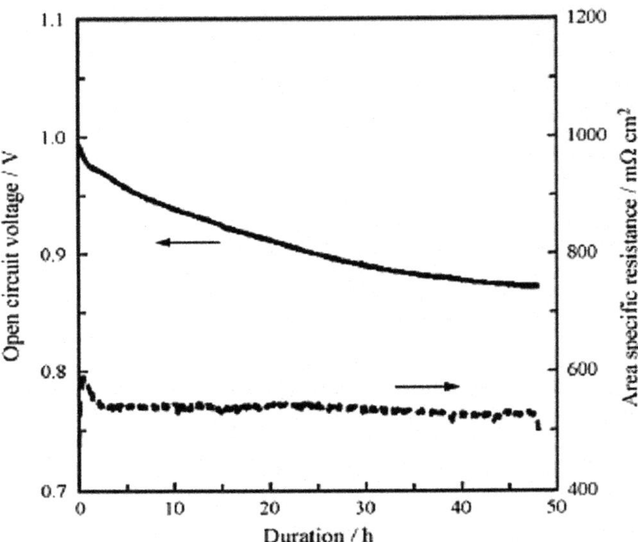

FIGURE 3.42 OCV decay (solid line) for tests carried out at 80°C, 30%RH under H_2|Air. *(Reprinted with permission from Sugawara et al. [87])*

simultaneous evolution of these two features in the CV profile is very similar to that observed in the presence (and due to the adsorption) of sulfate ions on Pt and thus correlates to the sulfate found in the effluent water. I-V polarization curves (Fig. 3.45) showed a loss in activity after the OCV hold followed by a subsequent partial recovery of performance after flushing out the sulfate ions with condensed water. Evidently, the degradation of the membrane releases sulfate ions that adsorb on the Pt catalyst surface and lower its specific activity. It was also postulated that the sulfate ions so released facilitate the formation of peroxide and further exacerbate the membrane degradation. For the duration of their study, the loss in electrochemical area was not found to be significant, but over longer periods of time that might also be a contributing factor since there is evidence that Pt dissolves and diffuses into the membrane to form a band.

Typically, the mechanism invoked to explain membrane degradation during OCV holds involves the formation of hydrogen peroxide at the anode catalyst surface (as a result of hydrogen and oxygen crossover) or as an intermediate of the ORR in a 2-electron process that can be expressed as follows:

$$O_2 + 2H^+ + 2e^- = H_2O_2 \qquad E^0 = 0.672\ V \tag{3.21}$$

Supporting evidence for the formation of H_2O_2 has been found in RRDE experiments, where small currents associated with peroxide reduction are routinely measured. Furthermore, peroxide currents have been reported to be

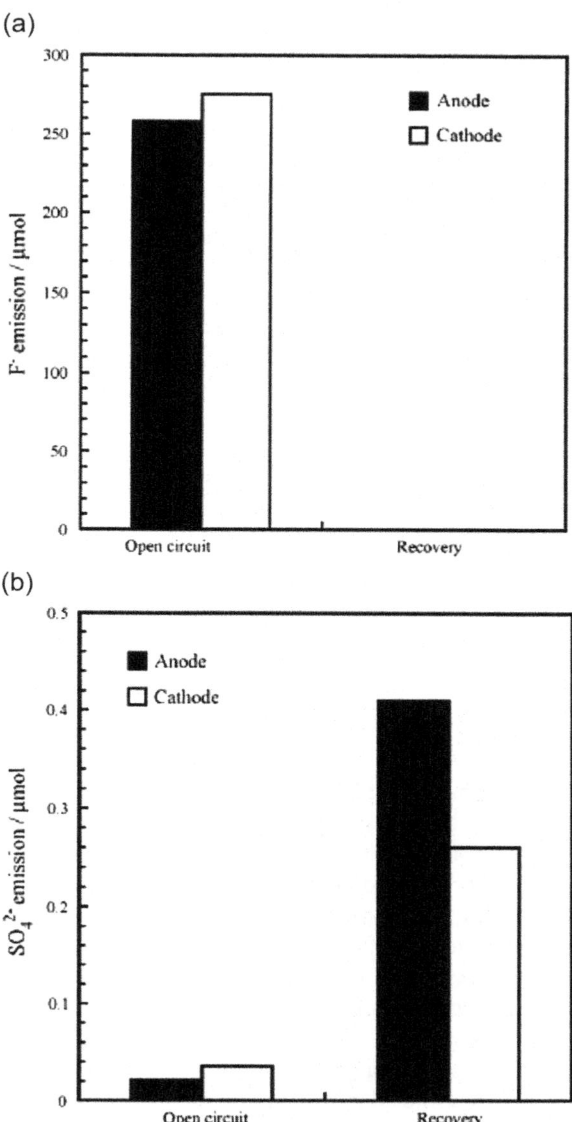

FIGURE 3.43 Ion chromatography studies of the effluent water for total fluoride (a) and sulfate (b) emission under 48 h of open circuit conditions as well as during the 2 h recovery operation. *(Reprinted with permission from Sugawara et al. [87])*

a function of the sulfate ion concentration as reported by Tsujita et al. [201] and illustrated in Fig. 3.46.

A common observation in OCV hold tests is that a Pt band is formed in the membrane after a period of time using conventional Pt/C catalysts. In the OCV

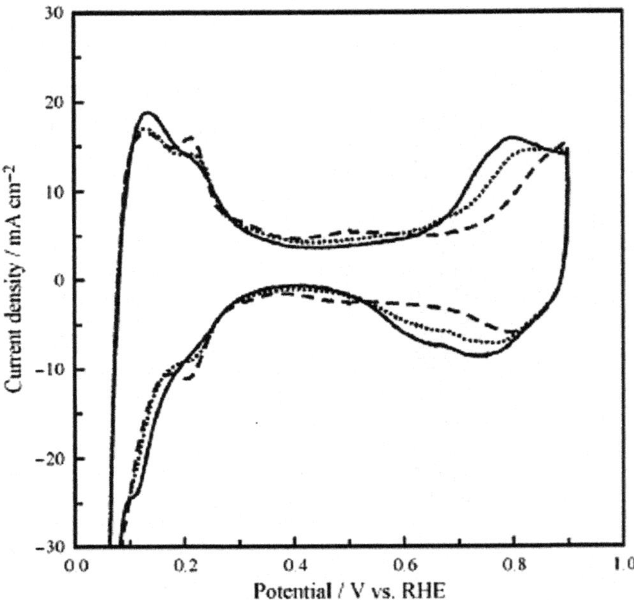

FIGURE 3.44 CVs of the cathode measured before and after OCV hold durability test operation as well as after recovery operation. CVs were measured on the cathode at 50 mV/s, 80°C, 0.5 SLPM and 100% RH. Anode was fed with H_2 at 0.5 SLPM, 100% RH. *(Reprinted with permission from Sugawara et al. [87])*

study conducted by Ohma et al. [207,208], three sample MEAs were evaluated under OCV conditions for 110 h using 10% $H_2|O_2$, 100% $H_2|O_2$ and 100% $H_2|$Air as the (Anode|Cathode) reactant flows. They found that the location of the band is a function of the partial pressures of hydrogen and oxygen on the anode and cathode. The band is formed closer to the anode when the partial pressure of hydrogen is lower; in contrast it forms in the vicinity of the cathode if air is used on the cathode side instead of O_2. In addition, fluoride ion elution rate (FER) measured from the effluent water (using IC) from the anode and cathode show higher rates on the side consistent with proximity to the location of the Pt band. In a follow-up paper, Ohma et al. [208] also used micro-Raman spectroscopy to analyze molecular level structural changes in the membrane to confirm that the membrane in the neighborhood of the Pt band was significantly degraded. Figure 3.47 illustrates the relative intensity profiles exhibiting a dip corresponding to degradation in the membrane in the vicinity of the Pt band.

Based on the understanding gained from these and other literature results, the following mechanism is proposed. Pt dissolves at the cathode when the potential is close to 1 V and Pt ions diffuse through the ionomer and into the membrane due to the concentration gradient. At the Pt nuclei sites in the membrane, oxidation of crossover H_2 (from the anode) and

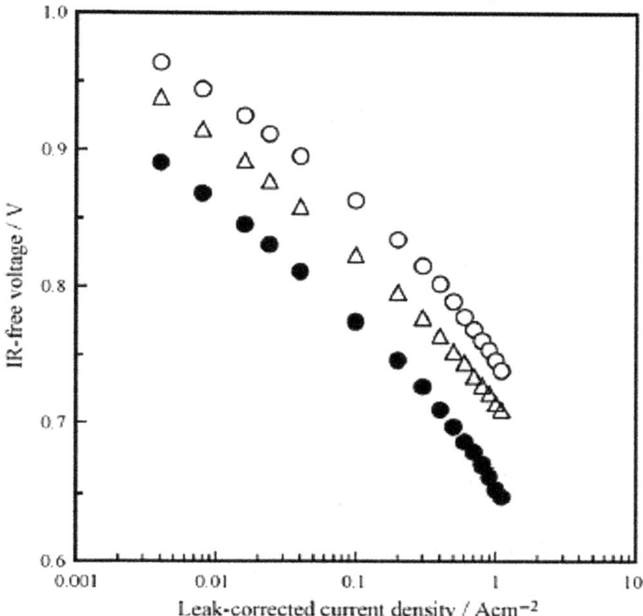

FIGURE 3.45 Tafel plots showing the loss in activity after the OCV hold durability test and after the recovery operation. Solid circles = after OCV; hollow triangles = after recovery; hollow circles = before OCV. *(Reprinted with permission from Sugawara et al. [87])*

reduction of cross-over O_2 (from the cathode) take place. Pt deposits and re-dissolves in a repetitive fashion in the membrane and travels across the membrane until it finally forms a stationary band at a location where the mixed potential at the Pt band is closer to 0. The process of reduction of Pt ions can be written as:

$$Pt^{2+} + H_2 = Pt + 2H^+ \tag{3.22}$$

The hypothesis for membrane degradation under OCV involves the formation of hydroxyl radicals in the vicinity of the Pt band. Hydroxyl radicals can attack both the main and side chain of the polymer to generate decomposition products of fluoride and sulfate that are observed in the effluent water. This process can be expressed as:

$$H_2O_2 + M^{2+} = M^{3+} + {}^{\bullet}OH + OH^- \tag{3.23}$$

As the membrane slowly decomposes, the crossover of gases continually increases contributing to a voltage drop and eventually may lead to catastrophic failure defined as a level of cross-over that is unacceptable (~50 mA/cm^2). Other factors that contribute to the degradation rate are humidity and

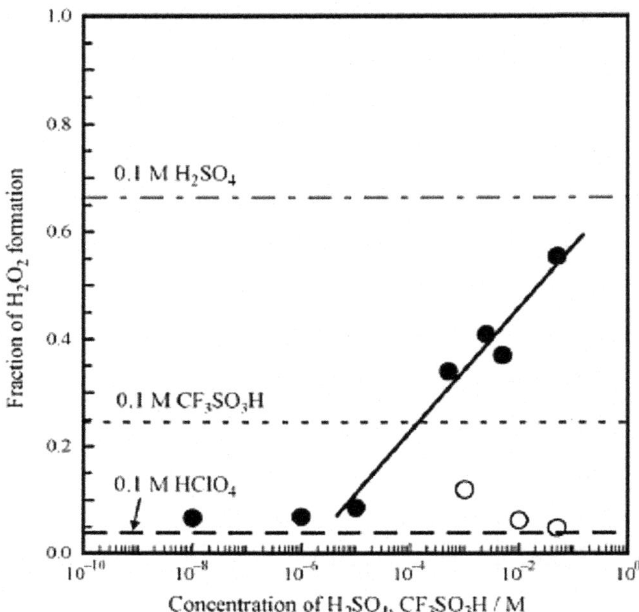

FIGURE 3.46 Fraction of peroxide formation at 0.60 V vs. RHE in 0.1 M $HClO_4$ with the introduction of different concentrations of H_2SO_4 and CF_3SO_3H. *(Reprinted with permission from Tsujita et al. [201])*

temperature. Higher humidity dilutes the generated peroxide and retards the decomposition of the membrane while elevated operating temperatures for obvious kinetic reasons accelerate the decomposition of the membrane.

4.2.2. Load Cycling

During what is considered to be normal operation of a vehicle, the automotive fuel cell stack has to endure about 300,000 cycles composed of diverse load profiles under varying conditions. The conditions endured by the stack are governed by the driving style of the vehicle operator, the driving conditions (gradients, traffic, highway passing, etc.), the environmental conditions (temperature, altitude, etc.) and vehicle system control management. A number of automotive drive cycles (FUDS cycle, US06, NEDC, and Japanese 10–15 mode drive cycles, etc.) are available to simulate statistically representative usage of a vehicle and are used to determine the mileage of ICE vehicles. These drive cycles can be transformed into power-time profiles based on vehicle specs and further into current-time and voltage-time profiles (based on the I-V performance of a specific stack) and applied towards the estimation of degradation rates of PEMFCs under simulated driving conditions (Fig. 3.48(a)).

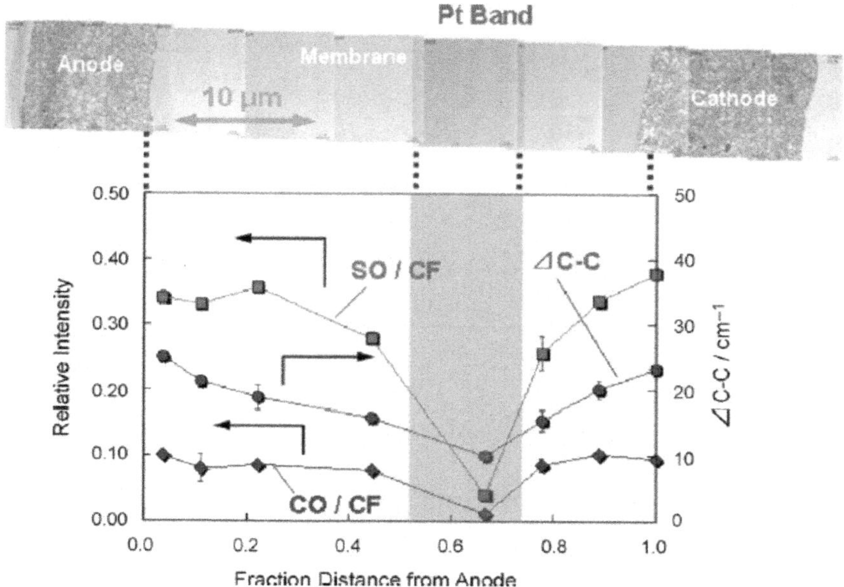

FIGURE 3.47 FWHM of C-C, and relative intensity ratios of the C-O/ C-F and S-O/ C-F (from micro-Raman spectra) as a function of normalized distance from the anode for sample 1 [208].

The largest amplitude of a cycle is about 0.60–0.95 V where 0.95 V represents idling and 0.60 V peak power. The upper limit of the voltage is determined fundamentally by the choice of catalyst and membrane (thickness/ permeability) which determine the OCV, and in part by the design of the vehicle system controls. The lower potential limit is usually selected with the goal of maintaining an electrical efficiency at peak power that does not fall below ~55%.

A generic cycle profile is illustrated in Fig. 3.48(b), representing a superset of all possible profiles. The elements of the profile that can be modulated are:

i. duration at low potential hold,
ii. duration of the ramp-up from low to high potential (or ramp-up rate),
iii. duration at high potential hold,
iv. duration of the ramp-down from high to low potential (or ramp-down rate),
v. magnitude of low potential, and
vi. magnitude of high potential.

The high and low potential magnitudes can have infinitely different values lying between the upper and lower limit in the range 0.6–1.0 V.

In order to gain both a practical estimate and a fundamental understanding of performance/activity losses of cathode catalysts under load cycling, it is worthwhile and indeed preferable to select and study the effect of simplified

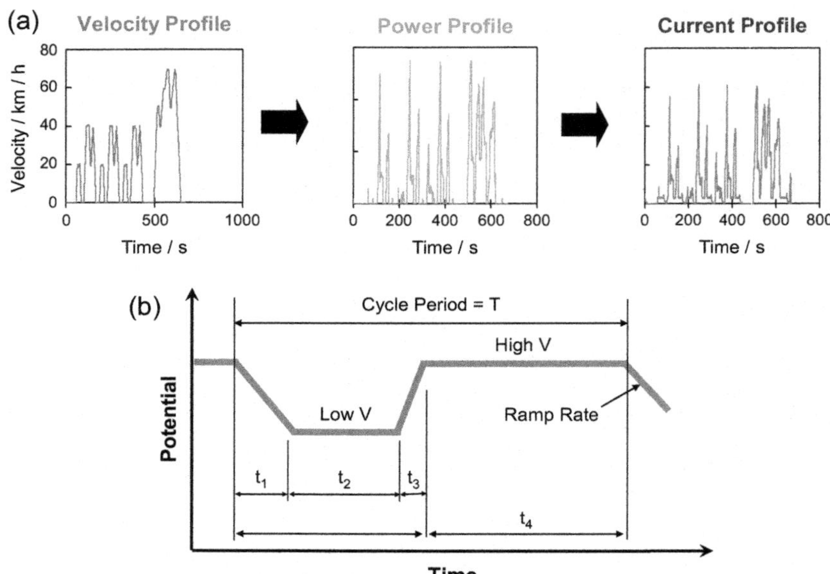

FIGURE 3.48 (a) Conversion of typical drive cycle to a voltage-current profile and (b) generic profile for load cycling that depicts the various elementary sections of the cycle that can be varied and affect electrode degradation under automotive conditions of PEMFC operation.

cycles with careful control of the elementary parameters mentioned above. Such characterization of Pt dissolution/degradation under elementary potential cycles may allow us to determine the specific cycle parameters and operating conditions that cause the greatest damage, and provide pathways to improve stack life. Before investigating the effect of cycle profile and mechanism of degradation, we investigated the effect of parameters such as the effect of potential span, oxygen partial pressure, temperature and humidity. The results from these experiments will allow us to select representative conditions that can be used as a baseline for studies involving a variable cycle profile. Some prerequisites for selection of representative conditions are reasonable potential range corresponding to vehicle operation and operating conditions that selectively isolate the catalyst degradation from membrane degradation.

Effect of Partial Pressure Potential cycling studies are typically conducted in fuel cells with pure H_2 flow maintained on the anode flow fields. This essentially allows us to investigate the cathode catalyst layer degradation with the anode potential maintained close to RHE of 0 V and suffering minimal degradation. The flows on the cathode side may be N_2, O_2, or air. The use of air on the cathode permits one to carry out studies under fairly realistic fuel cell conditions where the cell produces power, and water and heat are generated. But there are also several disadvantages, in that it is difficult to set the potential of the cathode to a precise fixed value over time; one can typically only set the cell voltage (rather than the resistance-free voltage) to a specified value making it difficult to investigate kinetics. In addition, for long term unmonitored tests under air, there may be issues caused by catastrophic failure of the membrane. On the other hand, with N_2 on the cathode, it is possible to set the potential precisely for measurements of kinetics and conduct some tests outside the normal fuel cell operating potential range. Thus, a number of researchers choose to carry out studies employing N_2 on the cathode. It thus becomes necessary to verify whether the degradation rate of the catalyst is influenced by the gas composition on the cathode.

Figure 3.49 shows the loss in ECA for two cells operated under the same conditions except for the cathode gas composition (N_2 or air). Based on the degradation rates, no significant effect of the substitution of gases is observable. This is despite the fact that the TEM micrographs (Fig. 3.50) indicate that the Pt dissolved from the cathode has traveled deep into the membrane when air is used, whereas it gets deposited close to the cathode/membrane interface when N_2 is used. These tests were both conducted at 100% RH, therefore, the results may differ for lower RH studies since the cathode water content would be higher when air is employed raising the degradation rate.

Recently, Sugawara et al. [209] have attempted to measure the CV simultaneously with the ORR polarization curve. Using the shielding technique, they found that the oxide peak profile in the extracted CVs was not affected significantly by the presence of oxygen in the atmosphere, lending support to the hypothesis that oxygen from the reactant gas does not play a critical role in the oxide coverage and hence in the degradation rates. Based on this evidence, we report on cycling experiments that were conducted with the cathode under nitrogen flow.

Effect of Potential Cycling Span Figure 3.51 shows a plot of a CV ($H_2|N_2$) superimposed over the plot of an I-V ($H_2|Air$) polarization curve. It is apparent that over the range of potentials where cycling can occur, the currents attributed to oxide adsorption (as well as the nature of the species and structure) vary considerably. At the low potential of ~0.60 V Pt is often considered to be almost oxide free, while at 0.95 V the coverage is quite high. In the range 0.80–0.95 V, the surface is always covered with a certain amount of oxide species.

In our attempt to unravel the potential regimes that are the most deleterious for Pt dissolution, we applied the waveforms shown in Fig. 3.52(a). We varied the lower potential E_L while keeping the upper potential constant at $E_H = 0.95$ V (Fig. 3.52(b)) and also varied the upper potential E_H while maintaining the lower potential at $E_L = 0.60$ V (Fig. 3.52(c)). We observed that the losses in ECA and mass activity i_m are the highest (~70%) for cycling in the widest potential span

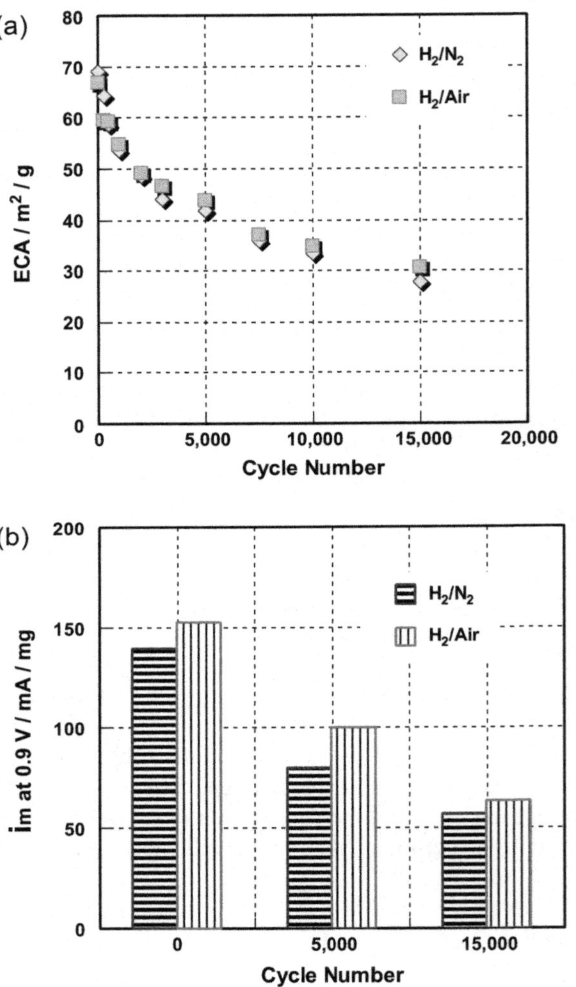

FIGURE 3.49 ECA (a) and mass activity (b) for MEAs cycled under anode|cathode flows of $H_2|N_2$ and $H_2|$Air. Potential cycling was carried out for 15,000 cycles, 80°C, 100 %RH with intermittent interruptions for diagnostics.

of 0.60–0.95 V. In restricted potential regimes such as 0.60–0.80 V and 0.95–0.80 V where the Pt surface is either covered with oxide species or relatively free of them, the losses are suppressed to ~20%. Therefore, it is when the Pt is subject to potential cycles that cover the entire span of oxide-free and oxide-covered potential regimes that the losses are found to reach precipitous levels. Thus a choice of 0.60–0.95 V as the potential span for many of the durability experiments reported in our studies was influenced by both practical and fundamental aspects.

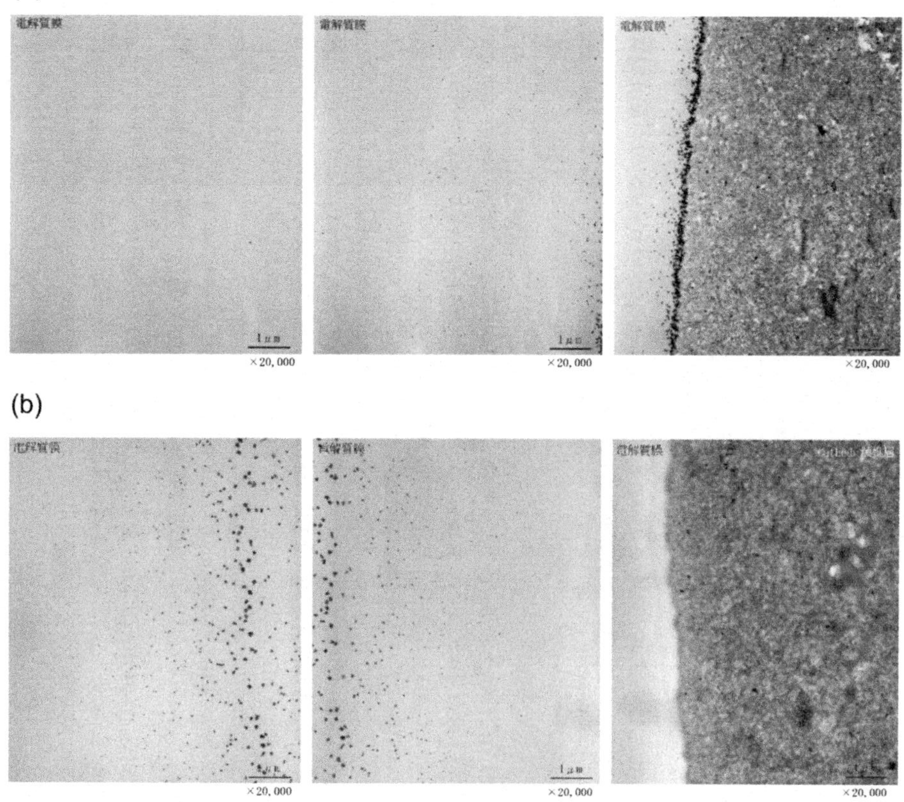

FIGURE 3.50 TEM for degraded MEAs extracted from cells after cycling under $H_2|N_2$ (a) and $H_2|Air$ (b) for 15,000 cycles. A Pt band at the cathode-membrane interface is observed for the case of $H_2|N_2$ cycling and a band closer to the center of the membrane for the $H_2|Air$ case.

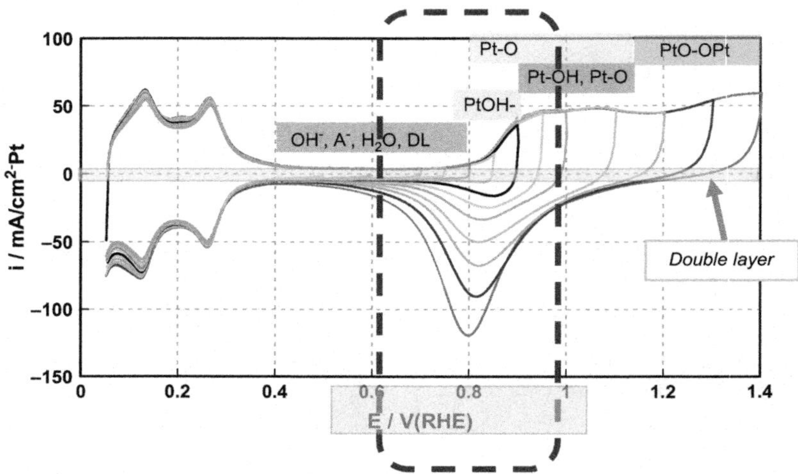

FIGURE 3.51 CVs of Pt/C obtained at different potential ranges (0.05–0.60 to 0.05–1.4 V) illustrating the signature pseudo-capacitive peaks due to HUPD and oxide species superimposed over an apparent constant double layer current. Typical automotive operation includes cycling in the potential window of 0.60–0.95 V.

Effect of Temperature Although it is not feasible to operate an automotive PEMFC at a given fixed temperature, the effect of temperature on the degradation rate is vital; first, in order to obtain values that can be used in empirical predictive models and secondly to gain insight into the mechanism itself. In our studies (Fig. 3.53), we examined the effect of temperature in the range 50–90°C at 100% RH where conventional PFSA membranes are stable. Clearly, we see that the degradation rate of the surface area is higher at elevated temperatures. We obtained fits to the curves based on the equation used by Bett et al. [60,61]. At this time the relevance of the fit parameters is not fully understood and cannot be readily correlated to the mechanism of degradation. Based on the fit values, we obtained the activation energy from Arrhenius plots to be 40 kJ/mol.

Effect of Humidity Practical automotive PEMFCs are typically operated at less than 100% inlet RH. On examining the effect of humidity on the loss in ECA as illustrated in Fig. 3.54, we find that the catalyst degradation is lower at lower RH; the Pt band at the interface is the strongest for 100% RH. The implications of this result for electrocatalyst degradation are enormous; catalyst degradation can be significantly suppressed if one could operate a fuel cell at extremely low RH. Essentially, dissolution and diffusion of Pt is to be suppressed under low RH conditions. State-of-the-technology PEMFCs for automotives are operated at moderate humidity for the purpose of maintaining ionomer/membrane conductivity and durability with the concomitant cost of complex water management issues. If the search for a membrane that has reasonable protonic conductivity at low RH is successful, the catalyst degradation problem would be

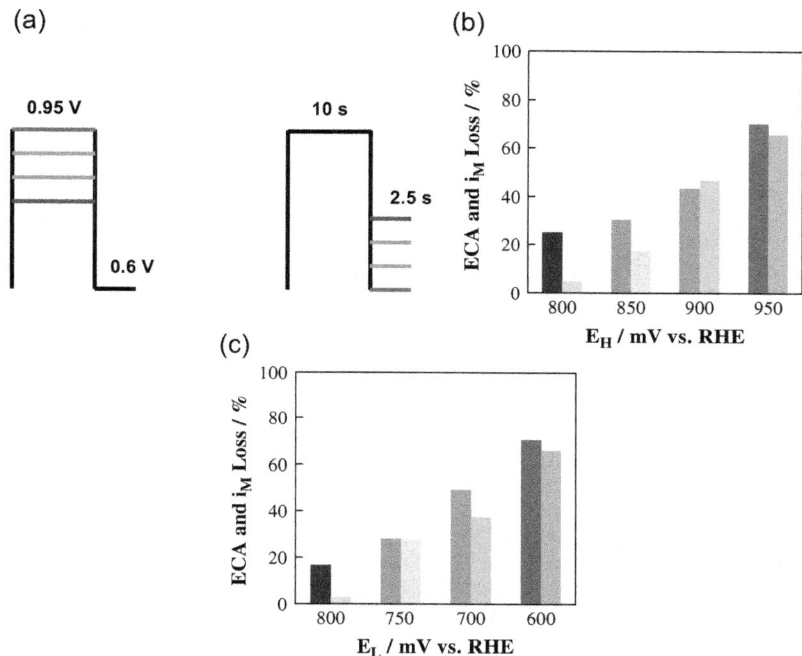

FIGURE 3.52 (a) Square wave profiles employed in cycling experiments (25 cm^2 subscale cells, catalyst loading of 0.35 mg/cm^2) conducted under $H_2|N_2$ (Anode|Cathode) at 80°C with hold times of 10 s at E_H and 2.5 s at E_L. (b) Effect of upper potential hold "E_H" (lower potential fixed at 0.60 V) on cycling durability (c) Effect on lower potential hold "E_L" (upper potential fixed at 0.95 V) on cycling durability. *(Uchimura et al. [11])*

significantly ameliorated. We note that this effect also applies to the cycling that occurs in the start-up/shut-down regime of 1.0–1.5 V; drying the stack before shut-down would help in significantly lowering the losses on start-up [154]. Based on these results, we have employed the harshest conditions of 100% RH in the cycling experiments performed in this section.

Effect of Cycle Profile Before undertaking a study on the effect of cycle profile, one needs to clarify the method of analysis of the degradation losses. For a given cycle profile, the degradation is usually higher per unit time compared to a steady state hold at a fixed potential in the same range. In other words, significant contributions to degradation arise from the 'ramp-up' or up-sweep and 'ramp-down' or down-sweep legs of a cycle. Figure 3.55 is a schematic of square and triangular cycle wave profiles having periods of 5 s and 10 s. Figures 3.56(a) and (b) shows the degradation per cycle and per unit time for respective cycles. As we can observe in Figs 3.56(a) and (b), in the selected time range of 5–10 s, the 'degradation/cycle' are almost identical while the 'degradation/h' increases at higher cycling frequencies. The degradation rate/

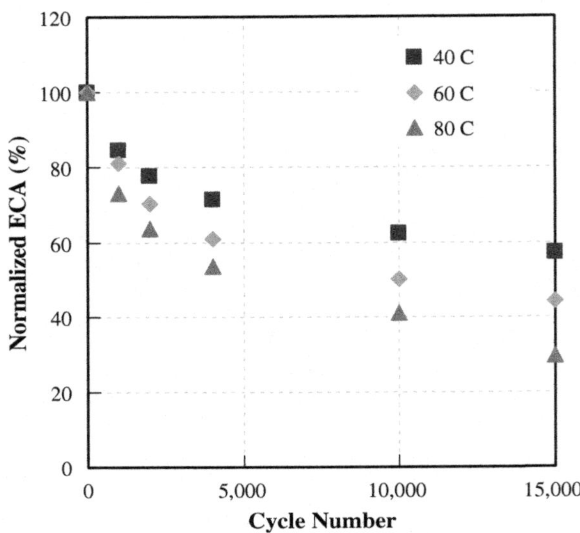

FIGURE 3.53 Effect of temperature on the degradation of the catalyst surface area over 15,000 cycles. Square wave profiles were employed in all cycling experiments (25 cm^2 subscale cells, catalyst loading of 0.35 mg/cm^2) conducted under $H_2|N_2$ (Anode|Cathode) at 100% RH with hold times of 10 s at E_H and 2.5 s at E_L.

cycle is obviously the appropriate parameter to be studied in order to understand the effect of cycle profile on degradation.

We note from Fig. 3.56(a) that degradation rate was significantly higher for the square wave than the triangular wave. The higher degradation rate for square wave cycles could be due to the faster ramp-up and ramp-down rates as well as the 'hold time' at the upper (E_H) and lower potentials (E_L). The time duration at the high (0.95 V) and low (0.60 V) potential hold for square waves was investigated; increasing upper potential hold times resulted in increasing losses until about ~30 s, whereas the losses for lower potential hold times reach a maximum in ~3 s correlating to the short times for reduction of oxide species. As the hold time at low potential was systematically lowered to <0.5 s degradation tended to diminish (to values corresponding to a constant potential hold) since the oxides formed during the upper potential hold were not completely reduced and the surface was maintained in a passivated state.

Having already observed the higher losses for square wave profiles as compared to triangular profiles, we refined the study to examine the losses for symmetric and asymmetric triangular cycle profiles with the objective of deconvoluting the effect of the ramp rate. Figure 3.57 depicts the effect of slow and fast anodic ramp rates ((70/70), (35/7,500), and (7,500/35)); the degradation was the highest for the fast anodic/slow cathodic sweep.

(a)

Normalized ECA / %

Cycle Number

■ 30% ◆ 70% ▲ 100%

(b)

100% RH 70% RH 30% RH

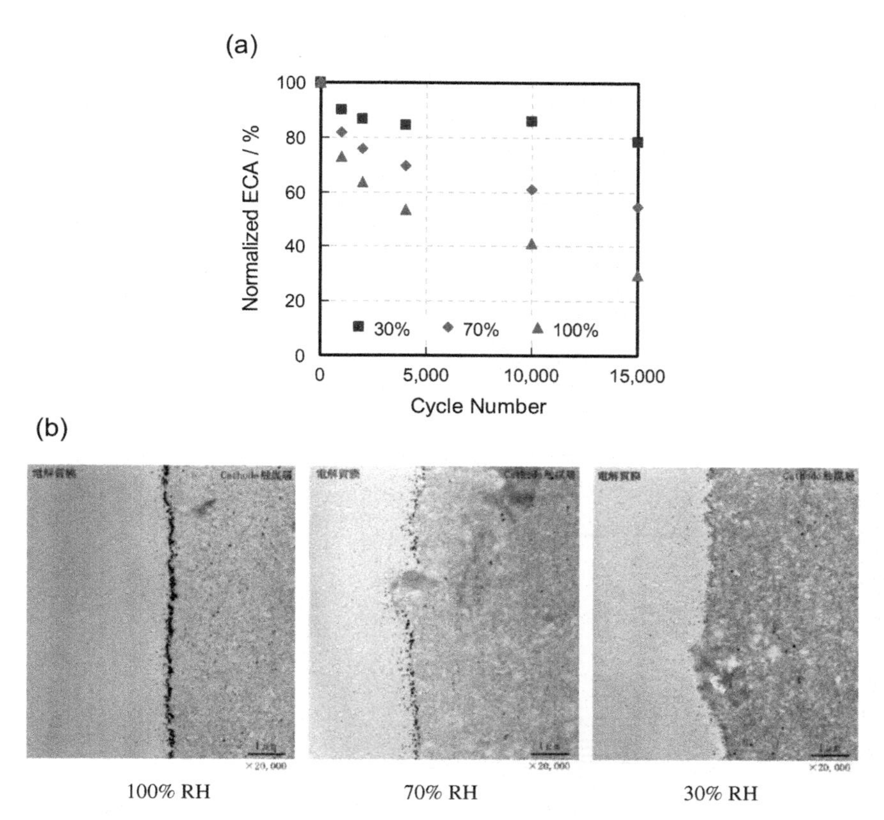

FIGURE 3.54 (a) Plot of normalized ECA vs. cycle number at 30%, 70% and 100% RH; (b) TEM images of the cathode-membrane interface corresponding to 100%, 70%, and 30% RH. Square wave profiles were employed in all cycling experiments (25 cm^2 subscale cells, catalyst loading of 0.35 mg/cm^2) conducted under $H_2|N_2$ (Anode|Cathode) at 80°C with hold times of 10 s at E_H and 2.5 s at E_L.

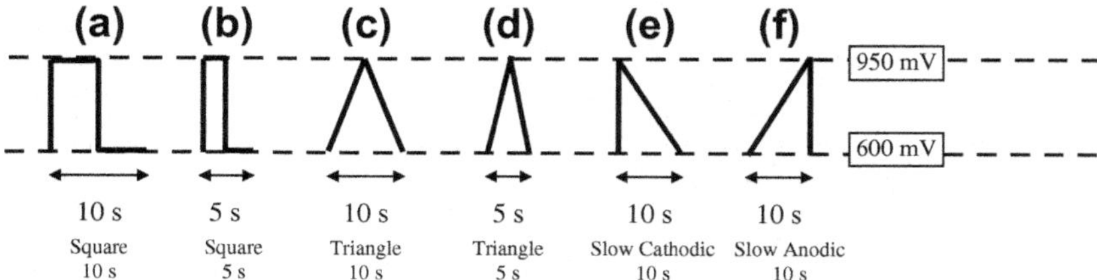

FIGURE 3.55 Schematic of voltage cycle profiles used in this study with upper potential limit of 0.95 V and lower potential limit of 0.60 V. (a) square 10 s; (b) square 5 s; (c) triangle 10 s; (d) triangle 5 s (e) asymmetric triangle, slow cathodic 10 s; (f) asymmetric triangle, slow anodic, 10 s.

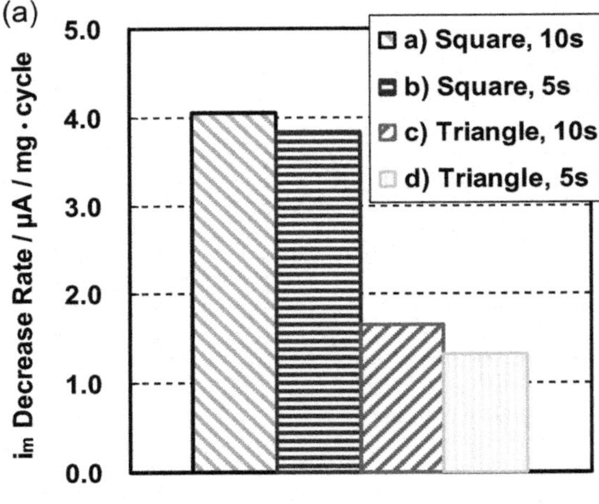

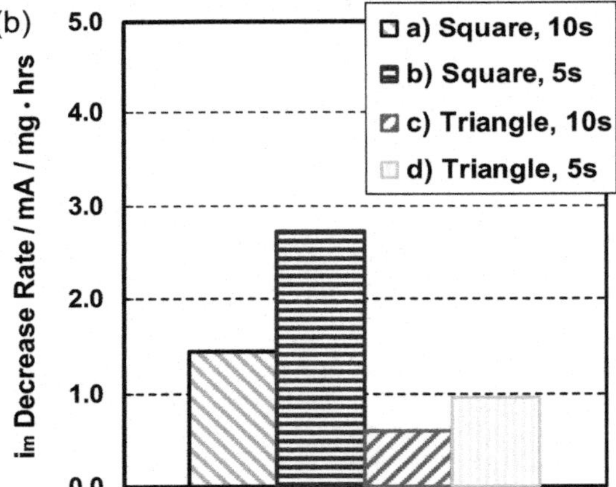

FIGURE 3.56 (a) Plot of mass activity loss/cycle for symmetric square and triangular cycle profiles for cycle periods of 10 s and 5 s. (b) Plot of mass activity loss/h for symmetric cycle profiles for cycle periods of 10 s and 5 s. Data collected at 80°C, 100% RH in 25 cm^2 subscale cells having a cathode catalyst loading of 0.35 mg/cm^2. *(Uchimura and Kocha [10])*

The actual profile shape in automotive drive cycles is complex, but portions of it can be simulated with cycle profiles that incorporate elements of both square and triangular waveforms similar to the generic profile shown in Fig. 3.48(b). It is fairly difficult to determine and isolate the contributions to degradation of each of the elements of such a wave profile, since the degradation that is taking place is dependent on the state of the Pt surface that is governed by the previous element. Another aspect is our inability to maintain a constant 'total cycle time' for variable ramp rates while at the same time maintaining a constant hold time. Figure 3.58 illustrates the cycle profiles and corresponding losses in surface area as the anodic ramp rate is systematically increased. We find that the faster anodic ramp rates result in greater degradation

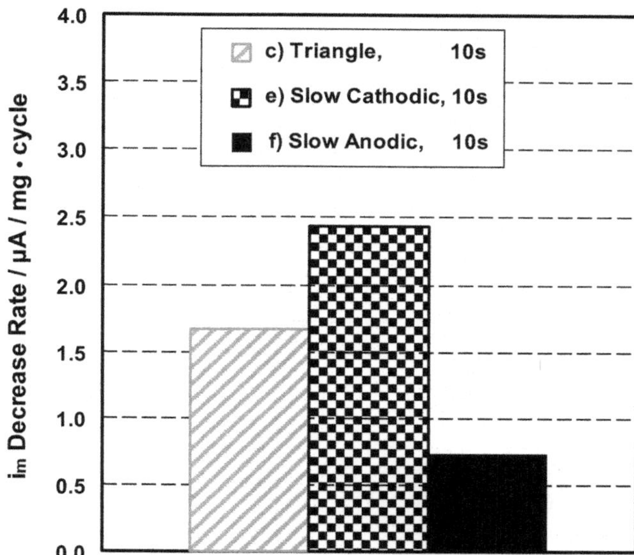

FIGURE 3.57 Mass activity loss per cycle for (c) asymmetric triangular profiles, 10 s (e) slow cathodic, 10 s and (f) slow anodic, 10 s. Data collected at 80°C, 100% RH in 25 cm^2 subscale cells having a cathode catalyst loading of 0.35 mg/cm^2. *(Uchimura and Kocha [10])*

of the surface area. A more complex trend is observable when variable cathodic ramp rates are systematically studied. The above results on the impact of cycle profile on degradation rate have practical implications in automotive applications; in a hybrid FCV, the battery may be used to augment or absorb power (as part of a control system strategy) in an attempt to lower the degradation rates.

Dissolution Mechanism It has been reported sporadically in the literature that the onset of oxides shifts positively for catalysts with enhanced activity. Figure 3.59, for example, illustrates that the signature 'onset of oxides' in CVs for poly-Pt disk is shifted positively vs. Pt black; likewise, the 'onset of oxides' for Pt black is more positive than for baseline Pt/C. A positive shift in the onset of oxide species implies a more oxophobic surface, generally attributed to a smaller number of edge and corner sites. We note that the specific activity for the three materials measured in perchloric acid follow the trend: poly-Pt disk (~1,500 $\mu A/cm^2$)>Pt black (~500 $\mu A/cm^2$)>Pt/C (~200 $\mu A/cm^2$) corroborating the particle size effect. It is not clear at this time if the particle size effect depends to an equal extent on the total number of atoms to produce a 'bulk-like' effect or only on the 'degree of extension' of a film.

Figure 3.60(a) illustrates a similar trend in the shift of oxide onset for Pt-alloy/C in comparison to Pt/C. Figure 3.60(b) further demonstrates that the Pt-alloy/C has a lower degradation rate than Pt/C under typical potential cycling in

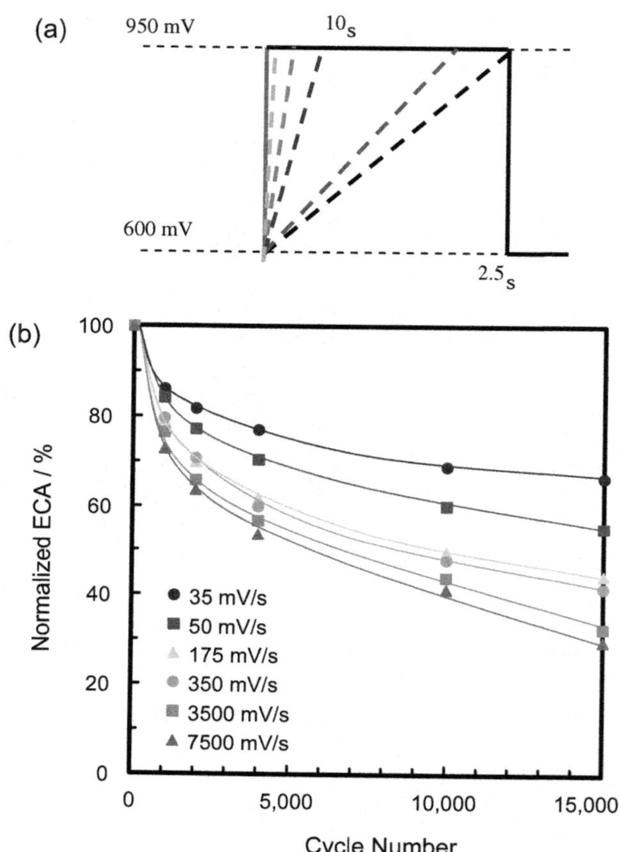

FIGURE 3.58 Effect of anodic ramp rate (for the profiles shown in (a)) on the loss in surface area of the cathode catalyst shown in (b). Data collected at 80°C, 100% RH under $H_2|N_2$ in 25 cm^2 subscale cells having a cathode catalyst loading of 0.35 mg/cm^2.

MEAs of PEMFCs. Thus we find a correlation between the positive shift of the onset of oxides on Pt to the enhancement of specific activity and durability. This is a remarkably fortuitous finding, in that, there does not appear to be a conflict in realizing a catalyst with the simultaneous improvement in both specific activity and durability.

If we examine a simple cyclic voltammogram of Pt/C for a triangular voltage sweep, we observe that the typical reduction peak for the oxide species is more negative than the oxidation peak (Fig. 3.59). The surface species evidently goes through a different series of states during the forward and backward sweeps. This is especially true when the potential is positive enough in the anodic sweep that place-exchange of Pt and O in the topmost surface

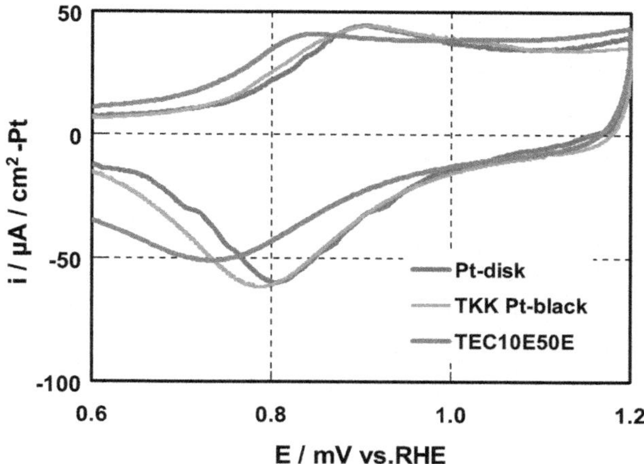

FIGURE 3.59 Shift in the onset of oxide formation observed in the CVs for Pt/C < Pt-black < poly-Pt. All tests were conducted in 0.1 M $HClO_4$ at 25°C, N_2 atmosphere in RDE set-up [215].

layers takes place. Wagner and Ross carried out LEED spot profile analysis of Pt(100) and Pt(111) single crystal surfaces subject to positive potentials >1 V and reported that spot widths changed [210]. They ascribed this phenomenon to the formation of randomly stepped surfaces caused by Pt and oxygen undergoing a place-exchange mechanism. In the place-exchange process, after half or more than half a monolayer is adsorbed on the Pt surface, the Pt and O atoms 'turnover' or 'flip' or 're-organize' due to repulsive forces to accommodate even more oxygen. During the reduction sweep, the Pt atoms displaced from their surface lattice locations, in the surface structures so formed, do not revert to their original positions causing the observed irreversibility. STM data reported by Itaya and co-workers [211] on Pt(111) revealed that after cycling of the samples, randomly oriented monoatomic or diatomic islands were observed on the terraces. The authors also attributed these islands to Pt adatoms produced during the anodic sweep that did not go back to their original positions in the lattice on reversing the potential sweep. Figure 3.61 is a schematic that describes the atomic level surface features before and after cycling.

Based on our understanding of literature and our results on potential cycling in MEAs and liquid electrolytes, we propose that during the first complete cyclic sweep, irregular island/mesa-like structures are formed and the displaced low-coordinated Pt atoms do not revert to their original positions during the reduction sweep. During the following anodic sweep, the atoms at the edges of these islands are obviously more active and susceptible to attack and dissolution. When the anodic sweep is extremely fast, the coverage of oxide species on the Pt surface is slightly lower (due to less time for oxide formation); this results in more bare Pt atoms being exposed and subject to the high potential at the peak of the cycle, accelerating dissolution. In addition, the oxidation stage

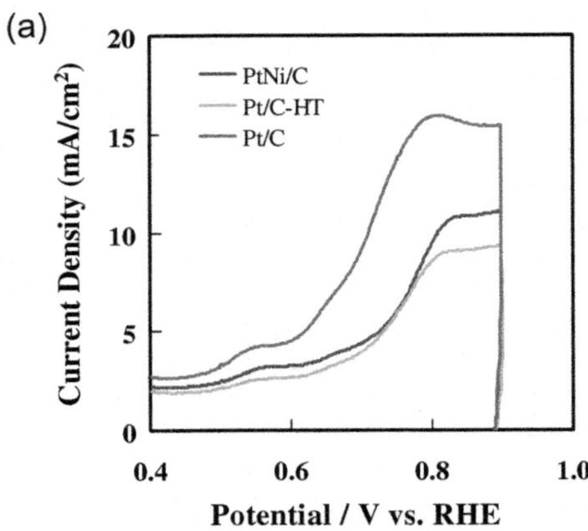

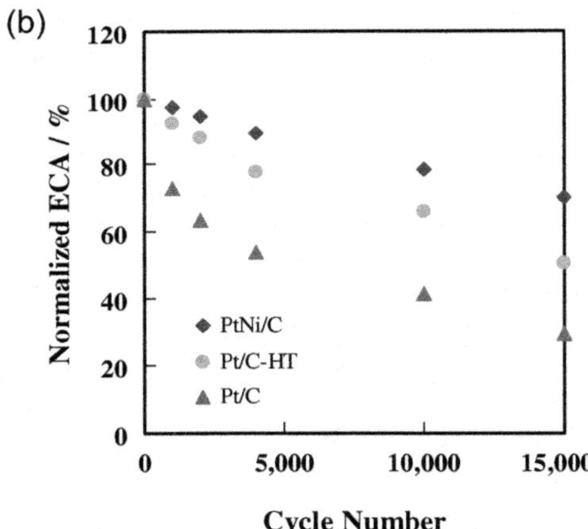

FIGURE 3.60 (a) Positive shift in the onset of oxides in the CVs Pt/C< Pt-HT/C<Pt-alloy/C is shown. (b) Plot of the ECA for the three catalysts before and after cyclic durability tests for 15,000 cycles in the range 0.60–1.0 V in $HClO_4$ shows that the heat treated catalyst Pt/C-HT and Pt-alloy catalyst PtCo/C exhibit improved resistance to potential cycling.

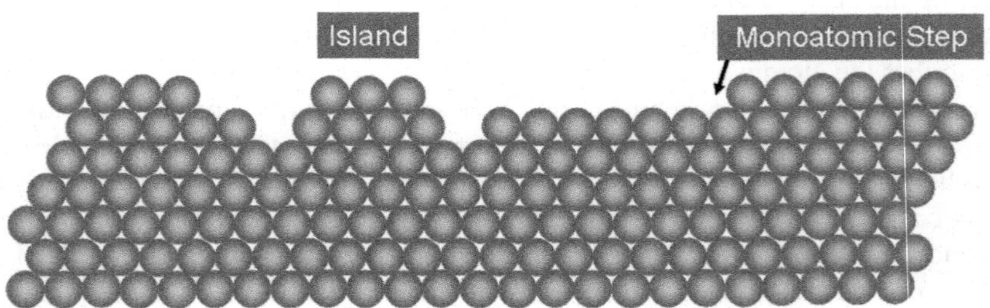

FIGURE 3.61 Schematic derived from STM observations and supported by LEED spot profile intensities illustrating monatomic steps that exist on clean uncycled, unoxidized surfaces as well as irreversibly formed islands after cycling to high potentials. Surfaces with islands possess a larger number of less-coordinated atoms.

in a fast anodic sweep very likely results in a structure that has greater number of smaller islands (less time for diffusion to islands) and is also more irreversible (observed from CVs measured as a function of scan rate). In slower anodic sweeps, corresponding to slow triangular waves, there is more time for Pt ions to diffuse to and stick to islands that already exist, therefore lowering the dissolution rate. In the case of cycling within the oxide regime only, larger islands form on the surface and the inner atoms of the island are protected/passivated from further attack and dissolution.

For binary and ternary alloy catalysts of Pt or heat-treated Pt, we find significantly higher durability than for Pt/C under cycling. Stamenkovic and co-workers have recently reported on the extraordinarily high activity of PtNi(111) versus Pt(111) and also on the higher durability of both PtNi(111) and Pt(111) skin structures [99]. The skin structures have been found to exhibit better tolerance to dissolution under cycling. They attribute this result to the contraction of the topmost surface layer of Pt atoms. Intuitively, such surfaces in which OH or oxide onset is suppressed are also more resistant to corrosion. For high surface area, heat-treated Pt and Pt-alloys on carbon, we suggest a similar mechanism where the surface is more stable due to:

i. heat treatment resulting in less relaxed surface structures; and
ii. alloying element causing a shift in d-band center and shift in onset of oxides.

In any case, the results demonstrating the trend where high specific activity and suppression of Pt dissolution occur together is especially encouraging for improvements in catalyst durability. At this stage of research, the hypothesis proposed above is only partially supported by data and more studies on the nature, species and structure of the oxides on Pt are needed for verification and substantiating the results. Such in-depth research into fundamental phenomena is being rekindled and being pursued in-house as well as in collaboration with laboratories and universities all over the world.

Based on the results obtained so far at both the fundamental and practical level, we attempt to predict the future pathway for obtaining a high activity, high durability catalyst layer that meets the cost targets for automotive commercialization. Figure 3.62 is a plot of mass activity versus number of monolayers of Pt. The curves represent lines of constant specific activity based on measurements for traditional nanoparticles Pt/C and Pt-alloy/C as well as bulk or poly-Pt disk and Pt-alloy catalysts. At this time, the bulk materials have shown superior specific activity and dissolution-resistance compared to nanoparticles. The question arises 'Can we harness these high specific activity materials by applying them as extended thin films (ETFECS) that also possess high mass activity?' In principle it appears that about three monolayers of Pt-extended thin-film should suffice to achieve the mass activity that is currently obtainable from 2–4 nm nanoparticles Pt/C or Pt-alloy/C. Keeping in mind the place exchange mechanism, we have arbitrarily assumed that three monolayers

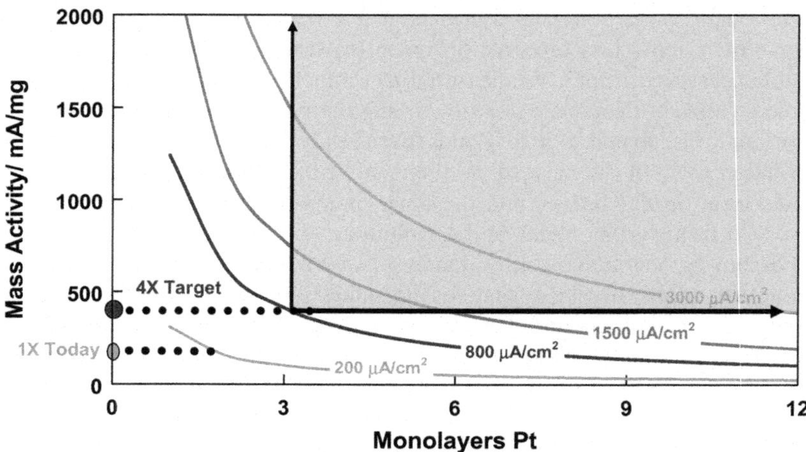

FIGURE 3.62 Mass activity vs. monolayers of Pt plotted for 4 curves of constant specific activity corresponding to state-of-the-art carbon supported nanoparticles Pt/C (green = 200 μA/cm^2), DOE target value for carbon supported nanoparticle Pt-alloy/C (blue = 800 μA/cm^2), bulk films of polycrystalline Pt (red = 1,500 μA/cm^2) and bulk polycrystalline Pt-alloy (orange = 3,000 μA/cm^2).

is the minimum thickness required for catalyst stability. If we can obtain the high specific activity that is now measured for bulk Pt-based materials while utilizing a film constituted of only a few monolayers, we should be able to achieve even higher mass activities than the commercialization requirements. Recent results indicate that ETFECS can be prepared with ECA exceeding 30 m^2/g and mass activities higher than conventional Pt/C catalysts. A major challenge in accomplishing this task is likely to be the ability to use an extended film in an electrode structure that permits high mass-transport of reactants and water and allows high oxygen limiting currents similar to that obtainable in state-of-the-art 10 μm Pt/C catalyst layers.

4.2.3. Contamination

Contamination of PEMFCs is an important topic since components of the fuel cell, such as high surface area carbon blacks and the PFSA ionomer, can act as filters and easily absorb impurities from degradation/corrosion products generated by the cell components as well as from the fuel and atmospheric air. Pt is readily susceptible to the adsorption of various contaminants that may reversibly or irreversibly occupy the sites necessary for the slow 4e$^-$ ORR unless the contaminant is oxidized or flushed off from its surface. The severity of the effect of impurities on Pt/C and the reversibility of the losses incurred depend on the specific contaminant. The main focus of this section is the effect of commonly found contaminants present in ambient air on the cathode Pt catalyst. Some such contaminants are sulfur compounds such as SO_2 and H_2S, and nitrogen compounds such as NO_x. The sources of these contaminants are pollutants produced by ICE vehicles as well as from industrial plants and power generation plants and from regions that have hot springs. In Japan, for example, the average concentrations of SO_2 and NO_2 derived from monitoring roadside pollution are 4 ppb and 27 ppb.

Experimental set-ups to evaluate the effect of contaminants essentially consist of the source of a specific impurity that is mixed in with an inert gas such as N_2 or air and introduced into the fuel cell at a location after the humidifier so as not to contaminate the test-stand itself. Special precautions have to be taken to use extremely high purity hydrogen (99.9995%), oxygen (99.9999 %) and nitrogen (99.9999 %) for obvious reasons. Several kinds of diagnostic experiment are typically conducted to quantify and understand the mechanism of contamination and recovery methods if any. They are:

i. steady-state hold experiments at a constant current (of say 1 A/cm^2) under H_2|Air (w and w/o contaminant); these conditions allow the evaluation of impurity effects under somewhat realistic fuel cell conditions that are interrupted intermittently by diagnostic I-V polarization curves and CVs;

ii. steady-state holds at consecutive constant potentials in the range (0.6–0.95 V) under $H_2|N_2$ followed by CVs; these experiments are designed to investigate the kinetics of the adsorption and de-sorption of impurities that may manifest as features in the CVs; and

iii. contaminating a sample followed by a series of I-Vs or CVs until it is de-contaminated.

We here briefly discuss the key results of a study on the effect of contaminants including SO_2, H_2S, NO_2, NO and NH_3 on the durability of cathode catalyst carried out by Nagahara et al. [212]. Figure 3.63 shows the effect of 0.5 ppm SO_2, 0.6 ppm H_2S, 2 ppm NO, 2 ppm NO_2 and 5 ppm of NH_3 on the iRFree voltage of a cell at 1 A/cm^2 versus the cumulative amount of contaminants injected into the air stream. This confirms that the loss in performance due to the sulfur compounds is quite severe compared to the nitrogen compounds.

Further investigation was carried out into the degradation caused by SO_2. Figure 3.64 shows the impact of 50 ppb of SO_2 on the performance at 1 A/cm^2; a loss of 96 mV was observed after 45 hours of operation. Simply by following this durability test with an H_2|Air polarization curve in the absence of contaminant injection, the performance recovered by ~57 mV. Subsequently, by initiating a high humidity operation at 1 A/cm^2 for 2 hours an almost complete recovery of performance was observable. The complete I-V polarization curves (H_2|Air and $H_2|O_2$) taken before and after the durability test (not shown here) showed a voltage loss over the entire current regime indicating that the losses were kinetic and not related to mass-transport. Cyclic voltammograms showed a positive shift in the onset of oxide after the durability test and a movement

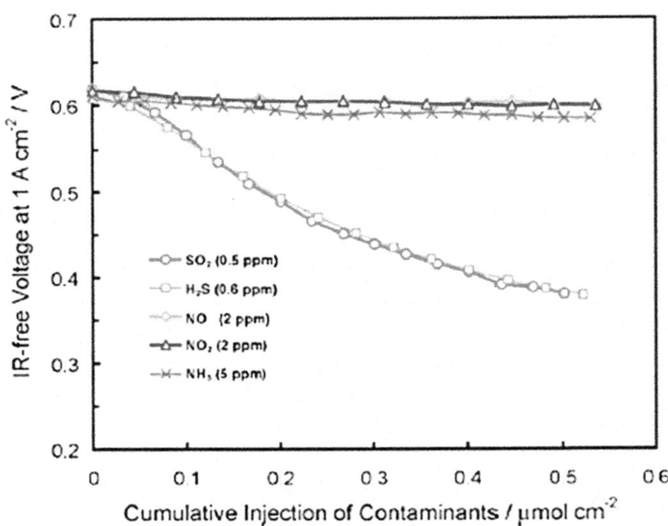

FIGURE 3.63 Impact of various atmospheric impurities including SO_2, H_2S, NO, NO_2 and NH_3 on the iRFree voltage of PEMFCs measured at 1 A/cm^2. *(Reprinted with permission from Nagahara et al. [212])*

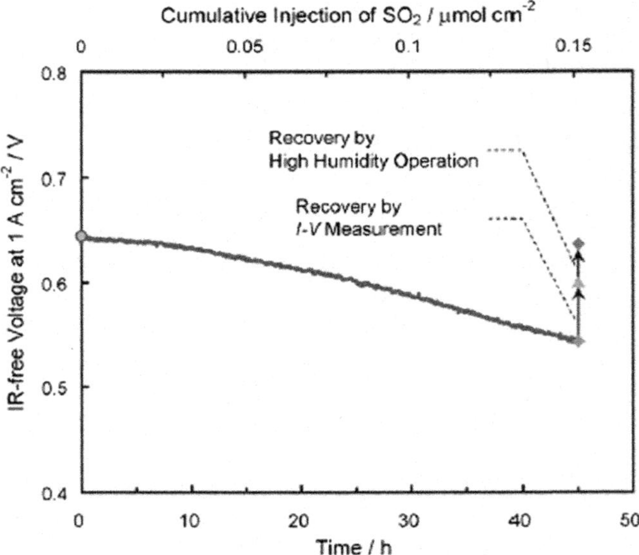

FIGURE 3.64 Negative impact of 50 ppb of SO_2 on the iRFree voltage with time over 50 h followed by partial recovery on carrying out an I-V polarization curve and subsequently by a high humidity operation. *(Reprinted with permission from Nagahara et al. [212])*

back towards negative potentials after recovery operations hinting at the interaction of oxide species with the contaminant.

In order to investigate the mechanism of contamination in more detail constant potential durability tests were conducted at various potential holds. Figure 3.65 shows the cyclic voltammograms for two extreme cases of potential holds at 0.5 V and 0.9 V as a function of hold time in the range 0–180 min.

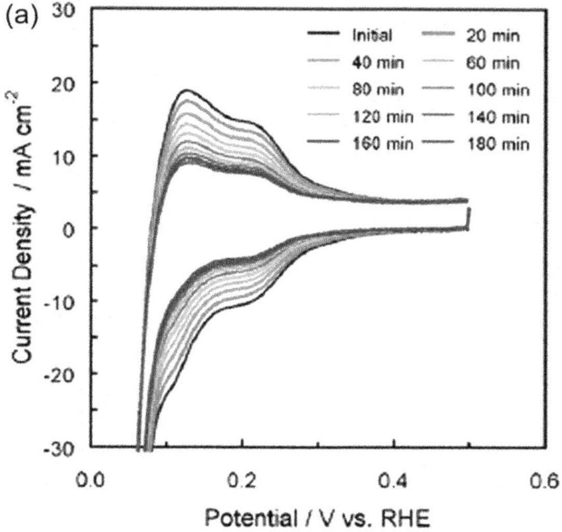

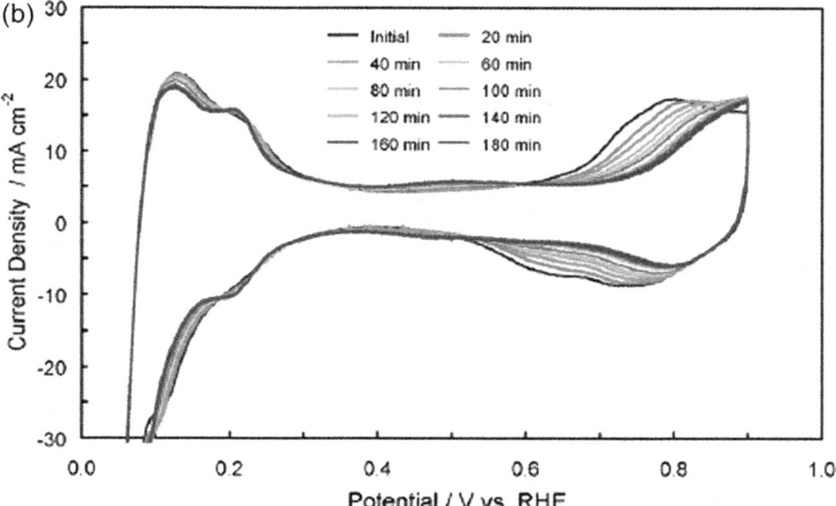

FIGURE 3.65 Cyclic voltammograms for two extreme cases of potential holds at 0.5 V and 0.9 V as a function of hold time of 0–180 min. *(Reprinted with permission from Nagahara et al. [212])*

Holding at 0.5 V resulted in a significant reduction in HUPD peaks whereas holding at 0.9 V only showed the positive shift of the onset of oxide formation with little effect on the HUPD peaks. These results indicate that a potential hold of 0.9 V and higher can oxidize the Pt-S to sulfate, thus freeing up active sites of Pt for the ORR.

$$\text{Pt-S} + 4\text{H}_2\text{O} \leftrightarrow \text{SO}_4^{2-} + 8\text{H}^+ + 6\text{e}^- + \text{Pt} \qquad (3.24)$$

It should be noted that in a practical fuel cell it will not be possible to reach potentials higher than the OCV, which is around 0.95 V on the cathode electrode. The authors postulate a mechanism in which sulfur dioxide is adsorbed strongly on Pt causing a loss in active sites and corresponding kinetic losses. The adsorbed sulfur can be oxidized by excursions to potentials above 0.90 V to sulfates and a high humidity operation can flush out the sulfates resulting in a complete recovery of performance. These results indicate that occasional low doses of SO_2 can be handled by PEMFC stacks without suffering serious permanent degradation.

Briefly, we also illustrate the impact of trace concentrations of chloride ions on the cyclic voltammograms of Pt measured in sulfuric acid in RDE experiments. Cl^- ions are often present in small quantities in the electrolyte or Pt catalyst itself, through contamination of glassware in half-cells, from the atmosphere, or generated by the decomposition of electrolytes ($HClO_4$) used. Figure 3.66 shows that the onset of oxides is shifted positively and the charge due to oxides is lower as chloride ions adsorb on the surface of Pt. Not shown is

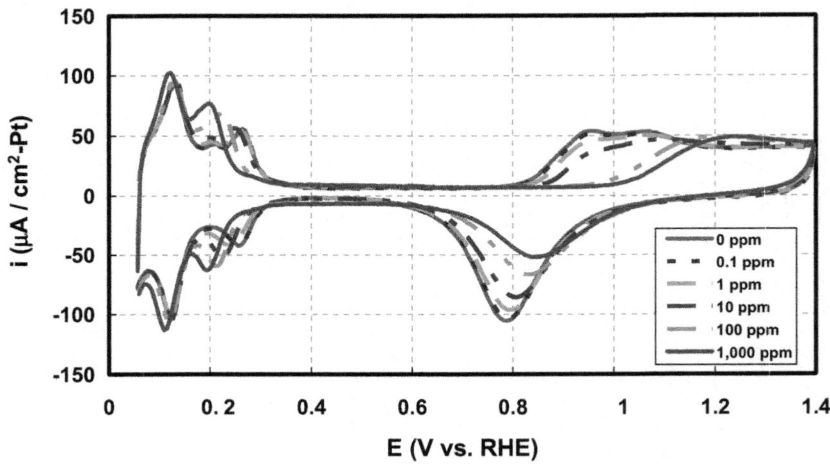

FIGURE 3.66 Changes in the HUPD and oxide regimes of Pt CVs when various trace concentrations of Cl^- ions are introduced into the electrolyte in RDE experiments conducted at 25°C in H_2SO_4 for scan rates of 50 mV/s.

the well-known effect of loss in ORR activity easily observable in I-V curves. Electrodes contaminated with chloride impurities lose catalytic activity that is completely recoverable when the impurity is eliminated. In fuel cells, operation at high current densities for short periods of time cleanses the electrode of the contamination and recovers performance.

4.2.4. Start-up Shut-down

In the literature review section, we discussed the thermodynamics and kinetics of carbon corrosion, the characterization of carbon support materials in acid electrolytes as well as PEMFCs and the scope and requirements for new and improved materials that might replace carbon blacks. Since the primary mechanism of carbon support degradation in automotive PEMFCs is a consequence of the phenomenon that takes place during uncontrolled start-up and shut-down, we discuss in this section the details of the phenomenon, and the basic principles governing the approach to its mitigation.

Figure 3.67(a) illustrates the scenario of start-up of a single cell after it has been shut-down for a sufficiently long period so that both chambers are filled with air. The following processes are likely to transpire once the reactant flows are stopped:

i. residual H_2 in the lines and anode flow fields permeate through the membrane to the cathode side and also leak to the atmosphere through seals and exit valves;
ii. over a period of time, due to cooling/condensation etc. the pressure in the flow fields drops and ambient air leaks in.

Both anode and cathode compartments and all the lines will eventually be filled with air. During the subsequent start-up of the stack, H_2 is turned on and enters the anode replacing and pushing out the resident air; a 'H_2-Air front' comes into being for a brief period of time. The passage of such a H_2-Air front through the anode has been found to lower the electrolyte phase potential and cause the cathode potential to rise to 1.5 V; carbon support corrosion on the cathode electrode ensues and unprecedented electrode degradation takes place. As the carbon support corrodes, the electrode structure falls apart and Pt particles simply agglomerate into larger particles of low surface area. The in-plane transport of protons has a negligible influence beyond a hundredth of a centimeter in relation to several centimeters range of gas flow path. With this assumption, we can simplify the description of the phenomenon to a H_2-Air fuel cell (left-half) driving an Air-Air fuel cell (right-half). A quasi steady-state model can be developed with a knowledge of the values of the ORR, OER, COR and HOR kinetics. At the location of the front itself the potential changes precipitously and therefore pseudo-capacitance (which can supply protons and electrons to partially limit COR) effects need to be included for completeness.

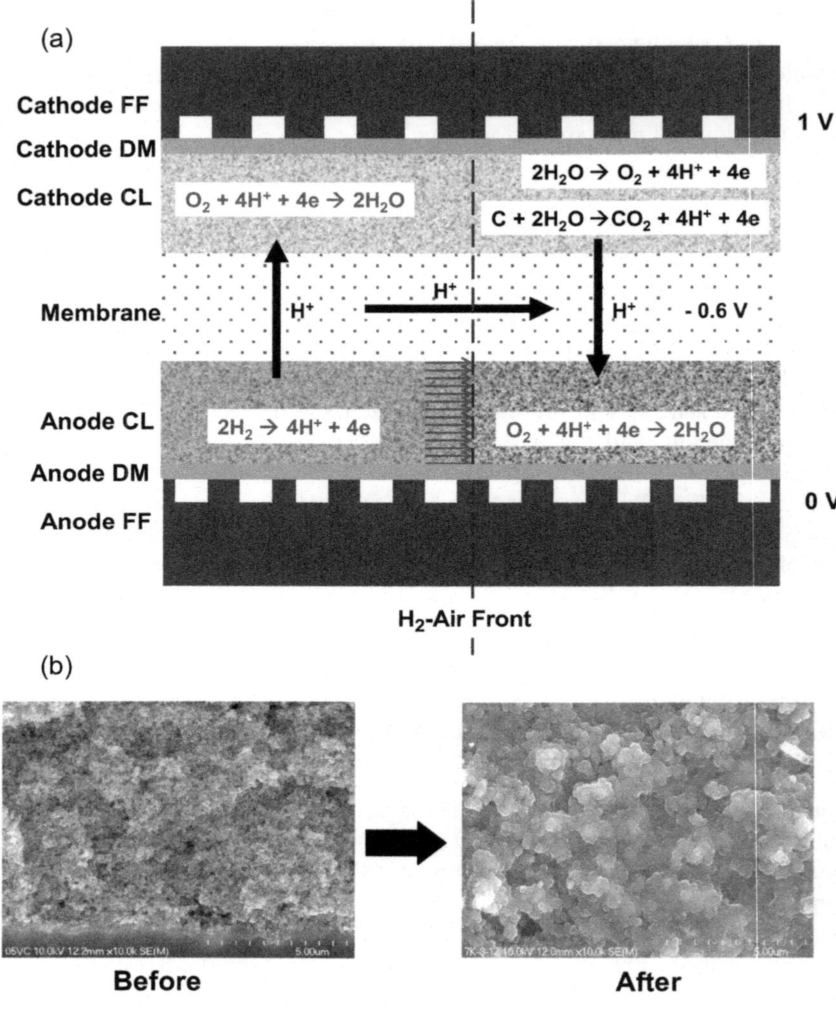

FIGURE 3.67 (a) Schematic based on start-up/shut-down phenomenon that occurs in PEMFCs. Components of a single cell as well as reactions occurring in various sections of the cell during the passage of a H_2-Air front through the anode during start-up is illustrated. (b) SEM micrographs of the cathode catalyst layer at the cathode exit before and after start-up/shut-down cycling illustrating the coarsening of the porous carbon microstructure.

If no special procedures are taken to ameliorate the condition, the cathode can lose half its surface area in just 50–100 cycles of start-up/shut-down. Figure 3.67(b) shows SEM images showing the coarsening of the cathode catalyst layer structure after only 50 cycles. Interestingly, little or no damage occurs on the anode catalyst layer itself due to the front. This phenomenon can also manifest locally:

i. when areas or sections of anode flow fields are blocked due to water build-up and water droplets,
ii. where diffusion media tents,
iii. where local fuel starvation occurs,
iv. near the perimeter of the CCM where the anode and cathode catalyst layers overlap partially, and
v. in the vicinity of membrane pinholes.

4.2.4.1. Start-up/Shut-down Mitigation

Based on our understanding of the phenomenon described above, a number of partial solutions to mitigate the losses encountered during start-up/shut-down are now available and reported in patents and literature [150–152]. We briefly elucidate the fundamental principles that can be applied without entering into engineering details, much of which is proprietary information. In practice, a combination of multiple mitigation strategies outlined below would be employed with the goal of minimizing the complexity and cost of system controls. A three-fold approach that involves modifying the operating conditions, cell component materials and cell design is outlined below.

Operating Conditions Reactant control: The deleterious effects of the H_2-Air front lasts for the duration of time that the front is progressing through the anode flow fields. One can increase the velocity of the front by using high flow rates of H_2 at the anode on start-up and using high flow rates of air purge at shut-down. It has been found that front times of <0.5 seconds lower but do not eliminate the degradation. A high flow purge would require the operation of a blower or compressor that would have to be powered by a battery and has limits. Dilution of the concentration of H_2, or oxygen in air prior to start-up, will also result in lower degradation rates. A residual, trickle flow of hydrogen at the anode, after shut-down, may prevent occurrence of the front. Lower temperatures as well as lower humidity can significantly lower the rates of carbon corrosion as well as Pt dissolution.

Electronic control: By the use of a shorting resistor or some form of an external load, one can limit the voltage spike that the cathode encounters during passage of the front. Further, by shorting the stack while hydrogen is still available or flowing in the anode, the current drawn by the load will consume some or all of the oxygen in the cathode preparing the stack for shut-down. Figure 3.68 illustrates the degree of degradation of an MEA in a PEMFC when it is subjected to potential cycles that vary from 1.0 V to higher potentials and demonstrates that degradation is suppressed for lower potential cycles above OCV.

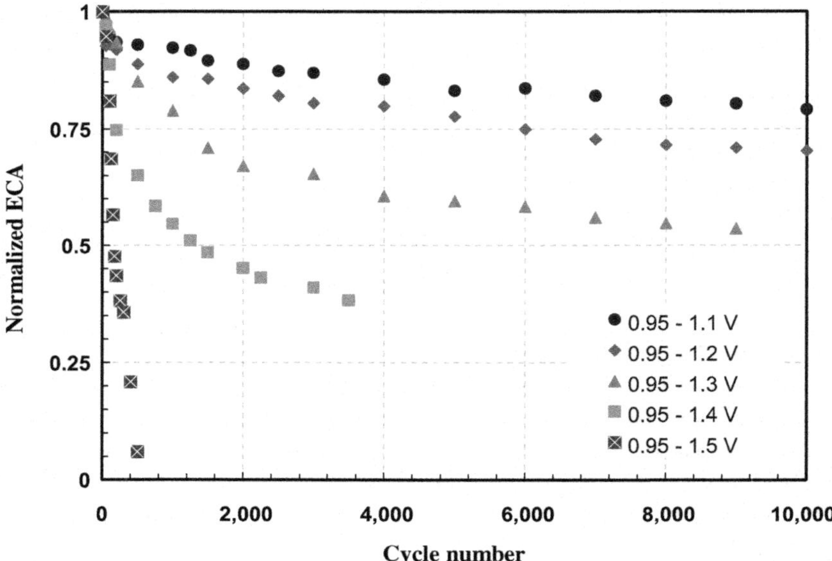

FIGURE 3.68 Effect of the upper potential E_H on degradation losses in the voltage regime between OCV and start-up/shut-down potential peaks where carbon corrosion occurs. Data collected at 100% RH, 20 mV/s under $H_2|N_2$ in 25 cm^2 subscale cells that have cathode catalyst loadings of 0.35 mg/cm^2.

Materials The use of a non-carbon support (or a corrosion-resistant graphitized carbon support) that does not corrode at high potentials would eliminate support and structural losses, but some residual losses due to potential cycling of Pt on the cathode will still remain. A diffusion media material that is easily and rapidly dried (with an air purge) can help incrementally in combination with the approach of lowering the RH. The protonic conductivity of the membrane and ionomer also influences the degradation rates. If one can suppress the ORR reaction rate on the anode, the COR on the cathode will in turn be suppressed; this can be achieved by using a lower loading of Pt (< 0.05 mg/cm^2) on the anode or finding a selective HOR anode catalyst that has poor ORR kinetics. Similarly, an ORR catalyst on the cathode that is also preferentially active to the OER will result in the suppression of the COR on the cathode and protect the carbon support.

Cell Design The use of flow fields that facilitate high reactant velocities will enable a fast H_2-Air front passage, and also prevents water from plugging channels reducing local H_2 starvation related degradation. Seals that have a low permeability for gases and valves that are leak resistant will also be helpful in preventing ambient air from entering the stack prematurely. The ratio of dead space in the manifolds of the anode and cathode also plays an important role depending on the applied strategy.

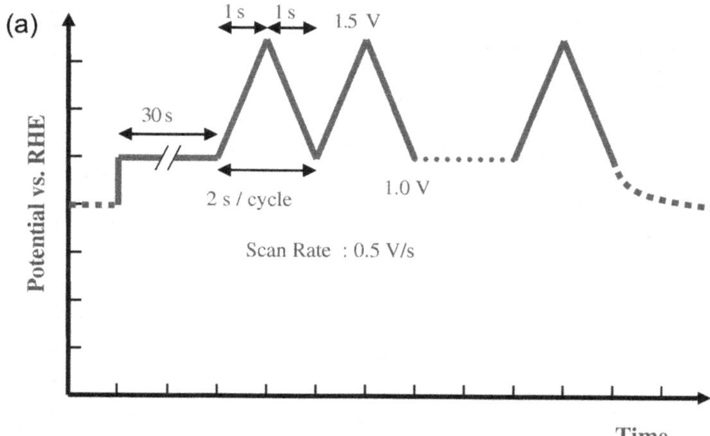

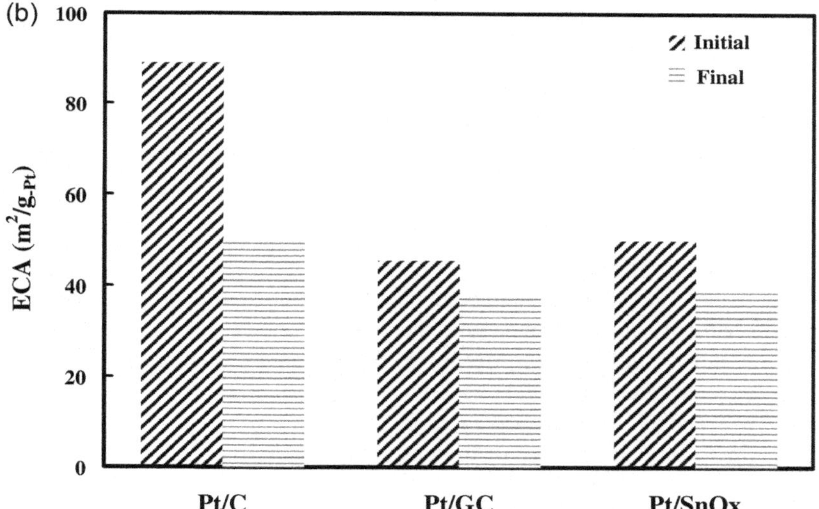

FIGURE 3.69 (a) Protocol for screening new catalyst supports for corrosion resistance. Testing carried out in RDEs using 0.1 M $HClO_4$ at 80°C under N_2 atmosphere at 500 mV/s. (b) Comparison of ECA loss for baseline Pt/C, Pt on graphitized carbon support–Pt/GC, and Pt on a non-carbon support–Pt/SnOx subjected to protocol from (a). [215]

Although start-up/shut-down can be simulated realistically using highly modified, safety-enabled fuel cell test stands, a simple protocol (Fig. 3.69(a)) will suffice for the evaluation of new catalysts or support candidates that show promise. This protocol (1.0–1.5 V, 2 s/cycle, 50–500 mV/s, triangular sweeps, 1,000–10,000 cycles) can be applied not only in subscale PEMFCs but also in half-cell RDE measurements for screening new support material.

Figure 3.69(b) demonstrates the application of this protocol as a successful screening tool to evaluate several catalysts supported on different carbons as well as SnO_2 supports.

4.3. Durability of PEMFC Stacks in Vehicles

In the last decade a considerable amount of work has been carried out by automotive companies on the durability of fuel cells in subscale cells, short-stacks, entire stacks (200–400 cells) on power train test benches, and stacks installed in operational vehicles. To our knowledge, no degradation studies have been reported focused on the detailed comparison of studies from test benches to stacks that are installed and tested in operational fuel cell vehicles. In a very recent report, Shimoi and co-workers from Nissan have reported on an extensive experimental study of degradation rates modeled in terms of start-stop, load cycling and idling modes evaluated in power train test benches. The studies were complemented with predictions of the degradation rate of stacks in actual fuel cell vehicles (Nissan X-TRAIL FCV) tested on both public roads and proving grounds. We briefly summarize these results below.

4.3.1. Data Collection for Estimation of Degradation

Degradation rates were estimated by carrying out independent tests under each of three operating modes described below for the bench testing of fuel cell stacks.

1. Estimation of degradation under start-up/shut-down: The schematic in Fig. 3.67 elucidates the phenomena of start-up/shut-down. In order to systematically take into account the various possible scenarios that can occur during start-up and shut-down, independent experiments were conducted to measure degradation:
   - **i.** for repeated start-stops in a short interval of time;
   - **ii.** after complete change of the gas composition to air; and,
   - **iii.** for different concentrations of oxygen in the cathode.

   A separate test-rig was designed to determine oxygen concentration in the anode and cathode as a function of 'standing time'. 'Standing time' refers to the non-operational time after shut-down during which the cathode gas composition slowly changes due to crossover, leaks, etc. These data were collected and used to estimate degradation rates along with a frequency distribution of the standing times of normally operated ICE vehicles.
2. Estimation of degradation under load cycling: Three load cycling patterns with different high and low potentials were defined and durability tests independently conducted under each of these conditions. The collected data were combined with the expected frequency (statistical occurrence) of each cycle pattern in normally operated ICE vehicles to estimate the degradation rate.

3. Estimation of degradation under idling: What one generally finds is that practical vehicles do not undergo continuous, extended, uninterrupted periods at the high potential idling at high temperatures. Thus, to simulate and reproduce realistic vehicle operating conditions and avoid over-estimation, data were collected for idling conditions in combination with some load cycles. The collected data were combined with expected idle times for normally operated ICE vehicles to estimate degradation rates.

4.3.2. Prediction of Stack Degradation

Using the data from stacks tested for each of the three modes of degradation and the statistical occurrence of these operating modes from ICE vehicles, the stack degradation was estimated. Figure 3.70 illustrates the contribution of the three operating modes to the total fuel cell stack degradation. The results (predicted degradation) as shown in Fig. 3.70 indicate that degradation attributed to start-up/shut-down cycles was the largest at 44% while the other two modes contributed ~28% each. Although the losses incurred during load cycling operation are intrinsically higher, the cumulative time under idling is longer resulting in comparable total losses for the two modes.

4.3.3. Comparison of Estimated and Actual Degradation

A number of X-TRAIL FCVs have been driven on both public roads and proving grounds for 2–3 years in the US, Japan and Europe. It is fairly complex to gather vast amounts of data from stacks in fuel cell vehicles and reduce them

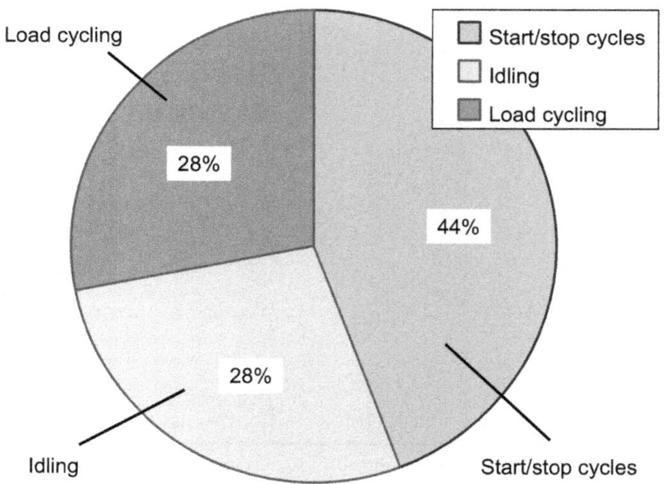

FIGURE 3.70 Estimation of losses in stacks that are attributed to the three major degradation modes of load cycling, start-stop and idling. *(Shimoi et al. [213])*

to manageable levels amenable to analysis. Briefly, fuel cell stack performance data from vehicles were recorded each time (trigger) a specified high current value was exceeded. The overall performance degradation as a function of:

i. operational period in years;
ii. operational time in hours;
iii. mileage; and
iv. number of start-stop cycles

revealed that no one parameter exclusively exhibited a strong correlation with the total degradation, implying that multiple factors were responsible. Figure 3.71 shows the comparison of estimated and actual vehicle degradation indicating good predictive capability of the estimation model. We note that the losses due to start-stop cycles were lower than estimated in Fig. 3.70; this is a result of a lower frequency of extended standing times between cycles rather than the total number of cycles.

MEAs were randomly selected from degraded stacks and I-V performance, ECA, SEM and EPMA diagnostics conducted on them to verify that the assumptions of the three modes of degradation were accurate. Based on the insights gained from this study, a new generation of PEMFC stack (with twice the power density of the previous generation) has been developed that incorporates a reduction in Pt loading by 50% while simultaneously showing a substantial improvement in durability and consequently a lowering of cost [213].

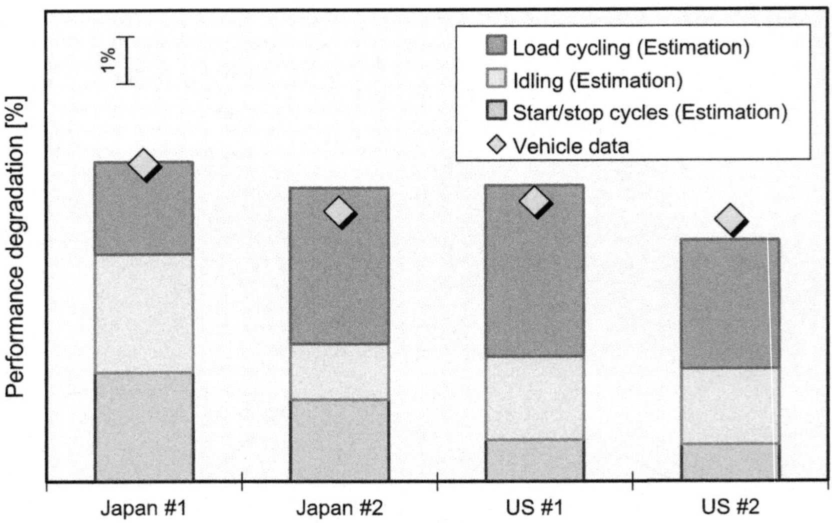

FIGURE 3.71 Comparison of estimated vehicle degradation (yellow diamonds) vs. actual degradation, sub-divided into load cycling, idling and start-up/shut-down cycles. *(Shimoi et al. [213])*

5. SUMMARY AND FUTURE CHALLENGES

Significant progress has been made over the last decade in enhancing the performance, durability and cost of PEMFC stacks for automotive use. Stack lifetimes of >3,000 h are now achievable with the right combination of materials, stack design and operating conditions, and the target lifetime of 5,000 h/10 years is considered achievable by 2015. Three pathways for improved catalytic materials include: supported nanoparticles of Pt-alloys, core-shell catalysts, and extended thin film electrode structures or ETFECS. Both the core-shell and thin extended films are based on the principle of using the minimal number of monolayers of Pt or Pt-alloys on the surface in order to deliver the targeted activity and durability. An interesting aspect in the development of new Pt-alloy catalysts as well as thin film catalysts is their potential ability to simultaneously exhibit a high specific activity together with high resistance to dissolution. System controls to limit potential spikes during start-up/shut-down, and to protect the carbon support have advanced to a reasonable level. Development of highly corrosion-resistant support materials is needed to allow further system simplification and lower costs. Implementing the new advanced catalysts and corrosion-resistant supports in an electrode structure that is optimized for utilization and mass-transport is a challenge that must be met in the coming years through innovative nanoscale engineering.

ACKNOWLEDGMENTS

I wish to thank all the members of the Advanced Electrocatalysis Group (NML-YF0:Fuel Cell Laboratory) especially Masanobu Uchimura, Shigemasa Kuwata, Jianbo Zhang, Ikuma Takahashi, Yoshiki Nagahara, Yosuke Suzuki, and Seiho Sugawara. I would also like to acknowledge Akihiro Iiyama and Kazuhiko Shinohara for their kind support during my stay (2005–2008) in Japan.

NOMENCLATURE

d	particle diameter, nm
ρ_{Pt}	density of Pt, 21.5 g/cm^3
M_{Pt}	molecular weight of Pt, 195 g/mol
L	catalyst loading, mg/cm^2
n	moles
t	time, s
T	temperature,°C
P	pressure, kPa
R	universal gas constant, J/mol K
Q	charge density, μC/cm^2
F	Faraday's constant, 96,487, C/equiv
E	thermodynamic reversible potential, V
V	voltage, V
η	overpotential, V
RHE	reversible hydrogen electrode, 0 V

θ_{ox}	surface coverage with oxide species
m	ORR reaction order at constant potential
γ	ORR reaction order at constant overpotential
ECA	electrochemical area, m^2/g
I	current density, A/cm^2
i_s	(area) specific activity, $\mu A/cm^2$
i_m	mass (specific) activity, mA/mg
i_L	limiting current, A/cm^2
Sr	scan rate, mV/s

REFERENCES

[1] N. Garland, J. Marcinkoski, DOE Hydrogen Program Record, http://www.hydrogen.energy.gov/pdfs/8019_fuel_cell_system_cost.pdf (2008).

[2] S. Satyapal, Hydrogen Program Overview, 2009 DOE Hydrogen Program and Vehicle Technologies Program, http://www.hydrogen.energy.gov/pdfs/review09/program_overview_2009_amr.pdf (2009).

[3] S. Satyapal, Fuel Cell Project Kickoff, US DOE, EERE Fuel Cell Technologies Program, http://www1.eere.energy.gov/hydrogenandfuelcells/pdfs/satyapal_doe_kickoff.pdf (2009).

[4] D. Papageorgopoulos, Fuel Cell Technologies 2009 DOE Hydrogen Program & Vehicle Technologies Program, Merit Review and Peer Evaluation Meeting, http://www.hydrogen.energy.gov/pdfs/review09/fc_0_papageorgopoulos.pdf (2009).

[5] S.S. Kocha, Principles of MEA preparation, in: W. Vielstich, A. Lamm, H. Gasteiger (Eds), Handbook of Fuel Cells-Fundamentals, Technology and Applications, John Wiley & Sons, New York, 2003. Vol. 3 Chapter 43, pp. 538–565.

[6] J. Matthey, Platinum today, Johnson Matthey Precious Metal Marketing http://www.platinum.matthey.com/prices/price_charts.html (2007).

[7] M. Kromer, T. Rhodes, M. Guernsey, Update on Pt availability and assessment of Pt leasing strategies for fuel cell vehicles – AN12, DOE Merit Review, http://www.hydrogen.energy.gov/pdfs/review08/an_12_kromer.pdf (2008).

[8] US DOE, Fuel cell targets, http://www1.eere.energy.gov/hydrogenandfuelcells/mypp (2007).

[9] S. Grot, W. Grot, Platinum recycling technology development, 2008 DOE Hydrogen Program Review, http://www.hydrogen.energy.gov/pdfs/review08/fcp_3_grot.pdf (2008).

[10] M. Uchimura, S. Kocha, The impact of cycle profile on PEMFC durability, ECS Trans. 11 (2007) 1215–1226.

[11] M. Uchimura, S. Sugawara, Y. Suzuki, J. Zhang, S.S. Kocha, Electrocatalyst durability under simulated automotive drive cycles, ECS Trans. 16 (2008) 225–234.

[12] US DOE, DOE cell component accelerated stress test protocols for PEM fuel cells, http://www1.eere.energy.gov/hydrogenandfuelcells/fuelcells/pdfs/component_durability_profile.pdf (2007).

[13] A. Iiyama, K. Shinohara, S. Igushi, A. Daimaru, Membrane and catalyst performance targets for automotive fuel cells, John Wiley & Sons, Ltd. Vol. 5 (2009).

[14] M.F. Mathias, R. Makharia, H.A. Gasteiger, J.J. Conley, T.J. Fuller, C.J. Gittleman, S.S. Kocha, D.P. Miller, C.K. Mittelsteadt, T. Xie, S.G. Yan, P.T. Yu, Two fuel cells in every garage? Electrochem. Soc. Interface 14 (2005) 24–35.

[15] US DOE-EERE, Technical Plan – Fuel Cells, Multi-Year Research, Development and Demonstration Plan: Planned Program Activities for 2005–2015, (2007).

[16] H.A. Gasteiger, S.S. Kocha, B. Sompalli, F.T. Wagner, Activity benchmarks and requirements for Pt, Pt-alloy, and non-Pt oxygen reduction catalysts for PEMFCs, Appl. Catal. B: Environmental 56 (2005) 9–35.

[17] M. Pourbaix, Atlas of electrochemical equilibrium in aqueous solutions, 1st ed., Pergamon Press, New York, 1966.

[18] P. Bindra, S.J. Clouser, E. Yeager, Pt dissolution in concentrated phosphoric acid, J. Electrochem. Soc. 126 (1979) 1631.

[19] S. Mitsushima, S. Kawahara, K.-i. Ota, N. Kamiya, Consumption rate of pt under potential cycling, Journal of The Electrochemical Society 154 (2007) B153–B158.

[20] X. Wang, R. Kumar, D.J. Myers, Effect of voltage on Pt dissolution, Electrochemical and Solid-State Letters 9 (2006) A225–A227.

[21] P.J. Ferreira, G.J. la O, Y. Shao-Horn, D. Morgan, R. Makharia, S. Kocha, H.A. Gasteiger, Instability of Pt/C electrocatalysts in proton exchange membrane fuel cells–a mechanistic investigation, Journal of The Electrochemical Society 152 (2005) A2256–A2271.

[22] X. Wang, R. Kumar, D.J. Myers, Effect of voltage on platinum dissolution–relevance to polymer electrolyte fuel cells, Electrochemical and Solid-State Letters 9 (2006) A225–A227.

[23] S. Mitsushima, Y. Koizumi, S. Uzuka, K. Ota, Dissolution of Pt in acidic media, Electrochim. Acta 54 (2008) 455–460.

[24] V. Komanicky, K.C. Chang, A. Menzel, N.M. Markovic, H. You, X. Wang, D. Myers, Stability and dissolution of platinum surfaces in perchloric acid, J. Electrochem. Soc. 153 (2006) B446–B451.

[25] R.M. Darling, J.P. Meyers, Kinetic model of platinum dissolution in PEMFCs, J. Electrochem. Soc. 150 (2003) A1523–A1527.

[26] R.M. Darling, J.P. Meyers, Mathematical model of platinum movement in PEM fuel cells, J. Electrochem. Soc. 152 (2005) A242–A247.

[27] C. Puglia, A. Nilsson, B. Hernnais, O. Karis, P. Bennich, N. Martensson, Physisorbed, chemisorbed and dissociated oxygen on Pt(111) studied by different core level spectrosocpy methods, Surf. Sci. 342 (1995) 119–133.

[28] S.J. Clouser, J.C. Huang, E. Yeager, Temperature dependence of the Tafel slope for oxygen reduction on Pt in concentrated phosphoric acid, J. Appl. Electrochem 23 (1993) 597.

[29] A. Damjanovic, V. Brusic, Electrode kinetics of oxygen reduction on oxide-free platinum electrodes, Electrochim. Acta 12 (1967) 615–626.

[30] A. Damjanovic, V. Brusic, J.O.M. Bockris, Electrode kinetics of oxygen reduction on platinum electrodes, J. Phys. Chem. 71 (1967) 2471.

[31] D.B. Sepa, M.V. Vojnovic, A. Damjanovic, Reaction intermediates as a controlling factor in the kinetics and mechanism of oxygen reduction at platinum electrodes, Electrochim. Acta 26 (1981) 781–793.

[32] D.B. Sepa, M.V. Vojnovic, L.M. Vracar, Effect of pH on Pt in acid solutions, Electrochim. Acta 32 (1987) 129.

[33] E. Yeager, M. Razaq, D. Gervasio, A. Razaq, D. Tryk, Dioxygen reduction in various acid electrolytes, J. Serb. Chem. Soc. 57 (1992) 819–833.

[34] D.C. Johnson, D.T. Napp, S. Bruckenstein, A ring-disk electrode study of the current/potential behaviour of platinum in 1.0M sulphuric and 0.1M perchloric acids, Electrochim, Acta 15 (1970) 1493–1509.

[35] B.E. Conway, T.C. Liu, Characterization of electrocatalysis in the ORR at platinum by evaluation of behavior of surface intermediate states at the oxide film, Langmuir 6 (1990) 268–276.

[36] S.G. Roscoe, B.E. Conway, State of the surface oxide films at Pt anodes and 'Volcano' behavior in electrocatalysis for anodic chlorine evolution, J. Electroanal. Chem. 224 (1987) 163–188.

[37] H. Angerstein-Kozlowska, B.E. Conway, A. Hamelin, L.J. Stoicoviciu, Elementary steps of electrochemical oxidation of single crystal planes of Au. Part II A chemical and structural basis of oxidation of the (111) plane, J. Electroanal. Chem 228 (1987) 429–453.

[38] S. Gottesfeld, B.E. Conway, Real conditions of oxidized Pt electrodes. Part 2. Resolution of reversible and irreversible processes by optical and impedance studies, J. Chem. Soc., Faraday Tans 1: Phy. Chem. in condensed phases 69 (1973) 1090–1107.

[39] J.L. Ord, F.C. Ho, Anodic oxidation of platinum: An optical study, J. Electrochem. Soc. 118 (1971) 46.

[40] G.C. Allen, P.M. Tucker, X-ray photoelectron spectroscopy of adsorbed oxygen and carbonaceous species on platinum electrodes, J. Electroanal. Chem. 50 (1974) 335–343.

[41] J.S. Hammond, N. Winograd, XPS spectroscopic study of potentiostatic and galvanostatic oxidation of Pt electrodes in sulfuric and perchoric acid, J. Electroanal Chem. 78 (1977) 55–69.

[42] K.S. Kim, N. Winograd, R.E. Davis, X-ray photoelectron spectroscopic studies on platinum, J. Am. Chem. Soc. 93 (1971) 6296.

[43] M. Peuckert, H.P. Bonzel, Characterization of oxidized platinum surfaces by X-ray photoelectron spectroscopy, Surf. Sci. 145 (1984) 239–259.

[44] J.P. Hoare, The electrochemistry of oxygen, Interscience, New York, 1968.

[45] B.E. Conway, Electrochemical oxide film formation at noble metals as a surface-chemical process, Prog. Surf. Sci. 49 (1995) 331–452.

[46] D.A. Harrington, Simulation of anodic Pt oxide growth, J. Electroanal. Chem. 420 (1997) 101–109.

[47] M. Alsabet, M. Grden, G. Jerkiewicz, Comprehensive study of the growth of thin oxide layers on Pt electrodes under well–defined temperature, potential, and time conditions, J. Electroanal. Chem. 589 (2006) 120–127.

[48] G. Jerkiewicz, G. Vatankhah, J. Lessard, M.P. Soriaga, Y.-S. Park, Surface-oxide growth at platinum electrodes in aqueous sulfuric acid reexamination of its mechanism through combined cyclic-voltammetry, electrochemical quartz-crystal nanobalance, and Auger electron spectroscopy measurements, Electrochim. Acta 49 (2004) 1451–1459.

[49] A.R. Kucernak, G.J. Offer, The role of adsorbed hydroxyl species in the electrocatalytic carbon monoxide oxidation reaction on platinum, Phys. Chem. Chem. Phys. 10 (2008) 3699.

[50] B.E. Conway, G. Jerkiewicz, Surface orientiation dependence of oxide film growth at platinum single crystals, J. Electroanal. Chem. 339 (1992) 123–146.

[51] Z. Nagy, H. You, Applications of surface X–ray scattering to electrochemistry problems, Electrochim. Acta 47 (2002) 3037–3055.

[52] G. Tremiliosi-Filho, G. Jerkiewicz, B.E. Conway, Characterization and significance of the sequence of stages of oxide film formation at platinum generated by strong anodic polarization, Langmuir 8 (1992) 658–667.

[53] V.I. Birss, M. Chang, J. Segal, Platinum oxide film formation-reduction: An *in-situ* mass measurement study, J. Electroanal. Chem. 355 (1993) 181–191.

[54] V.I. Birss, M. Goledzinowski, The unusual reduction behavior of thin, hydrous platinum oxide films, J. Electroanal. Chem. 351 (1993) 227–243.

[55] M. Farebrother, M. Goledzinowski, G. Thomas, V.I. Birss, Early stages of growth of hydrous platinum oxide films, J. Electroanal. Chem. 297 (1991) 469–488.

[56] G.A. Ragoisha, N.P. Osipovich, A.S. Bondarenko, J. Zhang, S. Kocha, A. Iiyama, Characterisation of the electrochemical redox behaviour of Pt electrodes by potentiodynamic electrochemical impedance spectroscopy, J. Solid State Electrochem. (2008).

[57] J. Tafel, B. Emmert, Ueber die ursache der spontanen depression des kathodenpotntials bei der eletrolyse verduennter schwefelsaeure, Zeit. Physik. Chem. (1905) 349–373.

[58] D.A.J. Rand, R. Woods, A study of the dissolution of platinum, palladium, rhodium and gold electrodes in 1 M sulphuric acid by cyclic voltammetry, J. Electroanal. Chem. 35 (1972) 209–218.

[59] V.A.T. Dam, F.A. de Bruijn, The stability of PEMFC electrodes: Pt dissolution vs potential and temperature investigated by quartz crystal microbalance, J. Electrochem. Soc. 154 (2007) B494–B499.

[60] J.A.S. Bett, K. Kinoshita, P. Stonehart, Crystallite growth of Pt dispersed on graphitized carbon black, J. Catal. 35 (1974) 307–316.

[61] J.A.S. Bett, K. Kinoshita, P. Stonehart, Crystallite growth of Pt dispersed on graphitized carbon black II. Effect of liquid environment, J. Catalysis 41 (1976) 124–133.

[62] K.F. Blurton, H.R. Kunz, D.R. Rutt, Surface area loss of Pt supported on graphite, Electrochim. Acta 23 (1978) 183–190.

[63] G.A. Gruver, The corrosion of carbon black in phosphoric acid, J. Electrochem. Soc. 125 (1978) 1719.

[64] J.F. Connolly, R.J. Flannery, B.L. Meyers, Recrystallization of supported platinum, J. Electrochem. Soc. 114 (1967) 241–243.

[65] K. Kinoshita, K. Routsis, J.A.S. Bett, C.S. Brooks, Morphological changes in the electrodes of phosphoric acid fuel cells, Electrochim. Acta 18 (1973) 953.

[66] P. Stonehart, P. Zucks, Sintering and recrystallization of small metal particles. Loss of surface area by Pt-black fuel cell electrocatalysts, Electrochim. Acta 17 (1972) 2333–2351.

[67] K. Kinoshita, J.T. Lundquist, P. Stonehart, Potential cycling effects on platinum electrocatalyst surfaces, Journal of Electroanal. Chem. 48 (1973) 157–166.

[68] E. Yeager, Electrocatalysis for O_2 reduction, Electrochim. Acta 29 (1984) 1527–1537.

[69] P. N. J. Ross, Catalyst deactivation, in E. E. Petersen, A. T. Bell, (Eds) Marcel Dekker: New York, 1987, pp 165–178.

[70] J. Aragane, T. Murahashi, T. Odaka, Change of Pt distribution in the active components of phosphoric acid fuel cell, J. Electrochem. Soc. 135 (1988) 844–850.

[71] J. Aragane, H. Urushibata, T. Murahashi, Effect of operational potential on performance decay rate in phosphoric acid fuel cell, J. App. Electrochem. 26 (1996) 147–152.

[72] L.J. Bregoli, The influence of platinum crystallite size on the electrochemical reduction of oxygen in phosphoric acid, Electrochim. Acta 23 (1978) 489–492.

[73] K. Kinoshita, Particle size effects for oxygen reduction on highly dispersed platinum in acid electrolytes, J. Electrochem. Soc. 137 (1990) 845–848.

[74] V.M. Jalan, E.J. Taylor, Importance of interatomic spacing in catalytic reduction of oxygen in phosphoric acid, J. Electrochem. Soc. 130 (1983) 2299–2302.

[75] S.S. Kocha, R. Makharia, H.A. Gasteiger, Voltage cycling durable catalysts. US20070003822 2005.

[76] M.S. Wilson, J.A. Valerio, S. Gottesfeld, Low platinum loading electrodes for polymer electrolyte fuel cells fabricated using thermoplastic ionomers, Electrochim. Acta 40 (1995) 355–363.

[77] M.S. Wilson, F.H. Garzon, K.E. Sickafus, S. Gottesfeld, Surface area loss of supported platinum in polymer electrolyte fuel cells, J. Electrochem. Soc. 140 (1993) 2872–2877.
[78] T. Tada, High dispersion catalysts including novel carbon supports, in: W. Vielstich, A. Lamm, H. Gasteiger (Eds), Handbook of Fuel Cells-Fundamentals, Technology and Applications, John Wiley & Sons, Ltd, 2003. Vol. 3, Chapter 38, pp. 481–488.
[79] J. St-Pierre, D.P. Wilkinson, S. Knights, M.L. Bos, Relationships between water management, contamination and lifetime degradation in PEFC, J. New Mat. Electrochem. Systems 3 (2000) 99–106.
[80] T. Patterson, Fuel cell technology topical conference proceedings, in: G.J. Igweand, D. Mah (Eds), AIChE Spring National Meeting, New York, 2002, p. 313.
[81] B. Merzougui, S. Swathirajan, Rotating disk electrode investigations of fuel cell catalyst degradation due to potential cycling in acid electrolytes, J. Electrochem. Soc. 153 (2006) A2220–A2226.
[82] D.A. Blom, J.R. Dunlap, J.R. Nolan, L.F. Allard, Preparation of cross-sectional samples of proton exchange fuel cells by ultramicrotomy for TEM, J. Electrochem. Soc. 150 (2003) A414–A418.
[83] K.L. More, R. Borup, K.S. Reeves, Identifying contributing degradation phenomena in PEM fuel cell membrane electrode assemblies via electron microscopy, ECS Trans. 3 (2006) 717–733.
[84] K.L. More, K.S. Reeves, Microstructural Characterization of PEM Fuel Cell MEAs, Microscopy and Microanalysis. 11(2) 2104 (2005).
[85] R.L. Borup, J.R. Davey, F.H. Garzon, D.L. Wood, M.A. Inbody, PEM fuel cell electrocatalyst durability measurements, J. Power Sources 163 (2006) 76–81.
[86] J. Healy, C.X. Hayden, T.K. Olson, R. Waldo, A. Brundage, J. Abbot, Aspects of chemical degradation of PFSA ionomer used in PEMFCs, Fuel Cells 5 (2005) 302–308.
[87] S. Sugawara, T. Maruyama, Y. Nagahara, S.S. Kocha, K. Shinohara, K. Tsujita, S. Mitsushima, K.-i. Ota, Performance decay of PEMFCs under OCV conditions induced by membrane degradation, J. Power Sources 187 (2009) 324–331.
[88] T. Xie, K. Teranishi, K. Kawata, S. Tshushima, S. Hiroi, Degradation mechanism of PEMFC under open circuit operation, Electrochem. Solid-State Lett. 9 (2005) A475–A477.
[89] C.H. Paik, G.S. Saloka, G.W. Graham, Influence of cyclic operation on PEM fuel cell catalyst stability, Electrochemical and Solid-State Letters 10 (2007) B39–B42.
[90] M. Uchimura, S.S. Kocha, The impact of oxides on activity and durability of PEMFCs, Annual AIChE Meeting Utah, November 6, 2007. 2007.
[91] W.M. Vogel, J.M. Baris, The reduction of oxygen on platinum black in acid electrolytes, Electrochim. Acta 22 (1977) 1259–1263.
[92] V.M. Jalan, D.A. Landsman, Noble metal-refractory metal alloys as catalysts and method for making. 4186110, 1980.
[93] D.A. Landsman, F.J. Luczak, Noble metal-chromium alloy catalysts and electrochemical cell. 4316944, 1982.
[94] F. J. Luczak, D.A. Landsman, Method for making ternary fuel cell catalysts containing platinum, cobalt and chromium. 4613582, 1986.
[95] D. Landsman, F.J. Luczak, Catalyst studies and coating technologies, Vol. 4, John Wiley & Sons Ltd, West Sussex, England, 2003.
[96] F.J. Luczak, Determination of d-band occupancy in pure metals and supported catalysts by measurement of the LIII x-ray absorption threshold, J. of Catal. 43 (1976) 376–379.
[97] K. Kinoshita, Electrochemical oxygen technology, John Wiley & Sons, New York, 1992, p. 176–180.
[98] S.S. Kocha, K. Deluca, F.J. Luczak, A novel ternary cathode catalyst for phosphoric acid fuel cells, 193rd Electrochemical Society Meeting, San Diego, CA, 1998, 34.

[99] V.R. Stamenkovic, B.S. Mun, M. Arenz, K.J.J. Mayrhofer, C.A. Lucas, G. Wang, P.N. Ross, N.M. Markovic, Trends in electrocatalysis on extended and nanoscale Pt-bimetallic alloy surfaces, Nature Materials 6 (2007) 241–247.

[100] S. Mukerjee, S. Srinivasan, Enhanced electrocatalysis of oxygen reduction on platinum alloys in proton exchange membrane fuel cells, J. Electroanal. Chem. 357 (1993) 201–224.

[101] S. Mukerjee, S. Srinivasan, M.P. Soriaga, J. McBreen, Role of structural and electronic properties of Pt and Pt alloys on electrocatalysis of oxygen reduction, J. Electrochem. Soc. 142 (1995) 1409–1422.

[102] D. Thompsett, Pt alloys as oxygen reduction catalysts, in: W. Vielstich, H. Gasteiger, A. Lamm (Eds), Handbook of Fuel Cells-Fundamentals, Technology and Applications, John Wiley & Sons Ltd, Chichester, UK, 2003. Vol. 3, Chapter 41, pp. 467–480.

[103] S.S. Kocha, H.A. Gasteiger, The use of Pt-alloy catalyst for cathodes of PEMFCs to enhance performance and achieve automotive cost targets, 2004 Fuel Cell Seminar, San Antonio, TX, 2004.

[104] P. Yu, M. Pemberton, P. Plasse, PtCo/C cathode catalyst for improved durability in PEMFCs, J. Power Sources 144 (2005) 11–20.

[105] R. Makharia, S.S. Kocha, P.T. Yu, M.A. Sweikart, W. Gu, F.T. Wagner, H.A. Gasteiger, Durable PEMFC electrode materials: Requirements and benchmarking methodologies, ECS Trans. 1 (2006) 3–18.

[106] S. Hidai, M. Kobayashi, H. Niwa, Y. Harada, M. Oshima, Y. Nakamori, T. Aoki, E. Ikenaga, Degradation mechanism of platinum cobalt alloy catalysts revealed by X-ray photoemission spectroscopy, Electrochemical Society, Vienna, 2009.

[107] M. Kobayashi, S. Hidai, H. Niwa, Y. Harada, M. Oshima, Y. Horikawa, T. Tokushima, S. Shin, Y. Nakamori, T. Aoki, Co oxidation accompanied by degradation of Pt-Co alloy cathode catalysts in polymer electrolyte fuel cells, Phys Chem Chem Phys. 2009 Oct 1; 11 (37): 8226–8230.

[108] V. Stamenkovic, T.J. Schmidt, P.N. Ross, N.M. Markovic, Surface segregation effects in electrocatalysis: Kinetics of oxygen reduction reaction on polycrystalline Pt_3Ni alloy surfaces, Journal of Electroanal. Chem. (2003) 191–199, 554–555.

[109] V.R. Stamenkovic, B. Fowler, B.S. Mun, G. Wang, P.N. Ross, C.A. Lucas, N.M. Markovic, Improved oxygen reduction activity on $Pt_3Ni(111)$ via increased surface site availability, Science 315 (2007) 494–497.

[110] K.C. Neyerlin, R. Srivastava, C. Yu, P. Strasser, Electrochemical activity and stability of dealloyed Pt-Cu and Pt-Cu-Co electrocatalysts for the oxygen reduction reaction, J. Power Sources 186 (2009) 261–267.

[111] P. Strasser, Electrocatalyst stability at the nanoscale, Electrochemical Society, Vienna, 2009.

[112] M. Debe, Advanced MEAs for enhanced operating conditions, amenable to high volume manufacture, 2005 DOE Hydrogen Program Review (2005).

[113] M. Debe, Advanced cathode catalysts and supports for PEM fuel cells, 2008 DOE Hydrogen Program Review, http://www.hydrogen.energy.gov/pdfs/review08/fc_1_debe.pdf (2008).

[114] M. Debe, Advanced cathode catalysts and supports for PEM fuel cells, 2009 DOE Hydrogen Program Review, http://www.hydrogen.energy.gov/pdfs/review09/fc_17_debe.pdf (2009).

[115] M.K. Debe, Nstf catalysts, in: W. Vielstich, A. Lamm, H. Gasteiger (Eds), Handbook of fuel cells-fundamentals, technology and applications, John Wiley & Sons, Ltd, New York, 2003. Vol. Chapter 45.

[116] M.K. Debe, A.K. Schmoeckel, S.M. Hendricks, G.D. Vernstrom, G.M. Haugen, R.T. Atanasoski, High voltage stability of nanostructured thin film catalysts for PEM fuel cells, ECS Trans. 1 (2006) 51–55.

[117] S.S. Kocha, K.C. Neyerlin, T. Olson, B. Pivovar, Electrochemical characterization and implementation of extended surface Pt catalysts, Paper presented at the Electrochemical Society – 218th ECS Meeting Abstracts 2010, MA, 2010. 1747.

[118] J. Zhang, F.H.B. Lima, M.H. Shao, K. Sasaki, J.X. Wang, J. Hanson, R.R. Adzic, Pt monolayer on nonnoble metal-noble metal core-shell nanoparticle electocatalysts for O_2 reduction, J. Phys. Chem. B 109 (2005) 22701–22704.

[119] J. Zhang, Y. Mo, M.B. Vukmirovic, R. Klie, K. Sasaki, R.R. Adzic, Pt monolayer electrocatalysts for oxygen reduction; Pt monolayer on Pd(111) and on carbon-supported Pd nanoparticles, J. Phys. Chem. B 108 (2004) 10955–10964.

[120] J. Zhang, R. Sasaki, E. Sutter, R.R. Adzic, Stabilization of platinum for oxygen reduction electrocatalysts using gold clusters, Science 315 (2007) 220–222.

[121] K. Sasaki, Y. Mo, J.X. Wang, M. Balasubramanian, F. Uribe, J. McBreen, R.R. Adzic, Pt submonolayers on metal nanoparticles/novel electrocatalysts for H_2 oxidation and O_2 reduction, Electrochim. Acta 48 (2003) 3841–3849.

[122] S. Ball, S.L. Burton, E. Christian, A. Davies, J. Fisher, R. O'Malley, S. Passot, B. Tessier, B.R.C. Theobald, D. Thompsett, Activity and stability of Pt monolayer core shell catalysts, 240th Meeting of the Electrochemical Society, Vienna, 2009.

[123] A. Ruban, B. Hammer, P. Stoltze, H.L. Skriver, J.K. Norskov, Surface electronic structure and reactivity of transition and noble metals, J. Mol. Catal. A: Chem 115 (1997) 421–429.

[124] X. Wang, N. Kariuki, J.T. Vaughey, J. Goodpastor, R. Kumar, D.J. Myers, Bi-metallic Pd-Cu oxygen reduction electrocatalysts, J. Electrochem. Soc. 155 (2008) B602–B609.

[125] R.G. Gonzalez-Huerta, J.A. Chavez-Carvayar, O. Solorza-Feria, Electrocatalysis of oxygen reduction on carbon supported Ru-based catalysts in a polymer electrolyte fuel cell, J. Power Sources 153 (2006) 11–17.

[126] P.J. Sebastian, F.J. Rodrinue, Development of $Mo_xRu_ySe_z$ (Co)n electrocatalysts by screen printing and sintering for fuel cell applications, Sur. Eng. 16 (2000) 43.

[127] R. Bashyam, P. Zelenay, A class of non-precious metal composite catalyst for fuel cells, Nat. Lett. 443 (2006) 63–66.

[128] F. Jaouen, F. Charreteur, J.P. Dodelet, Fe-based catalysts for oxygen reduction in PEMFCs, J. Electrochem. Soc. 153 (2006). A689.

[129] C. Medard, M. Lefevre, J.P. Dodolet, F. Jaouen, G. Lindbergh, Oxygen reduction by Fe-based catalysts in PEM fuel cell conditions: Activity and selectivity of the catalysts with two Fe precursors and various carbon supports, Electrochim. Acta 51 (2006) 3202–3213.

[130] K. Ota, S. Ishihara, S. Mitsushima, K. Lee, Y. Suzuki, N. Horibe, T. Nakagawa, N. Kamiya, Improvement of cathode materials for polymer electrolyte fuel cell, J. New Mater. Electrochem. Syst. 8 (2005) 25–35.

[131] R. Jasinski, A new fuel cell cathode catalyst, Nature 201 (1964) 1212–1213.

[132] H. Jahnke, M. Schonborn, G. Zimmerman, Pre-adsorption of M-N4 on carbon support, Top. Cur. Chem. 61 (1976) 133–181.

[133] V.S. Bagotsky, M.R. Tarasevich, K.A. Radyushkina, O.E. Levina, S.I. Andrusyova, Electrocatalysis of the oxygen reduction process on metal chelates in acid electrolytes, J. Power Sources 2 (1977) 233–240.

[134] S. Gupta, D. Tryk, I. Bae, W. Aldred, E. Yeager, Heat-treated polyacrylonitrile-based catalysts for oxygen electroreduction, J. Appl. Electrochem. 19 (1989) 19–27.

[135] J.P. Dodelet, Oxygen reduction reaction in PEM fuel cell conditions: Heat treated non-precious metal N4-macrocycles and beyond, Springer Science + Business Media. Inc., New York, 2006. Vol. Chapter 3.

[136] F. Jaouen, J.P. Lefevre, M. Cai, Heat treated Fe/N/C catalysts for O_2 electroreduction: Are active sites hosted in micropores? J. Phys. Chem. B 110 (2006) 5553.

[137] J.P. Lefevre, J.P. Dodelet, Fe-based electrocatalysts made with microporous pristine carbon black supports for the reduction of oxygen in PEM fuel cells, Electrochim. Acta 53 (2008) 8269.

[138] J.P. Lefevre, E. Proietti, F. Jaouen, J.P. Dodelet, Major improvments in oxygen reducion activity of iron based catalysts in PEM fuel cells, Science 324 (2009) 71–74.

[139] K. Kinoshita, J.A.S. Bett, Electrochemical oxidation of carbon black in concentrated phosphoric acid at 135°C, Carbon 11 (1973) 237–247.

[140] K. Kinoshita, J.A.S. Bett, Potentiodynamic analysis of surface oxides on carbon blacks, Carbon 11 (1973) 403–411.

[141] P.L. Antonucci, F. Romeo, M. Minutoli, E. Alderucci, N. Giordano, Electrochemical corrosion behavior of carbon in PA, Carbon 26 (1988) 197–203.

[142] K. Kinoshita, Carbon electrochemical and physicochemical properties, John Wiley & Sons, New York, 1988.

[143] M. Wissler, Graphite and carbon powders for electrochemical applications, J. Power Sources 156 (2006) 142.

[144] S.S. Kocha, D.J. Yang, J.S. Yi, Characterization of gas crossover and its implications in PEM fuel cells, AIChE Journal 52 (2006) 1916–1925.

[145] L.M. Roen, C.H. Paik, T.D. Jarvi, Electrocatalytic corrosion of carbon supports in PEMFC cathodes, Electrochem. Solid-State Lett. 7 (2004) A19–A22.

[146] P.T. Yu, W. Gu, R. Makharia, F. Wagner, H. Gasteiger, The impact of carbon stability on PEM fuel cell start-up and shutdown voltage degradation, ECS Trans. 3 (2006) 797.

[147] Y. Shao, J. Wang, R. Kou, M. Engelhard, J. Liu, Y. Wang, Y. Lin, The corrosion of PEMFC catalyst supports and its implications for developing durable catalysts, Electrochim. Acta 54 (2009) 3109–3114.

[148] R. J. Balliet, C. A. Reiser, System and method for shutting down a fuel cell power plant US20040001980, 2004.

[149] L.V.L. Dine, M.M. Steinbugler, C.A. Reiser, G.W. Scheffler Procedure for shutting down a fuel cell system having an anode exhaust recycle loop US20020098393, 2001.

[150] C.A. Reiser, D.Yang, R.D. Sawyer, Procedure for shutting down a fuel cell system using air purge US6858336, 2005.

[151] C.A. Reiser, D.J. Yang, R.D. Sawyer, Procedure for starting up a fuel cell system using a fuel purge, US7410712, 2005.

[152] C.A. Reiser, L. Bregoli, T.W. Patterson, J.S. Yi, J.D. Yang, M.L. Perry, T.D. Jarvi, A reverse-current decay mechanism for fuel cells, Electrochemical and Solid-State Letters 8 (2005) A273–A276.

[153] P.T. Yu, S.S. Kocha, L. Paine, W. Gu, F.T. Wagner, The effects of air purge on the degradation of PEMFC during startup and shutdown procedures, 2004 Annual Meeting AIChE, New Orleans. LA, 2004. April 25–29.

[154] S.S. Kocha Mitigating fuel cell start up/shut down degradation US20060240293 2005.

[155] N. Takeuchi, T.F. Fuller, Modeling of transient state carbon corrosion for PEMFC electrode, ECS Trans. 11 (2007) 1021–1029.

[156] N. Takeuchi, T.F. Fuller, Investigation of carbon loss on the cathode during PEMFC operation, ECS Trans. 16 (2008) 1563–1571.

[157] N. Takeuchi, T.F. Fuller, Modeling and investigation of design factors and their impact on carbon corrosion of PEMFC electrodes, J. Electrochem. Soc. 155 (2008) B770–B775.

[158] S.D. Knights, K.M. Colbow, J. St-Pierre, D.P. Wilkinson, Aging mechanisms and lifetime of PEFC and DMFC, J. Power Sources 127 (2004) 127–134.

[159] A.I. Medalia, Morphology of aggregates: I Calculation of shape and bulkiness factors; application to computer-simulated random flocs, J. Colloid. Interfac. Sci. 24 (1967) 393–404.

[160] A.I. Medalia, F.A. Heckman, Morphology of aggregates - II. Size and shape factors of carbon black aggregates from electron microscopy, Carbon 7 (1969) 567–582.

[161] K.H. Kangasniemi, D.A. Condit, T.D. Jarvi, Characterization of vulcan electrochemically oxidized under simulated PEMFC conditions, J. Electrochem. Soc. 151 (2004) E125–E132.

[162] P. Stonehart, Carbon substrates for phosphoric acid fuel cell cathodes, Carbon 22 (1984) 423–431.

[163] E. Passalacqua, P.L. Antonucci, M. Vivaldi, A. Patti, V. Antonucci, N. Giordano, K. Kinoshita, The influence of Pt on the electro–oxidation behavior of carbon in PA, Electrochim. Acta 37 (1992) 2725–2730.

[164] A.M. Chaparro, N. Mueller, C. Atienza, L. Daza, Study of the electrochemical instability of PEMFC electrodes in aqueous solutions by means of membrane inlet mass spectrometry, J. Electroanal. Chem. 591 (2006) 69–73.

[165] J. Willsau, J. Heitbaum, The influence of platinum concentration on the corrosion of carbon in gas diffusion electrodes - A DEMS study, J. Electroanal. Chem. 161 (1984) 93–101.

[166] K. Kinoshita, Differential thermal analysis of PtO_2/carbon, Thermochimica Acta 20 (1977) 297–308.

[167] M. Cai, M.S. Ruthkosky, B. Merzougui, S. Swathirajan, M.P. Balogh, S.H. Oh, Investigation of thermal and electrochemical degradation of fuel cell catalysts, J. Power Sources 160 (2006) 977–986.

[168] K. Lee, J.J. Zhang, H.J. Wang, D.P. Wilkinson, Synthesis and catalyzation of nanotubes for fuel cells, J. Appl. Electrochem. 36 (2006) 507.

[169] P. Serp, M. Corrias, P. Kalck, Review of nanotubes for catalysis applications, Appl. Catal. A 253 (2003) 337.

[170] E. Antolini, Carbon supports for low temperature fuel cell catalysts, App. Cat. B: Environmental 88 (2009) 1–24.

[171] R.L. Borup, J.P. Meyers, B. Pivovar, Y.S. Kim, R. Mukundan, N. Garland, D.J. Myers, M. Wilson, F. Garzon, D.L. Wood, P. Zelenay, K. More, K. Stroh, T.A. Zawodizinski, J. Boncella, J. McGrath, M. Inaba, K. Miyatake, M. Hori, K.-i. Ota, Z. Ogumi, S. Miyata, A. Nishikata, Z. Siroma, Y. Uchimoto, K. Yasuda, K.-i. Kimijima, N. Iwashita, Scientific aspects of polymer electrolyte fuel cell durability and degradation, Chem. Rev. 107 (2007) 3904–3951.

[172] Y. Shao, G. Yin, J. Zhang, Y. Gao, Comparative investigation of the resistance to electrochemical oxidation of carbon black and carbon nanotubes in aqueous sulfuric acid solution, Electrochim. Acta 51 (2006) 5853–5857.

[173] G. Girishkumar, T.D. Hall, K. Vinodgopal, P.V. Kamat, Single wall carbon nanotube supports for portable direct methanol fuel cells, J. Phys. Chem. B 110 (2005) 107–114.

[174] J. Wang, G.M. Swain, Fabrication and evaluation of Pt/diamond composite electrodes for electocatalysis: Preliminary studies of the ORR, J. Electrochem. Soc. 150 (2003) E24–E32.

[175] A.C.C. Tseung, K.Y. Chen, Hydrogen spill-over effect on Pt/WO_3 anode catalysts, Catalysis Today 38 (1997) 439–443.

[176] A.C.C. Tseung, S.C. Dhara, Loss of surface area by Pt and supported Pt black electrocatalysts, Electrochim. Acta 20 (1975) 681–683.

[177] M. Haruta, M. Date, Advances in the catalysis of Au nanoparticles, Appl.Catal. A: General 222 (2001) 427–437.

[178] T. Ioroi, Z. Siroma, N. Fujiwara, S. Yamazaki, K. Yasuda, Sub-stoichiometric titanium oxide-supported Pt electrocatalyst for polymer electrolyte fuel cells, Electrochem. Commun. 7 (2005) 183–188.

[179] Turner, J.A., US DOE Annual Merit Review May 10th, 2011 http://www.hydrogen.energy.gov/pdfs/review11/fc084_turner_2011_0.pdf.

[180] A. Ishihara, J.-H. Kim, S. Doi, S. Mitsushima, N. Kamiya, K. Ota, Non-Pt cathode based on Ta for PEMFC, ECS Trans. 3 (2006) 255–261.

[181] L. Liu, A. Ishihara, S. Mitsushima, N. Kamiya, K. Ota, Zirconium oxides for PEFC cathodes, Solid-State Lett. 8 (2005) A400–A402.

[182] Y. Shibata, A. Ishihara, S. Mitsushima, N. Kamiya, K. Ota, Effect of heat treatment on catalyst for ORR of TaO_xN_yTi prepared by electrodeposition, Electrochem. Solid-State Lett. 10 (2007) B43–B46.

[183] I.C. Halalay, B. Merzougui, M.K. Carpenter, S. Swathirajan, G.C. Garabedian, A.M. Mance, M. Cai, Supports for fuel cell catalysts 20060246344, 2006.

[184] B. Merzougui, I. C. Halalay, M. K. Carpenter, S. Swathirajan Conductive matrices for fuel cell electrodes. 0251954, 2006.

[185] F. Garzon, E. Brosha, B. Pivovar, T. Rockward, T. Springer, F. Uribe, I. Urdampilleta, J. Valerio, Freedom car fuel contaminants; effect on PEMFCs, 2006 Annual DOE Fuel Cell Program Review, 2006.

[186] R. Mohtadi, W.K. Lee, J.W.V. Zee, SO_2 contamination, J. Power Sources 138 (2004) 216–225.

[187] J.M. Moore, P.L. Adcock, J.B. Lakeman, G.O.J. Mepsted, Effect of battlefield contaminants on PEMFCs, J. Power Sources 85 (2000) 254.

[188] K.E. Swider, D.R. Rolison, The chemical state of sulfur in carbon supported fuel cell electrodes, J. Electrochem. Soc. 143 (1996) 813–819.

[189] K.E. Swider, D.R. Rolison, Reduced poisoning of Pt electrocatalyst supported on desulfurized carbon, Electrochemical and Solid-State Letters 3 (2000) 4–6.

[190] J. St-Pierre, PEMFC contamination model: Competitive adsorption followed by an electrochemical reaction, J. Electrochem. Soc. 156 (2009) B291–B300.

[191] Y.S. Kim, R. Mukundan, F. Garzon, B. Pivovar, Sub-freezing fuel cell effects, 2005 HFCIT Program Review (2005).

[192] K. Wipke, S. Sprik, J. Kurtz, T. Ramsden, J. Garbak, Second-generation fuel cell stack durability and freeze capability from national FCV learning demonstration (2005).

[193] E.L. Thompson, J. Jorne, H.A. Gasteiger, Oxygen reduction reaction kinetics in subfreezing PEM fuel cells, J. Electrochem. Soc. 154 (2007) B783–B792.

[194] C. Chacko, R. Ramasamy, S. Kim, M. Khandelwal, M. Mench, Characteristic behavior of polymer electrolyte fuel cell resistance during cold start, J. Electrochem. Soc. 155 (2008) B1145–B1154.

[195] S. Kim, M. Khandelwal, C. Chacko, M.M. Mench, Investigation of the impact of interfacial delamination on polymer electrolyte fuel cell performance, J. Electrochem. Soc. 156 (2009) B99–B108.

[196] M.R. Tarasevich, V.S. Vilinskaya, Eleckrokhimiya 9 (1973) 96.

[197] A. Damjanovic, On the kinetics and mechanism of ORR on oxide covered Pt, J. Electrochem. Soc. 138 (1991) 2315–2320.

[198] Y. Suzuki, S. Sugawara, N. Horibe, S.S. Kocha, K. Shinohara, MEA performance modeling for breakdown of catalyst layer polarization components, 214th Electrochemical Society Meeting, Honolulu, HI, 2008.

[199] N.M. Markovic, H.A. Gasteiger, B.N. Grgur, P.N. Ross, Oxygen reduction reaction on Pt(111): effects of bromide, J. Electroanal. Chem. 467 (1999) 157–163.

[200] M. Uchimura, S.S. Kocha, The influence of Pt-oxide coverage on the ORR reaction order in PEMFCs, 214th Electrochemical Society Meeting, Honolulu, HI, 2008.

[201] K. Tsujita, S. Sugawara, S. Mitsushima, K. Ota, Effect of sulfate on peroxide generation, Fall Meeting of the Electrochemical Society of Japan, Tokyo, 2007, 9.

[202] M.C. Lefebvre, R.B. Martin, P.G. Pickup, Characterization of ionic conductivity profiles within proton exchange membrane fuel cell gas diffusion electrodes by impedance spectroscopy, Electrochem.Solid-State Lett. 2 (1999) 259–261.

[203] B. Pivovar, Y.S. Kim, The membrane-electrode interface in PEFCs: I A method for quantifying membrane-electrode interfacial resistance, J. Electrochem. Soc. 154 (2007) B739–B744.

[204] D. Malevich, E. Halliop, B.A. Peppley, J.G. Pharoah, K. Karan, Effect of relative humidity on electrochemically active area and impedance response of PEM fuel cells, ECS Trans. 16 (2008) 1763–1774.

[205] D. Malevich, E. Halliop, B.A. Peppley, J.G. Pharoah, K. Karan, Investigation of charge transfer and mass-transport resistances in PEMFCs with microporous layers using EIS, J. Electrochem. Soc. 156 (2009) B216–B224.

[206] J. Zhang, S. S. Kocha, Intricacies in the measurement of surface oxides of Pt using different electrochemical techniques. (unpublished reports), (2009).

[207] A. Ohma, S. Suga, S. Yamamoto, K. Shinohara, Membrane degradation behavior during OCV hold test, J. Electrochem. Soc. 154 (2007) B757–B760.

[208] A. Ohma, S. Yamamoto, K. Shinohara, Membrane degradation mechanism during OCV test, J. Power Sources 182 (2008) 39–47.

[209] S. Sugawara, K. Tsujita, S.S. Kocha, K. Shinohara, S. Mitsushima, K. Ota, Simultaneous electrochemical measurement of ORR kinetics and Pt oxide formation/reduction, 214th Elecrochemical Society Meeting, Honolulu, HI, Wednesday, October 16, 2008 (2008).

[210] F.T. Wagner, N. Ross, LEED spot profile analysis of the structure of electrochemically treated Pt(100) and Pt(111) surfaces, Surf. Sci. 160 (1985) 305–330.

[211] K. Itaya, S. Sugawara, K. Sashikata, In situ scanning tunneling microscopy of platinum (111) surface with the observation of monatomic steps, J. Vac. Sci. Technol. A 8 (1990) 515–519.

[212] Y. Nagahara, S. Sugawara, K. Shinohara, The impact of air contaminants on PEMFC performance and durability, J. Power Sources 182 (2008) 422–488.

[213] R. Shimoi, A. Takahashi, A. Iiyama, Development of fuel cell stack durability based on actual vehicle test data: Current status and future work, SAE International 2009-01-1014 (2009).

[214] S.S. Kocha, Performance measurement of MEAs, 214th Meeting of the ECS - HTM Working Group, 2008.

[215] S. S. Kocha, Accelerated stress tests for non-carbon supports, http://www.1eere.energy.gov/hydrogenandfuelcells/pdfs/durability_working_group_minutes_oct_2010.pdf, Durability working group meeting, October 10, 2010, Las Vegas, NV.

Chapter 4

Gas Diffusion Media and their Degradation

Ahmad El-kharouf and Bruno G. Pollet
PEM Fuel Cell Research Group, Centre for Hydrogen and Fuel Cell Research, School of Chemical Engineering, The University of Birmingham, Birmingham, United Kingdom

1. INTRODUCTION

Gas diffusion media (GDM), also called gas diffusion layers (GDLs), are important sub-components in PEMFC membrane electrode assemblies (MEAs) (Fig. 4.1).

The GDL, with its porous nature, plays an essential role in assisting the hydrogen oxidation (HOR) and oxygen reduction (ORR) reactions in the catalyst layers (anode and cathode) by allowing the reactants (hydrogen and oxygen/air) to diffuse from the flow field channels (on the flow field plates) to the active sites on the electrocatalyst. The GDL also facilitates water management in the catalyst layer and the polymeric proton exchange membrane (PEM), by allowing water vapor together with the reactants to diffuse out to ensure sufficient humidification for the PEM. At the same time, the GDL facilitates the liquid water produced on the cathode side to flow out of

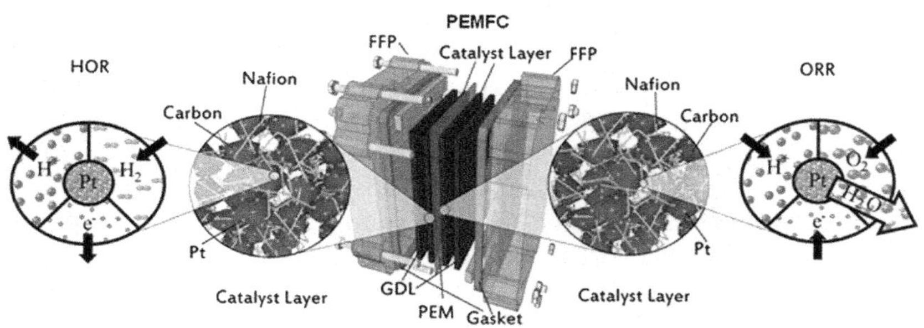

FIGURE 4.1 Explosive view of a Membrane Electrode Assembly (MEA).

Polymer Electrolyte Fuel Cell Degradation. DOI: 10.1016/B978-0-12-386936-4.10004-1

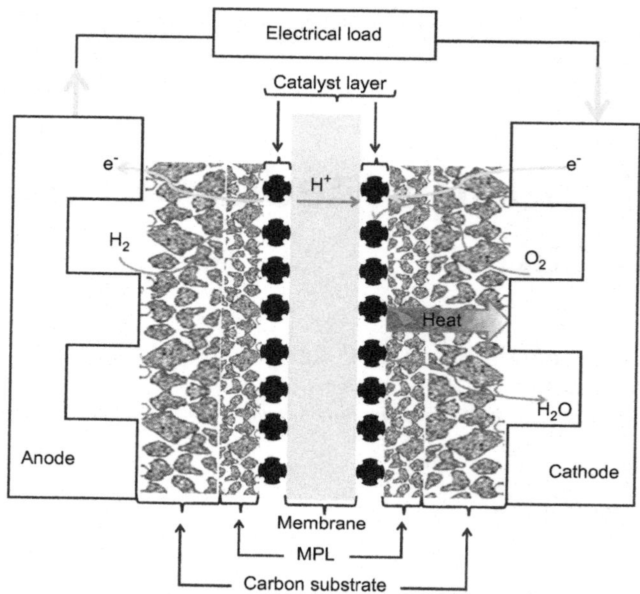

FIGURE 4.2 GDL structure and work principle in a single stack PEM fuel cell.

the fuel cell to prevent water flooding, which would block electrocatalyst active sites [1]. GDLs are electrically 'connected' to the catalyst ink and offer a supporting structure for the catalyst layers. In addition, the GDL is an electrically conducting medium that transfers electrons between the catalyst layer and the flow field or bipolar plates (BPPs) (Fig. 4.2).

In PEMFC electrode fabrication, the catalyst ink ((Pt/C/Nafion®) is either applied to:

i. the GDL to form a gas diffusion electrode (GDE) or a catalyst coated substrate (CCS); or
ii. the PEM to form a catalyst coated membrane (CCM).

The CCSs are usually prepared by spreading, spraying, deposition, ionomer impregnation, electrodeposition (continuous and pulsed) and sputtering, whereas the CCMs are fabricated by impregnation reduction, evaporation deposition, sputtering, dry spraying and decaling. For further details on the above methods, the reader is invited to refer to the excellent reviews on the analysis of PEM fuel cell design and manufacturing by Mehta and Cooper [1] and Litster and McLean on PEM fuel cell electrodes [2].

GDLs are made of porous electrically conductive materials, mainly carbon. The carbon substrates are typically water-proofed (e.g. by adding a hydrophobic material) to prevent the blockage of pores with water, as this can disrupt the diffusion of reactants to the catalyst layers during PEM fuel cell operation [1]. A microporous layer (MPL) made of carbon and a hydrophobic agent (see later), is often added onto the GDL surface between the catalyst layer and the GDL to enhance water removal from the catalyst layer, minimizing the electrical contact resistance with the adjacent catalyst layer, and preventing the catalyst ink from leaking into the GDL, thereby increasing the catalyst utilization and reducing the tendency of electrode flooding [3]. Furthermore, it has been reported that the presence of MPL in PEM fuel cell electrodes improves their performances and enhances their durability [4].

2. FABRICATION OF WOVEN AND NON-WOVEN GDLs

There are two types of porous carbon substrate GDLs, namely: carbon papers (non-woven) and carbon cloths (woven). Carbon paper GDLs are used because of their high permeability for gases and electrical conductivity. They are fabricated in four stages:

- **i.** a pre-pregging step (continuous strands are aligned with spools and a surface treatment is followed by a resin bath and the formation of a layered structure);
- **ii.** a moulding step;
- **iii.** a carbonization step; and
- **iv.** a graphitization step.

Carbon cloths are also fabricated in four stages:

- **i.** a carbonaceous fiber production step (made from *meso*-phase pitch spun by melt spinning, centrifugal spinning, blow spinning, etc.);
- **ii.** a fiber oxidation step;
- **iii.** a cloth formation step by weaving or knitting; and
- **iv.** a graphitization step [1].

After fabrication, GDLs can be produced in two further steps: either a one-stage process or a two-stage process is required. Conventional GDLs are usually fabricated via a one-stage process whereby the carbon substrate is often coated with a hydrophobic material to produce a high hydrophobicity GDL. In the two-stage process, a MPL is added onto the GDL produced in the one-stage process. The MPL is coated on either one side or both sides of the GDL by depositing a blend of carbon powder with a hydrophobic material [5]. Figure 4.3 shows a diagram of the GDL fabrication steps.

GDLs vary in their properties and therefore their performances based on their fabrication process and materials used. For example, GDLs vary not only in their structures (as paper and woven cloth), but also in their densities and thicknesses. Studies have shown that MEAs using carbon cloths exhibit higher power densities than those using carbon papers [5,6]. However, generally MEA manufacturers prefer using carbon paper ('industry standard'

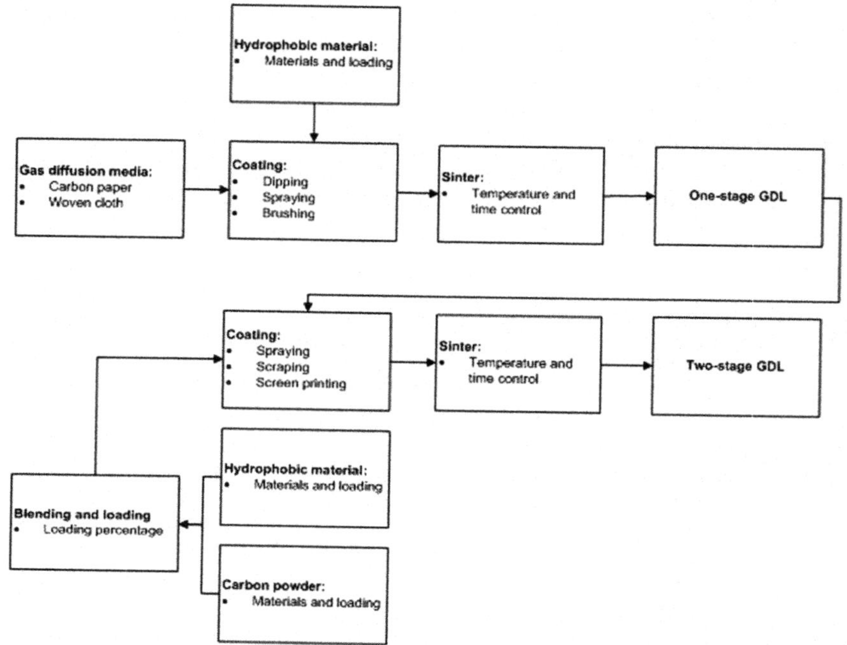

FIGURE 4.3 GDL fabrication process [5].

GDL: Toray TGP-H-60) due to its low cost and ease of applying MPLs and even the catalyst layer directly onto it [3]. Wetting carbon substrates by applying hydrophobic materials produces not only an increase in hydrophobicity of GDLs but also make their surfaces smoother, especially when applying the catalyst layers onto them in turn reducing contact resistances. However, excessive loading (wetting) results in a significant decrease in GDL electrical conductivities and in blockage of the GDL pores causing a reduction in reactant flows to the catalyst active sites to an unacceptable level (to maintain the required reaction rates). Studies show that MEAs perform best at a specific MPL loading. For example, polytetrafluoroethylene (PTFE) is the most commonly used material in GDLs, wetting with a loading ranging from 15 to 20 wt% yielding optimum performances [7–10]. Here, the hydrophobic material can be applied onto the GDL using various coating methods such as, brushing, spraying or dipping.

As stated earlier, MPL consists of a carbon black powder mixed with PTFE. Many MPL formulations of carbon powder types and PTFE loadings and their effect on MEA performance have been reported in the literature. For example, Table 4.1 lists some of the materials and their loadings used in various studies. Here, the MPL were applied either by printing, spraying, brushing, or painting on either or both sides of the GDL.

TABLE 4.1 MPL Composition and Loadings

Carbon Powder	Hydrophobic Agent	Optimum Performance	Reference
Acetylene Black, Black Pearls 2000, Composite Carbon and Acetylene Black with 10% Black Pearls 2000	PTFE: 30 wt%	Composite carbon with PTFE 30 wt%. Carbon loading 0.5 mg/cm^2	[11]
Vulcan XC-72	PTFE: 10–40 wt%	Vulcan XC-72 with PTFE 20–30 wt%. MPL loading 3.5 mg/cm^2	[9,12]
Carbon Black	PTFE: 24, 35, 45 wt%	Carbon black with PTFE 35 wt% Carbon loading 2 mg/cm^2	[8]

3. MANUFACTURERS OF GDLs

Worldwide, a few companies are focusing on GDL research and development, and fabrication. Table 4.2 lists the main GDL manufacturers for PEM fuel cell (as well as direct methanol fuel cell (DMFC)) applications the production of carbon GDLs is in the form of roll goods. The ready availability of low-cost carbon media in rolls would lead to considerations of highspeed manufacturing processes for the production of MEAs, given that PEM membrane material is available as roll goods.

4. GDL PROPERTIES AND THEIR CHARACTERIZATION

GDL properties and characteristics have a crucial effect on the PEM fuel cell performance. The properties are interrelated and affected by one another. GDL characterization and properties measurement are essential for assessing the GDL performance and degradation mechanism(s) (see later). In addition, studying GDL properties is essential for PEM fuel cell designers and developers as different characteristics are required in various applications e.g. stationary and automotive. GDL properties and their characterization methods are discussed in this section and summarized in Table 4.3.

The test methods, applied to most woven and non-woven substrates (in the absence and presence of MPL), cover the measurements of thickness, electrical conductivity and in-/through-planes air permeability of GDL materials as a function of applied vertical pressure.

TABLE 4.2 GDL Manufacturers

Manufacturers	Country	Company information	Products information
Ballard www.ballard.com	Burnaby, British Columbia, Canada	Ballard Power Systems Inc. is a world leading company in clean energy hydrogen fuel cells.	Ballard produces a wide range of GDL materials, both woven and non-woven/fiber and paper based with and without MPLs.
Sigracet www.sglgroup.com	Sigracet has 43 production sites in North America, Europe and Asia	SGL Group – The Carbon Company – is one of the world's leading manufacturers of carbon-based products.	SGL Group works in four fields of carbon – coarse-grain graphite, fine-grain graphite, expanded natural graphite as well as carbon fibers and carbon fiber composites. A wide range of carbon paper GDLs are produced with different thicknesses, PTFE loading, and MPL presence.
Toray www.toray.com	Tokyo, Japan	Toray Group is an integrated chemical industry group working on nanotechnology using organic synthesis chemistry, polymer chemistry and biotechnology as its core technology. In addition to the foundation businesses of fibers and textiles and plastic chemicals.	Toray carbon fiber paper 'TGP-H' has been used in PEMFC as a diffusion media for many years. Four sizes of TGP-H are available with variation in paper thickness and PTFE loading.
Freudenberg Group [Fuel Cell Components Technology (FCCT)] www.freudenbergfcct.com	Germany	Freudenberg FCCT is supplying GDL materials based on Freudenberg's unique non-woven technology.	FCCT develops fuel cell components with a focus on the development of stack seals, GDL, filters and

TABLE 4.2 GDL Manufacturers—cont'd

Manufacturers	Country	Company information	Products information
			humidifiers. Applications: PEMFC (automotive, stationary – co-generation, back-up power etc.) as well as for DMFC.
(E-TEK) BASF –The Chemical Company www.jonnycoder.net/aces/home.php	USA	Since 1990, E-TEK has focused on developing materials and manufacturing technologies for advancing fuel cells.	E-TEK develops and manufactures catalysts, GDLs, GDEs, and MEAs for PEM fuel cells of all types, sizes and temperature ranges.
CeTech www.ce-tech.com.tw	Taichung (Central Taiwan), Taiwan	CeTech was established in September, 2006 based on the notion of creating 'Clean Energy Technology'	The use of carbon materials to produce GDLs for fuel cells.

GDLs in PEM fuel cells are used under high compressive loads (>100 psi); therefore, the GDL materials have to be evaluated for the following:

i. compressive elastic and plastic deformation;
ii. compressed electrical properties; and
iii. air permeability under compression.

It is important to understand how GDLs behave under stress, as they are under load in the PEM fuel cell stack. However, if a substrate is compressible it will distort into the flow fields when under pressure; this would block or distort the gas flow dramatically. The *in-situ* lateral compressive loads may be cyclic and the GDL materials exposed to load cycles during the manufacturing of MEAs. Hence, thermal expansion, swelling and creep need to be considered.

Furthermore, in a PEM fuel cell, the GDL has a compressive load applied; this results from the cell compression and is necessary to seal cells and improve conformity to give a good electrical connection. Therefore, it is usually considered important to understand how a GDL behaves under loaded conditions as well as in the unloaded state. There has to be a balance between the GDL

TABLE 4.3 GDL Properties and Characterization Methods

GDL Properties	Characterization Methods
Surface Morphology and Fibers Structure	Scanning Electron Microscopy (SEM) and Transmission Electron Microscopy (TEM)
Porosity	–Mercury Porosimetry –Immersion Method
Gas and Water Permeability	In-house Built Systems
Bulk and Contact Resistance	–Electrochemical Impedance Spectroscopy (EIS) –4-wire Kelvin Impedance
Hydrophobicity	Sessile Drop Method
Mechanical Strength	Mechanical Compression

mechanical properties, particularly relating to compressibility. There is no prescribed figure for this parameter as the optimum depends upon the PEM fuel cell geometry, GDL properties and flow field design. If a substrate is too compressible it will distort into the flow field when under pressure and potentially reduce or block the gas flow area in the flow field. The other extreme is premature fracture due to lack of GDL compressibility. Additionally, the in-cell lateral compressive loads may be cyclic. Hence, consideration of thermal expansion, swelling and creep needs to be made. Table 4.4 lists some of the basic GDL properties as reported in manufacturers' datasheets. The wide variation is required to meet the needs of the fuel cell designs, operating conditions and applications.

PEM fuel cell performance is also strongly dependent upon the transport of the reactants and the reaction products. The transport characteristics in the catalyst layer and GDL are determined by its porosity permeability and hydrophobicity. PTFE is used to give the electrodes and GDLs hydrophobic characteristics. GDL conductivity can be influenced by changes to substrate thickness as well as its composition; the latter includes:

- Fiber type and length;
- Carbon fill;
- Binders; and
- Topcoat.

Gas permeability is another important characteristic of the substrate and plays an important role within PEMFCs. If gases are unable to penetrate the substrate, performance is severely affected.

TABLE 4.4 GDL Manufacturers, Trademarks and Main Properties

GDL Manufacturers	Trademark	Type	Thickness /μm	Area Weight/ g/m^2	Throughplane Resistance/ mΩ·cm^2
Ballard	1071HCB	carbon cloth	356	123	7.7
	2002HD	carbon cloth	298	96	7.2
	EP40	carbon paper	200	38	48
	EP40T	carbon paper + PTFE	200	43	13
	GDS3215 (D13)	carbon paper	210	60	14
	P50	carbon paper	170	50	6.4
	P70	carbon paper	230	75	7.4
	P50T	carbon paper+ PTFE	180	62	11.7
	P70T	carbon paper+ PTFE	255	85	13.4
	GDS1120	carbon paper	210	79	14.5
	GDS2120	carbon paper	260	101	14
	P75T	carbon paper + PTFE	255	88	13.4
	GDS 22100	carbon paper + PTFE +MPL	240	185	17
	MGL 200	molded graphite laminates	200	100	11.2
	MGL 400	molded graphite laminates	400	190	8.8
	GDS 2050-L	carbon paper	260	–	22
	GDS 2050-A	carbon paper	200	–	12

(*Continued*)

TABLE 4.4 GDL Manufacturers, Trademarks and Main Properties—cont'd

GDL Manufacturers	Trademark	Type	Thickness /μm	Area Weight/ g/m^2	Throughplane Resistance/ mΩ·cm^2
Toray	TGP-H-030	carbon paper	110	–	
	TGP-H-060	carbon paper	190	–	
	TGP-H-090	carbon paper	280	–	
	TGP-H-120	carbon paper	370	–	
Freudenberg	T10	carbon paper	210	105	20
	I6	carbon paper	210	115	10
	C2	carbon paper	250	130	12
	C4	carbon paper	250	130	11
	I3 C1	carbon paper	270	1458	15
	I2 C6	carbon paper	250	135	11
	I2 C8	carbon paper	230	135	10
E-TEK	LT 1200 N	non-woven web	185	75	
	LT 1200 W	woven web	275	200	
Sigracet	GDL 10 BA	carbon paper	400	85	12
	GDL 10 BC	carbon paper	420	135	16
	GDL 24 BA	carbon paper	190	54	10
	GDL 24 BC	carbon paper	235	100	12
	GDL 25 BA	carbon paper	190	40	10
	GDL 25 BC	carbon paper	235	86	12
	GDL 34 BA	carbon paper	280	86	11
	GDL 34 BC	carbon paper	315	140	14
	GDL 35 BA	carbon paper	300	54	12
	GDL 35 BC	carbon paper	325	110	15

TABLE 4.4 GDL Manufacturers, Trademarks and Main Properties—cont'd

GDL Manufacturers	Trademark	Type	Thickness /μm	Area Weight/ g/m^2	Throughplane Resistance/ mΩ·cm^2
CeTech	N0S1005	carbon paper	180	50	<7
	N1S1007	carbon paper + MPL	210	80	<15
	GDS210	carbon paper	210	40	<6
	GDS240	carbon paper + MPL	240	90	<15
	GDS340	carbon paper	340	55	<6
	GDS370	carbon paper + MPL	370	105	<15
	W0S1002	carbon cloth	360	125	<5
	W1S1005	carbon cloth + MPL	410	180	<13

4.1. Surface Morphology and Fiber Structure

Carbon fibers used in the fabrication of GDLs are either of non-woven or woven structures for carbon papers and carbon cloths respectively. The fiber structure has an important effect on the pore size distribution, electrical resistance and the GDL performance within the PEM fuel cell. Moreover, the morphology of the MPL has an effect on the catalyst layer morphology, which in turn has an effect on the active area of the catalyst and the PEM fuel cell performance.

Scanning Electron Microscopy (SEM) and Transmission Electron Microscopy (TEM) are used to study: the surface morphology, the fiber structure of the GDL, and the GDL/MPL degradation.

4.1.1. Fiber Structure

The most common fiber structure in GDLs is the one used in carbon papers due to its manufacturing simplicity. Figure 4.4(a) shows the fiber structure of a commercial carbon paper where PTFE loading can be seen between the fibers giving a more rigid structure. Another fiber structure for carbon papers is shown in Fig. 4.4(b): the fibers present a 'spaghetti' shape giving more flexibility to the carbon paper. Figure 4.4(c) shows a SEM image of a carbon cloth – this

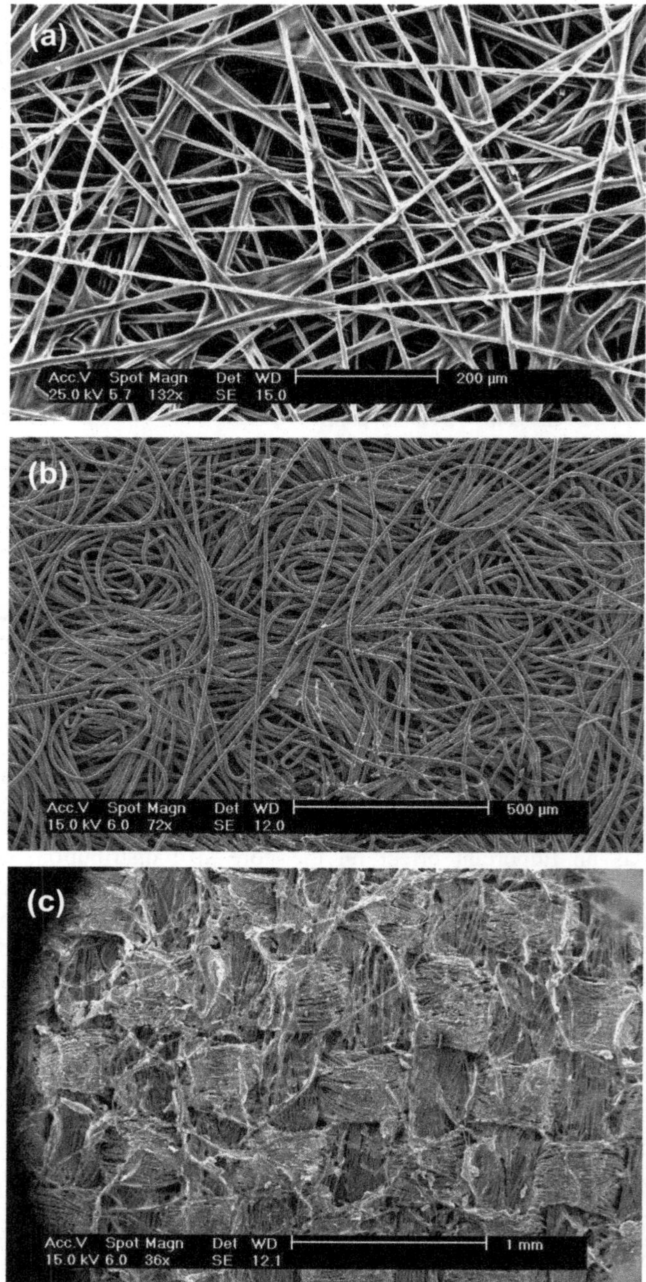

FIGURE 4.4 (a) Carbon paper GDL fiber structure, straight fibers. (b) Carbon paper GDL fiber structure, spaghetti fibers. (c) Carbon cloth GDL fiber structure.

structure offers high flexibility to the GDL and has shown high performances when used in PEM fuel cells.

This difference in GDL fiber structures has an effect on the PEM fuel cell performance. Generally speaking carbon cloths are recognized to exhibit better performances than carbon papers; however, their production cost is expensive relative to other GDLs. Wang et al. studied the difference between carbon paper and carbon cloth [13]. They found that carbon cloth GDLs perform better in high relative humidity (RH) conditions as the fiber structure has a smaller tortuosity and a rougher back surface assisting water droplets to detach from the surface. Carbon papers, on the other hand, have higher tortuosity and smoother surfaces leading to stagnation of water produced at the cathode side, hence they operate better in low RH conditions [13].

Many investigations have shown that the mechanical rigidity of the GDL has an effect on the PEM fuel cell performance. Flexible carbon cloth extrudes into the flow field channels when assembled under load in a stack. Carbon cloth has 43–125% higher intrusion than carbon paper [14]. Moreover, Kandlikar et al. reported that carbon paper GDL intrusion can be non-uniform across the channels due to its heterogeneous structure [15]. However in most cases, GDL intrusion into flow field channels causes a pressure drop across the channels and an uneven distribution of the reactants, leading to poor PEM fuel cell performance [16]. Thus, this property has a direct effect on the design of the flow field [17], as narrow channels can only be used with carbon cloths; however wider channels can be used with carbon papers.

Finally, it should be emphasized that an improved knowledge of the heat treatment history and the types of fiber precursor used for commercial GDLs would help to explain the above observations.

4.1.2. Microporous Layer (MPL)

The MPL morphology is determined by the size and the loading of the carbon powder particles; for example, finer carbon powder results in a smoother surface with smaller pores. Wang et al. studied the effect of the carbon powder size on the MPL and its effect on the PEM fuel cell performance. They found that the carbon powder size and loading, as well as the hydrophobic agent concentration, need to be optimized for the various applications (stationary and automotive) [11]. Figure 4.5 shows SEM images for some commercial GDLs.

4.1.3. Structural Characterization

The microstructure of the GDL is important as it controls the porosity and tortuosity. This strongly influences both the diffusion and mass flow from the flow fields to the catalyst layers. As previously stated, the GDLs are usually

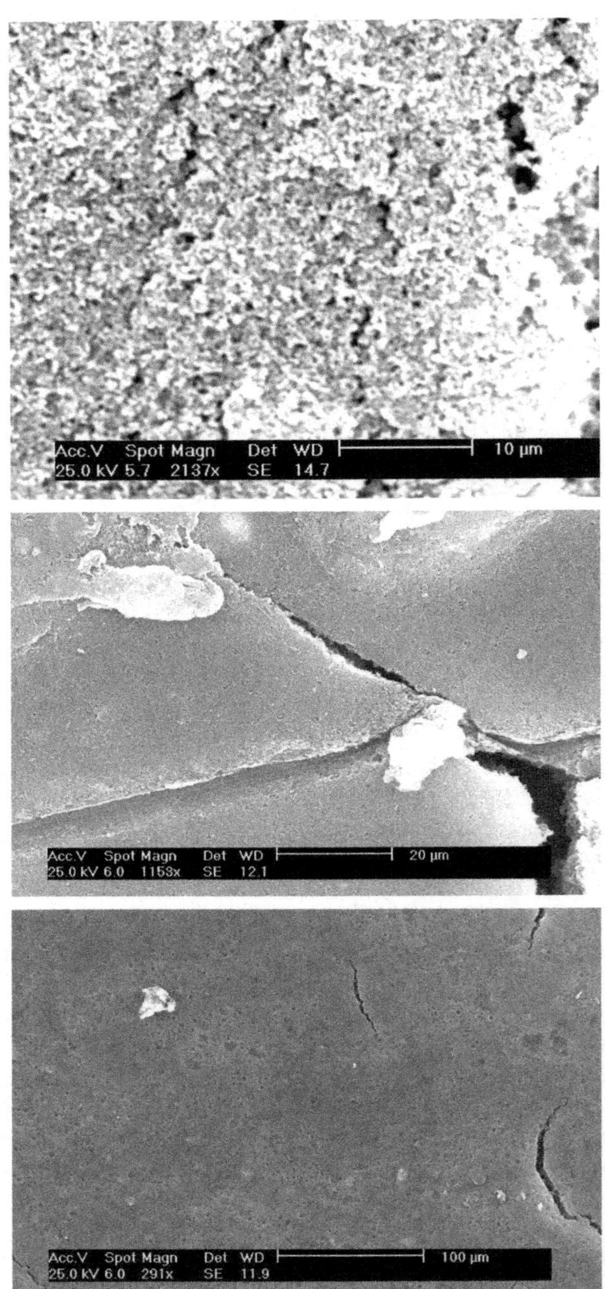

FIGURE 4.5 SEM images for MPL in commercial GDLs.

examined by SEM and TEM in order to investigate the structure of the carbon fibers and the distribution of MPL. SEM is also used to study:

i. the surface morphology and the fiber structure of the GDL; and
ii. the effect of degradation on the MPL surface and the GDL fibers.

4.2. Porosity and Gases and Water Transport

The reactant flows through the GDL, by both diffusion and convection, are dependent upon the geometry, the material parameters, and the flow field plate design [19,20]. The GDL porosity has a direct impact on the 'effective' diffusion coefficient (D_{eff}) [18] and permeability [19], and therefore the reactants and water transport through the GDL. *Note that molecular diffusion is proportional to the flux of the reactants and the gradient in species concentrations.* GDL bulk porosity is measured by using two methods: mercury porosimetry and the immersion method [18].

As briefly explained earlier, a GDL can consist of a single layer (one-stage GDL) or multi-porous layers (two-stage GDL, see above). As carbon substrates have relatively larger pore sizes compared to MPLs, in most cases a GDL is a multi-porous layer [18]. Figure 4.6(a) shows a commercial GDL pore size distributions measured by mercury porosimetry. The figure clearly shows the change in distribution pore size diameters when increasing the hydrophobic agent loading (Fig. 4.6(a)) and when adding a MPL (Fig. 4.6(b)). It can be observed that increasing the hydrophobic agent loading (to a one-stage GDL) and adding a MPL decreases the GDL porosity leading to a new range of pore size diameters (lower than the original ones).

Permeability is a property of a solid allowing a fluid to go through it in the presence of a differential pressure (Δp). Permeability is an important property for PEM fuel cell performance as it describes the flow of the reactants through the GDL. In-plane and through-plane permeability values depend greatly on several GDL parameters e.g. thickness, density, hydrophobic agent loading, fiber structures, and the presence and type of MPL. GDLs gas permeability (in-plane and through-plane) measurement methods reported in the literature are based on a similar and simple principle in which one side of a GDL is subjected to a controlled gas flow and a pressure drop is measured across it. Furthermore, in-plane permeability of a GDL is an important factor as it influences the gas flow over the flow field plate (FFP) and through the GDL. There is no standard method *per se* for GDL in-plane permeability measurements and experiments reported in the literature are usually carried out using different FFP designs and either air [21,22] or nitrogen [23] as gases. Figures 4.7(a) and 4.7(b) show in-plane and through-plane permeability measurement kits respectively.

In in-plane and through-plane permeability measurements, the input flows are controlled and varied, and the drops in pressure across the GDL (Δp) are measured. Using the experimental values and the Darcy-model, the in-plane

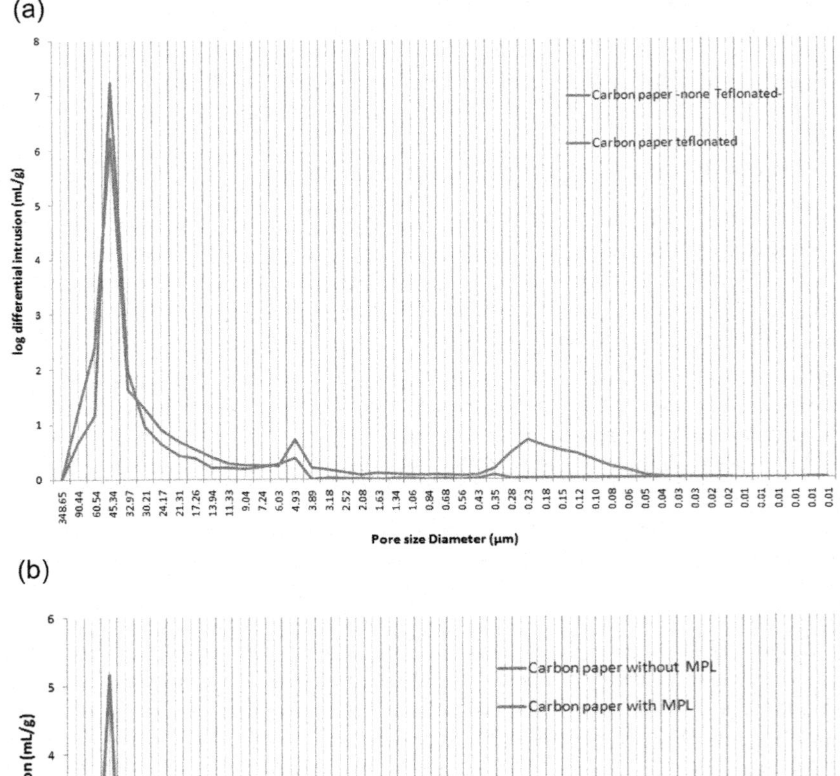

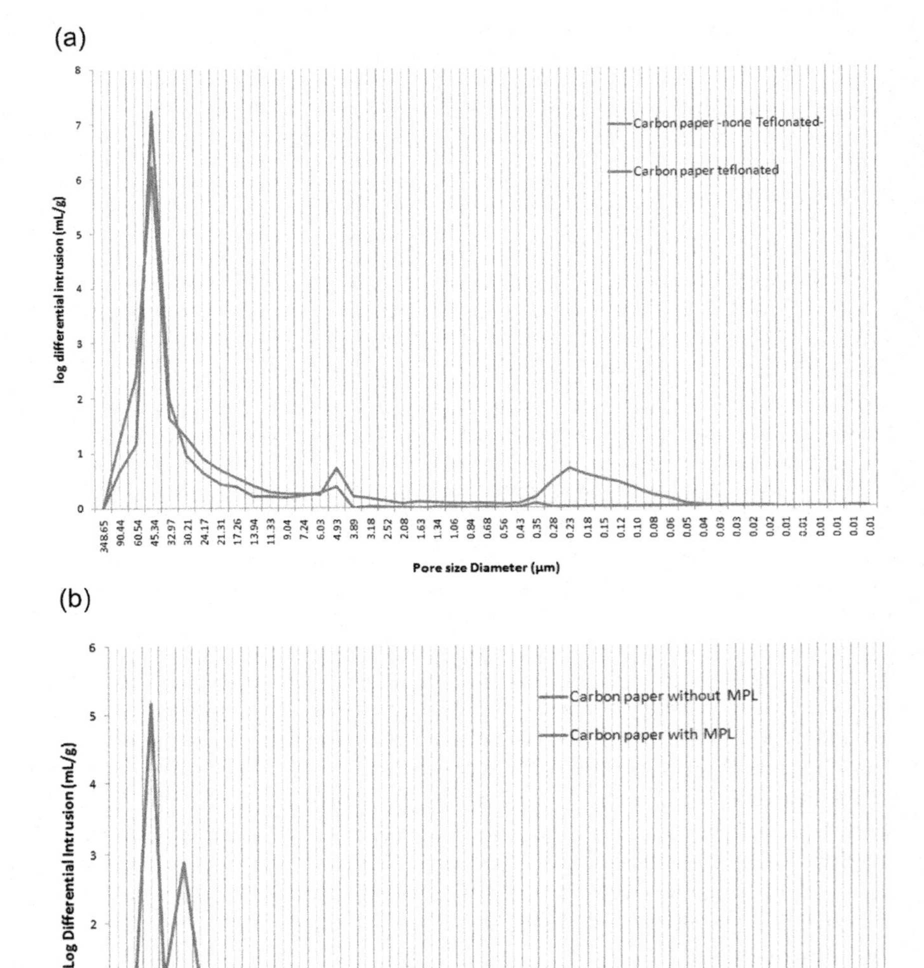

FIGURE 4.6 (a) GDLs pore size distribution change with increasing the hydrophobic agent loading. (b) GDLs pore size distribution change when adding a MPL.

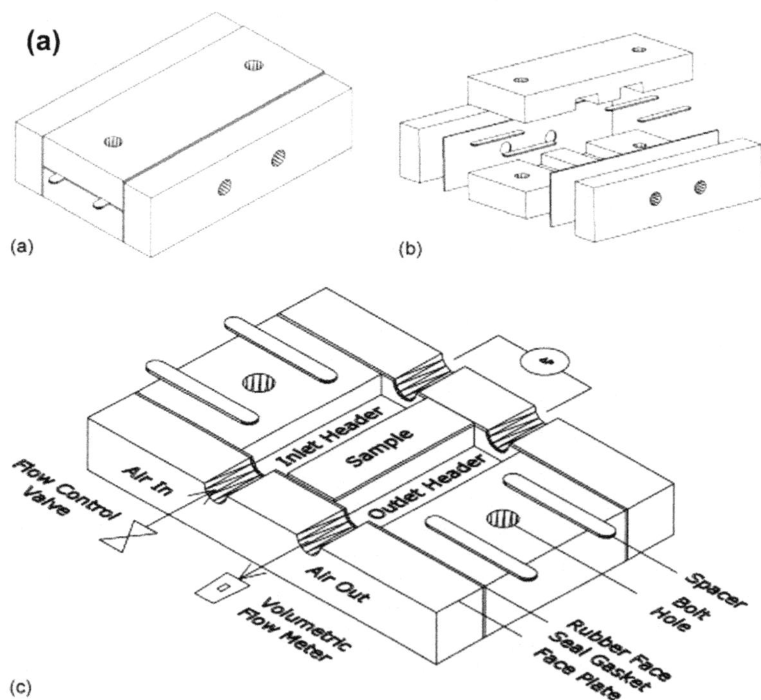

FIGURE 4.7 (a) Schematic of in-plane permeability measurement [21].

and through-plane permeability of a GDL sample at constant airflow can be calculated using equations (4.1) and (4.2):

$$\frac{dp}{dr} = -\frac{\eta}{k}u_r \quad p(r) = p_i - \frac{\eta}{k}\frac{Q}{2\pi h}ln\left(\frac{r}{R_i}\right) \tag{4.1}$$

$$u_r = \frac{Q}{2\pi h} \quad \Delta p := p_i - p_a = \frac{\eta}{k}\frac{Q}{2\pi h}ln\left(\frac{R_a}{R_i}\right) \tag{4.2}$$

where:
k is the air permeability in mm^2
η is the viscosity of air at 298.15 K (18.2 x 10^{-6} Pa.s)
Q is the airflow in cm^3/min
h is the thickness of sample in μm
u_r is the gas velocity at radius r in m/s
Δp is the pressure differential/100 in Pa
p_i is the input pressure/100 in p_a
r, R_a are the distance between the inlet channel and the outlet channel
R_i is the radius of the inlet channel

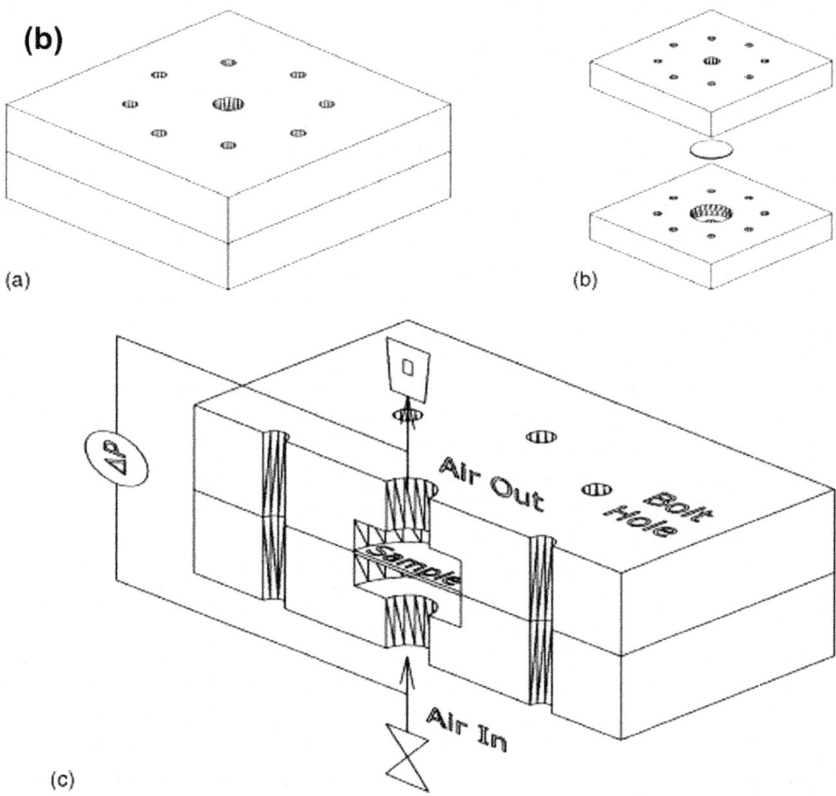

FIGURE 4.7 (b) Schematic of through-plane permeability measurement [21].

Water transport is another crucial function of the GDL. Recently, Tamayol and Bahrami studied water permeability using various commercial GDLs [24]. In their study, the threshold water pressure for permeation through different types of GDLs was measured using a bespoke membrane pressurized filtration cell system as shown in Fig. 4.8. The pressure was controlled by controlling the water level over the GDL. Water permeability was calculated from the measured mass of the collected water after a set period of time. It was found that PTFE loading and thickness increase the permeation threshold, but this is not affected by moderate compression of the GDL. They also reported that thickness has no effect on water permeability [24].

4.3. Electrical Conductivity and Contact Resistance

The interface contact resistance between MEA sub-components can have an influence on the performance of the PEM fuel cell. Electron transport through

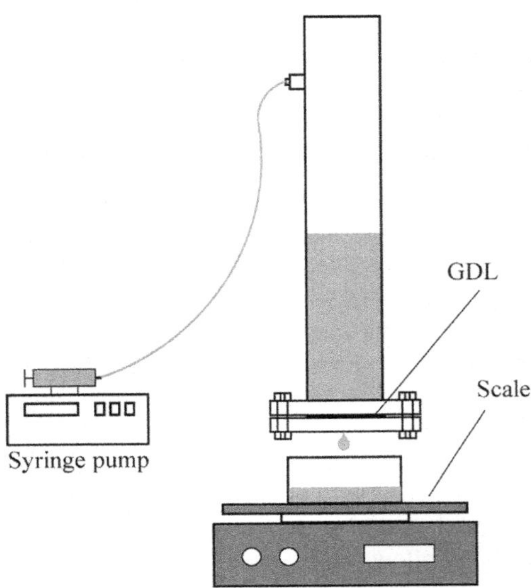

FIGURE 4.8 Water permeability testing setup as reported by Tamayol and Bahrami [24].

the GDL is affected by the GDL thickness, gas channel width and electronic conductivity [25]. In addition, the transfer of electrons between the GDL, the catalyst layer and the FFP/BPP affects the transport of electrons and has a crucial role in PEM fuel cell performance [18]. Many studies have focused on studying the current distribution across the MEA active area and techniques for measuring it, as it is of great importance for PEMFC stack developers [26–28].

In assessing the GDL electron transport capability, three parameters are commonly measured, namely in-plane, through-plane and contact resistances. Zhou and Liu studied the effect of the electrical resistance of the GDL on the PEM fuel cell performance by considering the anisotropic nature of the GDL structure. They found that GDL through and in-plane resistances can be neglected as they have little effect on the PEM fuel cell overall performance [29]. However, Nitta et al. studied the contact resistance between the GDL and the catalyst layer (CL) [30]. They observed that the contact resistance between the GDL-CL is of one order of magnitude higher than that between the GDL and the graphite (Poco) FFP. Furthermore, they also found that all contact resistances were also dependent upon the compression, therefore non-uniform compression of the GDL may result in an uneven current distribution along the MEA [30]. The same authors added that the GDL compressive modulus affects the contact resistance between the GDL and the FFP, and they concluded that carbon cloth has lower contact resistances compared to carbon paper [30]. Higier and Liu studied the difference in contact resistances under the channel and the land of the FFP [31]. They concluded that the variations in contact resistance depend strongly upon the channel width as it becomes more significant in wide channels and negligible in narrow channels. Higier and Liu also explain that the reason for this significant difference is due to an increase in contact resistance between the GDL and the CL caused by 'insufficient' compressions [31].

Potential techniques for measuring contact resistance have been reported [25–31], and include:

i. A highly conductive, thin, flat layer which is added to the substrate surface such that the contact resistance is removed completely. The layer completely coats the substrate surface while not penetrating into the bulk. Assumptions are made that the bulk resistance of the added layer is sufficiently low to prevent the added thickness significantly affecting the overall resistance, and the added layer contact resistance may also be considered as negligible.

ii. The resistance can be measured between two points on the same side of the GDL sample. If the distance between the points is then doubled, the bulk component of the resistance should also double, while the contact resistance remains the same. This allows the contact resistance to be measured. AC impedance is a common method of 'deconvoluting' contact and bulk resistance. For poorly conducting FFP materials, it is likely that the contact resistance is more significant. If this is the case then the measured resistance will increase more rapidly for carbon substrates with a high contact resistance. Contact resistance is also measured using simply a 4-wire Kelvin resistance measurement vs. clamping pressure. A non-linear decrease in the overall resistance with pressure increase is often observed as shown in Fig. 4.9.

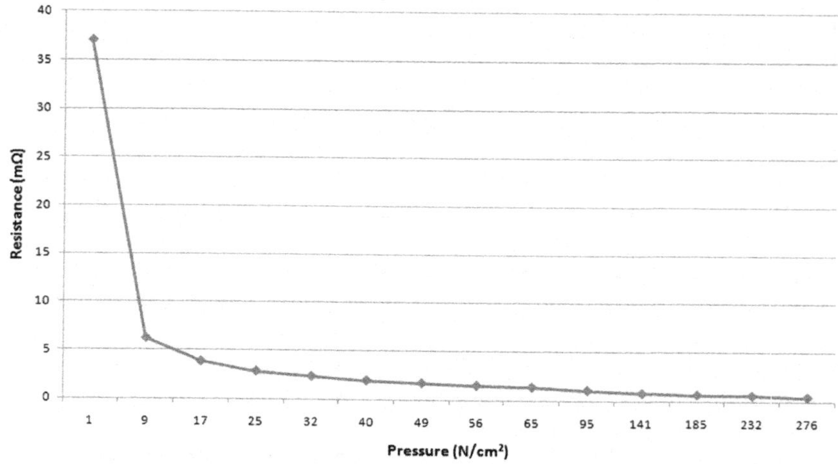

FIGURE 4.9 GDL resistance versus pressure.

4.4. Thermal Conductivity and Contact Resistance

Thermal conductivity in GDLs is an important property for predicting the temperature distribution within the MEA. As is the case for electrical resistance, thermal resistance is also dependent on clamping pressure; for example, both through-plane thermal and contact resistances decrease with an increase in compression or clamping pressure. It was shown that the presence of PTFE in the GDL leads to an increase in thermal conductivity at low clamping pressures (but to a decrease at higher compressions) [32]. Recently, Karimi et al. showed that the presence of MPL in the GDL decreases both thermal conductivity and contact resistance [32]. Furthermore, the humidification of the GDL increases its thermal conductivity [33]. Again, contact resistance has a dominant contribution to the thermal losses under compressive loads [34].

4.5. Hydrophobicity

As mentioned earlier in this chapter, GDL hydrophobicity is an important parameter for water management in the PEM fuel cell. GDL hydrophobicity depends upon two main factors: the hydrophobic agent loading and its pore diameter. Park and Popov studied numerically the effect of these parameters on the cathode GDL and reported that the presence of MPL decreases the average pore diameter and increases the hydrophobic area leading to enhanced water management and oxygen diffusion [35]. Here, the hydrophobicity of the GDL is measured by measuring water droplet contact angle using the sessile drop method (Fig. 4.10).

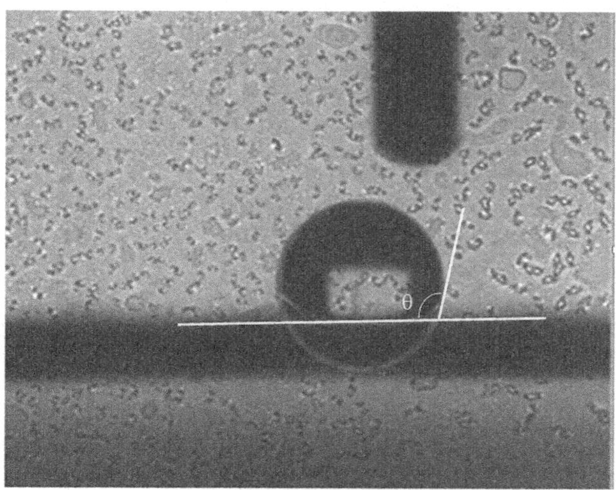

FIGURE 4.10 GDL water contact angle (θ).

4.6. Mechanical Characteristics

GDLs are often subjected to high compression forces as part of the MEA when assembled, and thus, GDLs need to be rigid and stable enough to withstand these mechanical stresses. The GDL mechanical stability is usually studied by applying an incremental pressure to it, and measuring the changes in thickness. Various investigations have focused on GDL compression optimization for maximum PEM fuel cell performance [36], but to the authors' knowledge, no studies in the literature have reported the effect of GDL mechanical resistance on reactant flows. As mentioned previously, clamping pressure decreases electrical and thermal resistances; however, it also affects the reactant flows through the GDL [37] and causes high intrusion into the flow field channels for woven GDLs (Fig. 4.11). Many investigations have reported that the optimum clamping pressure varies with different GDL types under various operating conditions [38,39]. Ge et al. found that the effect of clamping pressure on GDL is higher at high current densities [38].

4.7. Other GDL Characteristics

Other characteristics such as GDL thickness and density have an effect on GDL performance. For example, GDL thickness needs to be optimized as very thin

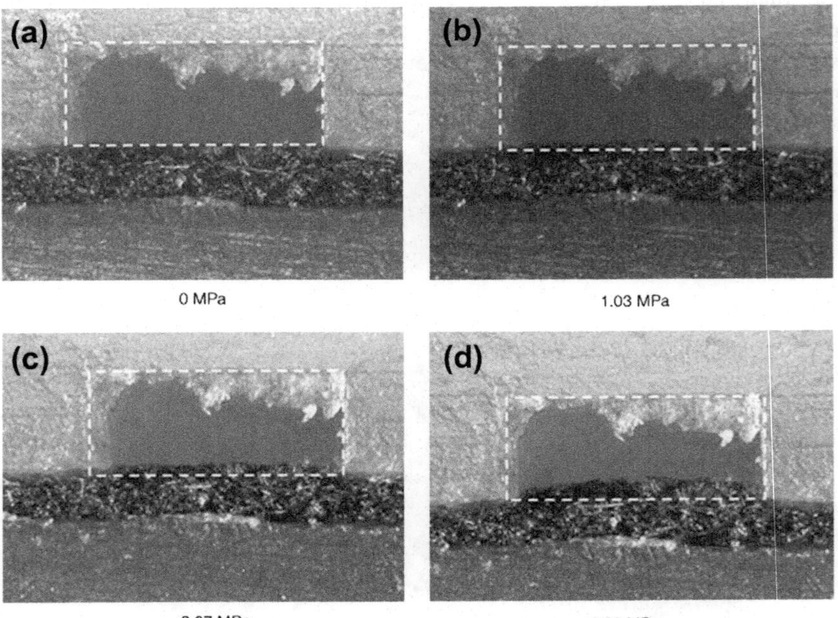

FIGURE 4.11 Effect of clamping pressure on GDL intrusion [15].

GDLs suffer from poor performance at high current densities due to high contact resistances, high mass transfer losses and water pooling. Thick GDLs increase activation losses, through-plane resistances, and mass transfer losses [10,40]. That is why commercial GDLs thickness vary between 100–420 μm (see Table 4.4).

5. THE DEGRADATION MECHANISMS OF GDLs

As described in previous sections, GDL roles and characteristics are important and affect the overall performance and lifetime of the PEM fuel cell [41]. GDL degradation can be detected by changes in its characteristics and properties, and therefore its performance in the PEM fuel cell. Very limited literature is available on GDL degradation, as the focus in PEM fuel cell studies so far has been on catalyst and membrane development. Due to the difficulty in separating the GDL degradation from MEA sub-components in *in-situ* experiments, GDL degradation studies are often performed by *ex-situ* testing; in other words, by simulating a PEM fuel cell environment. Therefore, a good understanding of the GDL properties and testing methods as presented in the previous sections is essential for GDL degradation studies. *In-situ* studies are usually conducted to study the effect of the changes in properties on the PEM fuel cell performance.

Two main changes have been reported in GDL degradation, namely:

- **i.** wetting behavior changes due to loss of the hydrophobic agent and carbon surface changes; and
- **ii.** changes in the structure of the GDL due to carbon corrosion and mechanical stress [42].

In addition, changes in MPL have been observed due to losses of PTFE/C and carbon oxidation [43]. GDL degradation can be categorised into:

- **i.** electrochemical;
- **ii.** mechanical; and
- **iii.** thermal degradation.

Extreme PEM fuel cell operating conditions enhance and accelerate the degradation of GDL and other MEA sub-components. High cell voltages (and potentials), low relative humidities, and high temperatures accelerate GDL degradation [41]. In addition, dynamic loading and potential cycling can also accelerate degradation.

5.1. Electrochemical Degradation

Platinum (Pt) is still the only electrocatalyst which provides sufficiently high activity for the ORR. To reduce the amount of precious metal used, Pt is usually supported on carbon in the form of dispersed nanoparticles, allowing for high-catalyst surface areas at low and ultra-low catalyst loadings. However, carbon-supported catalysts are highly susceptible to catalyst nanoparticle agglomeration and/or dissolution. Another major issue of high-performance, low-loaded PEMFC electrodes resides in the thermodynamic instability of carbon under PEMFC cathode conditions [1]. Above 0.207 V vs. SHE, carbon is oxidized to carbon dioxide following Eqn 4.3 (here, the oxidation to carbon monoxide (Eqn 4.4) is thermodynamically not favored [44] (Eqn 4.5)). The GDL suffers from carbon loss by oxidation and loss of structure rigidity due to the electrochemical activity of the PEM fuel cell.

$$C + 2H_2O \rightarrow CO_2 + 4H^+ + 4e^- \quad E^\circ = 0.207 \text{ V vs. SHE} \tag{4.3}$$

$$C + H_2O \rightarrow CO_2 + 2H^+ + 2e^- \quad E^\circ = 0.518 \text{ V vs. SHE} \tag{4.4}$$

$$CO + H_2O \rightarrow CO_2 + 2H^+ + 2e^- \quad E^\circ = -0.103 \text{ V vs. SHE} \tag{4.5}$$

Oxidation of carbon support, also known as carbon corrosion, can lead to performance decrease due to accelerated loss of ECSA (electrochemical surface area) and modification of pore morphology as well as surface characteristics. Hence, carbon loss from the GDL results in significant changes in its properties and to PEM fuel cell performance. Potential cycling, particularly at high cell potentials, results in loss of carbon material [41,45]. Aoki et al. reported that after electrochemically treating the cathode GDL in an *ex-situ* electrochemical corrosion test, the GDL experienced a loss in hydrophobicity and the corroded GDL maintained the same oxidant utilization and diffusion losses, whereby cathode flooding was observed at high current densities [46].

Chen et al. studied electrochemical degradation of the GDL under simulated PEMFC conditions at potentials in the range of 1–1.4 V. At high potentials (> 1.2 V) significant loss in carbon led to a thinning of the GDL fibers. In addition, the authors observed changes in the GDL morphology, resistance, gas permeability and contact angle. Chen et al. also reported that the ohmic resistance, charge transfer and mass transfer resistances increased significantly, together with an obvious drop in the overall PEMFC performance [47]. Recently, Hiramitsu et al. [48] found that carbon oxidation in the GDL resulted in an increase in diffusion overpotentials.

Kumar et al. reported a decrease in GDL rigidity and an increase in strain under PEM fuel cell clamping pressure due to electrochemical degradation. Carbon paper suffers from loss of structure rigidity with aging due to the weak fiber interface, in comparison to carbon cloths which possess a more stable structure. This loss in rigidity results in an increase in GDL intrusion into the flow field channels, which affects the PEM fuel cell performance [49].

Low relative humidities (RHs) [41,45] and high temperatures [52] enhance carbon corrosion. Experiments showed an increase in the MPL pore sizes due to the loss of the carbon particles in the MPL. The carbon fibers are often treated by PTFE to protect them, but with time, they lose their hydrophobicity and mass transport losses increase due to water accumulation in the GDL [41]. On the other hand, it has been reported that carbon loss increases from 8% to 36% when the temperature increases from 120 to 150°C [41].

It is important to mention that graphitized carbon has a higher oxidation resistance [49], hence it can be used in PEM fuel cells to mitigate oxidation losses. Further studies are required to investigate the effect of graphitized carbon on GDL properties. Hiramitsu et al. showed that homogeneous hydrophobic coating of the GDL protects the GDL carbon fibers from oxidation and controls the diffusion overpotential under long term operation of the PEM fuel cell [48].

5.2. Mechanical Degradation

GDL mechanical degradation occurs due to clamping pressures (compressions) applied to PEM fuel cell assembling, erosion by the reactant and water flows, and water freezing at low temperatures. High compression results in GDL deformation and changes in thickness due to the breakage and displacement of fibers under high pressures. The variations in GDL thickness are irreversible and result in a change in electrical and thermal resistance, together with a decrease in porosity. Sadeghi et al. observed high hysteresis in experiments involving pressure cycling whereby hysteresis became minimal after approximately five cycles of compression and the deformation became permanent in the GDL fiber structure [34]. Bazylak et al. reported the breakage of fibers and deterioration of the hydrophobic coating after applying high clamping pressures on the GDL. This resulted in a change in the structure and affected the water pathway through the GDL as water took other preferential pathways [51]. Figure 4.12 shows SEM images for the effect of pressure cycling on a carbon paper GDL structure.

GDLs are subjected to an uneven distribution of pressure due to GDL intrusion into the flow field channels on the FFP due to the uneven pressure applied by the bolts on the edges of the cell [15]. Maher et al. studied the mechanical stress on the MEA prior to and after PEM fuel cell operation under various conditions [52]. They reported that the variations in temperature and RH resulted in local bending in the MEA, and therefore produced a non-uniform distribution of stress across it (formation of cracks was observed) [52]. Another factor which contributes to the mechanical stress on the MEA and GDL is the heterogeneity of both the GDL and CL [36].

As the GDL is in contact with the reactant flows, it has been observed that these flows may erode the GDL surface, resulting in a loss in GDL hydrophobicity (due to the loss of the PTFE from both the fibers and the MPL) [53]. This mechanism is enhanced and accelerated at increased flow rates and temperatures, where it has been observed to be higher on the cathode side as the flow of air is higher than that of hydrogen on the anode side. The loss in hydrophobicity was reported to occur in short periods of operation of 100–150 hours [53] which resulted in an increase in GDL electrical resistance and porosity. Wu et al. studied the *in-situ* accelerated degradation of GDLs and the effect of elevated temperatures and flow rates on the PEM fuel cell performance. They found that

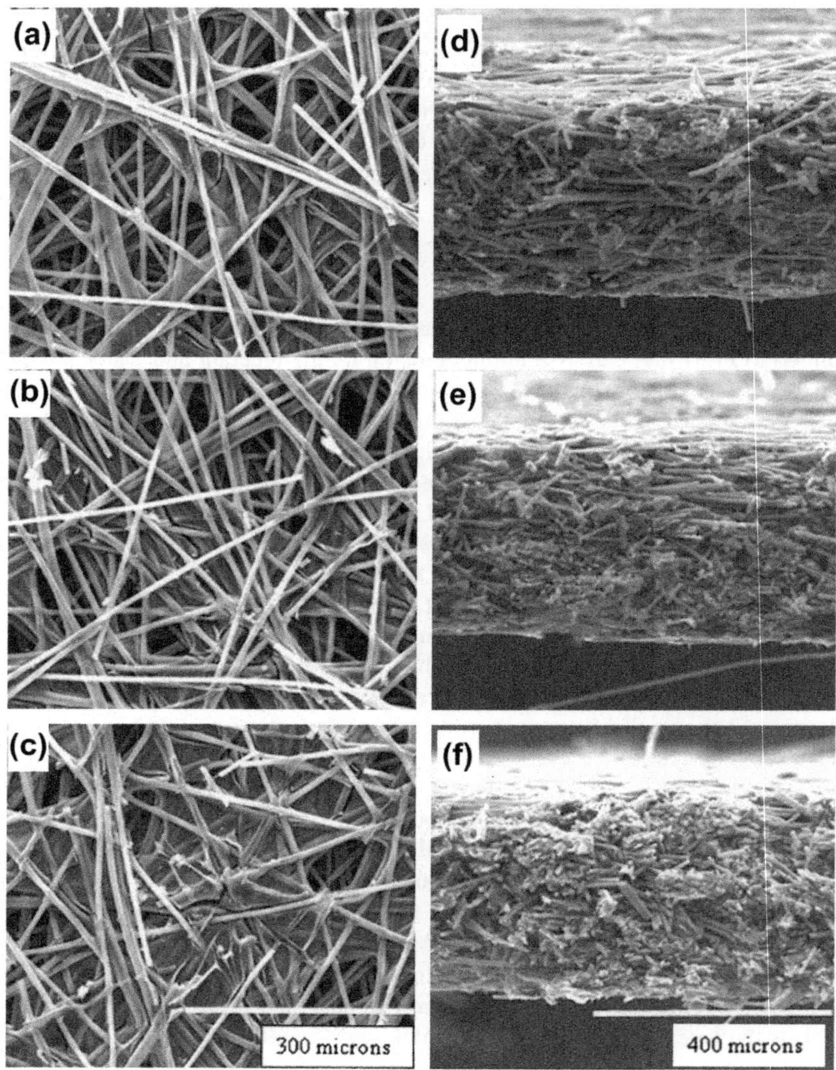

FIGURE 4.12 Effect of pressure cycling on carbon paper GDL structure [14].

when using the degraded GDL, its performance is comparable to that of a fresh one at high potentials and low current densities; however, at high current densities, concentration losses increase significantly [54].

Water transport through the GDL also contributes to the GDL degradation. Lin et al. studied MPL degradation in a simulated PEM fuel cell under water flooding conditions. They observed that the MPL was washed out from the GDL surface and moved into the GDL structure. Under their conditions, however, they found that the performance of the GDL after degradation was better than a GDL without MPL (as the MPL material maintained the ability of the GDL to transport the produced water and prevent flooding and concentration losses in the PEM fuel cell) [55].

In automotive applications, PEM fuel cells can be subjected to sub-zero conditions at start-up conditions. The GDL under such conditions suffers from deformation in its structure and a drop in its performance. This type of degradation can be seen as thermal degradation affecting the GDL mechanical integrity. A few studies have addressed the effect of freezing conditions on GDL performance, and the formation of ice between the GDL/CL/membrane interfaces. The presence of ice applies extra stress to the brittle GDL when clamped within the fuel cell structure, causing it to 'crush' [56]. Yan et al. [57] reported the effect of freeze/thaw cycles on the MEA properties and performance. They observed significant damage to the GDL when operated under freezing conditions caused by the backing layer (Teflon) and the binder structure on the GDL carbon paper surface. However, Pelaez and Kandlikar [58] and Lee and Merida [51] reported no effect on the GDL properties due to freeze/thaw cycling in *ex-situ* experiments. In contrast to the above findings, Lim et al. observed, by performing *in-situ* experiments on three types of GDL (carbon paper, carbon felt and carbon cloth), a drop in PEM fuel cell performance under freeze/thaw cycling conditions [59]. They found that the carbon felt GDL showed the smallest degradation in performance, due to its high stiffness which prevented gaps forming between the GDL and the membrane, which prevented the increase in contact resistances [59]. The *ex-situ* testing also showed that the carbon felt GDL surface was the less damaged than the carbon paper GDL, which exhibited the highest damage (Fig. 4.14) [59].

GDL degradation due to freeze/thaw cycling has a more significant effect on GDLs that contain MPLs. Lee et al. reported damage to the catalyst layer when MPL was present whereby the MPL lost its effect in decreasing the contact resistance and water management after 40 cycles [60].

It can be concluded that the effect of sub-freezing temperatures on the GDL is dependent upon the humidification level of the MEA and the presence of trapped pools of water within the MEA. Therefore, the freezing effect can be avoided by either purging the PEM fuel cell (removing any accumulated water), or using low RHs before the freezing conditions are experienced, as observed by Hou et al. [61]. Furthermore, this group also found that the MPL may be weakened by freeze/thaw cycling, making it prone to material loss from air flow through the GDL.

Finally, the GDL structure has a significant effect on its mechanical integrity and tolerance to mechanical degradation. Poornesh et al. found that the heterogeneous carbon fibers in the GDL contribute to surface rupture, hydrophobic coating deterioration, and breakage of the fiber, under PEM fuel cell loading and working conditions [62]. They concluded that using homogeneous fibers may enhance GDL durability [62].

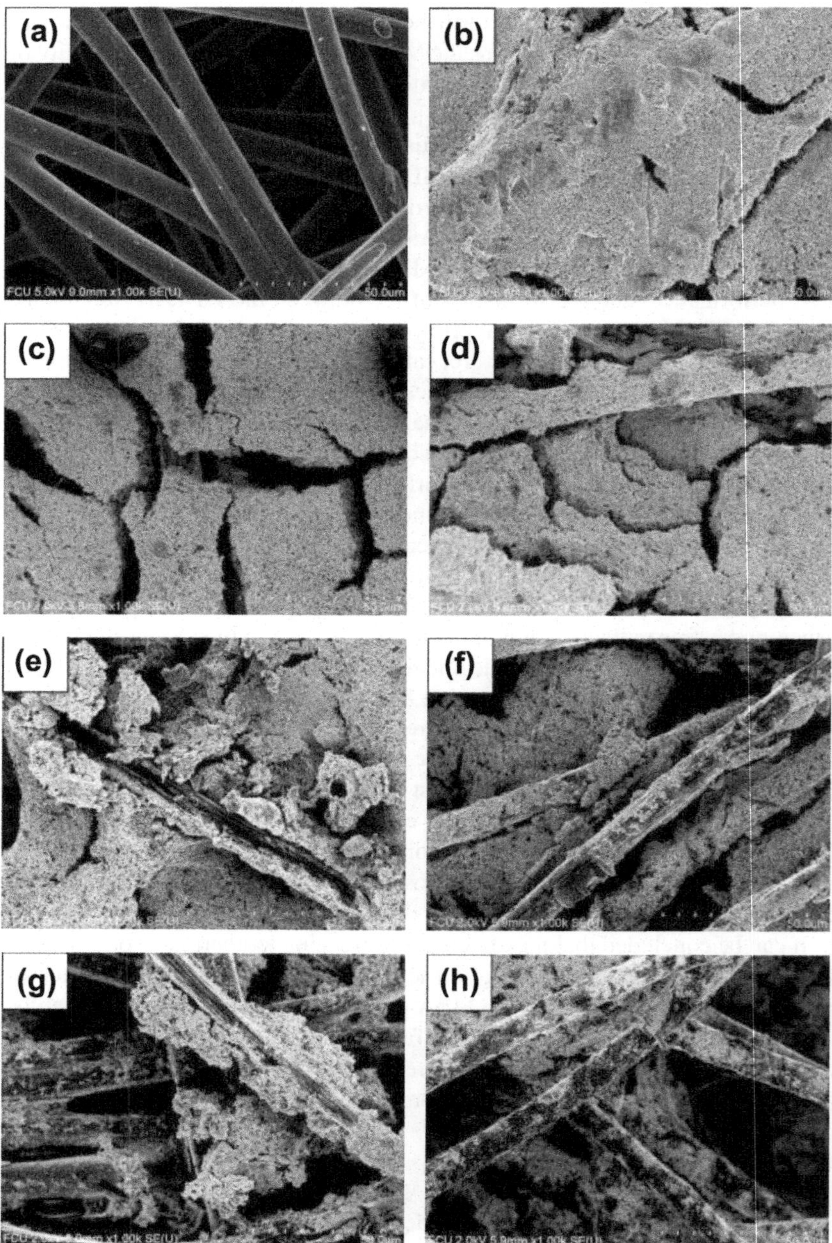

FIGURE 4.13 Comparison of GDL surfaces in MPL water washing at various times: (a) Carbon paper substrate, (b) GDL with MPL, (c) GDL-1HR (in 1 hour water washing), (d) GDL-5HR (in 5 hour water washing), (e) GDL-10HR (in 10 hour water washing), (f) GDL-50HR (in 50 hour water washing), (g) GDL-100HR (in 100 hour water washing), (h) GDL-200HR (in 200 hour water washing) [55].

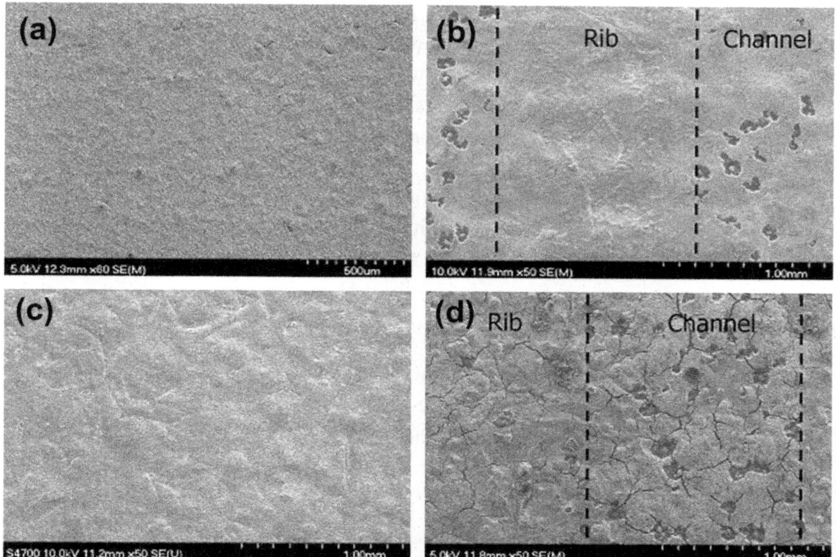

FIG 4.14 GDL surface degradation after 50 freeze/thaw cycles. (a) fresh MEA, (b) Tested MEA with a carbon cloth GDL, (c) Tested MEA with a carbon felt GDL, (d) Tested MEA with a carbon paper GDL [59].

5.3. Thermal Degradation

No significant studies have been carried out in the literature which address the effect of temperature on GDLs; there are merely a few observations. High temperature significantly affects the GDL maximum strain, resulting from weakening PTFE within the GDL [26] and enhancing carbon corrosion [41]. Moreover, it is thought that hot press (when fabricating MEAs) and freezing conditions weaken the GDL and MPL structures and lead to material loss during PEM fuel cell operation [29,54].

6. CONCLUSIONS

The main roles of a GDL are to:

i. allow gas transport to the catalyst layer;
ii. allow water transport away from the membrane at a rate that keeps the membrane hydrated but does not allow water accumulation; and
iii. conduct electrons to and from the catalyst layers.

However, a GDL suffers from electrochemical, mechanical and thermal degradation under PEM fuel cell operating conditions. This degradation results in changes in its properties, which affect the overall performance of the PEM fuel cell. The main effects on the PEM fuel cell performance are due to mass transfer losses caused by the loss of water transfer ability and changes in structure that affect the flows of reactants to the catalyst layer.

PEM fuel cell degradation studies have mainly focused on studying membrane and catalyst layer degradation and little attention has been given to GDL degradation. Further studies are required to develop a better understanding of these mechanisms, and methods for *in-situ* GDL degradation detection. Finally, the authors believe that GDL characteristics and durability tests need to be standardized.

REFERENCES

[1] V. Mehta, J.S. Cooper, Review and analysis of PEM fuel cell design and manufacturing, Journal of Power Sources 114 (1) (2003) 32–53.

[2] S. Litster, G. McLean, PEM fuel cell electrodes, Journal of Power Sources 130 (1–2) (2004) 61–76.

[3] C. Lim, C.Y. Wang, Effects of hydrophobic polymer content in GDL on power performance of a PEM fuel cell, Electrochimica Acta 49 (24) (2004) 4149–4156.

[4] H.K. Atiyeh, K. Karan, B. Peppley, A. Phoenix, E. Halliop, J. Pharoah, Experimental investigation of the role of a microporous layer on the water transport and performance of a PEM fuel cell, Journal of Power Sources 170 (1) (2007) 111–121.

[5] W.-M. Yan, et al., Effects of fabrication processes and material parameters of GDL on cell performance of PEM fuel cell. International Journal of Hydrogen Energy 32 (17) (2007) 4452–4458.

[6] T.R. Ralph, G.A. Hards, J.E. Keating, S.A. Campbell, D.P. Wilkinson, M. Davis, J. St-Pierre, M.C. Johnson, Low Cost Electrodes for Proton Exchange Membrane Fuel Cells, Journal of The Electrochemical Society 144 (11) (1997) 3845–3857.

[7] Y. Wang, S. Al Shakhshir, and X. Li, Development and impact of sandwich wettability structure for gas distribution media on PEM fuel cell performance. Applied Energy. In Press, Corrected Proof.

[8] Z. Qi, A. Kaufman, Improvement of water management by a microporous sublayer for PEM fuel cells, Journal of Power Sources 109 (1) (2002) 38–46.

[9] G. Velayutham, J. Kaushik, N. Rajalakshmi, K.S. Dhathathreyan, Effect of PTFE Content in Gas Diffusion Media and Microlayer on the Performance of PEMFC Tested under Ambient Pressure, Fuel Cells 7 (4) (2007) 314–318.

[10] M. Prasanna, H.Y. Ha, E.A. Cho, S.-A. Hong, I.-H. Oh, Influence of cathode gas diffusion media on the performance of the PEMFCs, Journal of Power Sources 131 (1–2) (2004) 147–154.

[11] X.L. Wang, H.M. Zhang, J.L. Zhang, H.F. Xu, Z.Q. Tian, J. Chen, H.X. Zhong, Y.M. Liang, B.L. Yi, Micro-porous layer with composite carbon black for PEM fuel cells, Electrochimica Acta 51 (23) (2006) 4909–4915.

[12] J.M. Song, S.Y. Cha, W.M. Lee, Optimal composition of polymer electrolyte fuel cell electrodes determined by the AC impedance method, Journal of Power Sources 94 (1) (2001) 78–84.

[13] Y. Wang, C.-Y. Wang, K.S. Chen, Elucidating differences between carbon paper and carbon cloth in polymer electrolyte fuel cells, Electrochimica Acta 52 (12) (2007) 3965–3975.

[14] V. Radhakrishnan, P. Haridoss, Effect of cyclic compression on structure and properties of a Gas Diffusion Layer used in PEM fuel cells, International Journal of Hydrogen Energy 35 (20) (2010) 11107–11118.

[15] S.G. Kandlikar, Z. Lu, T.Y. Lin, D. Cooke, M. Daino, Uneven gas diffusion layer intrusion in gas channel arrays of proton exchange membrane fuel cell and its effects on flow distribution, Journal of Power Sources 194 (1) (2009) 328–337.

[16] Y.-H. Lai, P.A. Rapaport, C. Ji, V. Kumar, Channel intrusion of gas diffusion media and the effect on fuel cell performance, Journal of Power Sources 184 (1) (2008) 120–128.

[17] L. Peng, J. Mai, P. Hu, X. Lai, Z. Lin, Optimum design of the slotted-interdigitated channels flow field for proton exchange membrane fuel cells with consideration of the gas diffusion layer intrusion. Renewable Energy 36 (5) (2011) 1413–1420.

[18] L. Cindrella, A.M. Kannan, J.F. Lin, K. Saminathan, Y. Ho, C.W. Lin, J. Wertz, Gas diffusion layer for proton exchange membrane fuel cells–A review, Journal of Power Sources 194 (1) (2009) 146–160.

[19] J.P. Feser, A.K. Prasad, S.G. Advani, On the relative influence of convection in serpentine flow fields of PEM fuel cells, Journal of Power Sources 161 (1) (2006) 404–412.

[20] J.G. Pharoah, On the permeability of gas diffusion media used in PEM fuel cells, Journal of Power Sources 144 (1) (2005) 77–82.

[21] J.T. Gostick, M.W. Fowler, M.D. Pritzker, M.A. Loannidis, L.M. Behra, In-plane and through-plane gas permeability of carbon fiber electrode backing layers, Journal of Power Sources 162 (1) (2006) 228–238.

[22] J.P. Feser, A.K. Prasad, S.G. Advani, Experimental characterization of in-plane permeability of gas diffusion layers, Journal of Power Sources 162 (2) (2006) 1226–1231.

[23] M.V. Williams, E. Begg, L. Bonville, H.R. Kunz, J.M. Fenton, Characterization of Gas Diffusion Layers for PEMFC, Journal of The Electrochemical Society 151 (8) (2004) A1173–A1180.

[24] A. Tamayol, M. Bahrami, Water permeation through gas diffusion layers of proton exchange membrane fuel cells. Journal of Power Sources 196 (15) (2011) 6356–6361.

[25] H. Meng, C.-Y. Wang, Electron Transport in PEFCs, Journal of The Electrochemical Society 151 (3) (2004) A358–A367.

[26] H. Sun, G. Zhang, L.-J. Guo, H. Liu, A novel technique for measuring current distributions in PEM fuel cells, Journal of Power Sources 158 (1) (2006) 326–332.

[27] M. Noponen, T. Mennola, M. Mikkola, T. Hottinen, P. Lund, Measurement of current distribution in a free-breathing PEMFC, Journal of Power Sources 106 (1–2) (2002) 304–312.

[28] M.M. Mench, C.Y. Wang, M. Ishikawa, In Situ Current Distribution Measurements in Polymer Electrolyte Fuel Cells, Journal of The Electrochemical Society 150 (8) (2003) A1052–A1059.

[29] T. Zhou, H. Liu, Effects of the electrical resistances of the GDL in a PEM fuel cell, Journal of Power Sources 161 (1) (2006) 444–453.

[30] I. Nitta, O. Himanen, M. Mikkola, Contact resistance between gas diffusion layer and catalyst layer of PEM fuel cell, Electrochemistry Communications 10 (1) (2008) 47–51.

[31] A. Higier, H. Liu, Effects of the difference in electrical resistance under the land and channel in a PEM fuel cell, International Journal of Hydrogen Energy 36 (2) (2011) 1664–1670.

[32] G. Karimi, X. Li, P. Teertstra, Measurement of through-plane effective thermal conductivity and contact resistance in PEM fuel cell diffusion media, Electrochimica Acta 55 (5) (2010) 1619–1625.

[33] O.S. Burheim, J.G. Pharoah, H. Lampert, Through-Plane Thermal Conductivity of PEMFC Porous Transport Layers, Journal of Fuel Cell Science and Technology 8 (2) (2011) 021013–11.

[34] E. Sadeghi, N. Djilali, M. Bahrami, Effective thermal conductivity and thermal contact resistance of gas diffusion layers in proton exchange membrane fuel cells. Part 2: Hysteresis effect under cyclic compressive load, Journal of Power Sources 195 (24) (2010) 8104–8109.

[35] S. Park, B.N. Popov, Effect of hydrophobicity and pore geometry in cathode GDL on PEM fuel cell performance, Electrochimica Acta 54 (12) (2009) 3473–3479.

[36] K. Han, B.K. Hong, S.H. Kim, B.K. Ahn, T.W. Lim, Influence of anisotropic bending stiffness of gas diffusion layers on the electrochemical performances of polymer electrolyte membrane fuel cells. International Journal of Hydrogen Energy 35 (22) (2010) 12317–12328.

[37] W.-k. Lee, C.-H. Ho, J.W. Vanzee, M. Murthy, The effects of compression and gas diffusion layers on the performance of a PEM fuel cell, Journal of Power Sources 84 (1) (1999) 45–51.

[38] J. Ge, A. Higier, H. Liu, Effect of gas diffusion layer compression on PEM fuel cell performance, Journal of Power Sources 159 (2) (2006) 922–927.

[39] X.Q. Xing, K.W. Lum, H.J. Poh, Y.L. Wu, Optimization of assembly clamping pressure on performance of proton-exchange membrane fuel cells, Journal of Power Sources 195 (1) (2010) 62–68.

[40] E. Sengul, S. Erkan, I. Eroglu, N. Bac, Effect of gas diffusion layer characteristics and addition of pore-forming agents on the performance of polymer electrolyte membrane fuel cells, Chemical Engineering Communications 196 (1–2) (2009) 161–170.

[41] W. Schmittinger, A. Vahidi, A review of the main parameters influencing long-term performance and durability of PEM fuel cells, Journal of Power Sources 180 (1) (2008) 1–14.

[42] J. Scholta, K. Seidenberger, and F. Wilhelm. Gas Diffusion Layer (GDL) degradation in polymer electrolyte fuel cells. in 2nd CARISMA international conference on progress in MEA materials for medium and high temperature polymer electrolytes fuel cells. 2010.

[43] J. Wu, X.Z. Yuan, J.J. Martin, H. Wang, J. Zhang, J. Shen, S. Wu, W. Merida, A review of PEM fuel cell durability: Degradation mechanisms and mitigation strategies, Journal of Power Sources 184 (1) (2008) 104–119.

[44] S. Maass, F. Finsterwalder, G. Frank, R. Hartmann, C. Merten, Carbon support oxidation in PEM fuel cell cathodes, Journal of Power Sources 176 (2) (2008) 444–451.

[45] R.L. Borup, J.R. Davey, F.H. Garzon, D.L. Wood, M.A. Inbody, PEM fuel cell electrocatalyst durability measurements, Journal of Power Sources 163 (1) (2006) 76–81.

[46] T. Aoki, A. Matsunaga, Y. Ogami, A. Maekawa, S. Mitsushima, K.-i. Ota, H. Nishikawa, The influence of polymer electrolyte fuel cell cathode degradation on the electrode polarization. Journal of Power Sources 195(8) 2182–2188

[47] G. Chen, H. Zhang, H. Ma, H. Zhang, Electrochemical durability of gas diffusion layer under simulated proton exchange membrane fuel cell conditions, International Journal of Hydrogen Energy 34 (19) (2009) 8185–8192.

[48] Y. Hiramitsu, H. Sato, K. Kobayshi, M. Hori, Controlling gas diffusion layer oxidation by homogeneous hydrophobic coating for polymer electrolyte fuel cells, Journal of Power Sources 196 (13) (2011) 5453–5469.

[49] R.J.F. Kumar, V. Radhakrishnan, P. Haridoss, Effect of electrochemical aging on the interaction between gas diffusion layers and the flow field in a proton exchange membrane fuel cell, International Journal of Hydrogen Energy 36 (12) (2011) 7207–7211.

[50] C. Lee, W. Mérida, Gas diffusion layer durability under steady-state and freezing conditions, Journal of Power Sources 164 (1) (2007) 141–153.

[51] A. Bazylak, D. Sinton, Z.-S. Liu, N. Djilali, Effect of compression on liquid water transport and microstructure of PEMFC gas diffusion layers, Journal of Power Sources 163 (2) (2007) 784–792.

[52] A.R. Maher, Sadiq Al-Baghdadi, A CFD study of hygro-thermal stresses distribution in PEM fuel cell during regular cell operation, Renewable Energy 34 (3) (2009) 674–682.

[53] D.L. Wood, S. Pacheco, J. Davey, R. Borup, Durability issues of PEMFC GDL and MEA under steady-state and drive-cycle operating conditions. in Fuel cell seminar. Los Alamos national laboratory.

[54] J. Wu, J.J. Martin, F.P. Orfino, H. Wang, C. Legzdins, X.-Z. Yuan, C. Sun, In situ accelerated degradation of gas diffusion layer in proton exchange membrane fuel cell: Part I: Effect of elevated temperature and flow rate, Journal of Power Sources 195 (7) (2010) 1888–1894.

[55] J.-H. Lin, W.-H. Chen, S.-H. Su, Y.-J. Su, T.-H. Ko, Washing Experiment of the Gas Diffusion Layer in a Proton-Exchange Membrane Fuel Cell, Energy & Fuels 22 (4) (2008) 2533–2538.

[56] Q. Guo, Z. Qi, Effect of freeze-thaw cycles on the properties and performance of membrane-electrode assemblies, Journal of Power Sources 160 (2) (2006) 1269–1274.

[57] Q. Yan, H. Toghiani, Y.-W. Lee, K. Liang, H. Causey, Effect of sub-freezing temperatures on a PEM fuel cell performance, startup and fuel cell components, Journal of Power Sources 160 (2) (2006) 1242–1250.

[58] J.A. Pelaez, S.G. Kandlikar, Effects of freezing and thawing on the structures of porous gas diffusion media in the Fifth International Conference on Nanochannels, Microchannels and Minichannels, Puebla, Mexico, 2007.

[59] S.-J. Lim, G.-G. Park, J.-S. Park, Y.-J. Sohn, S.-D. Yim, T.-H. Yang, B.K. Hong, C.-S. Kim, Investigation of freeze/thaw durability in polymer electrolyte fuel cells. International Journal of Hydrogen Energy 35 (23) (2010) 13111–13117.

[60] Y. Lee, B. Kim, Y. Kim, X. Li, Effects of a microporous layer on the performance degradation of proton exchange membrane fuel cells through repetitive freezing. Journal of Power Sources 196 (4) (2011) 1940–1947.

[61] J. Hou, H. Yu, S. Zhang, S. Sun, H. Wang, B. Yi, P. Ming, Analysis of PEMFC freeze degradation at -20 °C after gas purging, Journal of Power Sources 162 (1) (2006) 513–520.

[62] K.K. Poornesh, C.D. Cho, G.B. Lee, Y.S. Tak, Gradation of mechanical properties in gas-diffusion electrode. Part 2: Heterogeneous carbon fiber and damage evolution in cell layers, Journal of Power Sources 195 (9) (2010) 2718–2730.

Chapter 5

Bipolar Plate Durability and Challenges

Hazem Tawfik,[1,2] Yue Hung[1] and Devinder Mahajan[2]
[1]*Farmingdale State College,*
[2]*Stony Brook University and Brookhaven National Lab.*

1. INTRODUCTION

Bipolar plates constitute the backbone of a hydrogen fuel cell power stack; they isolate the individual cells, conduct current between cells, facilitate water and thermal management through the cell, and provide conduits for reactant gases as well as removing reaction products. Polymer electrolyte membrane (PEM) fuel cell commercialization and market penetration necessitate the mass production of bipolar plates; therefore, they are required to be made of materials with excellent manufacturability and suitable for cost-effective, high volume, automated production systems. Because membrane electrode assemblies (MEAs) with gas diffusion layers (GDL) are made from relatively light materials, with very small thickness typically between 500 and 600 microns, bipolar plates comprise more than 60% of the weight and 30% of the total cost of a fuel cell stack [1–4]. The weight, volume and cost of a fuel cell stack can be reduced significantly by improving the layout configuration of flow field design and the use of lightweight materials.

Poco graphite has been considered as the PEM fuel cell industry's reference standard for bipolar plates because of its excellent corrosion resistance and low interfacial contact resistance (ICR). However, due to the graphite's brittleness and lack of mechanical strength, combined with its relatively poor manufacturability and cost effectiveness for large production volume, Poco graphite bipolar plate material is deemed unsuitable for automotive application and commercialization. A number of materials are currently being developed and tested in laboratories around the world to produce cost-effective and durable bipolar plates for PEM fuel cells. Varieties of non-coated and coated metals, metal foams and non-metal graphite composites are being developed and reviewed as possible replacements for Poco graphite.

Polymer Electrolyte Fuel Cell Degradation. DOI: 10.1016/B978-0-12-386936-4.10005-3

The ideal characteristics of a bipolar plate's material are high corrosion resistance, high mechanical strength, low interfacial contact resistance, high contact angle, no reactant gases permeability and no brittleness. Currently, graphite composite is used in industry as the bipolar plate material due to its relatively high corrosion resistance. In addition, lower interfacial contact resistance and higher electrical conductivity can be obtained by having higher carbon to polymer ratio in the graphite composite. However, higher carbon ratios will increase the composite's brittleness and will elevate the volumetric power density when compared to thin metallic plate; which is the material favored in the automobile industry. Metals, on the other hand, provide higher mechanical strength, better durability to shocks and vibration, no permeability and more flexibility in fabrication. The main challenge of a metallic bipolar plate, however, is that corrosion-resistant metal, such as stainless steel, develops a passive oxide layer on the surface. Although this passive layer protects the bulk metal from the progression of corrosion, it also causes an undesirable effect of high interfacial contact resistance. This causes the dissipation of a considerable amount of electric energy into heat and a reduction in the overall efficiency of a fuel cell power stack. Figure 5.1 shows that the corrosion product formed on a stainless steel bipolar plate surface causes a voltage drop in the fuel cell performance. The corrosion product is not necessarily just chromium oxide; depending on the type of stainless steel used, it could be iron oxide, nickel oxide or others.

Various types of metals and alloys are currently being tested and evaluated by researchers working in the field of PEM fuel cells to develop bipolar plates that possess the combined merits of graphite and metals. The ideal characteristics of a bipolar plate's materials are high corrosion resistance and low surface

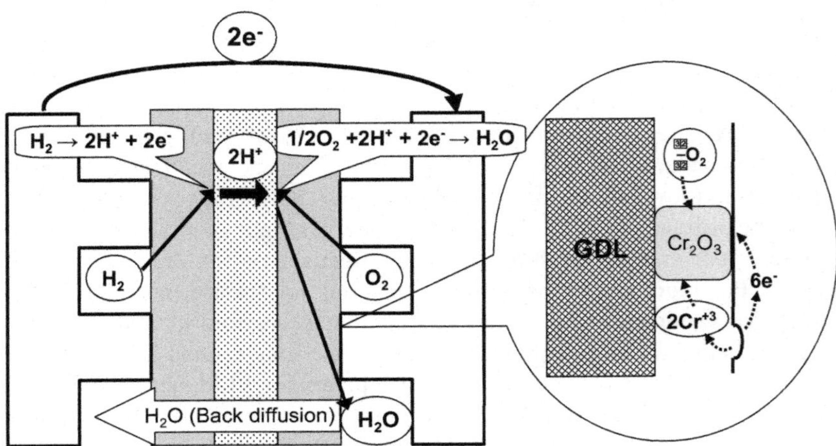

FIGURE 5.1 Corrosion occurs on stainless steel bipolar plate surface.

contact resistance, like graphite, and high mechanical strength, no permeability to reactant gases and no brittleness, like metals such as stainless steel, aluminum, titanium, etc. The key characteristics necessary for a bipolar plate for transportation applications are as follows [5–8]:

Article I. Electrochemically stable in the fuel cell environment

- High corrosion resistance with corrosion current at −0.1 V (SCE) and H_2 purge: < 1 μA/cm^2
- High corrosion resistance with corrosion current at 0.6 V(SCE) and Air purge: < 1 μA/cm^2

Article II. Possess steady low ohmic resistance throughout the operation

- ICR @140 N/cm^2: 10 milliohm.cm^2
- High surface tension with water contact angle close to 90° or larger than 90° – i.e. high dehydration or hydrophobic
- Light weight
- High mechanical strength: < 200 N/m^2
- High volume cost-effective manufacturability: $10/kW

2. LITERATURE SURVEY OF METALLIC BIPOLAR PLATE TECHNOLOGY

Considerable attention has been given to metallic bipolar plates due to their particular suitability for PEMFC automotive applications. As mentioned earlier, metals provide many mechanical advantages over graphite-based materials, and the only concerns in using them are their relatively low corrosion resistance and high interfacial contact resistance (ICR) – both of which can cause considerable power degradation. This section investigates a number of candidate materials and how far they are developed for potential application in fuel cell bipolar plate technology.

2.1. Non-coated Metals

Major concerns have focused on metal corrosion and corrosion by-products, as well as the decrease in ICR values once a surface passivation film forms. Candidate metallic materials for bipolar plate application, such as aluminum, stainless steel and titanium, have been tested by various researchers working in the field. For example, Hermann et al. [8] reported that aluminum, stainless steel, titanium and nickel bipolar plates exposed to an operating environment similar to that of a fuel cell (pH: 2–3; T ~ 80°C) were prone to corrosion or dissolution. A passive film on the surface of a bipolar plate increases electrical resistance and decreases a cell's electric power output. While this surface oxide layer protects the metal and stops the corrosion from progressing to the lower layers, it forms an electrically insulating film on the plate's top surface. As the thickness of the oxide layer increases, ICR also increases, accordingly causing a decrease in electric power output.

Moreover, for uncoated metals, metal ions and oxides could directly foul the electrolyte and tarnish the catalyst in the membrane electrode assembly (MEA), resulting in considerable adverse effects on cell performance. If unprotected metal bipolar plates are exposed to the corrosive environment inside a fuel cell, where conditions are typically relative humidity (> 90%), acidity (pH = 2–3) and temperature (60–80°C), metal dissolution will occur. The dissolved metal ions will diffuse into the membrane and then get trapped in the ion exchange sites inside the ionomer, resulting in lowered ionic conductivity as described by Mehta and Cooper [6]. A highly conductive corrosion-resistant coating with high bonding strength at the interfacial layer between base metal substrate and coating layer is required to solve this problem.

In addition, the compaction between the interfacial surfaces plays an important role in the ICR. Davies et al. [11] observed that under a constant compaction pressure of 220 N/cm^2 imposed on fuel cell plate experiments, the ICR values of various grades of stainless steel and other alloys decreased in the following order 321 > 304 > 347 > 316 > Ti > 310 > 904 > Incoloy 800 > Inconel 601 > Poco graphite. For materials of higher chromium content, the authors observed that the oxygen content in the surface film was not as prominent as illustrated for other grades of stainless steel, which suggests that the passive film was thinner in these relatively high Cr stainless steel samples. The results showed that the passive film decreased in thickness following the order 321 > 304 > 316 > 347 > 310 > 904 > Incoloy 800 > Inconel 601. The results also indicated that the performance of the bipolar plates is related to the thickness of the passive layer and its ICR value; as the thickness and ICR increase, more heat energy is generated and less electric energy output is produced.

2.1.1. Untreated Stainless Steel

Many researchers tend to use stainless steel as the benchmark material since it is known to have good corrosion resistance, and is commercially available at relatively low cost. Wang et al. [12,13] found that both austenitic (349™) and ferritic (AISI446) stainless steel with high Cr content showed good corrosion resistance and could be suitable for bipolar plate applications, though AISI446 requires some improvement in ICR due to the formation of a surface passive layer of Cr_2O_3. This group also verified that Cr in the alloy formed the passive film on the surface of the stainless steel. As the Cr content increased, the corrosion resistance improved, as is commonly known in corrosion studies, and these results agreed with the findings of Davies et al. [11]. However, a thick non-conductive surface passive layer of Cr_2O_3 will produce an undesirably high ICR. Wang and Turner [13] studied stainless steel samples of AISI434, AISI436, AISI441, AISI444, and AISI446. They noted that in both PEM fuel cell anode and cathode environments, passive films formed on AISI446 stainless steel were very stable. An increase in ICR between steel and the carbon backing material (gas diffusion layer) due to passive film formation was also reported. X-ray photoelectron spectroscopy (XPS) depth profiles indicated that the thickness of the passive film on AISI446 was estimated to be 2.6 nm for the film formed at 0.1 V in the simulated PEM fuel cell anodic environment, and 3.0 nm for the film formed at 0.6 V in the simulated cathodic environment. Wang and Turner [13] stated that ICR for AISI446 increased after passivation. The passive films on AISI446 were mainly chromium oxide, with iron oxides playing only a minor role. In the simulation of the PEMFC, the passive film formed on the cathode environment was thicker than that formed on the anode environment, with the former resulting in higher ICR. The authors also recommended that some modification of the passive film, which is dominated by chromium oxide, is required to improve the ICR of the steel.

Furthermore, Wang et al. [16] conducted potentiodynamic and potentiostatic tests to investigate the corrosion behavior of SS316L in simulated PEM fuel cell anode and cathode environments. This group indicated that at free corrosion potential, the corrosion resistance of SS316L in oxygen-containing environments is higher than that in hydrogen-containing environments, because the corrosion potential in the solution with O_2 is higher according to the Nernst equation. This finding was also confirmed in open circuit potential and potentiodynamic tests. However, the real PEM fuel cell operating conditions for anode and cathode are about −0.1 V (SCE) and 0.6 V (SCE), respectively. The negative anodic current can provide cathodic protection for SS316L and therefore less corrosion was observed by optical microscopy and scanning electron microscopy (SEM) in the anode environment [16]. The inductively coupled plasma optical emission spectrometry (ICP-OES) results confirm that 25 and 42 ppm of metal ion concentrations are at the anode and cathode, respectively, after 5,000 hours of potentiostatic tests. This group also suggested that SS316L must be coated or modified in order to be used as a bipolar plate material.

2.1.2. Nickel-based Alloys

Many types of alloy have been developed for bipolar plate applications where stainless steels, such as SS304 or SS316, do not provide adequate interfacial contact resistance. These alloys often contain Ni, Cr and Mo as stabilizing elements to achieve desirable corrosion properties. Wilson et al. [42] evaluated a number of stainless steels (SS316, E-Brite 26-1 & AL2205) and nickel alloys (AL201, AL400 & AL600) as bipolar plates. This group performed aggressive 125 hour immersion tests to evaluate the corrosion resistance of these materials. The acidic solution (pH 2 sulfuric acid at 80°C) used in this test was much more severe than actual fuel cell conditions. Their results showed that E-Brite had better corrosion resistance (less than 1 μm/yr) while nickel alloys AL600 and SS316 had less corrosion resistance in the pH 2 environment. To put this result in context, Wind et al. [10] reported that 316L stainless steel tested for 100 hours in a cell operation environment at 75°C as a bipolar plate material resulted in Ni contamination levels of 76 $\mu g/cm^2$. The lack of corrosion resistance is most likely due to the existence of Ni in the materials.

2.1.3. Porous Materials and Metal Foams

Kumara and Reddy [41] investigated three different porous materials; namely Ni-Cr metal (Fe: up to 8%, C: up to 2%, Cr: 30–54%, Ni: balance) foam with 50 PPI (pores per inch), SS316 metal foam with 20 PPI, and carbon cloth. They suggested that the MEA metal ion contamination can be minimized to a great extent by optimizing the fluid-flow in metal foams. Any metal ion products that are formed will not stagnate in the cell stack but are exhausted along with the water by-product. A single cell performance test showed that the metal foams, particularly Ni-Cr metal foam, perform better than the SS316 conventional channel design flow-field. In addition, the low performance of the single cell with carbon cloth was due to the high contact pressure in the cell stack which blocks the pores in the carbon cloth.

The same authors [41] indicated that with a decrease in permeability of the metal foam, the cell performance increased. This is because the decreasing permeability of the material can cause an increase in pressure drop across the flow-field. The cell performance could be further improved by optimizing the size, shape and distribution of pores in the metal foam. An additional advantage of these metal foams is that they could possibly be used as a catalyst support in the electrochemical reactions within the fuel cell, thereby eliminating the need to use carbon electrodes. However, corrosion and lifetime testing were not mentioned in this study. The untreated metal foam can corrode when in direct contact with the acidic membrane and would lead to severe MEA contamination.

2.2. Coated Metals

Metallic bipolar plates should be coated with protective coating to avoid corrosion or the formation of thick passive films that may result in elevated ICR. Coatings should be conductive and adhere to the base metal without exposing the substrate to corrosive media [7]. Two types of coatings, carbon-based and metal-based, have been investigated [6,7,17]. Carbon-based coatings include graphite, conductive polymer, diamond-like carbon and organic self-assembled monopolymers [8]. Noble metals, metal nitrides and metal carbides are some of the metal-based coatings that have been explored [10,18–22]. To maximize the adherence force between the coating and the substrate, the coefficient of thermal expansion of base metal and coating should be as close as possible to each other to eliminate the formation of micropores and microcracks in coatings due to unequal thermal expansion [19]. In addition, some coating processes are prone to pinhole defects and various techniques for coating bipolar plates are still under development [8]. Mehta and Cooper [6] presented an overview of carbon-based and metallic bipolar plate coating materials.

Woodman et al. [19] concluded that the coefficients of thermal expansion (CTE) for both coating and substrate, corrosion resistance of coating, micropores and microcracks have a combined effect in protecting bipolar plates from the PEM fuel cell's corrosive environment. The authors also argued that even though PEM fuel cells typically operate at temperatures of less than 100°C, vehicle service would impose frequent start-up and shut-down conditions, and temperature differentials of 75–125°C would be expected during typical driving conditions. A large difference in the CTE of the substrate and coating materials may lead to coating layer failure. One technique to minimize the CTE differential is to add intermediate coating layers with less CTE mismatch between the adjacent layers. However, materials such as Al, Cu, Sn, Ni and Ni phosphorous are very susceptible to electrochemical corrosion in the acidic solutions that are typical of PEMFC operating conditions. Also, coating techniques and surface preparation must be optimized to improve the bonding strength between the coating material and the substrate base plate to eliminate the possibility of separation, microcracks or pinholes.

2.2.1. Metal-based Coatings

2.2.1.1. Gold and Precious Metal Coatings

Noble metals such as gold and platinum have low ICR and high corrosion resistance and therefore their fuel cell performance when used as a bipolar plate coating is very similar to Poco graphite bipolar plates [9,10]. However, the high cost of these metals has prohibited their utilization for commercial use. Hentall et al. [9] machined current collectors from aluminum (Al) to the exact dimensions of graphite bipolar plates. The Al plates were then coated with gold (Au) by a solution process and used in a fuel cell. During initial warm-up, the data indicated that the Au-coated plates performance was very similar to graphite (1.2 A/cm^2 at 0.5 V) because Au-coated Al plates had a similar ICR with gas diffusion media (GDM) to graphite. However, the performance degraded quickly to 60 mA/cm^2 at 0.5 V. The analysis revealed that some of the Au coating was lifted from the plate and physically embedded in the membrane. Wind et al. [10] also indicated that Au-coated bipolar plate (SS 316L) clearly demonstrated no difference between the metal-based and graphite plates due to the similar ICR mentioned above. Woodman et al. [19] stated the coefficient of thermal expansion (CTE) for aluminum as approximately 24 μ in./in./°C (over 0–400°C) while the CTE for Au was approximately 14 μ in./in./°C over the same temperature range. The authors concluded that failure of the Au coating would be expected when it reaches the plastic deformation threshold.

In another study, coated titanium bipolar plates for PEM fuel cells were developed by Wang et al. [24]. In this work, surface-treated titanium (Ti) bipolar plates were used in PEM fuel cells to reduce electric energy losses due to the formation of passive layers. The coating materials were iridium oxide (IrO_2) and platinum (Pt). These coatings were applied on titanium plates by sintering and proprietary methods. The cost of sintering 20 g/m^2 iridium oxide on to the plates was about US$320 and the cost of 2.5 μm platinum coating for a 25 cm^2 active area single cell was about US$400. The authors indicated that the performance of iridium oxide (IrO_2) and platinum (Pt) coated single cells was close to the graphite bipolar plate single cells. They also suggested that titanium substrate with proper surface modification could be used as PEM fuel cell bipolar plate material.

Due to the high price of Au-coated and precious metal-coated bipolar plates, this technology will face considerable competition from other, less expensive, corrosion-resistant coatings for bipolar plates. Table 5.1 shows the high-volume material cost for the candidate materials [18], quoted from the London Metal Exchange. As indicated in Table 5.1, the price of gold is much higher than aluminum and nickel. However, as indicated earlier, Ni phosphorous coatings are very unstable in the PEM fuel cell environment even though their cost, compared to gold, might make them very attractive. Therefore, researchers have been developing new types of coating materials that are economical and affordable for bipolar plate application.

2.2.1.2. TiN Coating

For nitride coatings, Zhang et al. [27] studied the corrosion behavior of TiN-coated stainless steel as a bipolar plate material. The authors applied a TiN coating to the surface of SS304 bipolar plate by two different techniques; namely, magnetron sputtering (MS) and pulsed bias arc ion plating (PBAIP). Electrochemical measurement was performed in 0.5 M H_2SO_4 + 2 ppm HF solution at room temperature. The result shows that the corrosion current (I_{corr}) of uncoated SS304 was 2.6 $\mu A/cm^2$, whereas SS304/Ti_2N/TiN (MS) and SS304/TiN (PBAIP) were lower than 0.0145 $\mu A/cm^2$. The authors also indicated that the corrosion behaviors of the coating depend strongly on microstructural characteristics [27]. After 8 hours of potentiostatic test at 0.6 V(SCE), SEM images showed that corrosion occurred mainly on the grain boundary of the SS304/Ti_2N/TiN sample prepared by magnetron sputtering (MS). However, for the SS304/TiN sample prepared by PBAIP, the corrosion took place on the large particles because no trace of grain boundary was found on the film. In addition, the same authors evaluated ICR values of the coated

TABLE 5.1 Bipolar Plate Materials and High-volume Material Costs

Material	Material Cost ($/g)	Density (g/cm^3)
Graphite	0.105	1.79
Aluminum	0.003	2.7
Gold	47.8	19.32
Nickel	0.026	8.19

and uncoated samples before and after polarization. The results show that at a compaction force of 240 N/cm^2, the contact resistance of the original TiN-coated SS304 sample (~19 mΩ·cm^2) was lower than that of the as-received SS304 sample (~140 mΩ·cm^2). Also, after polarization at 0.6 V (SCE) for 8 hours, the contact resistance of the TiN-coated sample was increased slightly to 25 mΩ·cm^2.

Similar work was conducted by Li et al. [25] and Cho et al. [26]. They also observed a significant improvement in the TiN-coated SS316 bipolar plates. The TiN coating was applied using a hollow cathode discharge (HCD) ion plating method. Li et al. [25] demonstrated that the coating showed no significant degradation under typical fuel cell load conditions after four hours. SEM observations show that a very small portion of substrate areas had passivated during immersion tests of TiN coatings in an O_2 environment for 1,000 hours and in a H_2 environment for 240 hours, respectively, due to a coating defect. Cho et al. [26] also indicated that the water contact angle of graphite and TiN-coated SS316 was almost the same, and equal to 90°, while that of uncoated SS316 was 60°. These findings imply that graphite and TiN-coated SS316 bipolar plate can remove water more efficiently than uncoated SS316. This group also operated the TiN-coated single cell successfully for 700 hours with some power degradation (current density at 0.6 V decreased from 896 to 598 mA/cm^2 during 700 hours).

Overall, the results revealed that a TiN coating could offer higher corrosion resistance and electrical conductivity than uncoated SS316. Further improvements to the coating quality and evaluation of the long-term stability of TiN-coated SS316 bipolar plate under simulated and actual fuel cell operation conditions are required.

2.2.1.3. CrN/Cr_2N Coating

For almost a decade, Brady et al. [28,30–32] have studied Cr-nitride coatings on Ni-Cr and Fe-Cr base alloys as protection for PEMFC bipolar plates. They have studied the nitridation of commercially available austenitic 349™ alloys, AISI446, high Cr (30–35 wt%) Ni-Cr alloys and a ferritic high Cr (29 wt%) stainless steel (Hastelloy G-30® & G-35™) as bipolar plate candidate materials. Corrosion resistance in sulfuric acid solutions (simulated fuel cell environment) and ICR values were measured and evaluated. Thermal nitridation of high Cr stainless steel showed low ICR readings after 2,700 hours of single fuel cell operation with nitrided Hastelloy G-35™ anode and cathode plated. They also indicated that oxygen impurities in the nitriding environment play a significant role in the nitrided surface structures. Existence of oxygen during the thermal nitridation process can cause oxide formation on G-30® & G-35™ at the surface under some conditions. However, Brady et al. were able to conclude that the nitridation of Cr-bearing alloys can yield low ICR, good electrical conductivity and corrosion-resistant CrN or Cr_2N based surfaces that are promising for fuel cell bipolar plate application.

In addition to Brady's group, other researchers have focused their attention on CrN coating. Fu et al. [29] applied Cr-nitride films to stainless steel 316L by PBAIP. Different compositions of Cr-nitride films could be obtained by changing the flow rate of N_2 during the PBAIP process. Their work also confirmed that after potentiodynamic and potentiostatic tests in both the simulated anodic and cathodic environment at 70°C, the corrosion resistance of a Cr-nitrided SS310L sample was greatly enhanced in comparison to the uncoated SS316L sample. The interfacial contact resistance between the Cr-nitrided SS316L sample and the Toray carbon paper was about two orders of magnitude lower than that of the untreated SS316L. In addition, this group indicated that the contact angle of the Cr-nitrided sample with water was 95°, which is beneficial for water management in fuel cells.

2.2.1.4. Carbide-based Amorphous Metallic Coating Alloy

In recent years, additional work regarding both carbide and nitride coatings has been published. Bai et al. [34] investigated chromized coatings on carbon steel AISI 1020 as a bipolar plate material. A pack chromization process was employed to apply the coatings onto electrical discharge machining (EDM) activated and textured steel surfaces, and also non-EDM steel. The coating was deposited at a temperature of 700°C. Electron probe microanalysis (EPMA) line-scan and X-ray diffraction (XRD) results showed that the chromized coatings on carbon steels consist mainly of (Cr, Fe) carbides with minor amounts of (Cr, Fe) nitrides. The authors reported that the chromized coating on EDM activated low-carbon steel has a low corrosion current density of 5.78×10^{-8} A/cm^2 under 0.5 M H_2SO_4 solution and a low ICR value of 11.8 mΩ·cm^2 at 140 N/cm^2. This was a considerable improvement compared to the raw material, with an ICR value of 403.8 mΩ·cm^2. Moreover, the corrosion resistance of all chromized coatings is much better than that of raw materials, and the low ICR value of the coating was due to the increase in levels of conductive (Cr, Fe) carbides and nitrides.

Hung et al. [49] conducted a comprehensive study on the corrosion and contact resistance behavior of a Cr_3C_2 coating on SS316 and AL bipolar plates. The coating was deposited onto the substrate materials by a high velocity oxygen fuel (HVOF) thermal spray technique. Their results showed that the conductive Cr_3C_2 coating on either AL or SS316 substrates has a stable, low ICR value of 14–15 mΩ·cm^2 at 140 N/cm^2 even after 9.5 hours of potentiostatic testing at 0.6 V (SCE) and 80°C. The corrosion current of the coated SS316 sample and the coated AL sample, measured in pH 3+0.1 ppm HF solution at 80°C was 8.07 μA/cm^2 and 28 μA/cm^2, respectively. The higher corrosion current in the coated AL sample was due to the existence of a nickel binding agent in the coating, and the increase in surface roughness caused by the HVOF process. However, this group was able to operate the Cr_3C_2-coated aluminum bipolar plate in a single cell for 1,000 hours at 70°C with only slight power degradation. Optimizing the coating deposition technique is necessary, because this localized high temperature deposition method could introduce local deformations of the bipolar plate due to the mismatch of CTE between the coating and the substrate material. Natesan and Johnson [33] also reported that, other than fuel cell application, chromium carbide-coated alloy has better corrosion resistance than chromium-coated and uncoated SS310 in high temperature oxygen and sulfur environments.

Ren and Zeng [35] have applied titanium carbide (TiC) coatings onto 304SS with minimum porosity without pinholes using a high-energy micro-arc technique, with a metallurgical bonding between the coating and the substrate. In general, the coating produced by this technique is denser than that produced by the PVD process, and thus it has better protection on the substrate material. The open circuit potential of TiC-coated steel in 1 M H_2SO_4 solution at 25°C is 0.2 V higher than that of uncoated 304SS, and the corrosion current decreased from 8.3 A/cm^2 for the uncoated steel to 0.034 A/cm^2 for the coated steel. They also reported that the TiC coating did not experience any degradation under the condition of potentiostatic polarization at 0.6 V (SCE), and exhibited high stability during 30 days of immersion test. Long-term durability evaluation of single cells and power stacks with these carbide based coatings (TiC and Cr_3C_2) is necessary to determine their stability.

2.2.1.5. Nickel Alloy Coatings

Abo El-Enin et al. [36] tested several nickel alloys applied on an aluminum 1050 substrate. Different pretreatment methods were tested and the procedure of the optimum pretreatment method is:

1. Dip the specimen in a 12.5% NaOH for 3 min;
2. Electroless zinc plating (Chemical doping) for 2 min;
3. Dip quickly in the electroplating bath after rinsing with distilled water.

Energy depressive X-ray analysis (EDAX) was employed to determine the chemical composition of each alloy. The chemical composition for each alloy (in percentage) is approximately $Ni_{69}Co_{31}$, $Ni_{32}Co_{32}Fe_{36}$, $Ni_{88}Mo_9Fe_3$ and $Ni_{88}Fe_2Mo_9Cr$. The corrosion current of the different alloys under 10^{-4} M H_2SO_4+ 2 ppm HF solution at 90°C are as follows: (Ni-Co: 0.288 μA/cm^2), (Ni-Co-Fe: 0.882 μA/cm^2), (Ni-Mo-Fe: 12.31 μA/cm^2), (Ni-Mo-Fe-Cr: 1.782 μA/cm^2) and (annealed Ni-Mo-Fe-Cr: 0.398 μA/cm^2). The thickness of these coatings is 10–13 μm, and their electrical conductivities are lower than 0.66×10^6 S/cm. The result of corrosion testing seems to be promising for fuel cell bipolar plate application. However, the ICR and durability tests of these coating materials were not reported.

2.2.2. Zirconium-based Coating

Yoon et al. [23] conducted a coating investigation for a range of metals including: gold of various thicknesses (2 nm, 10 nm, and 1 μm), titanium, zirconium, zirconium nitride (ZrN), zirconium niobium (ZrNb), and zirconium nitride with a gold top layer (ZrNAu). These protective coatings were deposited on stainless steel substrates (304, 310, and 316) by electroplating and physical vapor deposition (PVD) methods. The authors performed electrochemical polarization and contact resistance tests to determine the corrosion resistance and ICR measurement of the coatings. The results show that Zr, ZrN, ZrNb, ZrNAu and 10 nm gold-coated samples satisfy the DOE target for corrosion resistance at anode potential (1 $\mu A/cm^2$) but only the Zr-coated sample satisfied both anode and cathode potential in a typical PEM fuel cell environment in the short-term. The authors reported that the interfacial contact resistance of the Zr-coated sample was 1,000 $m\Omega \cdot cm^2$ at 140 N/cm^2 which did not meet the DOE goal. However, the ICR of the Zr-coating was reduced by nitriding the Zr surface (ZrN) and reached 160 $m\Omega \cdot cm^2$. To further improve the ICR, the nitrided surface was then coated with a thin (10 nm) layer of gold (ZrNAu) which did significantly reduce the interfacial contact resistance (6 $m\Omega \cdot cm^2$) and meet the DOE contact resistance goal. Durability and single cell tests are required to determine the sustainability of these coating materials.

2.2.3. Conducting Polymer-based Coatings

Shine et al. [37] electrochemically coated 304 stainless steel with the conducting polymers polyaniline (PANI) and polypyrrole (PPY). Cyclic voltammetry was selected and used for the polymerization and deposition of these polymers onto the SS304 substrate because of the advantage of *in situ* deposition of the polymer. The authors reported that both PANI and PPY coatings with three cycles of cyclic voltammetry showed the best corrosion results. The corrosion current densities of bare SS304, PPY-and PANI-coated steels measured under H_2SO_4 solution (pH 3) were 10^{-5}, 10^{-6} and 10^{-7} A/cm^2, respectively. At a compaction pressure of 140 N/cm^2, the contact resistance of PPY- and PANI-coated steels were found to be approximately 0.8 $\Omega \cdot cm^2$, while the ICR of graphite plate in this study was 0.08 $\Omega \cdot cm^2$. However, the authors expected these polymers to be more conductive in an acidic environment because their conductivity is related to their degree of oxidation or reduction [37]. These coatings provided acceptable corrosion resistance but relatively high ICR in a fuel cell environment. Potentiostatic and long term single cell tests should be performed under simulated and actual fuel cell conditions.

2.2.4. Diamond-like Coating

Lee et al. [40] applied a PVD coating of aYZU001 diamond-like film onto 5052 Al alloy, and compared its performance to that of SS316L and graphite. Corrosion was measured using 0.5 M H_2SO_4 solution and the corrosion rates of uncoated Al, coated Al, SS316L and graphite were 1.16 μm/yr, 0.247 μm/yr, 0.1 μm/yr and 0.019 μm/yr, respectively. In addition, the coated Al, 316L stainless steel and graphite were fabricated into a single cell to measure ICR and to test cell performance. The authors reported that the contact resistance evaluation shows that at 30 kg force, the electrical resistances of coated

Al (24 mΩ) and uncoated SS31L (25 mΩ) plates were higher than that of graphite (14 mΩ). Also the electrical resistances of coated Al (25 mΩ) and uncoated SS316L (27 mΩ) were slightly raised after 4 hours of single cell testing, because the non-conductive passive film formed on these materials increased their electrical resistance. The authors also concluded that metallic bipolar plates, PVD coated 5052 aluminum and SS 316L, performed better than the graphite plate at a low voltage but experienced shorter cell life.

2.2.5. Other Proprietary Coatings

Andre et al. [15] have conducted a comprehensive study on corrosion behavior of stainless steel plate materials (316L and 904L) through *ex-situ* electrochemical investigations. The authors indicated that, although composition of 316L and 904L plays an important role for corrosion resistance, a passivation pre-treatment can heavily modify both structure and composition of the passive layer on the stainless steels. They treated the stainless steels (316L and 904L) bipolar plates with a patented, low cost, surface modification process. Inductively coupled plasma mass spectrometry (ICP-MS) analysis was performed on exhausted water samples to determine the cation release from the test samples. Exhausted water from the graphite bipolar plate after 1,000 hours of single cell test was analyzed as a bench mark. Their findings indicated that the lowest cation release during aging was observed for the patented surface treated 316L. For anodic operating conditions, surface treated 316L showed a great improvement in corrosion resistance. The total cation release from treated 316L during 500 hours of testing was 40 ppb – in comparison to 129 ppb for the bright annealed state 316L. However, the same surface modification applied on 904L did not show the same improvement, because the high nickel content may have caused a greater amount to dissolve during fuel cell operation. For cathodic operating conditions (0.8 V/SHE), the material remains in its passive state, and even as-received plates offer sufficient corrosion resistance, because total cation release from the 316L plate after 500 hours of testing was less than 60 ppb.

Wind et al. [10], Griffths et al. [52] and Hodgson et al. [54] also investigated a number of coating materials for metallic bipolar plate application over 500 hours of single cell operation. In one case, the coating material completed over 10,000 hours of operation. Lifetime single cell tests with these coatings show promising results. However, the composition of the coating materials and methods of coating technique were not reported.

2.3. Amorphous Alloys

Jayaraj et al. [38,39] investigated the corrosion behavior of Fe- and Ni-based amorphous alloys as alternative bipolar plate materials in comparison to stainless steel and graphite. The composition of the alloys were $Fe_{50}Cr_{18}$-$Mo_8Al_2Y_2C_{14}B_6$ (Fe-Al_2), $Fe_{43}Cr_{18}Mo_{14}C_{15}B_6Y_2Al_1N_1$ (Fe-Al_1N_1) and $Ni_{60}Nb_{20}Ti_{10}Zr_5Ta_5$ (Ni-Ta_5). These alloys were prepared by arc melting of high purity metals and Fe-Cr-N master alloy under an argon atmosphere. At 140 N/cm^2, Fe-based amorphous alloys, particularly the Fe-Al_1N_1 alloy (~13 $\Omega \cdot cm^2$), exhibited lower contact resistance than the Ni-based amorphous alloy Ni-Ta_5 alloy (~21 $\Omega \cdot cm^2$). This group also conducted corrosion test measurements by applying H_2 gas and pressurized air bubbled into a 1 M H_2SO_4 + 2 ppm F- solution at 80°C, to simulate the fuel cell environment. The corrosion test showed that both the Fe- and Ni-based amorphous alloys show better corrosion resistance than SS316L. The corrosion properties of Ni-based amorphous alloys were particularly superior to those of Fe-based under cathodic conditions. They also concluded that the Cr content of the alloy is an important factor in improving corrosion resistance.

Lafront et al. [14] have also investigated the corrosion behavior of bulk amorphous $Zr_{75}Ti_{25}$ alloy and 316L stainless steel (316L), using an electrochemical noise technique, in 12.5 ppm H_2SO_4 + 1.8 ppm HF solution (pH 3.2) at 25°C and 80°C. His group indicated that electrochemical noise analysis (ENA) was a non-destructive and non-interrupting method of monitoring the performance of bipolar plates. His group concluded that the bulk amorphous $Zr_{75}Ti_{25}$ alloy was a better candidate for bipolar plate material under anodic environment (bubbled H_2 solution), while the 316L alloy was better in the cathodic environment (bubbled O_2 solution).

2.4. Composite Plates

Composite plates can be categorized as either metal or carbon-based. A metal-based, composite, bipolar plate has been developed by the Los Alamos National Laboratory [43]. This design combines porous graphite, polycarbonate plastic and stainless steel in an effort to leverage the benefits of different materials. Polycarbonate and stainless steel provide rigidity and impermeability to the structure, while graphite resists corrosion and provides low ICR. Therefore, the plate design allows the use of porous graphite plates, which are not as time consuming or expensive to produce as nonporous graphite plates. Polycarbonate also provides chemical resistance and can be molded to any shape for gaskets and manifolds. This layered plate appears to be a very good alternative from stability and cost standpoints.

Carbon-based composite plates enjoy good chemical stability and high corrosion resistance. They also have low contact resistance due to the absence of passive film. However, their lack of mechanical strength and hydrogen permeability has caused difficulties in the production of thin bipolar plates, which are necessary for automobile application. Blunk et al. [44] used expanded graphite particles as conductive fillers for composite bipolar plates to reduce their electrical resistance and increase mechanical strength. A plate thickness of 0.5 mm can be achieved which does not limit H_2 permeation rates. However, the cost of expanded graphite particles is higher than the graphite powder that is typically used.

TABLE 5.2 Advantages and Disadvantages of Graphite, Graphite Composite and Metal Bipolar Plates

Bipolar Plate Material	Advantages	Disadvantages
Poco Graphite	Corrosion resistance Low ICR High power density	Porous Poor machinability Brittle Relatively expensive
Industrial Graphite Composite (Carbon powder + polymer resin)	Corrosion resistance Low ICR Good machinability and can be mass produced	Moderate electrical conductivity (higher electrical conductivity can be achieved by having higher carbon to polymer ratio; however, this will increase material's brittleness [44]) Low thermal conductivity
Metal	Not porous High electrical conductivity and thermal conductivity Good machinability and can be mass produced Durable Relatively inexpensive	Corrode in acidic environments High ICR

The advantages and disadvantages of graphite, graphite composite and metals are briefly summarized in Table 5.2. Also, the overall comprehensive testing and evaluation of various materials for metallic and non-metallic bipolar plates are compiled in Table 5.3 to provide a quick reference to the most up-to-date research findings in PEM fuel cell technology. The current bipolar plate specifications and the DOE technical and cost targets are also listed in Table 5.3.

3. METHODS AND APPROACHES

The main challenges facing metallic bipolar plates are their corrosion and electrical conductivity. For metal, its bulk resistance is much less than its interfacial contact resistance due to the possible non-conductive oxide layer formed on the metal surface. Therefore, two commonly used bipolar plate testing techniques, namely, the interfacial contact resistance test and the accelerated corrosion resistance test, are introduced in this section.

TABLE 5.3 Summary of Metallic Polar Plate Materials, Coatings and their Corrosion Current Density and Interfacial Contact Resistance

Plate Material	Coating Material (thickness)	Corrosion Current Density (DOE Target 1 μA/cm²) / Corrosion Rate	Contact Resistance (DOE Target 10 mΩ·cm²),	Cost (DOE 2010 Target $10/kw)	Ref.
316SS	Cr-nitride	316SS (~300 μA/cm²). Cr-Nitrided SS316 (1 μA/cm²), 0.5 M H_2SO_4 + 5 ppm F at 70°C	Before operation, 316SS (55 mΩ·cm²). Cr-Nitrided SS316 (10 mΩ·cm²)		[29]
Ni-Cr base alloys (Hastelloy G-30, G-35), Ferritic stainless steel (AL 29-4C)	Thermal nitridation	Anode current at 70°C, 1 M H_2SO_4 + 2 ppmF⁻ with hydrogen purge, Nitrided G-35 (0.5 μA/cm²), Nitrided AL29-4C (0.3 μA/cm²)	Before operation, G-30 & G-35 (30-75 mΩ·cm²). AL29-4C (>100 mΩ·cm²), Nitrided G-30 & G-35 (10 mΩ·cm²), Nitrided AL29-4C (>10 mΩ·cm²)		[32]
AISI446, 316LSS, 349TM, 2205	Nitrided AISI446	Anode potential CD at −0.1 V at 70°C, 1 M H_2SO_4 + 2 ppmF⁻ with hydrogen purge, AISI446 (−2.0~−1.0 μA/cm²), 2205 (−0.5~0.5 μA/cm²), 349 TM (−4.5~−2.0 μA/cm²), Nitrided AISI1446 (−1.7~−0.2 μA/cm²), Modified AISI446 (−9.0~−0.2 μA/cm²)	Before operation, AISI446 (190 mΩ·cm²), 2205 (130 mΩ·cm²), 349TM (110 mΩ·cm²), Nitrided AISI446 (6 mΩ·cm²), Modified AISI446 (4.8 mΩ·cm²), at 140 N/cm²	AISI446 (4.76 $/kW), 349TM (4.22 $/kW), 2205 (3.14 $/kW), Nitrided AISI446 (N/A), Modified AISI446 (N/A)	[20]

		Cathode potential CD at 0.6 V at 70°C, 1 M H_2SO_4 + 2 ppmF^- with air purge, AISI446 (0.3~1.0 μA/cm²), 2205 (0.3~1.2 μA/cm²), 349TM (0.5~0.8 μA/cm²), Nitrided AISI1446 (0.7~1.5 μA/cm²), Modified AISI446 (1.5~4.5 μA/cm²)			
Ni-50Cr alloy, 349TM SS	Thermal nitridation on Ni-50Cr (3–5 μm) and 349TM	Anode environment CD at −0.1 V, at 70°C, 1 M H_2SO_4 + 2 ppmF^-, with hydrogen purge, Nitrided Ni-50Cr (3–4 μA/cm²), Nitrided 349TM (15–20 mA/cm²),	Before operation, Ni-50CrL (~60 mΩ·cm²), nitrided Ni-50Cr (~10 mΩ·cm²), 349 (~100 mΩ·cm²), Nitrided 349 (~10 mΩ·cm²) at 150 N/cm²	–	[21]
		Cathode environment CD at 0.6 V, with air purge, 349TM (~0.25 mA/cm²)			
AISI446	Thermal nitridation on AISI446 (~1 μm)	Anode environment CD at −0.1 V, at 70°C, 1 M H_2SO_4 + 2 ppmF^-, with hydrogen purge, Nitrided AISI1446 (~−1 μA/cm²).	Before operation, Polarized 7.5 h at 0.6 V, Nitrided AISI446 (<40 mΩ·cm²), at 150 N/cm²	–	[22]
		Cathode environment CD at 0.6 V, with air purge, Nitrided AISI1446 (~0.6 μA/cm²)			

(Continued)

TABLE 5.3 Summary of Metallic Polar Plate Materials, Coatings and their Corrosion Current Density and Interfacial Contact Resistance—cont'd

Plate Material	Coating Material (thickness)	Corrosion Current Density (DOE Target 1 $\mu A/cm^2$) / Corrosion Rate	Contact Resistance (DOE Target 10 $m\Omega \cdot cm^2$),	Cost (DOE 2010 Target $10/kw)	Ref.
Ni-Cr alloy	Thermal nitridation (3–5 μm)	–	Before operation, 316L (~160 $m\Omega \cdot cm^2$), Ni–50CrL (~60 $m\Omega \cdot cm^2$) Nitrided Ni-50CrL (~10 $m\Omega \cdot cm^2$) at 140 N/cm^2	–	[17]
Ni-Cr alloy, AISI446	Thermal nitridation	–	Nitrided AISI446 (20 $m\Omega \cdot cm^2$), at ~150 N/cm^2	–	[28]
AISI434, 436, 441, 444, 446	–	Anode environment CD at −0.1 V, at 70°C, 1 M H_2SO_4 + 2 ppmF^-, with hydrogen purge, AISI446 (10–15 $\mu A/cm^2$), 444 (50 $\mu A/cm^2$), 436 (60 $\mu A/cm^2$), 434 (200 $\mu A/cm^2$), 441 (300 $\mu A/cm^2$) at 70°C, 1 M H_2SO_4 + 2 ppmF^-	Before operation AISI446>434>441>436>444 (between100–200 $m\Omega \cdot cm^2$) at 140 N/cm^2 (small difference)	–	[13]

		Cathode environment CD at 0.6 V, with air purge, AISI446 (10–15 μA/cm^2), 444 (20 μA/cm^2), 436 (20 μA/cm^2), 441 (60 μA/cm^2), 434 (100 μA/cm^2) at 70°C, 1 M H_2SO_4 + 2 ppmF$^-$	After passivation AISI446 (280 mΩ·cm^2 anode environment), (350 mΩ·cm^2 cathode environment) at 140 N/cm^2		
316LSS, Ni-Cr alloy	Thermal nitridation	–	Before operation, 316L (~160 mΩ·cm^2), Ni-50CrL (~60 mΩ·cm^2), Nitrided Ni-50CrL (~10 mΩ·cm^2) at 140 N/cm^2, After passivation, Nitrided Ni-50CrL (no increase)	–	[45]
349TM SS, 316L, 317L,904L	–	Anode environment CD at −0.1 V, at 70°C, 1 M H_2SO_4 + 2 ppmF$^-$, with hydrogen purge, 349TM >904L>317L>316L	Before operation (mΩ·cm^2) 316L>317L>904L>349 (160–100 mΩ·cm^2) at 140 N/cm^2	–	[12]
		Cathode environment at 0.6 V, with air purge, 349TM >904L>317L>316L	After passivation (mΩ·cm^2) 349 (200 mΩ·cm^2) at 140 N/cm^2		
316LSS	Electrochemical process	Electrochemical processed 316L (~0.030 mmpy), 0.5 M H_2SO_4	Before operation 316L (48 mΩ), Electrochemical processed 316L (~7–27 mΩ) at 15 kgf	–	[46]
304SS	–	–	–	–	[47]

(Continued)

TABLE 5.3 Summary of Metallic Polar Plate Materials, Coatings and their Corrosion Current Density and Interfacial Contact Resistance—cont'd

Plate Material	Coating Material (thickness)	Corrosion Current Density (DOE Target 1 $\mu A/cm^2$) / Corrosion Rate	Contact Resistance (DOE Target 10 $m\Omega \cdot cm^2$),	Cost (DOE 2010 Target $10/kw)	Ref.
316LSS	Electrochemical process	0.6 V Potential, 0.5 M H_2SO_4, 316L (60 $\mu A/cm^2$), Electrochemical processed 316L (15 $\mu A/cm^2$), 316LSS (0.1 mmpy), Electrochemical processed 316L (~0.030 mmpy)	–	–	[48]
316LSS, Aluminum 5052, Graphite	YZU001 On Aluminum 5052	Al (1.16 mmpy), Al-coated (0.247 mmpy), 316LSS (0.1 mmpy), Graphitel (0.019 mmpy), 0.5 M H_2SO_4	–	–	[40]
Low-carbon steel AISI 1020	Reforming pack chromization process	AISI 1020 (634 $\mu A/cm^2$), 1020-Cr (1.24 $\mu A/cm^2$), 1020-EMD-Cr (<1 $\mu A/cm^2$) 0.5 M H_2SO_4	AISI 1020 (403.8 $m\Omega \cdot cm^2$), 1020-Cr (39 $m\Omega \cdot cm^2$), 1020-EDM-Cr (<17 $m\Omega \cdot cm^2$),	–	[34]
Aluminum, graphite composite	Chromium carbide (0.1 mm)	–	–	–	[49]
Aluminum, graphite composite	Chromium carbide (0.1 mm)	–	–	–	[50]

316SS	TiN (2–4 μm)	TiN coating (0.25 μA/cm^2 with O_2 bubbled solution) & (0.32 μA/cm^2 with H_2 bubbled solution), 316SS (4.4 μA/cm^2 with O_2 bubbled solution) & (27.1 μA/cm^2 with H_2 bubbled solution), at 80°C, 0.01 M HCl + 0.01 M Na_2SO_4	–	–	[25]
316SS		Anode potential CD at −0.11 V, at 80°C, 0.01 M HCl + 0.01 M Na_2SO_4 bubbled with hydrogen, 316SS (~1.6 μA/cm^2),	–	–	[51]
316SS, Titanium	Ti-FC5, SS316FC6, SS316FC7		Before operation 316SS (~40 mΩ·cm^2), FC5,6,7 (~10–15 mΩ·cm^2) at 200 N/cm^2	–	[52]
321SS, 304SS, 347SS, 316SS, Ti, 310SS, 904LSS, Incoloy800, Inconel601, Poco Graphite	–	–	Before operation 321SS (100 mΩ·cm^2), 304SS (51 mΩ·cm^2), 347SS (53 mΩ·cm^2), 316SS (37 mΩ·cm^2), Ti (32 mΩ·cm^2), 310SS (26 mΩ·cm^2), 904SS (24 mΩ·cm^2), Incoloy800 (23 mΩ·cm^2), Inconel 601 (15 mΩ·cm^2), Poco Graphite (10 mΩ·cm^2) at 220 N/cm^2	–	[11]

(Continued)

TABLE 5.3 Summary of Metallic Polar Plate Materials, Coatings and their Corrosion Current Density and Interfacial Contact Resistance—cont'd

Plate Material	Coating Material (thickness)	Corrosion Current Density (DOE Target 1 μA/cm²) / Corrosion Rate	Contact Resistance (DOE Target 10 mΩ·cm²),	Cost (DOE 2010 Target $10/kw)	Ref.
			After 1200 h operation Ti (250 mΩ·cm²), SS316 (44 mΩ·cm²), SS310 (28 mΩ·cm²), Poco Graphite (10 mΩ·cm²) at 220 N/cm²		
310SS, 316SS, 904LSS	–	–	Before operation 904LSS< 310SS<316SS. After operation SS310<SS316	–	[53]
Titanium, 316SS, Poco Graphite	FC5 (1 μm) on Ti (proprietary)	–	Before operation 316SS(37 mΩ·cm²), FC5 (~13 mΩ·cm²), Graphite (10 mΩ·cm²) at ~220 N/cm²	–	[54]
Aluminum	Gold plated aluminum (2 μm)	Aluminum (~250 μmpy), Copper (>500 μmpy), Gold plated aluminum (~750 μm/year), 316LSS (<100 μmpy), Graphite (<15 μmpy), Silver (<15 μmpy), Gold (<15 μmpy), Nickel (>1000 μmpy), Phosphorous copper (~500 μmpy), Phosphorous	–	Graphite ($75/kg), conductive plastics ($5–$30/kg), Gold plated aluminum ($7/kg)	[19]

		nickel (<30 μmpy), Tin (>10000 μmpy), Titanium (<100 μmpy), Tungsten (<100 μmpy), Zinc (>2000 μmpy), 0.5 M H_2SO_4			
Aluminum	Multi layer coating, (Ni, Au) conductive polymers (polyaniline)	–	–	Graphite ($89/kW), Gold plated ($346/kW), Nickel plated, ($3.2/kW), Aluminum ($2.71/kW)	[18]
Fe- and Ni-base amorphous alloys, Fe-Al_2, Fe-Al_1N_1, Ni-Ta_5	–	Anode potential at −0.1 V at 80°C, 1 M H_2SO_4 + 2 ppmF^- with hydrogen bubbling, Fe-Al_2 (140 μA/cm^2), Fe-Al1N1 (48 μA/cm^2), Ni-Ta5 (52 μA/cm^2)	Before operation , (8–20 mΩ·cm^2)	–	[39]
316LSS, Fe based alloys – $Fe_{50}Cr_{18}Mo_8Al_2YC_{14}B_6$	–	Anode potential at −0.1 V at 75C, 1M H_2SO_4+2ppmF^- with hydrogen bubbling, Fe based alloy (2.48 μA/cm^2)	–	–	[38]
		Cathode environment at 0.6 V, at 75 °C, 1 M H_2SO_4+2ppmF^- with air bubbling, Fe based alloy (0.12 mA/cm^2)			

(Continued)

TABLE 5.3 Summary of Metallic Polar Plate Materials, Coatings and their Corrosion Current Density and Interfacial Contact Resistance—cont'd

Plate Material	Coating Material (thickness)	Corrosion Current Density (DOE Target 1 μA/cm^2) / Corrosion Rate	Contact Resistance (DOE Target 10 mΩ·cm^2),	Cost (DOE 2010 Target $10/kw)	Ref.
304SS	TiC on 304SS	Corrosion current density, Icorr, 304SS (8.3 μA/cm^2), 304SS/TiC (0.034 μA/cm^2), 1 M H_2SO_4	–	–	[35]
316SS, Graphite	TiN on 316SS (1 μm)	–	Before operation 316SS (34.2 mΩ·cm2), 316SS/TiN (32.7 mΩ·cm^2), Graphite (30.2 mΩ·cm^2) at 180 N/cm^2	–	[26]
304SS	TiN on 304SS	Corrosion current density, Icorr, 304SS (2.6 μA/cm^2), 304SS/TiN (0.145 μA/cm^2), 0.5 M H_2SO_4 +2 ppm HF	Before operation 304SS (~140 mΩ·cm^2), 316SS/TiN (19 mΩ·cm^2) at 240 N/cm^2	–	[27]
Titanium, 304SS	Plasma-polymerized HFP	–	–	–	[55]
304SS	NiAl (1 μm)	Corrosion current density, Icorr (49 μA/cm^2), 0.5 M H_2SO_4 at 25°C	–	–	[56]

304SS	Conductive polymers polyaniline (PANI) and polypyrrole (PPY)	Corrosion current density, Icorr, 304SS (10 μA/cm^2), PPY (1 μA/cm^2), PANI (0.1 μA/cm^2), 0.1 M H_2SO_4	Before operation 304SS (~100 mΩ·cm^2), PPY (~800 mΩ·cm^2), PANI (~800 mΩ·cm^2), Graphite (80 mΩ·cm^2) at ~140 N/cm^2	–	[37]
316LSS, $Zr_{75}Ti_{25}$	–	–	–	–	[14]
Ni-Cr metal foam, 316SS metal foam, 316SS channel, Carbon cloth	–	–	–	–	[41]
304LSS, 304LN, 316L, 316LN, 317L, 904L, E-brite, SAF2205, SAF2507, AL29-4-2, AL-6XN	–	–	–	–	[57]
316LSS	Gold, proprietary coatings	–	Oxide resistance –(19.6 to 668.36 mΩ/cm^2	–	[10]
Aluminum	Graphite overmolded	–	–	–	[58]
SS felt, Nickel foam, carbon paper, graphite	–	–	–	–	[59]
310SS	–	–	–	$6.44/kW	[60]
	–		–	–	[42]

(Continued)

TABLE 5.3 Summary of Metallic Polar Plate Materials, Coatings and their Corrosion Current Density and Interfacial Contact Resistance—cont'd

Plate Material	Coating Material (thickness)	Corrosion Current Density (DOE Target 1 μA/cm^2) / Corrosion Rate	Contact Resistance (DOE Target 10 mΩ·cm^2),	Cost (DOE 2010 Target $10/kw)	Ref.
SS316, E-Brite, AL600 (Nickel based alloy)		E-brite (<1 μm/year) better than SS316 and AL600			
Aluminum, 316LSS, Titanium	Gold plated aluminum and 316LSS,	–	Before operation 316SS (~110 mΩ·cm^2), Titanium (~70 mΩ·cm^2), Graphite (10 mΩ·cm^2) at ~140 N/cm^2	–	[9]
Fe-based alloys	–	–	Before operation Fe-based (~100 mΩ·cm^2), Ni-based (~10 mΩ·cm^2), Au-plated (~2 mΩ·cm^2) at ~140 N/cm^2	–	[61]
316SS	Sand blasted and etched	–	–	–	[62]
310SS, Chromium Carbide	–	–	–	–	[33]

3.1. Interfacial Contact Resistance (ICR) Measurement Setup

The interfacial contact resistance of a bipolar plate's candidate material can be determined by sandwiching the material between two gas diffusion layers (GDL) to simulate the actual cell conditions, and measuring the voltage drop (or the resistance, R = V/I) between the materials. Fig. 5.2 shows a schematic of an interfacial contact resistance test setup. The most important components of this setup are the two gold-plated copper terminals. A one centimeter squared contact area of the gold-plated copper terminals is preferred, in order to minimize the need for conversion from the measurement, since the unit of ICR in the field of PEM fuel cells is $m\Omega \cdot cm^2$. Gold-plated copper terminals are also preferred due to their known corrosion resistance, avoiding the formation of an oxide layer on the testing surfaces as this can negatively affect the accuracy of the ICR measurement. A current of 1A is applied across the two gold-plated copper plates by an external power supply to simulate the PEM fuel cell current density. The total resistance R1 can then be calculated by Ohm's law (R = V/I). This experimental setup has been widely used by many researchers [11,12,23,32,34,37,39,65].

Figure 5.3(a) shows a similar wiring diagram for the ICR test setup by measuring the voltages across the test fixture and across a shunt resistance. The voltage V1 across the one ohm shunt resistance (R) is identical to the current applied through this one ohm resistance (R) according to the Ohm's law. The total resistance R1 is equivalent to twice the interfacial contact resistance between the copper the terminal and the GDL. The bulk resistance of the sample material can be eliminated because it is much smaller than the interfacial contact resistance, particularly for metals. The total resistance R2, as shown in Fig. 5.3(b), is equivalent to twice the summation of the ICR value between the gold-plated copper terminal and the GDL added to the value of the ICR between the sample material and the GDL. Therefore, using Ohm's law,

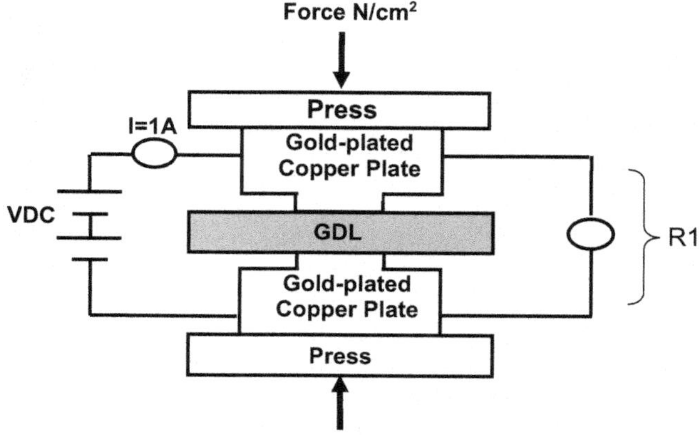

FIGURE 5.2 Schematic diagram of interfacial contact resistance (ICR) measurement.

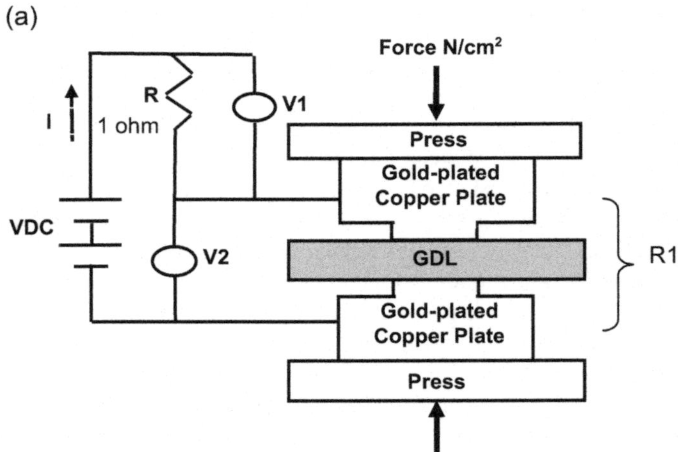

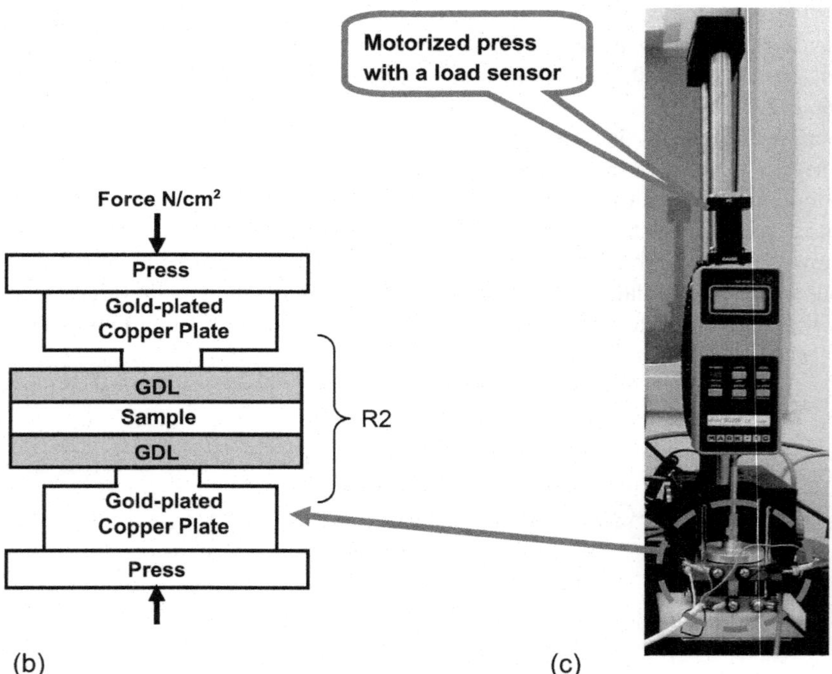

FIGURE 5.3 Schematic diagram of interfacial contact resistance (ICR) measurement setup, (a) test fixture schematic with GDL only, (b) test fixture schematic with combined sample and GDLs, (c) test fixture mounted on a motorized press with a load sensor.

both R1 and R2 can be measured by dividing V2 by V1 (see Figs. 5.3(a) and 5.3(b)). The ICR between the sample material and the GDL can then be calculated from the following equation:

$$ICR = (R2 - R1)/2 \qquad (5.1)$$

Since DOE has suggested an ICR target value for bipolar plate material of 10 $m\Omega \cdot cm^2$ at 140 N/cm^2, and the voltage drop is a strong function of clamping force, the pressure applied to the gold-plated copper terminals should be around 140 N/cm^2. Pressures ranging from 10 to 300 N/cm^2 can be applied during the measurement to obtain the relationship between pressure and ICR for each material. To automate the ICR measurement process, the test fixture can be mounted on a motorized press with a load sensor (Mark-10 Corp.)as shown in Fig. 5.3(c).

3.2. Accelerated Corrosion Resistance Rest Cell Setup

A conventional three-electrode system can be used for this test. Figure 5.4 shows a schematic diagram of a corrosion cell coupled with a counter electrode (CE), a reference electrode (Ref) and a working electrode (WE). The counter electrode is usually made of conductive and non-corrosive material such as platinum and graphite. The reference electrode can be either a saturated calomel electrode (SCE) or a standard hydrogen electrode (SHE) and the working electrode is the test sample. Figure 5.5 shows a commercially available corrosion test cell obtained from EG&G Princeton Applied Research. Different manufacturers produce conventional three-electrode corrosion test cells in different configurations. The cell is connected to a potentiostat to measure the corrosion current and corrosion rate of the bipolar plate candidate material [23].

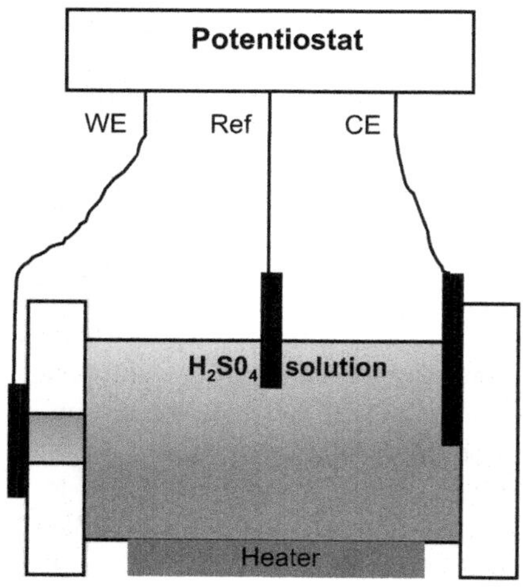

FIGURE 5.4 Schematic diagram of corrosion test cell and potentiostat.

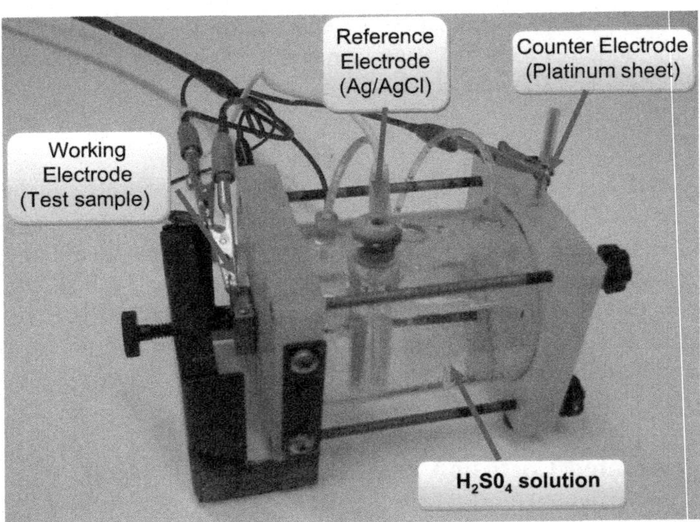

FIGURE 5.5 Corrosion test cell and testing solution.

Electrochemical experiments should be conducted in fuel cell simulated solution at 80°C (fuel cell operating temperature) to select the bipolar plate material with a low dissolution rate – i.e. low corrosion current. The fuel cell simulated solution is made of H_2SO_4. Antunes et al. [64] recently published a list of different concentrations of H_2SO_4 solutions that have been employed by many different researchers and developers. The most widely used solution in bipolar plate research is 0.5 M H_2SO_4 + 2 ppm F^- or 2 ppm HF. For accurate results, air and hydrogen are bubbled in the solution during the electrochemical experiments to simulate the actual fuel cell operating conditions.

Both potentiodynamic and potentiostatic techniques are employed to compare the corrosion resistance of different materials. In such tests, samples are stabilized in simulated fuel cell solution at the open circuit potential (OCP). For potentiodynamic testing, the potential is applied to the test sample and swept between potentials 0.5 V (SCE) below OCP and 1 V (SCE) above OCP at a scanning rate of 1 mV/s. For potentiostatic testing, the test sample is subjected to a constant voltage potential of 0.6 V (SCE) for approximately 8 hours [20,27,32]. These tests indicate the corrosion current of the material at different voltages. The potentiostatic test combined with ICR measurement allows researchers to observe the bipolar plate material behavior in a simulated fuel cell environment without having to build an actual fuel cell and testing it for a long period of time; an expensive and time consuming process.

Finally, considerable caution must be exercised when preparing the test solutions using concentrated sulfuric acid (H_2SO_4) and hydrofluoric acid (HF) because they are extremely dangerous and harmful substances for the human body if contacted directly. The basic safety requirements for solution preparation are as follows:

- Ventilation – Acidic solution preparation must be performed under a chemical fume hood to avoid breathing acidic vapor.
- Eye Protection – Wear chemical splash goggles together with a face shield.
- Body Protection – Wear a laboratory coat with a chemical splash apron that is made of natural rubber, neoprene, or viton material.
- Gloves – Wear gloves that are made of medium or heavyweight viton, nitrile, or natural rubber material to avoid direct contact with acidic solution. A second pair of nitrile exam gloves should be worn under the gloves for protection against leaks.

4. RESULTS AND DISCUSSION

4.1. Interfacial Contact Resistance (ICR) Measurements

Figure 5.6 shows a typical ICR and compression pressure relationship graph for materials, such as SS316 and graphite, that have good potential to act as bipolar plates. As the compression force increases, the ICR decreases due to the interface contact points between the material and the carbon paper increasing when pressed together. The increasing pressure enhances both

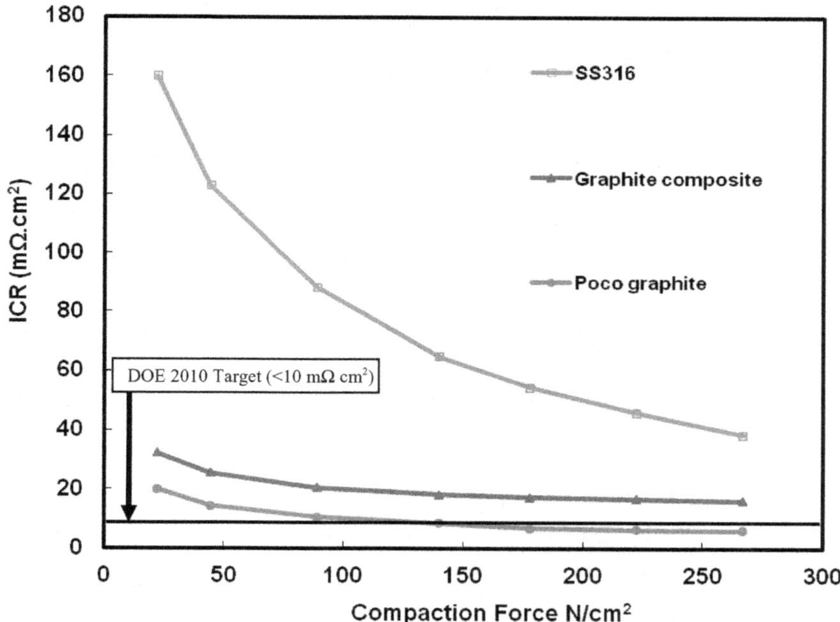

FIGURE 5.6 Interfacial contact resistance of graphite and SS316 bipolar plate materials.

electrical and thermal conductivity. The ICR of Poco graphite had the lowest ICR value of the samples tested. This chart also indicates that stainless steel had a higher ICR value than the commonly used graphite composite. The result of the stainless steel sample agrees with the trend observed by Davies et al. [11] and Wang et al. [12]. Stainless steel tends to form a passive film of chromium, iron and nickel oxides that inhibits electrical conductivity. However, the ICR can be improved by an increase in the Cr content of the alloy. This reduces the thickness of the surface passive film by hampering oxidation of the other elements of the stainless steel, such as iron, nickel etc. [12]. Improved ICR also can be obtained by modifying the surface thin film composition and the texture of stainless steel [28–32]. Further explanation of the effect of roughness on ICR and corrosion resistance can be found in Section 4.3 (Effect of Roughness). Despite the superiority of Poco graphite to metals in both corrosion and ICR, it is not a good candidate bipolar plate material, due to its brittleness, porosity and lack of durability.

4.2. Accelerated Corrosion Resistance Test

Corrosion behavior of bipolar plate material candidates can be obtained from their potentiodynamic polarization curves. Figure 5.7 shows polarization curves of uncoated SS316 in two different concentrations of H_2SO_4+HF solution. As

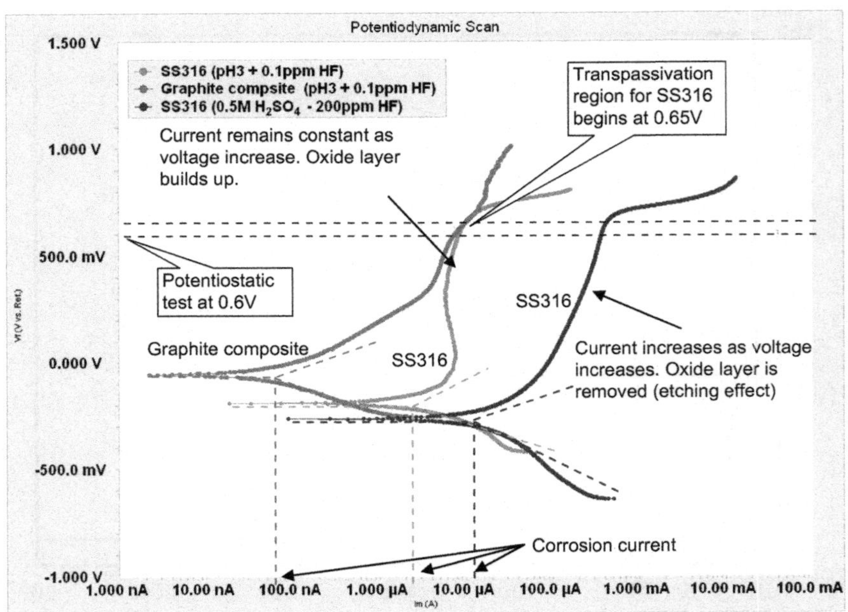

FIGURE 5.7 Potentiodynamic polarization curve of SS316 and graphite composite in two different concentration fuel cell simulated solution.

expected, the corrosion currents of uncoated SS316 obtained from the low concentration solution (pH 3 + 0.1 ppm HF) were much lower than those obtained from the high concentration solution (0.5 M H_2SO_4 + 200 ppm HF).

Performing corrosion tests in aggressive corrosive solutions may provide a good indication of the material's corrosion resistance. However, after hours of potentiostatic testing at 0.6 V (SCE), the interfacial contact resistance of the uncoated SS316 exposed to pH 3 + 0.1 ppm HF, would be higher than that of the uncoated SS316 sample exposed to 0.5 M H_2SO_4 + 200 ppm HF. The increase in ICR is due to the uncoated SS316 passivating below 0.65 V (SCE) at approximately 8 μA in a simulated fuel cell solution. As demonstrated in Fig. 5.7, the current remained relatively constant while the voltage increased from 0 to 0.65 V (SCE). Solutions of relatively low concentration and low voltages allowed the formation of oxide film on the material surface. This oxide film was built up within these operating voltages to prevent further corrosion on the surface. Also, due to the relatively low concentration of the solution, transpassivation was not reached at 0.6 V and therefore the passive layer remained intact during the potentiostatic test causing an increase in the interfacial contact resistance. In addition, the breakdown of the oxide film can occur due to transpassivation at 0.65 V or due to the highly acidic environment. This oxide film removal could result in lower ICR but cause MEA contamination and considerable power degradation if the material was used as a fuel cell bipolar plate.

On the other hand, Fig. 5.7 also shows that graphite did not show any signs of passive and transpassive stages. Therefore graphite composite has been widely used in the fuel cell industry due to its corrosion resistance as mentioned earlier. The combination of the ICR and accelerated corrosion tests can also predict the durability of the bipolar plate in the fuel cell environment. In particular, potentiostatic testing allows metal to passivate at fuel cell operating voltages. Many researchers have tested their bipolar plate samples using potentiostatic techniques in simulated fuel cell solution, and have measured the ICR value of the samples after the test to ensure the ICR value remained the same. Such methods can save researchers a tremendous amount of time, because an actual fuel cell is not needed in the test. The detailed results for corrosion current and ICR of different candidate materials are listed in Table 5.3 and Table 5.4 lists the detailed metallic bipolar plate durability test results from different researchers.

4.3. Effect of Roughness

The effect of surface roughness on corrosion and interfacial contact resistance (ICR) was also examined. Commercially available SS316 sheet was used to demonstrate the effect of roughness on corrosion resistance and ICR. The SS316 sheet was cut into one square inch samples, and these were polished with 1,000 grit sand paper and sand blasted with 80 mesh size aluminum oxide to obtain two different surface roughnesses. The polishing and sand blasting of

TABLE 5.4 Summary of Metallic Bipolar Plate Materials and their Durability Study

Material	Accelerated Durability/ Corrosion Test Method	ICR (Before Corrosion /Lifetime Testing)	ICR (After Corrosion/ Lifetime Testing)	Single Cell/ Stack Lifetime Testing	Ref
AISI446	Polarized at 0.6 V for 7.5 hours vs SCE	190 $m\Omega\cdot cm^2$ at 140 N/cm^2	~260 $m\Omega\cdot cm^2$ at 140 N/cm^2	–	[20]
AISI446 (Nitrided)	Polarized at 0.6 V for 7.5 hours vs SCE	6 $m\Omega\cdot cm^2$ at 140 N/cm^2	~16 $m\Omega\cdot cm^2$ at 140 N/cm^2	–	
Modified446	Polarized at 0.6 V for 7.5 hours vs SCE , Air Purge	4.8 $m\Omega\cdot cm^2$ at 140 N/cm^2	~8 $m\Omega\cdot cm^2$ at 140 N/cm^2	–	
Modified446	Polarized at −0.1 V for 7.5 hours vs SCE , H2 Purge	4.8 $m\Omega\cdot cm^2$ at 140 N/cm^2	~9.6 $m\Omega\cdot cm^2$ at 140 N/cm^2	–	
Ni-50Cr (Nitrided)	4,100 hours in simulated PEMFC corrosion test cell	10 $m\Omega\cdot cm^2$ at 140 N/cm^2	~10 $m\Omega\cdot cm^2$ at 140 N/cm^2	1,000 hours	[28]
Proprietary coating on SS316L	–	–	–	1,000 hours	[10]
Proprietary coating (FC5) on Ti	–	<10 $m\Omega\cdot cm^2$ at 200 N/cm^2	<10 $m\Omega\cdot cm^2$ at 200 N/cm^2	1,0000 hours	[52]
Proprietary coating (FC6) on SS316	–	<15 $m\Omega\cdot cm^2$ at 200 N/cm^2	–	3,000 hours	
Proprietary coating (FC7) on SS316	–	<10 $m\Omega\cdot cm^2$ at 200 N/cm^2	–	2,000 hours	

Superferritic SS (E-Brite)	–	–	–	200 hours	[42]
TiN on SS316	–	–	Ohmic & charge transfer resistance increased 20%	700 hours	[20]
SS316	–	–	Ohmic & charge transfer resistance increased 200%	200 hours	
TiN on SS316 (1kW stack)	–	–	Degradation rate 11%mV/1,000hour	1,000 hours	
SS310	–	–	–	3,500 hours	[60]
SS316	–	~50 mΩ·cm^2 at 140 N/cm^2	~70 mΩ·cm^2 at 140 N/cm^2	3,000 hours	[53]
SS310	–	~40 mΩ·cm^2 at 140 N/cm^2	~60 mΩ·cm^2 at 140 N/cm^2	3,100 hours	
SS904	–	~40 mΩ·cm^2 at 140 N/cm^2	–	1,100 hours	
SS349TM	Polarized at 0.6 V for 7.5 hours vs SCE	~120 mΩ·cm^2 at 140 N/cm^2	~200 mΩ·cm^2 at 140 N/cm^2		[12]
SS316L	–	–	Deteriorated continuously	300 hours	[46]
SS316L (Electrochemical processed)	–	–	Steady performance	300 hours	
Gold on stainless steel	–	–	Steady performance	300 hours	[63]

(Continued)

TABLE 5.4 Summary of Metallic Bipolar Plate Materials and their Durability Study—cont'd

Material	Accelerated Durability/ Corrosion Test Method	ICR (Before Corrosion /Lifetime Testing)	ICR (After Corrosion/ Lifetime Testing)	Single Cell/ Stack Lifetime Testing	Ref
SS304	Polarized at 0.6 V for 8 hours vs SCE	~140 mΩ·cm^2 at 240 N/cm^2	~112 mΩ·cm^2 at 240 N/cm^2	–	[27]
Ti_2N/TiN on SS304	Polarized at 0.6 V for 8 hours vs SCE	~19 mΩ·cm^2 at 240 N/cm^2	~37 mΩ·cm^2 at 240 N/cm^2	–	
TiN on SS304	Polarized at 0.6 V for 8 hours vs SCE	~19 mΩ·cm^2 at 240 N/cm^2	~25 mΩ·cm^2 at 240 N/cm^2	–	
Carbide-based coating on AL	–	–	–	1,000 hours	[49]
Hastelloy G35	–	30–75 mΩ·cm^2 at 100–200 N/cm^2	–	–	[32]
Hastelloy G35 (Nitrided)	Polarized at 0.84 V for 7.5 hours vs SHE	~10 mΩ·cm^2 at 140 N/cm^2	~20 mΩ·cm^2 at 140 N/cm^2	2,500 hours	
Ferritic stainless steel (AL 29–4C)	–	>100 mΩ·cm^2 at 100–200 N/cm^2	–	–	
Ferritic stainless steel (AL 29–4C) (Nitrided)	Polarized at 0.84 V for 7.5 hours vs SHE	~10 mΩ·cm^2 at 140 N/cm^2	~20 mΩ·cm^2 at 140 N/cm^2	–	

the samples removed the oxide layer from the sample surface. The existence of this layer could have negatively affected the ICR measurement. The roughness measurements (Ra) were about 0.2 μm for the polished samples and 3 μm for the sand blasted samples.

The ICR measurements were recorded immediately after the surface treatment, so that the effect of the oxide layer or the passive film on ICR was minimal. The existence of a passive film can complicate ICR measurements on smooth or rough surfaces, because passive film is an electric insulator and it greatly affects the ICR measurements. Without a passive film on the sample surfaces, Fig. 5.8 shows that the rougher surface (Ra ~3 μm) had a slightly lower ICR value than the smoother surface (Ra ~0.2 μm) on the carbon gas diffusion layer (GDL). This result agrees with the observations by Avasarala et al. and Kraytsberg et al. [66,67]. Figure 5.9 shows a SEM image of GDL carbon fibers used in this experimental work. The width of each carbon fiber was approximately 10 μm. Figure 5.10 illustrates the two different scenarios of carbon fibers contacting smoother and rougher surfaces. The lower ICR measurements for relatively high roughness can be explained by the fewer number of contact points and smaller area of contact between the GDL and the sample's surface. This results in higher pressure and better conductivity between the two surfaces at the contact points since pressure can be calculated as force divided by area. On the other hand, the smoother surface generates more contact points, and larger area of contact between the GDL and the

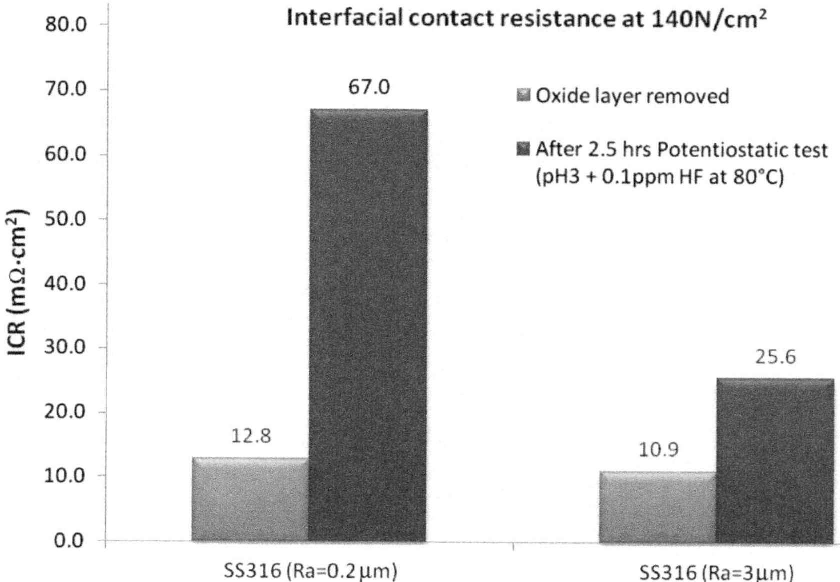

FIGURE 5.8 Interfacial contact resistance at 140 N/cm^2 on SS316 with roughness Ra = 0.2 μm and 3 μm.

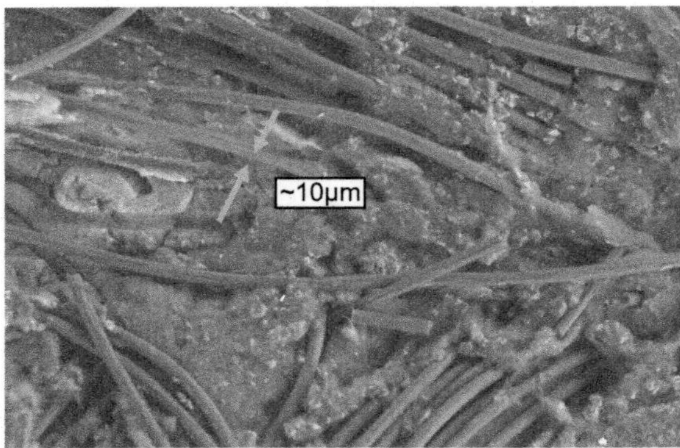

FIGURE 5.9 SEM image of gas diffusion layer (GDL) at 1,000x magnification.

sample. This reduces the pressure between the contact surfaces and results in a higher ICR, since the compression force remains the same in both cases.

However, a corrosion test in solution pH 3+0.1 ppm HF at 80°C shows that the rougher surface had the higher corrosion current, as shown by the potentiodynamic polarization curves (Fig. 5.11). This was attributed to a higher apparent surface area of the rougher surface per unit area exposed to the acidic solution allowing passive film to grow very sporadically on the rougher surface causing less protection to the surface, a higher dissolution rate and material degradation.

In addition, the ICR value of the SS316 sample (Ra = 0.2 μm) was higher than that of the rougher SS316 sample (Ra = 3 μm) after 2.5 hours of potentiostatic testing at 0.6 V (SCE) in pH 3 +0.1 ppm HF solution at 80°C, as

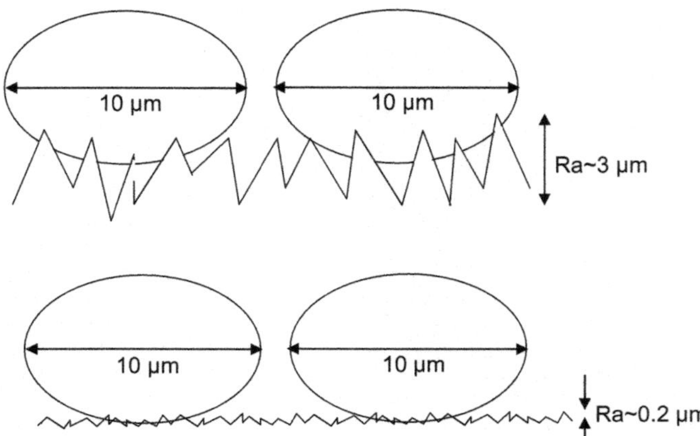

FIGURE 5.10 Illustration of carbon fiber in contact with rough (Ra~3 μm) and smooth (Ra~0.2 μm) surfaces.

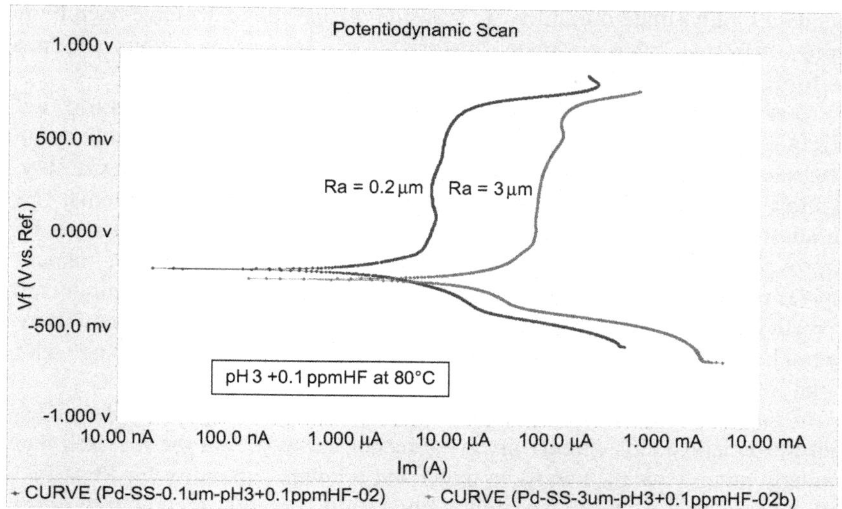

FIGURE 5.11 Potentiodynamic polarization curve of SS316 samples with roughness Ra = 0.2 μm and 3 μm.

shown in Fig. 5.8. This was due to a layer of dense and continuous passive film formed on the smoother surface, which protected the stainless steel sample from further corrosion and resulted in relatively low corrosion current as shown in Fig. 5.11. However, due to poor conductivity of the passive film, the electrical conductivity between the stainless steel and GDL surfaces decreased and the ICR of stainless steel increased, as shown in Fig. 5.8. A similar result was also observed in the SS316 sample with roughness (Ra = 3 μm). Although a passive film formed on the rougher surface after 2.5 hours of potentiostatic testing, causing an increase in ICR, this increase was much less than that of the smoother SS316 sample (Ra = 0.2 μm). This was attributed to a slowly grown, discontinuous passive film formed on the rough surface allowing some points of contacts between SS316 and GDL surfaces.

5. SUMMARY

The durability of bipolar plates is defined by the length of their lifetime in combating corrosion, while maintaining low ICR and not experiencing any drastic loss of power. This is one of the key requirements in fuel cell technology. Metallic plates have proven to surpass the mechanical strength of graphite composite plates, but, as indicated earlier, they are prone to corrosion in the fuel cell environment. Considerable research has been conducted to combat the metallic bipolar plate durability issue, in order to enhance the material's corrosion resistance and interfacial contact resistance. The accelerated corrosion test and interfacial contact resistance test are widely used to evaluate bipolar plate durability. In addition, actual fuel cells have been built using different bipolar plate materials to investigate their lifetime performance.

Cost and durability are still the two pronounced challenges for the PEM fuel cell industry. The cost of large supplies of fuel cell materials and high volume manufacturing processes must be reduced for PEM to reach economic viability, to allow it to penetrate the energy market and compete with other systems. The durability of the PEM fuel cell is another important parameter that must be improved to enhance the reliability of the two main components, namely bipolar plates and MEA. Further research and development must be conducted to rectify the four main bipolar plate corrosion mechanisms; corrosion failure by pin hole formation, electrocatalyst poisoning, membrane ion exchange with metal ion and passivation formation as noted by Kimble et al. [18].

In summary, the concept of replacing graphite with metallic bipolar plates looks promising. This chapter gives the reader an update on the research that can help attain a balance between metal and graphite suitability for PEM fuel cell applications. As more research is conducted to enable metals to perform well under fuel cell operating conditions, an ideal fuel cell system will emerge that can be deployed in the transportation sector.

REFERENCES

[1] X. Li, I. Sabir, Review of bipolar plates in PEM fuel cells: Flow-field designs, J. Power Source 30 (2005) 359–371.

[2] H. Tsuchiya, O. Kobayashi, Mass production cost of PEM fuel cell by learning curve, Int. J. Hydrogen Energy 29 (2004) 985–990.

[3] I. Bar-On, R. Kirchain, R. Roth, Technical cost analysis for PEM fuel cells, J. Power Source 109 (2002) 71–75.

[4] J. Larmine, A. Dicks, Fuel Cell Systems Explained (2000) P84–89.

[5] H. Tawfik, Y. Hung, D. Mahajan, Metal bipolar plates for PEM fuel cell – A review, J. Power Source 163 (2007) 755–767.

[6] V. Mehta, J. Smith Cooper, Review and analysis of PEM fuel cell design and manufacturing, J. Power Sources 144 (2003) 32–53.

[7] R.L. Borup, N.E. Vanderborgh, Design and testing criteria for bipolar plate material for PEM fuel cell application, Mater. Res. Soc. Symp. Proc. 393 (1995) 151–155.

[8] A. Hermann, T. Chaudhuri, P. Spagnol, Bipolar plates for PEM fuel cells: A review, Int. J. Hydrogen Energy 30 (2005) 1297–1302.

[9] P.L. Hentall, J.B. Lakeman, G.O. Mepsted, P.L. Adcock, J.M. Moore, New materials for polymer electrolyte membrane fuel cell current collectors, J. Power Sources 80 (1999) 235–241.

[10] J. Wind, R. Spah, W. Kaiser, G. Bohm, Metallic bipolar plates for PEM fuel cells, J. Power Sources 105 (2002) 256–260.

[11] D.P. Davies, P.L. Adcock, M. Turpin, S.J. Rowen, Bipolar plate materials for solid polymer fuel cells, J. Applied Electrochemistry 30 (2000) 101–105.

[12] H. Wang, M.A. Sweikart, J.A. Turner, Stainless steel as bipolar plate material for polymer electrolyte membrane fuel cells, J. Power Sources 115 (2003) 243–251.

[13] H. Wang, J.A. Turner, Ferritic stainless steels as bipolar plate material for polymer electrolyte membrane fuel cells, J. Power Sources 128 (2004) 193–200.

[14] A.M. Lafront, E. Ghali, A.T. Morales, Corrosion behavior of two bipolar plate materials in simulated PEMFC environment by electrochemical noise technique, Electrochimica Acta 52 (2007) 5076–5085.

[15] J. Andre, L. Antoni, J.-P. Petit, Corrosion resistance of stainless steel bipolar plates in a PEFC environment: A comprehensive study, Int. J. Hydrogen Energy (2010).

[16] Y. Wang, D.O. Northwood, Effects of O_2 and H_2 on the corrosion of SS316L metallic bipolar plate materials in simulated anode and cathode environments of PEM fuel cells, Electrochimica Acta 52 (2007) 6793–6798.

[17] M.P. Brady, K. Weisbrod, I. Paulauskas, R.A. Buchanan, K.L. More, H. Wang, M. Wilson, F. Garzon, L.R. Walker, Preferential thermal nitridation to form pin-hole free Cr-nitrides to protect proton exchange membrane fuel cell metallic bipolar plates, Scripta Materialia 50 (2004) 1017–1022.

[18] M.C. Kimble, A.S. Woodman, E.B. Anderson, Characterization of Corrosion-Protective Methods for Electrically conductive coatings on Aluminum, American Electroplaters and Surface Finishers Society 1999, AESF SUR/FIN '99 Proceedings, 6/21–24

[19] A.S. Woodman, E.B. Anderson, K.D. Jayne, and M.C. Kimble, Development of Corrosion-Resistant Coatings for Fuel Cell Bipolar Plates, American Electroplaters and Surface Finishers Society 1999, AESF SUR/FIN '99 Proceedings, 6/21–24.

[20] J.A. Turner, H. Wang, M.P. Brady, Corrosion Protection of Metallic Bipolar Plates for Fuel Cells, May 22–26, DOE Hydrogen Program Review.

[21] H. Wang, M.P. Brady, G. Teeter, J.A. Turner, Thermally nitrided stainless steels for polymer electrolyte membrane fuel cell bipolar plates Part 1: Model Ni–50Cr and austenitic 349^{TM} alloys, J. Power Sources 138 (2004) 86–93.

[22] H. Wang, M.P. Brady, K.L. More, H.M. Meyer III, J.A. Turner, Thermally nitrided stainless steels for polymer electrolyte membrane fuel cell bipolar plates Part 2: Beneficial modification of passive layer on AISI446, J. Power Sources 138 (2004) 79–85.

[23] W.S. Yoon, X. Huang, P. Fazzino, K.L. Reifsnider, M.A. Akkaoui, Evaluation of coated metallic bipolar plates for polymer electrolyte membrane fuel cells, J. Power Sources 179 (2008) 265–273.

[24] S.-H. Wang, J. Peng, W.-B. Lui, Surface modification and development of titanium bipolar plates for PEM fuel cells, J. Power Sources 160 (2006) 485–489.

[25] M. Li, S. Luo, C. Zeng, J. Shen, H. Lin, C. Cao, Corrosion behavior of TiN coated type 316 stainless steel in simulated PEMFC environments, Corrosion Science 46 (2004) 1369–1380.

[26] E.A. Cho, U.-S. Jeon, S.-A. Hong, I.-H. Oh, S.-G. Kang, Performance of a 1kW-class PEMFC stack using TiN-coated 316 stainless steel bipolar plates, J. Power Sources 142 (2005) 177–183.

[27] D. Zhang, L. Duan, L. Guo, W.-H. Tuan, Corrosion behavior of TiN-coated stainless steel as bipolar plate for proton exchange membrane fuel cell, Int. J. Hydrogen Energy (2010).

[28] M.P. Brady, P.F. Tortorelli, K.L. More, H.M. Meyer III, L.R. Walker, H. Wang, J.A. Turner, B. Yang, R.A. Buchanan, Cost-Effective Surface Modification for Metallic Bipolar Plates, DOE FY (2004). Progress Report.

[29] Y. Fu, G. Lin, M. Hou, B. Wu, H. Li, L. Hao, Z. Shao, B. Yi, Optimized Cr-nitride film on 316L stainless steel as proton exchange membrane fuel cell bipolar plate, Int. J. Hydrogen energy 34 (2009) 453–458.

[30] H. Wang, M.P. Brady, G. Teeter, J.A. Turner, Thermally nitrided stainless steels for polymer electrolyte membrane fuel cell bipolar plates Part 1: Model Ni-50Cr and austenitic 349™ alloys, J. Power Sources 138 (2004) 86–93.

[31] H. Wang, M.P. Brady, K.L. More, H.M. Meyer III, J.A. Turner, Thermally nitrided stainless steels for polymer electrolyte membrane fuel cell bipolar plates Part 2: Beneficial modification of passive layer on AISI446, J. Power Sources 138 (2004) 79–85.

[32] M.P. Brady, H. Wang, B. Yang, J.A. Turner, M. Bordignon, R. Molins, M. Abd Elhamid, L. Lipp, L.R. Walker, Growth of Cr-Nitrides on commercial Ni-Cr and Fe-Cr base alloys to protect PEMFC bipolar plates, Int. J. Hydrogen Energy 32 (2007) 3778–3788.

[33] K. Natesan, R.N. Johnson, Corrosion resistance of chromium carbide coatings in oxygen-sulfur environments, Surface and Coatings Technology 33 (1987) 341–351.

[34] C.-Y. Bai, M.-D. Ger, M.-S. Wu, Corrosion behaviors and contact resistances of the low-carbon steel bipolar plate with a chromized coating containing carbides and nitrides, Int. J. Hydrogen Energy 34 (2009) 6778–6789.

[35] Y.J. Ren, C.L. Zeng, Corrosion protection of 304 stainless steel bipolar plates using TiC films produced by high-energy micro-arc alloying process, J. Power Sources 171 (2007) 778–782.

[36] A. Sanaa, Abo El-Enin, O.E. Abdel-Salam, H. El-Abd, A.M. Amin, New electroplated aluminum bipolar plate for PEM fuel cell, J. Power Sources 177 (2008) 131–136.

[37] Shine Joseph, J.C. McClure, R. Chianelli, P. Pich, P.J. Sebastian, Conducting polymer-coated stainless steel bipolar plates for proton exchange membrane fuel cells (PEMFC), Int. J. Hydrogen Energy 30 (2005) 1339–1344.

[38] J. Jayaraj, Y.C. Kim, K.B. Kim, H.K. Seok, E. Fleury, Corrosion studies on Fe-based amorphous alloys in simulated PEM fuel cell environment, Science and Technology of Advanced Materials 6 (2005) 282–289.

[39] J. Jayaraj, Y.C. Kim, H.K. Seok, K.B. Kim, E. Fleury, Development of metallic glasses for bipolar plate application, Materials Science and Engineering A 449–451 (2007) 30–33

[40] S.-J. Lee, C.-H. Huang, Y.-P. Chen, Investigation of PVD coating on corrosion resistance of metallic bipolar plates in PEM fuel cell, J. Materials Processing Technology 140 (2003) 688–693.

[41] A. Kumara, R.G. Reddy, Materials and design development for bipolar/end plates in fuel cells, J. Power Sources 129 (2004) 62–67.

[42] M.S. Wilson, C. Zawodzinski, S. Møller-Holst, D.N. Busick, F.A. Uribe, T.A. Zawodzinski, PEMFC stacks for power generation, Proceedings of the 1999 U.S DOE Hydrogen Program Review.

[43] Los Alamos National Laboratory Home page (1998). Available from Website: http://www.ott.doe.gov/pdfs/contractor. Last retrieved 5 November 2001.

[44] R. Blunk, M. Hassan Abd Elhamid, D. Lisi, Y. Mikhail, Polymeric composite bipolar plates for vehicle applications, J. Power Sources 156 (2006) 151–157.

[45] M.P. Brady, K.L. More, P.F. Tortorelli, L.R. Walke, K. Weisbrod, M. Wilson, F. Garzon, H. Wang, I. Paulauskas and R.A. Buchanan, Cost-Effective Surface Modification for Metallic Bipolar Plates, DOE FY 2003 Progress Report.

[46] S.-J. Lee, J.-J. Lai, C.-H. Huang, Stainless steel bipolar plates, J. Power Sources 145 (2005) 362–368.

[47] S.-J. Lee, Y.-P. Chen, C.-H. Huang, Electroforming of metallic bipolar plates with micro-featured flow field, J. Power Sources 145 (2005) 369–375.

[48] S.-J. Lee, C.-H. Huang, J.-J. Lai, Y.-P. Chen, Corrosion-resistant component for PEM fuel cells, J. Power Sources 131 (2004) 162–168.

[49] Y. Hung, K.M. El-Khatib, H. Tawfik, Testing and evaluation of aluminum coated bipolar plates of PEM fuel cells operating at 70° C, J. Power Sources 163 (2006) 509–513.

[50] Y. Hung, K.M. El-Khatib, H. Tawfik, Corrosion-resistant lightweight metallic bipolar plates for PEM fuel cells, J. Applied Electrochemisty 35 (2005) 445–447.

[51] M.C. Li, C.L. Zeng, S.Z. Luo, J.N. Shen, H.C. Lin, C.N. Cao, Electrochemical corrosion characteristics of type 316 stainless steel in simulated anode environment for PEMFC, Electrochimica Acta 48 (2003) 1735–1741.

[52] G.M Griffths, E.E Farndon, D.R Hodgson, I.M Long, PEMcoat: A range of active coatings for metal bipolar plates, The Knowledge Foundation's 3rd Annual International Symposium, Washington, D.C., April 22–24, 2001.

[53] D.P. Davies, P.L. Adcock, M. Turpin, S.J. Rowen, Stainless steel as a bipolar plate material for solid polymer fuel cells, J. Power Sources 86 (2000) 237–242.

[54] D.R. Hodgson, B. May, P.L. Adcock, D.P. Davies, New lightweight bipolar plate system for polymer electrolyte membrane fuel cells, J. Power Sources 96 (2001) 233–235.

[55] A. Taniguchi, K. Yasuda, Highly water-proof coating of gas flow channels by plasma polymerization for PEM fuel cells, J. Power Sources 141 (2005) 8–12.

[56] S. K. Chen, H. C. Lin, and C. Y. Chung, Corrosion resistance study of stainless-steel bipolar plates with NiAl coatings, Feng Chia University and National Science Council of Republic of China under the Grant Numbers FCU-93GB27 and NSC 93-2218-E-35-006.

[57] J.S. Kim, W.H.A. Peelen, K. Hemmes, R.C. Makkus, Effect of alloying elements on the contact resistance and the passivation behaviour of stainless steels, Corrosion Science 44 (2002) 635–655.

[58] S. Ehrenberg, D. Baars, J. Kelly, and I. Kaye, One piece bi-polar plate with cold plate cooling, 2002, Dais Analytic Corporation, Rogers Corporation.

[59] A. Sergei Gamburzev, J. Appleby, Recent progress in performance improvement of the proton exchange membrane fuel cell (PEMFC), J. Power Sources 107 (2002) 5–12.

[60] F. J. Chimbole, J. P. Allen, R. M. Bernard, Results of PEMFC bipolar separator plate interconnect and current collector development at AEC, www.gencellcorp.com

[61] R. Hornung, G. Kappelt, Bipolar plate materials development using Fe-based alloys for solid polymer fuel cells, J. Power Sources 72 (1998) 20–21.

[62] C.E. Reid, W. R. Mérida, and G. McLean, Results and analysis of a PEMFC Stack using metallic bi-polar plates, Proceedings of 1998 Fuel Cell Seminar, Palm Spring, 1998.

[63] T. Matsuura, M. Kato, M. Hori, Study on metallic bipolar plate for proton exchange membrane fuel cell, J. Power Sources 161 (2006) 74–78.

[64] R.A. Antunes, M.C.L. Oliveira, G. Ett, V. Ett, Corrosion of metal bipolar plates for PEM fuel cells: A review, Int. J. Hydrogen Energy (2010).

[65] F. Barbir, PEM Fuel Cells (2005) P99–P107.

[66] B. Avasarala, P. Haldar, Effect of surface roughness of composite bipolar plates on the contact resistance of a proton exchange membrane fuel cell, J. Power Sources 188 (2009) 225–229.

[67] A. Kraytsberg, M. Auinat, Y. Ein-Eli, Reduced contact resistance of PEM fuel cell's bipolar plates via surface texturing, J. Power Sources 164 (2007) 697–703.

Chapter 6

Freeze Damage to Polymer Electrolyte Fuel Cells

Abdul-Kader Srouji[1] and Matthew M. Mench[2*]
[1]*Fuel Cell Dynamics and Diagnostics Laboratory, and Department of Mechanical and Nuclear Engineering, The Pennsylvania State University, PA, USA,* [2]*Electrochemical Energy Storage and Conversion Laboratory, and Department of Mechanical Aerospace and Biomedical Engineering, The University of Tennessee, TN, USA*

1. INTRODUCTION

The Department of Energy (DOE) 2015 technical target requires that fuel cell vehicles should deliver 50% of rated power in 30 seconds from a cold start at −20°C, with less than 62.5 J/We parasitic energy input for start-up and shut-down [1]. With those targets, they should also be able to start from −40°C after being soaked at this temperature for 8 hours. DOE targets for an 80 kW_e (net) fuel cell system for automotive applications are summarized in Table 6.1.

Dozens of studies have been performed to examine various aspects of fuel cell material compatibility and performance degradation in freezing environments. The damage resulting from a frozen soak, from freeze/thaw (F/T) cycling, and from frozen start have been examined. The purpose of this summary is to explore the damage resulting from freeze/thaw conditions. A detailed summary of damage observed due to some aspects of a frozen environment from various published studies is shown in Table 6.2. Generically, damage resulting from a frozen environmental condition is due to water generated at the cathode during sub-zero operation, or the existence of liquid water that resides in the membrane, porous media after shut-down. Liquid water that freezes in the channels and internal manifolds can hinder cold start and reactant flow, resulting in exacerbated degradation due to cell voltage reversal and carbon corrosion or local fuel starvation. An important result from accumulated studies is that, for conventional fuel cell materials and designs, no significant damage is observed from simply cycling the fuel cell material to subzero conditions without start-up operation or liquid water before freeze. This indicates that freeze-related damage can be eliminated through proper

Polymer Electrolyte Fuel Cell Degradation. DOI: 10.1016/B978-0-12-386936-4.10006-5

TABLE 6.1 **DOE Technical Targets for Automotive Applications: 80-kW$_e$ (net) Integrated Transportation Fuel Cell Power Systems Operating on Direct Hydrogen[a]. As Reported in [1]**

Characteristic	Units	2003 Status	2005 Status	2010	2015
Energy efficiency[b] at 25% of rated power	%	59	59	60	60
Energy efficiency at rated power	%	50	50	50	50
Power density	W / L	440	500	650	650
Specific power	W / kg	420	470[c]	650	650
Cost[d]	\$ / kW$_e$	200	110[e]	45	30
Transient response (time from 10% to 90% of rated power)	seconds	3	1.5	1	1
Cold start-up time to 50% of rated power					
at −20°C ambient temperature	seconds	120	20	30	30
at +20°C ambient temperature	seconds	60	<10	5	5
Start-up and shut-down energy[f]					
from −20°C ambient temperature	MJ	N/A	7.5	5	5
from +20°C ambient temperature	MJ	N/A	N/A	1	1
Durability with cycling	hours	N/A	~1,000[g]	5,000[h]	5,000[h]
Unassisted start from low temperatures[i]	°C	N/A	−20	−40	−40

[a] *Targets exclude hydrogen storage, power electronics and electric drive.*
[b] *Ratio of DC output energy to the lower heating value of the input fuel (hydrogen). Peak efficiency occurs at about 25% rated power.*
[c] *Based on corresponding data from the DOE report to account for ancillaries.*
[d] *Based on 2002 dollars and cost projected to high-volume production (500,000 systems per year).*
[e] *Status is from 2005 TIAX study and will be periodically updated.*
[f] *Includes electrical energy and the hydrogen used during the start-up and shut-down procedures.*
[g] *Durability with cycling is being evaluated through the Technology Validation activity. Steady-state stack durability is 20,000 hours.*
[h] *Based on test protocol issued by DOE in 2007.*
[i] *8-hour soak at stated temperature must not impact subsequent achievement of targets.*

design, materials, and operating protocol. Based on the summary of available studies shown in Table 6.2, not all environments, materials, or designs result in damage. In fact, there has been a seemingly high discrepancy between the results of different studies, which suggests there is still much to learn about the

TABLE 6.2 Summary of Observed PEFC Damage Due to Frozen Environments from Various Sources

Reference	Test Mode	PEM	CL	MEA	DM	Test conditions			Results
						T range (°C)	Number of cycles	Purge/no purge	
Wilson et al. 1994 [62]	*In-situ* F/T	Nafion 117	20 wt% Pt/C (0.16 mg/cm^2)	Decal process[a]	ELAT hydrophobic carbon cloth	−10/80	3	No purge (wet)	No performance loss
McDonald et al. 2004 [17]	*Ex-situ* F/T	Nafion 112	0.4 mg Pt/C /cm^2	N/A	None	−40/80	385	Dry state (λ<3)	No significant physical damage change in the molecular level
	In-situ F/T	Nafion 112	0.4 mg Pt/C /cm^2	N/A	Carbon paper		385	Dry state (λ<3)	No significant physical damage change in the molecular level
Liu 2006 [2]	*Ex-situ* F/T (immersion)	Nafion 112	N/A	N/A	None	−40/50	10	Immersed in water	Severe CL loss Severe deformation of MEA
		DSM	N/A	N/A	None		10	Immersed in water	No observable loss
	In-situ F/T	Nafion 112	N/A	N/A	N/A		40	N/A	No performance loss No ECSA loss
		DSM	N/A	N/A	N/A		40	N/A	No performance loss No ECSA loss

(Continued)

TABLE 6.2 Summary of Observed PEFC Damage Due to Frozen Environments from Various Sources—cont'd

Reference	Test Mode	PEM	CL	MEA	DM	Test conditions: T range (°C)	Test conditions: Number of cycles	Test conditions: Purge/no purge	Results
Patterson et al. 2006 [63,64]	*In-situ* F/T	N/A	N/A	N/A	N/A	−40/25	63	N/A	No performance loss
	Cold start-up	N/A	N/A	N/A	N/A	−15	N/A	N/A	End cell loss
Mukundan et al. 2006 [20,68]	*In-situ* F/T	Nafion 1135	20 wt% Pt/C (0.2 mg/cm^2)	Decal process[a]	Wet proofed carbon cloth	−40/80	100	No purge (wet)	No performance loss
					SGL 30DC		45	No purge (wet)	Mechanical failure of DM
		Nafion 1135	20 wt% Pt/C (0.2 mg/cm^2)		Wet proofed carbon cloth	−80/80	10	No purge (wet)	Performance loss HFR increase Interfacial delamination DM failure
Cho et al. 2003, 2004 [21,65]	*In-situ* F/T	Nafion 115	20 wt% Pt/C (0.4 mg/cm^2)	GDE[c]	Wet proofed carbon paper	−10/80	4	No purge (wet)	Performance loss, ohmic and charge transfer resistance increase ECSA loss
							4	Dry purge ($\lambda<2$)	No performance loss No ECSA loss

Gaylord 2005 [16]	Field test (stationary)	N/A	N/A	N/A	Carbon paper	Exposed to freezing	N/A	N/A	DM Fracture Membrane failure Severe CL delamination
Meyers 2005 [15]	*In-situ* F/T	Commercial MEAs (reinforced membrane)			N/A	−20/?	20	N/A	Membrane cracks CL delamination
Oszcipok et al. 2005, 2006 [66,67]	Cold start-up	Catalyst coated membrane			N/A	−10	10	Dry purge	Performance loss ECSA loss Hydrophobicity loss (MOL,DM)
		Catalyst coated membrane (0.4 mg Pt/cm^2)			Carbon cloth	−10	7	Partial purge	Significant performance loss
Yan et al. 2006 [14]	Cold start-up	Nafion 112,115,117	20 wt% Pt/C	GDE[d]	Carbon paper /cloth	−15	N/A		Interfacial delamination Membrane hole
Guo and Qi 2006 [18]	*Ex-situ* F/T	Commercial MEA with 30 μm membrane and 1.0 mg Pt/cm^2			None	−30/20	6	Dry purge (λ<4)	Negligible damage
								No purge (wet)	Severe damage Severe CL cracks
	In-situ F/T		Carbon paper				20	No purge (wet)	Severe CL cracks ECSA loss Negligible performance loss Easy flooding
								Dry purge (λ<4)	No physical damage No performance loss

(*Continued*)

TABLE 6.2 Summary of Observed PEFC Damage Due to Frozen Environments from Various Sources—cont'd

Reference	Test Mode	PEM	CL	MEA	DM	Test conditions: T range (°C)	Test conditions: Number of cycles	Test conditions: Purge/no purge	Results
Hou et al. 2006 [19]	*In-situ* F/T	Nafion 212	20 wt% Pt/C (0.8 mg Pt/ cm^2)	GDE	Carbon paper	−20/60	20		No performance loss No ECSA loss No physical damage
Alink et al. 2008 [31]	*In-situ* F/T		0.4 mg Pt/C /cm^2		Toray TGP-H-060	−40/60	120	Dry purge	Increase in porosity of MEA Decrease in electrode surface area more important at the cathode Micro-cavities on electrodes
			0.3 mg Pt/C /cm^2				62	No purge	Increase in porosity of MEA Serious detachment of electrode material Micro-cavities on electrodes
	Ex-situ F/T					−20/0.5	10		No damage to MEA No damage to DM
	Cold start-up		0.4 mg Pt/C /cm^2		Toray TGP-H-060	−40			
			0.3 mg Pt/C /cm^2				9		

Kim et al. 2007 [3]	*Ex-situ* F/T	Non-cracked 0.4 mg Pt/C /cm^2	Reinforced membrane 18 μm	Without DM/ MPL	−40/70	30	CL separation under channels
		Cracked 0.4 mg Pt/C /cm^2	Reinforced membrane 18 μm				Severe CL separation under channels
		Non-cracked 0.4 mg Pt/C /cm^2	Non-reinforced 18 μm				Nearly complete delamination of CL
		Non-cracked 0.4 mg Pt/C /cm^2	Reinforced 35 μm				Severe MEA damage, nearly complete delamination of CL under channels
		Non-cracked 0.4 mg Pt/C /cm^2	Reinforced membrane 18 μm	With DM/ MPL	−40/70	30	MEA largely intact. No sign of F/T damage
		Cracked 0.4 mg Pt/C /cm^2	Reinforced membrane 18 μm				Cracking with no delamination
		Non-cracked 0.4 mg Pt/C /cm^2	Non-reinforced 18 μm				Damage under channels

(*Continued*)

TABLE 6.2 Summary of Observed PEFC Damage Due to Frozen Environments from Various Sources—cont'd

Reference	Test Mode	PEM	CL	MEA	DM	Test conditions			Results
						T range (°C)	Number of cycles	Purge/no purge	
			Non-cracked 0.4 mg Pt/C /cm^2	Reinforced 35 μm					Frost heave damage and CL separation
Kim et al. 2008 [4]	*Ex-situ* F/T				Carbel-CL SGL 10BB SGL 25BC SGL 10BA	−40/70	30–100	Water submerged condition	Interfacial delamination DM/CL Deformation of stiff diffusion media

[a]*Decal printing (TBA+ form catalyst) and then hot pressing at 200°C.*
[b]*20% PTFE treatment with MPL.*
Abbreviation: F/T: Freeze/thaw thermal cycling DSM: Dimensionally stable membrane GDE : Gas diffusion electrode, PEM: Polymer electrolyte membrane.
[c]*Catalyst ink sprayed on DM and then hot pressing at 140°C.*
[d]*Sprayed on DM and then hot pressed.*

genesis and causes of freeze related damage. There are some commonalities between the studies. When freeze related damage was observed, it was generally observed at the following locations:

- The membrane in the form of pinholes on the surface.
- The catalyst layer (CL), in the form of local cracks as well as interfacial delamination between the CL|membrane and/or CL|DM interface, and loss of electrochemical surface area (ECSA).
- The diffusion media, via cracking of the microporous layer (MPL) and interfacial delamination with the CL, or loss of hydrophobicity.

The effect of different component characteristics on sustaining damage was studied [2,3,4]. One important result concludes that properly drying a cell before sub-zero cool down and freeze prevents observable physical damage. However, over-drying or non-uniform drying of the membrane during shut-down has been shown to cause an uneven stress distribution in the membrane and result in accelerated membrane degradation [5]. Although freeze-related damage can be eliminated by completely purging the cell, this mitigation method is generally too time consuming, parasitic, and potentially damaging to the membrane to be of use. Much more study of purge, evaporative removal, and knowledge of the locations and sources of freeze-related water damage is needed. From Table 6.2, it can be seen that damage is not uniformly observed and varies as a function of:

1. Material sets – Microporous layer (MPL)|CL combinations appear to impact results. This is a function of drainage at shut-down, interfacial contact, and distribution of compression pressure. Stiff or bonded DM appear to best mitigate damage because they can reduce interfacial liquid accumulation which has been shown to cause freeze delamination in some cases [3,4].
2. Cell design – The channel/land design also appears to impact damage. In particular, for a traditional channel/land configuration, low under-channel compression exacerbates damage. The wider the channel, the worse the damage appears to be. Figure 6.1 shows the result for freeze/thaw cycling to −30°C from a wet state with 2 mm wide lands. This testing qualitatively confirmed the results of computational simulation [6–8] which predicted that the major damage locations from a macroscopic perspective appear under the channel along various material interfaces in the cell.
3. Shut-down protocol – Obviously, the source of damage is water. Upon cool-down, condensation, diffusive flow, temperature gradient driven flow, and capillary transport take place. To avoid damage or hindered cold-start, residual water inside the CL should be removed. This can be accomplished with a variety of methods discussed in Section 4 of this review.
4. Location in stack – Anode end cells in particular have been shown by several groups to suffer aggravated damage compared to center and cathode end stack plates. This is apparently a result of greater heat transfer from end plates and concomitant phase-changed induced (PCI) motion [9].

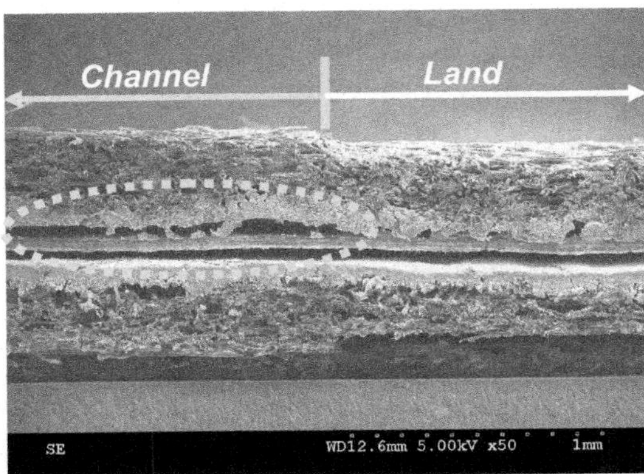

FIGURE 6.1 SEM image of cross-section of membrane electrode assembly after 100 freeze/thaw cycles to −30°C. Delamination damage is seen under the channel but not land locations due to the overburden pressure from the land [4].

Although many studies have been conducted to investigate freeze damage on different components of a polymer electrolyte fuel cell, there are significant variation in the results. Much, but likely not all, of this variation can be ascribed to the non-standardized testing procedures, cell designs, and materials used. There is still a clear need to resolve these discrepancies with fundamental understanding of the physicochemical mechanisms involved, so that optimized designs, materials, and protocols can be developed. The motivation of this review is to better understand and codify the existing data so that conflicting conclusions from the various studies can be better explained. Additionally, methods and potential concepts to mitigate freeze-induced damage are discussed. The operation of start-up from a frozen state and possible damage resulting exclusively from a frozen start is another topic that is not within the scope of this review.

2. COMPUTATIONAL MODEL EFFORTS

Several publications based on computational models for predicting the key parameters and conditions for freeze damage in PEFCs have been developed. These are based on porous media flow theories, and frost heave formation in thin cracks [6–8]. Models to predict the formation of ice, without the onset of damage, have also been developed, but are not within the subject of this review and are not summarized here. Damage resulting from ice formation can be separated into two different phenomenological categories:

1. damage due to ice formation and expansion from a liquid to solid state; and
2. damage due to an ice lens, which develops sufficient phase pressure to physically separate interfacial surfaces.

Damage due to the approximately 8% volume expansion of the ice phase is imagined by many to be the only possible mechanism for damage, but experimental and computational evidence now suggests that ice lens formation is responsible for much of the observed freeze damage. In fact, the porous media in fuel cells (CL and DM) are typically far from a full-saturation state, and another 8% expansion as the ice forms during a slow cooling at shutdown is not likely to cause severe morphological damage. In contrast, an ice lens can cause plastic or elastic delamination and deformation, and can form as a result of even slight interfacial water accumulation, or from water expulsion from the membrane under decreasing temperatures [6–8]. The presence of interfacial accumulation of liquid has been suggested or observed by several independent studies [10–12]. After ice nucleation, whether or not an ice lens continues to grow depends on the ice phase pressure, the overburden pressure, the hydraulic availability of liquid water flowing to the ice lens location, and the heat transfer rate from the porous media. The critical ice pressure for ice lens formation is referred to as the local overburden pressure P_{ovbd}. Overburden pressure is a function of the assembling pressure P_{assm} or the transmitted channel pressure P_{ch}, depending on whether the location is under the land or the channel respectively, the material tensile strength σ_{ts} and the shear stress τ_{sh} [6]. The possible cases are listed in Table 6.3. The impact of DM stiffness is discussed in the following section of this review.

The large overburden pressure restrains macroscopic ice lens formation. However, local delamination at the catalyst level could still occur due to high local ice-phase pressure. Figure 6.2 is a schematic representing potential locations of freeze/thaw damage according to a computational model, which included the impact of water motion in the ionomer and porous media during shut-down to a frozen state [6–8]. The results are in qualitative agreement with the observed F/T damage on SEM for materials which underwent *ex-situ* F/T cycling. Ice growth leading to damage most likely occurs under the channel and at the interface between CL|DM and CL|Membrane. The CL|Membrane ice growth is highly dependent on the freezing temperature depression properties of Nafion membrane. Non-freezing water flowing out of the membrane would immediately freeze upon contact with the catalyst layer. The maximum ice lens growth at this location would therefore depend on the initial water content of

TABLE 6.3 Possible Locations of Ice Lens Growth with their Respective Overburden Pressure Based on Data from [6]

Position	Under BP	Under CH
Within DM, CL or Nafion	$P_{ovbd} = P_{assm} + \sigma_{ts} + p_{ch}$	$P_{ovbd} = p_{ch} + \sigma_{ts} + \tau_{sh}$
At interface between bipolar plate(BP)/DM, DM/CL or CL/Nafion	$P_{ovbd} = P_{assm} + p_{ch}$	$P_{ovbd} = p_{ch} + \tau_{sh}$

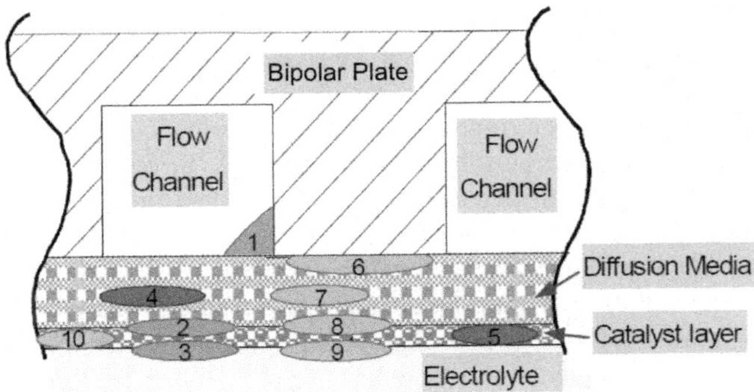

FIGURE 6.2 Schematic showing the potential locations of freeze/thaw damage [6].

the membrane, and the membrane type, as represented in Fig. 6.3. Reduction of the liquid water in contact with the ionomer at shut-down to a frozen state is the key to avoid damage, as the liquid contact is responsible for higher free-water content in the membrane, which is responsible for local degradation – as discussed. It should be noted that extensive *ex-situ* testing revealed that no damage was observed when F/T cycling in a purely gas phase (but vapor-saturated) environment [13].

3. MODES OF DEGRADATION

In this section, the various modes of observed freeze-related physicochemical damage are discussed. Where possible, the phenomena responsible for the damage are described. The section is divided into subsections based on

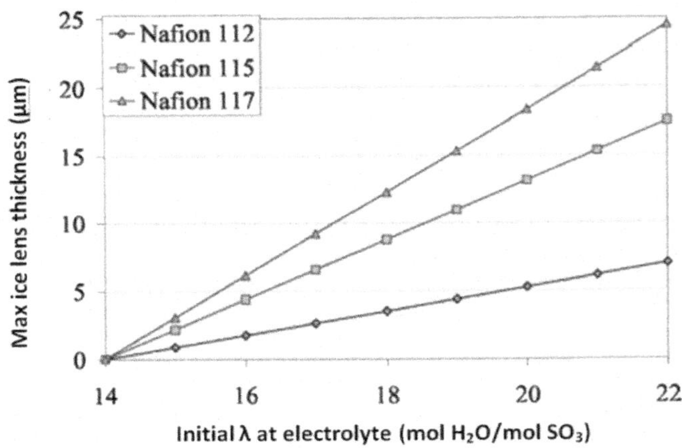

FIGURE 6.3 Maximum thickness of ice lens that could be formed by water expelled from Nafion during freezing, as a function of the initial water content and membrane type [6].

components for convenience, although it is likely that the observed damage is interrelated.

3.1. Membrane

Water freezing in the PEFC can damage the electrolyte membrane in different ways. Physical damage is observable with a scanning electron microscope (SEM). Increased roughness, cracks and pinholes were observed in a membrane after *in-situ* operation at sub-zero temperatures ranging from −5°C to −15°C [14]. The same was observed of cells in a fuel cell stack that were freeze/thawed 20 times at −20°C [15], and idle stacks during winter or long duration installation under freezing temperatures [16]. Figure 6.4 shows SEM images of damage to a Nafion membrane after being stored and operated at sub-zero ambient temperatures as low as −15°C [14]. The MEA was assembled by spraying the catalyst on wet-proofed carbon paper and then hot pressing the electrodes on the Nafion. After operation the electrodes were separated from the membrane and the polymer electrolyte was examined with an SEM. Increased membrane roughness is observed (Fig. 6.4(c)) after sub-zero operation compared to a membrane operated at room temperature (Fig. 6.4(b)). At higher magnification, (Figs. 6.4(d) and (e)) micro-cavities and pinholes at the cathode outlet region of the membrane are clearly visible after sub-zero operation. This type of damage can lead to performance loss through increased hydrogen crossover and loss of catalyst activity.

In the electrolyte, water content (λ) is defined as the number of moles of water per mole of sulfonic acid group in the electrolyte. Membrane hydration is elementary to ionic conductivity, but over-hydration and subsequent freeze can cause damage. A dried MEA with $\lambda < 4$ (water weight percent less than 6%), did not experience freeze-damage [17–20]. However, this state has difficulty in generating current from a frozen state because of low ionic conductivity [21,22]. Under sub-freezing conditions, Mukundan et al. [20] determined that $7<\lambda<12$ in Nafion® resulted in optimal conductivity. Later, Tajiri et al. [22] observed similarly low ohmic losses for $6.2<\lambda<14$ in a W.L. Gore Primea® MEA.

Water in the membrane can exist in a non-freezing state or a freezing state. Water in Nafion is subjected to a fluorocarbon environment and bonds strongly to cations and ion exchange sites which prevents it from freezing to −120°C [23]. Other water molecules, interacting weakly with ions and cations exchange sites, freeze on reaching −20°C. Free water inside the membrane behaves like bulk water on the surface of the membrane and freezes immediately at 0°C [23]. Since the strongly bonded water in the membrane has some freezing point depression, it should not cause damage as it does not freeze. Differential scanning calorimetry (DSC) has been used to measure the amount of unfrozen water in fine grained media as a function of temperature. In addition, the finer the

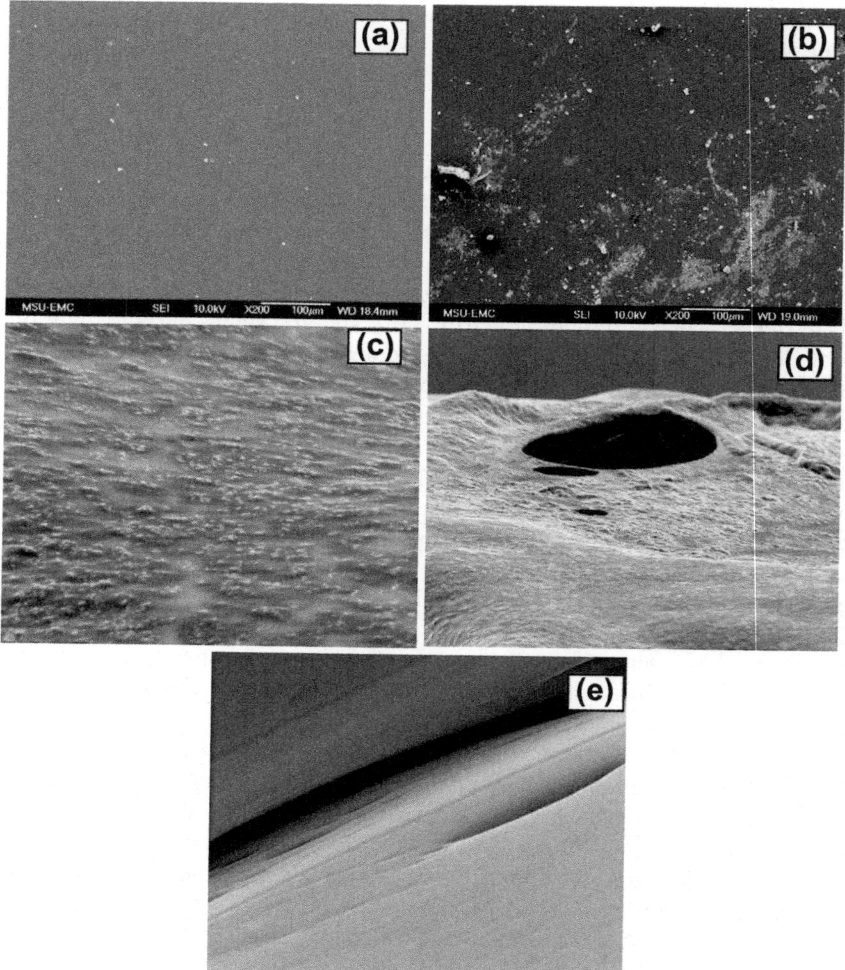

FIGURE 6.4 Effect of sub-zero temperature on membrane (a) Virgin Nafion membrane, (b) membrane after operation at room temperature, (c) membrane after operation at −15°C, (d) membrane from cathode outlet regions after operation at −15°C and (e) membrane from cathode outlet regions after operation at −15°C. Images from [14].

grains are, the greater the freezing point depression compared to free standing water. DSC data from the literature [24,25,26] characterizing water composition in Nafion has been extrapolated by He and Mench [6] and is shown in Fig. 6.5. It was shown that the weakly bonded water in Nafion pores sized 2 nm corresponds to a freeze point depression of 24.5 K. It was hypothesized by He et al. that this weakly bound water comes out of the membrane and results in freezing damage at the interface between the membrane and the catalyst layer. An

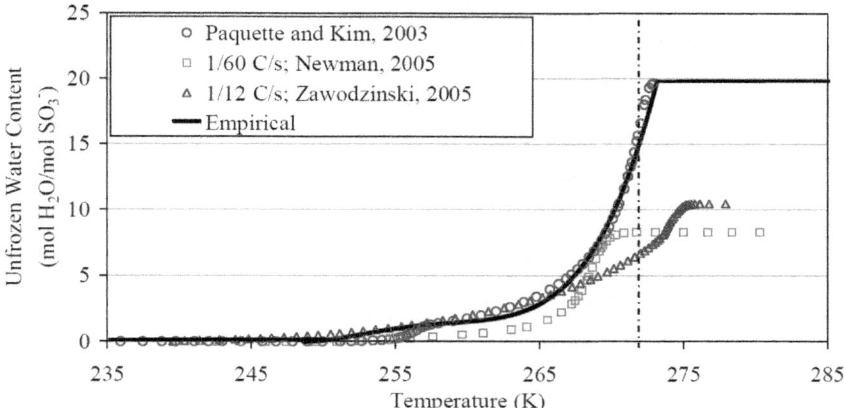

FIGURE 6.5 Unfrozen water versus temperature curves derived from Nafion DSC data of ref 24,25,26. [6].

experimental study by Pineri and co-workers appears to confirm this [27]. In the Pineri work, X-ray diffraction results indicate that some of the water desorbs out of the membrane below 0°C. Damage could also be a result of the membrane swelling and contraction with temperature. The Pineri result suggests that the damage is more likely a result of water outflow than membrane swelling. The end result is the same for either case, and indicates that *a major source of potential damage is excess membrane water resulting from liquid water in contact with the electrolyte at shut-down to a frozen state.*

A study by Liu measured the strain before failure of Nafion 112 membrane after dry and wet F/T cycles. The results are summarized in Table 6.4. The membrane after dry F/T cycles from −40°C to 80°C ruptures under significantly less strain than a membrane not subjected to F/T, indicating some internal change in structure. Wet F/T cycling did not show any additional

TABLE 6.4 Percent Elongation at Break of Nafion 112 Membranes Before/After Dry/Wet F/T Thermal Cycling Compiled from [28]

Material	Percent Elongation at break	
	Before Cycling	After 385 F/T Cycles (Dry)
Membrane (Machine direction)	1290	40
Membrane (Cross direction)	320	25
	Before Cycling	After 200 F/T Cycles (Wet)
Membrane (Machine direction)	>300	>300
Membrane (Cross direction)	>300	>300

potential, as the strain after 200 wet F/T cycles was the same as the strain of a new membrane [28]. The authors suggest that water in the membrane relieves structural change that can occur during freezing, because it makes chain movements more facile. It therefore prevents the membrane from becoming brittle. This effect may also contribute to the observed exacerbated damage from uneven dry-out during purge of large stack plates. Areas of high local dry-out can suffer from reduced plasticity in the membrane on subsequent purges.

3.2. Catalyst Layer Damage

Maintaining catalyst layer integrity throughout operation is of critical importance to a fuel cell performance, since other non-freeze related degradation modes commonly cause significant irreversible damage to the catalyst, membrane, and support structure [29]. The catalyst layer (or electrode) is a porous media covering both faces of the membrane, with a typical thickness range of 5–30 μm and porosity of 0.4–0.6. A surface morphology characterization of catalyst layer was performed by Hizir et al. [30]. Local catalyst cracks with large relative dimensions $\mathcal{O}(\mu m)$ compared to pores are often observed after F/T and sub-zero operations of fuel cells. Because the catalyst layer is between the membrane and the gas diffusion media, interfacial CL|membrane and CL|DM delamination is often observed due to ice lens formation. In addition, because the catalyst layer is a reaction site, any damage to will most likely lead to a loss in electrochemically active area. However, there exist conflicting results, as some researchers observed damage at the electrodes in the form of lost ECSA, physical cracking or pulverization of the electrode, while other studies did not show any damage. The work of Kim et al. [3,4] investigated those conflicting conclusions by studying the effect of fuel cell component structures, DM stiffness/thickness and membrane rigidity on the impact of freeze thaw (F/T) damage on the electrodes. Kim et al. determined that a stiff DM with a thin, reinforced membrane was the best configuration to mitigate damage from a freeze/thaw environment. DM thickness was not found to play a significant role in freeze/thaw damage.

3.2.1. *Electrode Cracking*

Extensive cracking of the electrode structure has been observed on MEAs subject to freeze-thaw cycles in both *ex-situ* and *in-situ* conditions. Figure 6.6 is an SEM image from the work of Guo and Qi and depicts the change in surface morphology of a commercial MEA frozen after being subjected to ambient temperature and RH (Fig. 6(a)) versus a similar MEA that was frozen after being hydrated in water at 80°C for 10 minutes (Fig. 6(b)). Each were cycled six times between 20 and −30°C and soaked at −30°C for 6 hours during every cycle [18]. In Fig. 6.6(a) the electrode is smooth and almost no damage can be seen as a result of the six freeze thaw cycles. However, severe damage is apparent in Fig. 6.6(b) with cracks and obvious separation of the catalyst

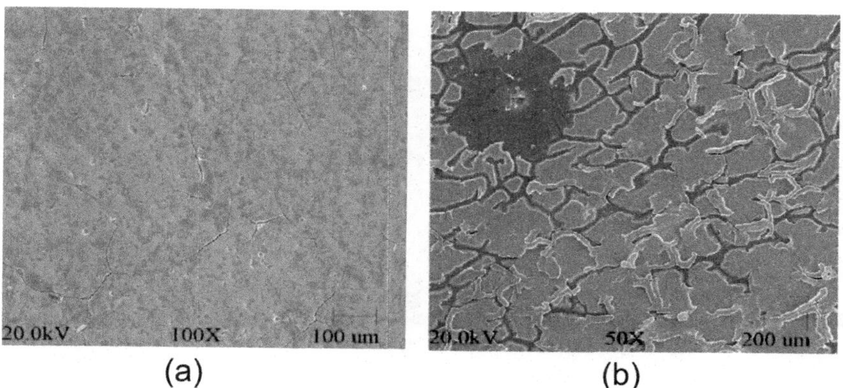

FIGURE 6.6 SEM of the cathode side of freestanding MEAs after six freeze-thaw cycles between 20 and −30°C: (a) MEA was only exposed to ambient temperature and relative humidity before going through the freeze-thaw cycles; 100x magnification; (b) MEA that was fully hydrated in water at 80°C for 10 min before going through the freeze/thaw cycles; 50x magnification. Images from [18].

surface. In the upper left corner, a total detachment of the catalyst material from the membrane can also be seen. Alink et al. also made a similar observation *ex-situ* with freeze/thaw cycling down to −40°C of a wet gas diffusion layer and MEA assembly without compressive forces [31]. In fact, without assembly compression the mechanical bond is very weak and repeated testing from various groups shows that detachment readily occurs.

Alink et al. also used a commercial type MEA for *in-situ* freeze/thaw cycling with compressive forces. One assembly was exposed to fully humidified reactants at the cathode and anode before being cycled. SEM showed catalyst damage and fracture, although catalyst layer segregation was not as noticeable as in the *ex-situ* experiment. This could be due either to the fact that an MEA in an assembled fuel cell is subjected to pressure from the backing plates, or the fact that the membrane water uptake is less when subjected to vapor instead of liquid [31] – or indeed both. This is in agreement with other research that explains how water drains and freezes in the catalyst layer below 0°C when membrane water content has reached a maximum [31]. This is also due to the fact that some water inside the membrane does not freeze, as explained in Section 2. Additionally, upon freezing in the electrode pores, water expands in volume and may generate micro-cracks on the surface of the catalyst [18,20,31].

3.2.2. Interfacial CL/Membrane & CL/DM Delamination

Delamination of the catalyst layer from both the membrane and the DM sides was shown to occur in several publications after the cell is subjected to sub-zero operation or brought to a frozen state without effective purging [3,4,14–16,20]. The catalyst layer and the DM are both porous media and permeable to gas. With pore sizes ranging from several nanometers to hundreds of micrometers,

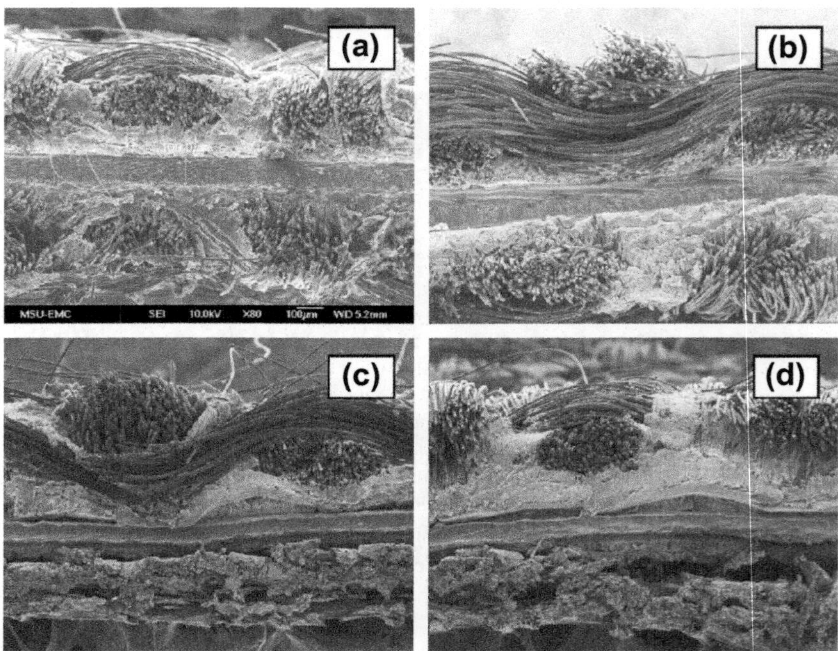

FIGURE 6.7 Effect of sub-zero temperature on MEA with cloth DM. (a) Virgin MEA, (b) MEA after operation at room temperature, (c) MEA after operation at −15°C. Images from [14].

water confined in those pores experiences a freezing point depression of only 2 to 4°C, which is not enough to prevent freeze damage [32]. The cathode side is more prone to separation as water is generated in the cathode catalyst layer by the oxygen reduction reaction. Figure 6.7 shows the evolution of a virgin MEA (Fig. 6.7(a)) upon operation at room temperature (Fig. 6.7(b)) where no delamination is observed from liquid water. Delamination on both the membrane and DM side became apparent after operation at −10°C (Fig. 6.7(c)) and −15°C (Fig. 6.7(d)). The delamination normally occurs under a channel location, where overburden pressure is low. For this reason, open or mesh flow field designs have an intrinsic advantage over conventional channel land design for limiting freeze/thaw damage.

To better understand CL delamination and how fuel cell components can help to promote or mitigate freeze damage, Kim et al. investigated the effect of DM stiffness, DM thickness and membrane rigidity on freeze/thaw damage in an *ex-situ* environment. The results are summarized below.

Effect of DM Stiffness

Frost heave formation and volume expansion of frozen water can induce shear force on the catalyst layer leading to interfacial delamination. Frost heaving is a phenomenon more complex than volume expansion of frozen water. Primary and secondary heave can occur whether or not an ice fringe exists, depending on the thermal and mass transport conditions. As shown in Fig. 6.8, the frozen fringe is the transition two-phase zone between 0% and 100% ice at the freezing front. Primary heave refers to frost heave with no frozen fringe, where no ice will penetrate into the unfrozen area. During secondary heave, the ice lens grows and penetrates into the frozen fringe [16]. Two test cells, one with flexible cloth DM and another with stiff carbon paper type DM, were F/T cycled 30 times from −40 to 70°C [4]. They are both shown in Fig. 6.9. Excessive surface damage and CL|DM delamination were observed on the catalyst layer of the cell assembled with cloth DM (Fig. 6.9(a)), while no cracks were observed on the cell using stiff carbon paper DM (Fig. 6.9(b)). A stiffer diffusion media more uniformly translates the compressive forces from under the land to under the channels and therefore provides more deformation resistance when subject to ice growth pressure. The stronger compression to the CL surface can also reduce interfacial water accumulation at shutdown. This also means that the channel width and channel/land ratio are important parameters in uniformly spreading the compressive forces; relatively wide channels will promote DM deformation. Figure 6.10 shows a calculated compression distribution from a common felt DM (SGL 10BB) onto the CL. To obtain the non-homogeneous compression pressure data required for the simulation, the DM thickness versus compression pressure data given in [33] is used to evaluate the non-homogeneous strain in the DM layer. Finally, using the DM strain data, the compression information under one land-channel configuration is extracted from the stress-strain data of the DM given in [34]. The DM thickness measurement was performed *ex-situ* for one set of land and channel (each of length 1 mm), and Fig. 6.10 shows the variation of the non-homogeneous compression pressure from mid-land to mid-channel location with a span of 1 mm. As can be seen, even for a stiff DM, the compression pressure on the catalyst layer drops off very sharply in the channel

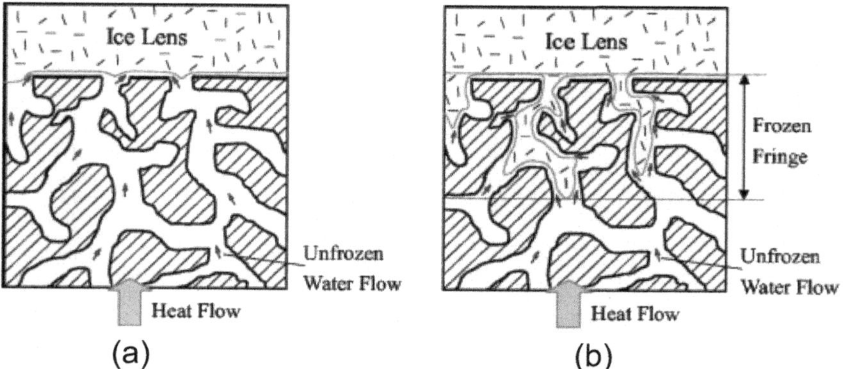

FIGURE 6.8 Comparison of (a) primary and (b) secondary frost heave. Images from [6].

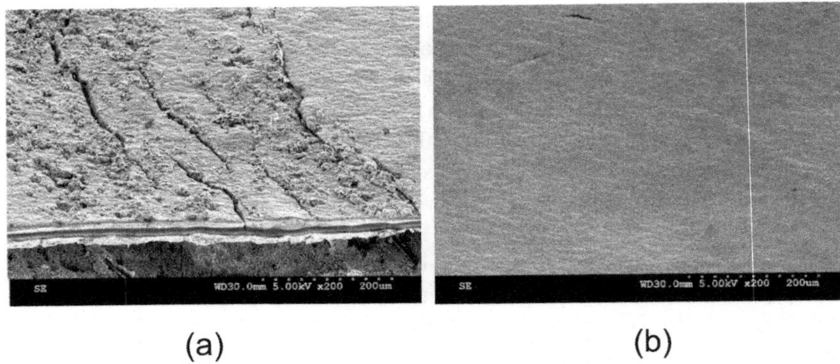

FIGURE 6.9 Surface images of MEAs cycled 30 times between −40 and 70°C with negligible cracks in the virgin catalyst layer and 18 μm reinforced membrane. Images shown correspond to locations under channel. (a) CARBEL-CL (cloth type) DM and (b) SGL 10BB (non-woven felt type) DM. Images from [4].

region. This is the reason delamination damage is most likely in this location. Based on modeling from S.He et al, the calculated ice-phase pressure rarely gets over 2 MPa, which is at the high range of the normal compression experienced under a land in a typical fuel cell.

Effect of DM Thickness

Kim et al. also compared the effect of carbon paper DM thickness on a 35 μm reinforced membrane known to be sensitive to damage [4] and an 18 μm

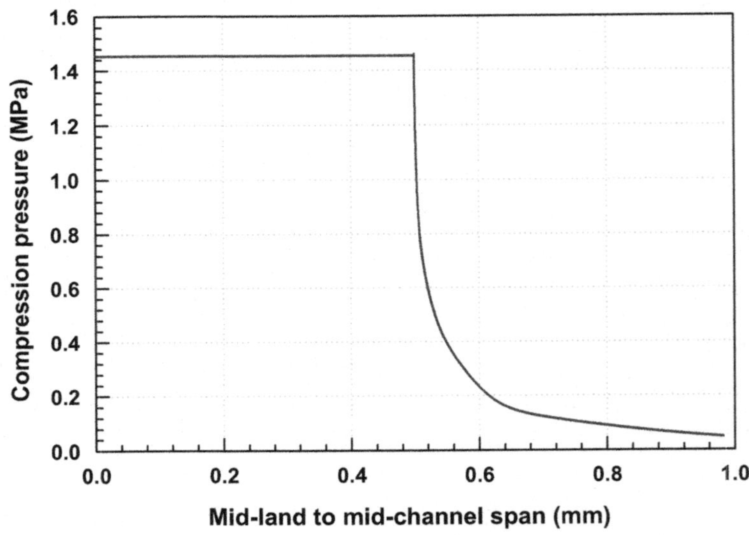

FIGURE 6.10 Calculated compression distribution from a common DM onto the CL.

reinforced membrane known to be less sensitive, as seen in Fig. 6.11. The non-woven felt type is stiffer due to its more three dimensional lattices. Both cells with 18 μm membrane showed no physical damage with either thin carbon paper DM (235 μm) of non-woven paper type (Fig. 6.11(a)) or thicker carbon (415 μm) non-woven felt type DM (Fig. 6.11(b)) respectively. This is a result of stiff DM applying some compressive force under the channels. Interestingly, the stiff DM did not prevent damage on the thicker membrane (Fig. 6.11(c)) and the thicker stiff DM showed as much damage (Fig. 6.11(d)). It was concluded that a thickness from 235 μm to 415 μm of non-woven type was not significant to mitigate the observed physical damage to the electrode surface [4].

Effect of Membrane Rigidity and Thickness

A test cell with 18 μm non-reinforced membrane and carbon paper DM was F/T cycled 30 times between −40°C and 70°C [3]. Fig. 6.12 shows interfacial delamination under the channel. Although the DM was stiff carbon

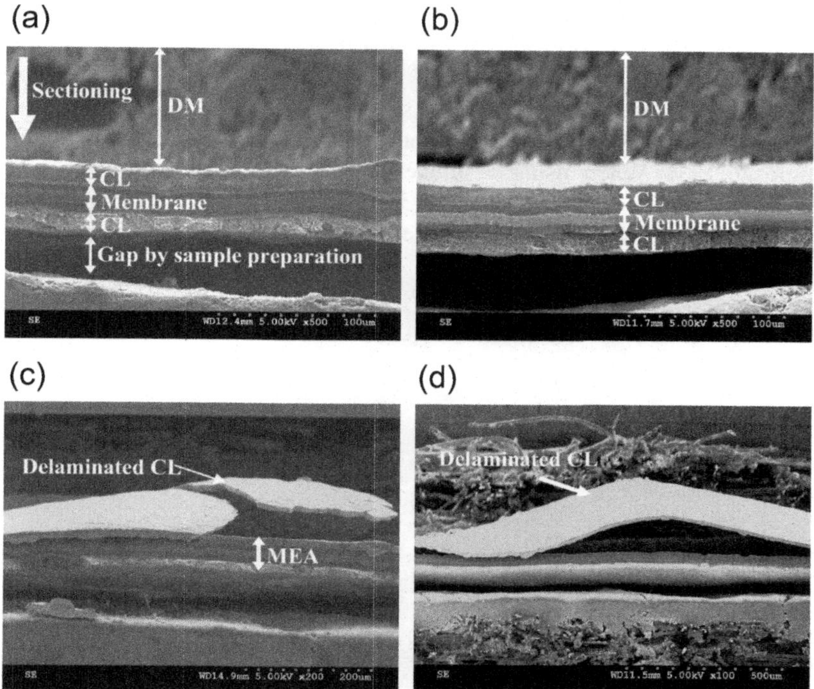

FIGURE 6.11 Cross-sectional images of MEAs with negligible virgin cracked catalyst layers, F/T cycled 30 times: (a) 18 μm reinforced membrane with SGL 25BC DM (thickness 235 μm); (b) 18 μm reinforced membrane with SGL 10BB DM (thickness 415 μm); (c) 35 μm reinforced membrane with SGL 25BC DM; (d) 35 μm reinforced membrane with SGL 10BB DM. Images from [4].

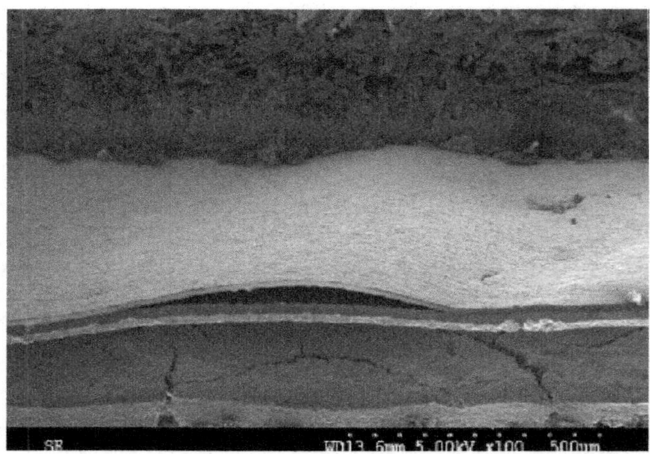

FIGURE 6.12 SEM image of F/T cycled non-cracked CL with 18 μm non-reinforced membrane under the channel location [3].

paper, the non-reinforced membrane promoted delamination. Membrane reinforcement is used to make a membrane mechanically stronger and more durable without significantly changing its conductive capabilities. The most common methods include adding a strong polymer such as expandable porous polytetrafluoroethylene or other fibers, resulting in a membrane composite [35]. When using a thicker 35 μm reinforced membrane with carbon paper DM, as shown in Fig. 6.13, frost heave damage is visible under the channels. Although the thicker membrane is reinforced, it is a bigger

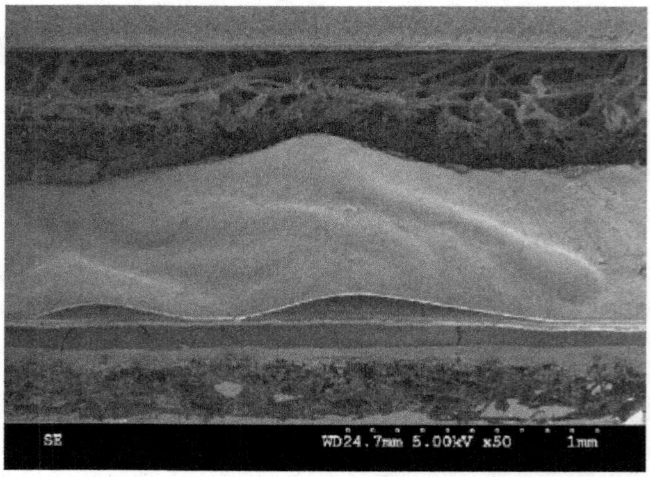

FIGURE 6.13 SEM image of F/T cycled non-cracked CL with 35 μm reinforced membrane under the channel location [3].

reservoir for water and by itself a source of water for damage in the CL [3]. Therefore, the best material combination to mitigate freeze-damage was found to be a non-cracked virgin catalyst layer on a reinforced, thin membrane, assembled with stiff diffusion media. Although this freeze-tolerable design reduced freeze damage under worst case scenarios of direct liquid contact with the ionomer at freeze, irreversible damage was still present, highlighting the importance of liquid removal from the catalyst layer before shut-down to a frozen state.

3.3. Loss of Electrochemical Surface Area

Besides physically observable damage, performance is directly relevant to the electrochemical surface area (ECSA) at the electrodes, which can be measured *in-situ* with cyclic voltammetry [18,20,32,36]. Even without major observable morphological damage, ECSA loss has been observed in F/T testing. After 20 freeze/thaw cycles (20 to −30°C at fully humidified state) Guo and Qi observed ECSA loss at both electrodes. ECSA decreased by 23% at the cathode and by 15% at the anode, as seen in Fig. 6.14. This difference could be due to storage of generated water in the cathode CL due to previous operations. However, the short term performance of the cell did not show much change [18]. Hou et al. investigated freeze degradation using 20 freeze/thaw cycles between −20°C and 60°C. The cell was operated at 60°C, purged by gases at 25°C with 58% RH after each operation, and then frozen to −20°C. Cyclic voltammetry (performed only at the cathode) showed that values of ECSA fluctuated between 45.4 and 51.2 m^2/g_{cat} and did not decrease progressively after each cycle [19]. Interestingly, this fluctuation did not alter the performance curves after each freeze/thaw cycle. Although the ECSA measurement fluctuation could be from the experimental device, cyclic voltammetry is a transient test and it is possible to have detected structure alteration or liquid water transients blocking the triple-phase boundary which would not affect a steady-state performance test. Mukundan et al. [20] also performed ECSA measurements to compare the durability of their Los Alamos National Lab (LANL)-made MEAs to MEAs from W.L. Gore. The W.L. Gore MEAs showed >50% loss in the catalyst surface area after five cold starts at −10°C, while LANL-prepared MEA showed negligible loss. In this study, catalyst layer morphology is obviously important for durability at freezing conditions; as previously discussed, water in smaller pores may not freeze at conditions in which water in larger pores does. This loss in ECSA after cold starts was not observed at the anode, clearly because water is generated at the cathode side. Ge and Wang [32] and Srouji [36] made the same observation regarding lack of damage at the anode from cold starts. Ge and Wang have also recorded 1 to 3% of Pt area loss at the cathode per cold start performed at −10°C, although each cold start was interrupted by a thaw and operation at 70°C, making the cell go to a freeze down process before each cold start [32]. Srouji recorded 4.4% of Pt area loss at the cathode after 25 consecutive cold starts at −10°C. The protocol

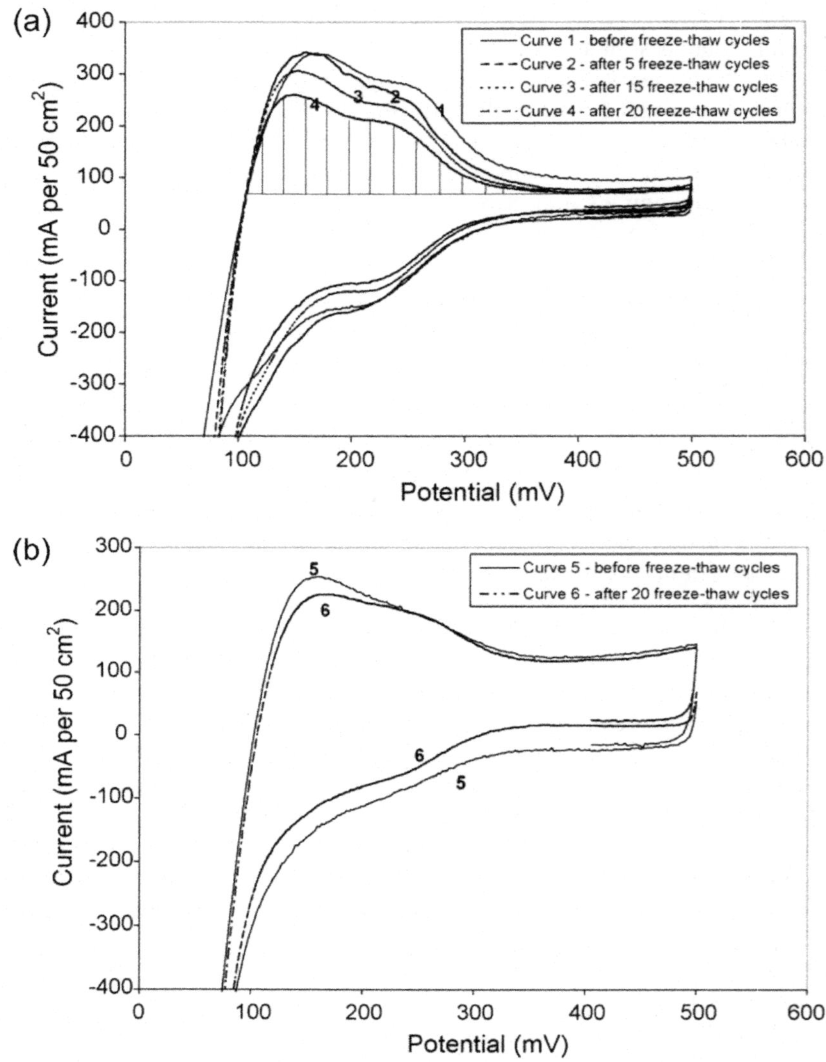

FIGURE 6.14 Cyclic voltammograms of (a) cathode and (b) anode of an MEA after 0–20 freeze/thaw cycles at fully humidified state. Images from [18].

developed for rapid consecutive cold starts with known initial membrane water content is described in detail by Chacko et al. [37], and is capable of isolating sub-zero operation damage from the water generated at the cathode from residual water damage resulting from the freeze down process itself. A challenge in cyclic voltammetry studies is to correlate ECSA loss with performance loss during steady-state operation.

3.4. DM Fracture and Loss of Hydrophobicity

Some failure of the gas diffusion media has been observed as an apparent consequence of F/T cycles. Mukundan et al. [20] observed DM failure after 10 F/T cycles down to −80°C. However, Yan et al. [14] witnessed no damage to DM after exposure to normal conditions but noticed an increase in porosity and darkness in color after sub-zero temperature exposure. They attributed this phenomenon to ice forming in the DM. Results from neutron imaging studies have shown that the saturation of the DM materials rarely can exceed 30%, so that the additional 8% volume expansion from freezing should be tolerable in the overall structure. However, some damage can occur if the water is locally confined by an enclosed pore structure or surrounding ice. Although the DM provides a stiff support for the MEA and a hydrophobic barrier, it does store a considerable amount of water in the CL after shut down [31]. Evidence of reduced hydrophobicity from exposure to freezing conditions with high liquid saturation has also been observed. The damage to the DM in freeze conditions needs to be investigated more for a more complete understanding.

4. METHODS OF FREEZE DAMAGE MITIGATION

There are various concepts for preventing the fuel cell system damage caused by freezing, as well as a damage-free rapid start-up in sub-freezing conditions through good energy/power management. Pesaran et al. [38] categorized a review of solutions into two strategies: 'Keep Warm' where the system uses energy during vehicle parking and 'Thaw and Heat at Startup' which consumes energy mostly at vehicle startup. A summary of the various approaches is shown in Fig. 6.15. Intellectual properties have been developed for most if not all of them, and a compilation of 160 patents for freeze damage mitigation are listed and summarized in reference [38]. In that same milestone report, it is concluded that the correct use of insulation around the stack components can delay stack freeze by several days after it is shut down.

Residual water reduction and evaporation during shut-down before the fuel cell is frozen can be achieved by several methods, including:

1. convective purge;
2. vacuum purge;
3. capillary drainage;
4. thermally driven drainage; and
5. combinations of the above.

No more than 62.5 J/We should be consumed during cold start-up, based on the DOE goals for parasitic losses. The ultimate goal is a non-parasitic shut-down with no damage.

Clearly, the key to shut-down is proper removal of liquid which is in contact with the ionomer without producing overly dry areas of the membrane. Several

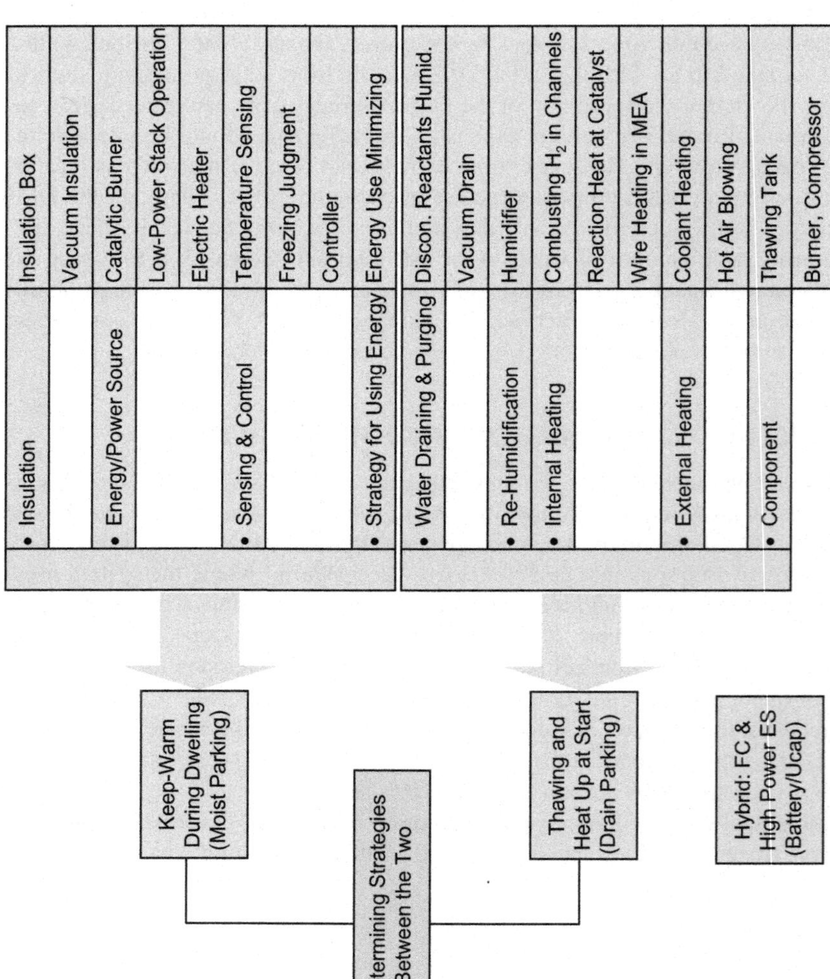

FIGURE 6.15 Method and technology chart for fuel cell start from subfreezing environment [38].

studies agree that an MEA equilibrated to 80~95% relative humidity is better suited for rapid cold start, since this provides some storage for generated water during start-up and assures limited liquid-phase ionomer contact at freeze. However, this is difficult to achieve in practice without an exceedingly long purge, high parasitic losses, or distributed stresses which can lead to degradation and the fact that it is progressively more difficult to remove water with decreasing temperature, to maintain a dryer than saturated state [39].

A suggested optimal purge strategy is to keep purging until the water in the channels and diffusion media is removed, while water is still largely present in the membrane [5]. An MK 9 series 10 cell stack used in this study had an optimal purge duration of 88 seconds with dry air and H_2 at 89 L/min and 25 L/min respectively; both at 70°C and 1.6 bar. However, this can be difficult to achieve in full size stack plates. Cho and Mench showed that for certain conditions, the water content in the cell is not correlated with high frequency resistance (HFR) during purge, and is not a good metric of water removal from the cell. Figure 6.16 shows this [40]. For this plot, data were taken using neutron imaging to record total liquid water content, and HFR, to record average membrane resistance. Different combinations of anode and cathode inlet relative humidity were used to purge a 250 cm^2 full size fuel cell stack plate. Each test began from the same initial conditions. All comparative purges were operated at the same flow rates relative to each other. As can be seen from the figure, a full humidity (100/100% RH anode/cathode) purge results in water removal from the cell, indicating significant accumulation in the channels can

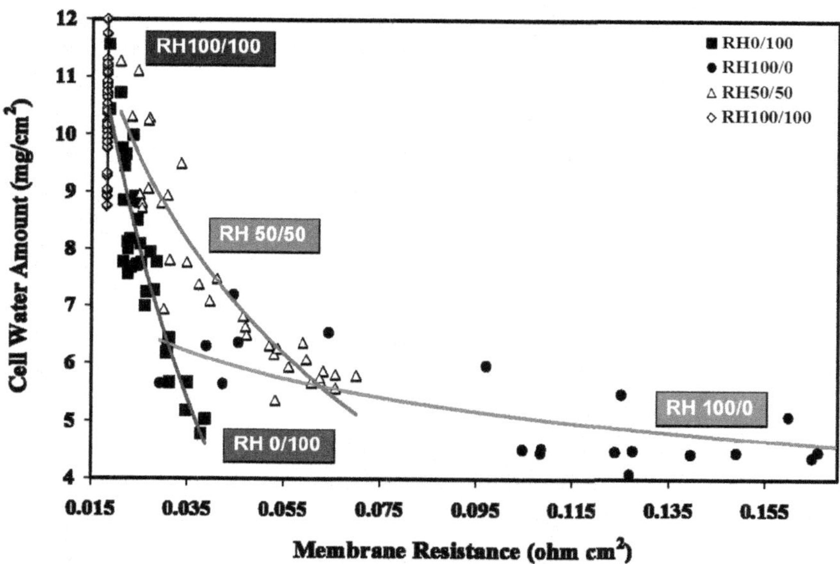

FIGURE 6.16 Cell water amount from neutron radiography with respect to membrane resistance at different operating conditions during purge [40].

be removed due to non-evaporative effects such as shear. Due to back diffusion and the initial water distribution, there is a sharp difference in the membrane dry-out compared to a dry anode or cathode purge. As discussed, membrane non-uniformities in water content have been shown to exacerbate damage and should be avoided. Recent work has shown a novel composite purge approach can most efficiently remove water content while preventing membrane dry-out [41,42], as shown in Figs 6.17(a) and (b). The characteristic water removal behavior during gas purge was analyzed using neutron radiography (NR) and HFR, as shown in Figs 6.17 (a) and (b). NR is used for quantifying the total

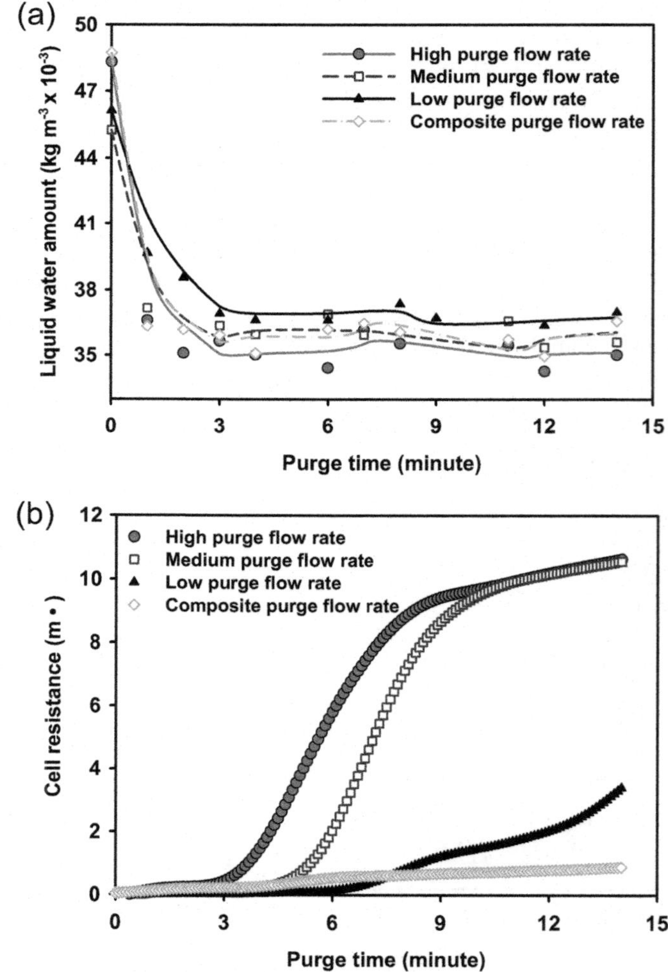

FIGURE 6.17 Water removal behavior of fuel cell during purge: (a) variation of water amount in the cell and (b) variation of total cell resistance [41].

amount of water residing in all the components of the fuel cell, whereas HFR is utilized to indicate variation of water content in the membrane. Therefore, by comparing both data sets during purge, water removal behavior can be understood in detail. As shown in Figs 6.17 (a) and (b), a high flow rate purge was very fast and efficient for decreasing the residual water in the cell, but increased the cell resistance substantially, raising issues of possible degradation of the membrane and high energy consumption. For a relatively low flow rate purge, the cell resistance did not increase severely, but water removal from the cell was not efficient. However, in the case of a composite purge with mixed purge flow rates (high flow rate for 1 min., medium flow rate for 3 min., and low flow rate for 10 min.), the water removal rate from the cell was almost identical to the medium flow rate case, but with reduced membrane resistance increase (−91%) and less energy consumption (−24%). More details of this can be found in ref. [41].

A typical convective method of removing residual water during shut-down is purging with hot dry gas, which is effective in rapidly evaporating residual liquid water from inside the DM or CL and the channels. The convenience of this method depends on the reactant gas flow field patterns. It often leads to non-uniform water distribution, which can result in rapid degradation of the MEA. For example serpentine flow field patterns have more water content near the outlet and suffer from dry out at the gas inlets. This results in mechanical stress causing physical degradation. Serpentine flow fields also suffer from water accumulation around the 180° turns, as shown in Fig. 6.18. This accumulation would tend to damage the cell upon freeze or prevent proper start-up via channel blockage. However, parallel flow fields have less resistivity to fluid motion and hence mitigate non-uniformity [43]. This general effect has been observed consistently in both small and full size stack designs, leading to a modern design paradigm that seeks to straighten the flow field as much as

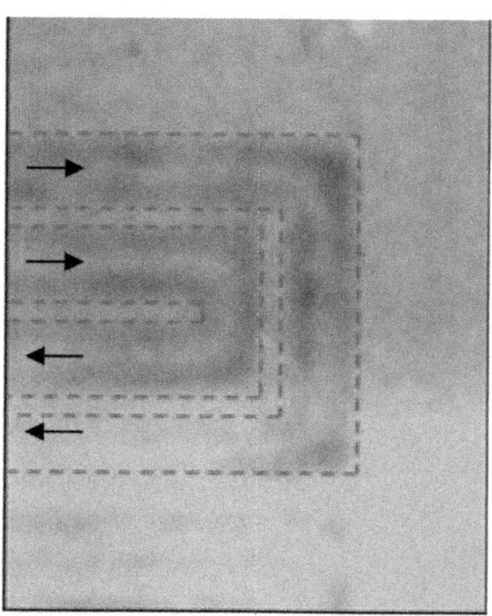

FIGURE 6.18 Neutron radiograph showing a tendency for water accumulation at corners and switchbacks in the fuel cell flow channel [43].

possible and manage water content through thermal or other transport mechanisms to eliminate these effects and reduce water content.

Dry purging should be done with careful attention to the purge gas temperature. An MK 513 series single cell of Ballard Power Systems Inc. [5,44] experienced freeze damage after a dry hot purge (dry N_2 purge was conducted right after operation for one minute at 85°C on both sides). On the other hand, no damage occurred when the cell was cooled down to ambient temperature and then purged with cold dry N_2. Although no reason behind this observation was disclosed, it's important to note that the MK 513 series cell has very long, parallel flow channels. The hot purge may have over-dried the MEA near the inlet and then cooled and wetted the MEA near the outlets, leading to freeze-damage. However, a cold purge induces less gradients of moisture leading to slower evaporation but less damage and a more uniform water distribution. The key point is removal of liquid water in contact with the catalyst layer without inducing damage from uneven stress caused by drying.

Although water removal is necessary for freeze damage mitigation, it is very important not to over-dry the MEA. A dehydrated membrane will have a very low electric conductivity and cold start-up will not be possible. HFR measures the ionic resistance and therefore can be used as a diagnostic tool to determine optimal purge duration. HFR is not affected when the purging process removes residual liquid water from the channels and DM. Cell resistance starts to increase when water removal is initiated at the membrane level. An optimum strategy is to stop the purge at the inflection point of the resistance versus time curve [5] as show in Fig. 6.19 for an Mk9 10-cells stack.

Vacuum purging was proposed as a method for drying out the DM and MEA of a cell [45], before shut-down in a freezing environment, since water is more easily drained at higher temperatures because of better evaporation. It's

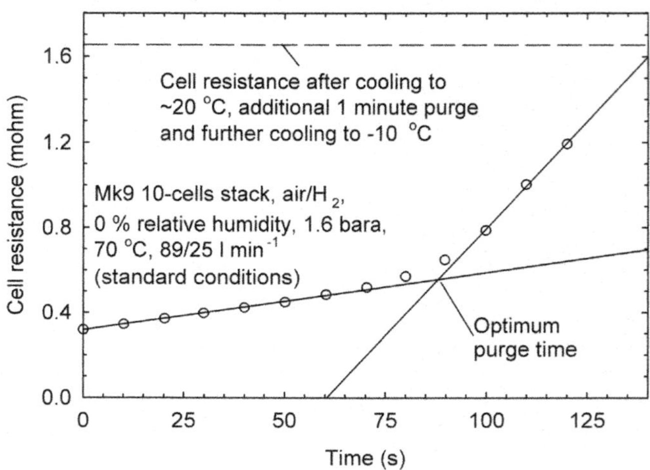

FIGURE 6.19 HFR change with purge time [5].

preferable to start vacuum purging as soon as the cell is shut down from its operating temperature. Vacuum drying at a higher temperature dictates the need for a smaller vacuum pump which is typically already onboard a vehicle for other purposes. Although this approach is shown to result in damage mitigation, and may be appropriate in certain niche applications, in general it is not generally believed to be practical in operating systems.

Temperature-gradient driven water transport is an attractive non-parasitic water drainage method during fuel cell shut-down. The use of engineered temperature gradients within the stack has been demonstrated to prevent freeze damage [46,47]. There are two basic modes of temperature-gradient driven flux of water that are relevant at shut-down; thermo-osmotic transport in the membrane, and phase-change-induced (PCI) flux through the open voids. Thermo-osmosis in the membrane is the water flux observed when water with different temperatures is separated by the membrane [44,46,48–54]. Thermo-osmotic water flux in fuel cell membranes is from the cold to the hot side, and depends on the difference in entropy between water stored in the membrane and water external to the membrane [53]. Unbound water transport is thermodynamically favored in the direction with increasing entropy [52,53]. Kim et al. further investigated water flux through the membrane and concluded that water flux is proportional to temperature difference as shown in Fig. 6.20, and inversely proportional to membrane thickness as seen in Fig. 6.21. An Arrhenius rate law was determined to capture this transport mode.

PCI flow occurs with the presence of a temperature gradient and gas phase in the CL, MPL or main DM and dominates once irreducible saturation is

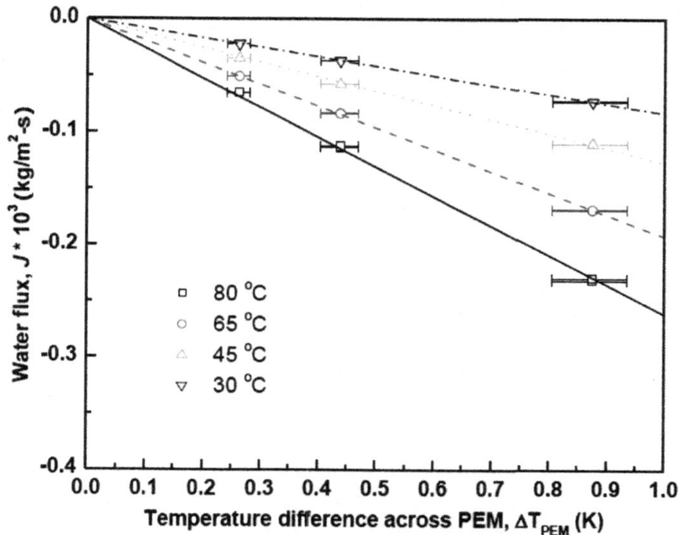

FIGURE 6.20 Thermo-osmotic water flux in Nafion 112 membrane [55].

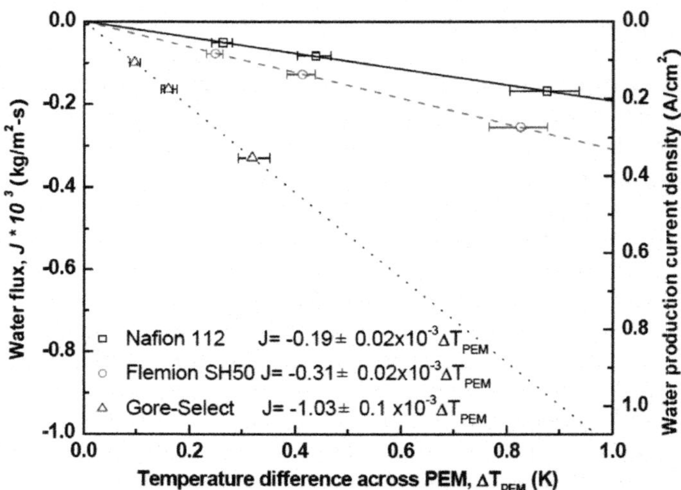

FIGURE 6.21 Comparison of thermo-osmotic water flux of membranes [55].

attained in the porous media [54–56]. PCI flow is strongly dependent on average membrane temperature and temperature gradients [54–56]. The effect of DM/CL thermal mass was negligible. M. Khandelwal and Mench showed that thermo-osmotic flow can either assist or oppose PCI flow depending on the hydrophobic properties of the membrane [57]. In fuel cell media, it generally opposes the PCI flow. Thus, a residual water content in the warmer electrode can result under significant temperature gradients, which would tend to occur near the end plates of a stack. Therefore, to minimize water in the cathode CL, thermo-osmosis flux across the membrane is very important to help freeze durability. Both PCI flow and thermo-osmosis in various membranes and DM material sets have been experimentally investigated and quantified, and it was determined that both modes of transport can be well-correlated using Arrhenius rate laws as shown in Fig. 6.22. Although the type of reinforcement in the membrane has some impact, thermo-osmosis is fairly constant for perfluorosulfonic type membranes, but significantly less than regular concentration-based diffusion. Therefore, this mode of transport is not normally critical during operation, given the existing high range of uncertainty in published diffusivity values. However, during shut-down to a frozen state, thermo-osmosis can become important, as it can counteract the PCI flow, which moves liquid toward the cold location. The result of the interaction can be a residual frozen water saturation in the warmer-side catalyst layer of the MEA, as has been shown via recent modeling of this effect [57]. In general, the PCI flow is much more significant for even the small temperature gradients expected during shut-down between MEA components [56]. Several fuel cell manufacturers have also investigated this effect, and Ballard suggested

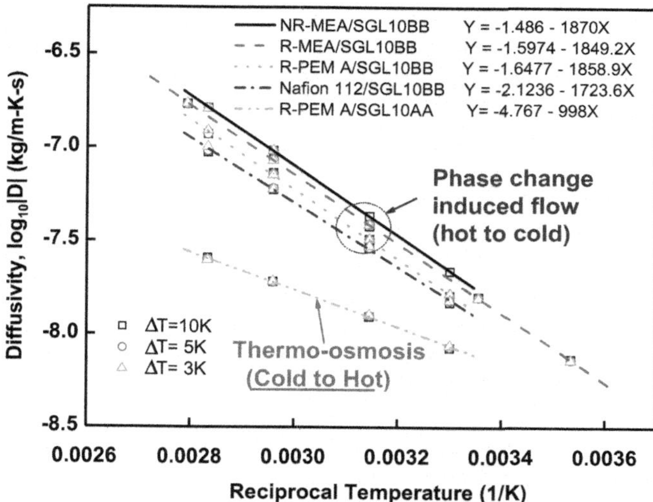

FIGURE 6.22 Correlated thermo-osmosis and PCI flow relationships based on an Arrhenius rate law. Data from different temperature gradients all conveniently collapse into a single curve for a given material set [54,55].

a unique stack design to promote internal temperature gradients near end-plate locations to avoid damage and promote reliable start-up from a frozen condition [46].

4.1. Damage Mitigation via Material Choice and Design

Although the various presented methods of mitigation are useful, in principle no action would be needed at shut-down if the operational overhead of liquid water was reduced to a value below that at which damage occurs. That is, if proper materials and design to reduce the liquid water overhead can be chosen, the required parasitic purge can be reduced. *Work by Turhan et al. has shown that water content in a fuel cell can be reduced by as much as 50% with little performance change, just by changing the DM thickness and channel/land design [58].* The following parameters have been determined from accumulated research to be key controlling parameters in the water content in the fuel cell porous media and flow channels:

1. The operating conditions: It should be noted that high current does not necessarily result in higher water content. In fact, the high channel flow rates and heat produced by inefficiency often reduce water content as current increases. Low current conditions often have the greatest total stored water content in the fuel cell.
2. The thermal boundary conditions and heat transport: PCI flow plays a critical role in water distribution, as proven by various studies. Water

distribution and storage can be controlled through manipulation of this boundary condition via coolant channel design or material selection.

3. The material choices: Tremendous shifts in water content at similar operating conditions have been observed depending on the thickness and type of diffusion media and other components.
4. The channel/land interface: The shape and surface energy (e.g. contact angle) have been shown to be critical in the drainage of liquid from accumulation under the lands, as described in [58]. This impact should not be overlooked in terms of expected water content and freeze effects. In general, liquid flow across this interface is dominated by capillary action, so that the interface shape, roughness, contour, and surface energy are important aspects of drainage. A hydrophilic interface is preferred to allow drainage from the DM into the channel.
5. Manifold design: The ability for water to drain from the internal channel structure into the main manifold is a key factor. In this location, even a small amount of water can impede the ability to properly start-up from a frozen state. Thus, it is critical that this location remain free of accumulation at shut-down.
6. Channel design: As described, there is a general desire to reduce the number of flow switchbacks and flow deceleration points to avoid water accumulation. Thus, the general design paradigm from this result is to straighten the flow path as much as possible, and maintain water balance through other means such as boundary temperature control.
7. Channel shape and surface energy: It has been shown by many researchers in the fuel cell and micro-fluidics field, that there is a clear relationship between water retention in the channels and channel shape and surface energy. However, using hydrophobic channels is not a particularly good solution since it restricts water removal from the DM and can result in operational instabilities related to the creation of multiple slugs of water [59].

Many of these parameters are not included in modern computational models. Thus, there is still a tremendous discrepancy between the water distribution predicted and that which is observed in practice [60]. Clearly, much additional research is needed before this can be fully resolved.

4.2. Comments on Proper Conditions for Experimental Testing of Freeze/Thaw

As discussed, there are discrepancies between the results in literature for seemingly similar testing. The issues which result in these discrepancies include:

1. Differences in the experimental configuration or materials. As discussed, the stiffness of the DM, membrane thickness and type, as well as channel to land width ratio and compression play a strong role in the development of freeze/thaw damage.

2. For single cells, it is imperative that precise thermal boundary conditions are maintained. Single-cell testing has traditionally taken place using a heating cartridge to maintain temperature, but neutron imaging has shown this to result in very different internal water distribution than if coolant channels are used. For accurate testing at the single cell level, it is imperative that coolant-based or other type boundary temperature control is used that is superior to cartridge heaters, which provide inconsistent and non-representative heating and cooling behavior.
3. The shut-down procedure used in the laboratory is obviously critical and should be carefully considered and controlled in terms of thermal boundary conditions. One of the main differences between single cell and stack cell freeze/thaw testing is that in a single cell, both sides are colder than the center of the cell at shut-down. This results in PCI flow removal of water from both electrodes, and mitigation of freeze damage compared to an in-stack cell, where the temperature gradient is in one direction on both sides of the membrane. In-stack cell testing can be simulated by using dual coolant controlled boundary conditions. If separate coolant flows are used, a temperature gradient representative of any particular location in the fuel cell stack can be simulated.
4. The initial conditions before shut-down and purge to a frozen state should also be carefully maintained. That is, the same shut-down procedure, executed on two similar cells with a different operational history, will result in a different final condition before freeze. This discrepancy can be eliminated by ending a cycle with a pre-shut-down step. By operating at a selected known condition for a significant period of time and then initiating shut-down, the previous operational history effects can be effectively erased, and the cells will be shut-down from a precise initial condition. For example, for a $50\,cm^2$ active area single cell, operation at 0.6 V for 30 minutes before initiating the purge protocol should eliminate any differences that might arise from operational conditions before purge.

5. SUMMARY AND FUTURE OUTLOOK

This chapter has examined the results of published studies that examine physicochemical degradation in polymer electrolyte fuel cells resulting from a shut-down to a frozen state. Damage caused from a frozen start-up is out of the scope of this publication, but deserves additional attention in the literature. Ultimately, no damage as a result of freezing to −40°C was observed for any common fuel cell materials if there was no contact with liquid water. This result indicates that damage-free shut-down to a frozen state is possible through proper engineering of operational and shut-down protocol, materials, and design. Achieving a damage-free frozen condition is difficult, however, because the time for purge should be short, and parasitic losses should be minimized.

Four to six years ago, the major automotive manufacturers reported successful start-up of their respective fuel cell vehicles in relation to the issue of freeze. It seems that their approach was through good systems engineering leading to a multitude of patents. In fact, sustaining a freezing environment is not the challenge holding up fuel cell vehicle commercialization. On the other hand, the 2007 DOE fuel cell technical plan reports that clear results of degradation rate over a 5,000 hour lifespan (150,000 miles equivalent) of an automotive stack have not been declared, although it is estimated to be $< 20\%$. The ultimate goal is 5% performance degradation at the end of life of a stack subjected to the full range of external environmental conditions (-40 to 40°C).

From a summary of the existing literature, damage to the fuel cell components is a result of water expansion upon freezing, and a frost-heave delamination mechanism unrelated to the expansion process. Electrochemical surface area (ECSA) reduction has been commonly measured as a function of frozen conditions. Physical damage to PEFC components were identified to include membrane|CL delamination, CL|DM delamination, and local pore damage in porous layer (CL and DM), and some membrane cracking. Loss of DM hydrophobicity and some morphological changes have also been observed, including some instances of DM punch-through from ice formation.

A key source of freeze damage is now known to be the result of liquid water contact with the ionomer in the CL and membrane at shut-down. After a frozen condition is reached, the excess water uptake in the membrane can cause significant local delamination damage along the CL interface.

Key factors which influence the degree of damage include the compression distribution on the MEA, membrane type and thickness, diffusion media stiffness, and shut-down conditions. Designs which limit areas of low compression are better suited for a frozen environment. Stiff diffusion media materials and thinner membranes with reinforced structure offer the greatest resistance to damage by limiting expansion and contraction forces, potential interfacial accumulations of water, and membrane-based sources of water under a frozen state.

The ideal shut-down condition appears to be one in which the membrane phase is slightly and uniformly under-humidified, ensuring a low level of liquid accumulation and contact with the ionomer. Overly drying the membrane results in a poor cold start and potential membrane damage from internal stress generation.

In order to achieve the desired shut-down condition of a slightly and uniformly under-humidified membrane, a simple, low temperature dry purge is effective, but too time consuming and parasitic to achieve desired levels of performance in practical operating systems. Various different purge approaches and damage mitigation techniques have been developed. Among the most promising is the use of controlled temperature gradients to assist liquid water drainage via a phase change induced flow. Engineering design of typical stack components or coolant flow can induce sufficient gradients during shut-down to assist water removal from porous media into the channels, which can then be flushed by a short blast purge under low temperature conditions that will not harm the membrane.

Many of the discrepancies between the experimentally observed phenomena can be attributed to the different materials, channel/land configurations and operational protocols. It is critical in freeze/thaw testing to achieve proper thermal boundary conditions and initial conditions for shut-down to a frozen state to assure reliable data and interpretation. A key difference between single cell and in-stack data is the thermal boundary conditions, which can control the final distribution of liquid going into a frozen state. A dual coolant system can be used to achieve near isothermal controlled boundary conditions, as well as to simulate accurate conditions for in-stack cells with a single laboratory cell.

Although much work has been done to identify and explain freeze damage in PEFCs, there is still work to be done. The role of the CL|DM interface has been shown to be critical, yet little is known about the *in-situ* nature of this interface, particularly under dynamic operating conditions. Understanding the nature of materials and design so that the residual liquid water overhead in the fuel cell can be reduced before shut-down is perhaps more critical to achieve desired performance levels. If cells can be designed to have greatly reduced stored water content during operation, less is obviously required of the shut-down. Finally, a more complete knowledge of the nature of condensation and evaporation in fuel cell media is required for accurate modeling. Currently, models are constructed based on thermodynamic driving force of saturation pressure gradients and no information on the potentially important effects of surface energy or morphology are included.

ACRONYMS

CL	Catalyst Layer
CV	Cyclic Voltammetry
DM	Diffusion Media
DSC	Differential Scanning Calorimetry
ECSA	Electrochemical Surface Area
F/T	Freeze/Thaw
GDL	Gas Diffusion Layer
HFR	High Frequency Resistance
LANL	Los Alamos National Lab
MEA	Membrane Electrolyte Assembly
MPL	Microporous Layer
PEFC	Polymer Electrolyte Fuel Cell
PEM	Polymer Electrolyte Membrane
RH	Relative Humidity
SEM	Scanning Electron Microscopy

REFERENCES

[1] US Department of Energy, Hydrogen Fuel Cells and Infrastructure Technologies Program. (www1.eere.energy.gov/hydrogenandfuelcells/mypp/pdfs/fuel_cells.pdf), 2005.

[2] H. Liu, Dimensionally Stable High Performance Membrane, Proceedings of 2006 DOE Annual Program Review, DOE, Washington DC, 2006.

[3] S. Kim, M.M. Mench, Physical Degradation of Membrane Electrode Assemblies Undergoing Freeze/Thaw Cycling: Micro-Structure Effects, J. Power Sources 174 (2007) 206–220.

[4] S. Kim, B.K. Ahn, M.M. Mench, Physical Degradation of Membrane Electrode Assemblies Undergoing Freeze/Thaw Cycling: Diffusion Media Effects, J. Power Sources 179 (2008) 140–146.

[5] J. St-Pierre, J. Roberts, K. Colbow, S. Campbell, A. Nelson, PEMFC Operational and Design Strategies for Sub Zero Environments, J. New Mater. Electrochem. Syst. 8 (2005) 163–176.

[6] S. He, M.M. Mench, One-Dimensional Transient Model for Frost Heave in Polymer Electrolyte Fuel Cell – Part I: Physical Model, J. Electrochem. Soc. 153 (2006) A1724–A1731.

[7] S. He, S.H. Kim, M.M. Mench, One-Dimensional Transient Model for Frost Heave in Polymer Electrolyte Fuel Cell – Part II: Parametric Study, J. Electrochem. Soc. 154 (2007) B1024–B1033.

[8] S. He, J.H. Lee, M.M. Mench, One-Dimensional Transient Model for Frost Heave in Polymer Electrolyte Fuel Cell – Part III: Heat Transfer, Microporous Layer, and Cyclic Effect, J. Electrochem. Soc. 154 (2007) B1227–B1236.

[9] M. Khandelwal, S. Lee, M.M. Mench, One-dimensional Thermal Model of Cold-Start in a Polymer Electrolyte Fuel Cell Stack, J. Power Sources 172 (2007) 816–830.

[10] C. Hartnig, I. Manke, R. Kuhn, N. Kardjilov, J. Banhart, W. Lehnert, Cross–sectional insight in the water evolution and transport in polymer electrolyte fuel cells, Appl. Phys. Lett. 92 (2008) 1–3. 134106.

[11] E.E. Kimball, J.B. Benziger, and Y.G. Kevrekidis, Effects of GDL Structure with an Efficient Approach to the Management of Liquid Water in PEM Fuel Cells, Fuel Cells, (2010) DOI: 10.1002/fuce.200900110.

[12] T. Swamy, E.C. Kumbur, M.M. Mench, Characterization of Interfacial Structure in PEFCs: Water Storage and Contact Resistance Model, J. Electrochem. Soc. 157 (2010) B77–B85.

[13] S. Kim, Ph.D. Thesis, The Pennsylvania State University, 2008.

[14] Q. Yan, H. Toghiani, Y. Lee, K. Liang, H. Causey, Effects of Sub-Freezing Temperatures on a PEM Fuel Cell Performance, Startup and Fuel Cell Components, J.Power Sources 160 (2006) 1242–1250.

[15] J.P. Meyers, Fundamental Issues in Subzero PEMFC Startup and Operation, Proceedings of Fuel Cell Operations at Sub–Freezing Temperature Workshop, DOE, Washington DC, (2005), (http://www1.eere.energy.gov/hydrogenandfuelcells/fc_freeze_workshop.html).

[16] R. Gaylord, Stationary Application and Freeze/Thaw, Proceedings of Fuel Cell Operations at Sub-Freezing Temperature Workshop, DOE, Washington DC, (2005), (http://www1.eere.energy.gov/hydrogenandfuelcells/fc_freeze_workshop.html).

[17] R.C. McDonald, C.K. Mittelsteadt, E.L. Thompson, Effects of Deep Temperature Cycling on Nafion® 112 Membranes and Membrane Electrode Assemblies, Fuel Cells 4 (2004) 208–213.

[18] Q. Guo, Z. Qi, Effect of Freeze-Thaw Cycles on the Properties and Performance of Membrane-Electrode Assemblies, J. Power Sources 160 (2006) 1269–1274.

[19] J. Hou, H. Yu, S. Zhang, S. Sun, H. Wang, B. Yi, P. Ming, Analysis of PEMFC Freeze Degradation at −20°C After Gas Purging, J. Power Sources 162 (2006) 513–520.

[20] R. Mukundan, Y.S. Kim, F. Garzon, B. Pivovar, 2006 DOE Hydrogen Program Annual Progress Report (2006) 926–929.
[21] E.A. Cho, J.J. Ko, H.Y. Ha, S.A. Hong, K.Y. Lee, T.W. Lim, I.H. Oh, Characteristics of the PEMFC Repetitively Brought to Temperatures Below 0°C, J. Electrochem. Soc. 150 (2003) A1667–A1670.
[22] K. Tajiri, Y. Tabuchi, C.Y. Wang, Isothermal Cold Start of Polymer Electrolyte Fuel Cells, J. Electrochem. Soc. 154 (2007) B147–B152.
[23] G. Xie, T. Okada, The State of Water in Nafion 117 of Various Cations Forms, Denki Kagakuoyobi Butsuri Kagaku 64 (1996) 718–726.
[24] J.W. Paquette, K.J. Kim, Behavior of Ionic Polymer-metal Composites Under Subzero Temperature Conditions, Proc. ASME Intl. Expo, 205 ASME (2003).
[25] J. Newman, Investigating Failure in Polymer-Electrolyte Fuel Cells, US DOE Annual Program Review Proc. (2005).
[26] T.A. Zawodzinski, Membranes and MEAs at Freezing Temperatures, DOE Fuel Cell Operations at Sub-Freezing Temperatures Workshop, (2005).
[27] M. Pineri, G. Gebel, R.J. Davies, O. Diat, Water Sorption-desorption in Nafion® Membranes at Low Temperature, Probed by Micro X-ray Diffraction, J. Power Sources 172 (2007) 587–596.
[28] H. Liu, Dimensionally Stable High Performance Membrane, Proceedings of 2008 DOE Annual Program Review, DOE, Washington DC, 2008.
[29] R. Borup, J. Meyers, B. Pivovar, Y.S. Kim, R. Mukundan, N. Garland, D. Myers, M. Wilson, F. Garzon, D. Wood, P. Zelenay, K. More, K. Stroh, T. Zawodzinski, J. Boncella, J.E. McGrath, M. Inaba, K. Miyatake, M. Hori, K. Ota, Z. Ogumi, S. Miyata, A. Nishikata, Z. Siroma, Y. Uchimoto, K. Yasuda, K.-i. Kimijima, N. Iwashita, Scientific Aspects of Polymer Electrolyte Fuel Cell Durability and Degradation, Chem. Rev. 107 (2007) 3904–3951.
[30] F.E. Hizir, S.O. Ural, E.C. Kumbur, M.M. Mench, Characterization of Interfacial Morphology in Polymer Electrolyte Fuel Cells: Micro-porous Layer and Catalyst Layer Surfaces, J. Power Sources 195 (2010) 3463–3471.
[31] R. Alink, D. Gerteisen, M. Oszcipok, Degradation Effects in Polymer Electrolyte Membrane Fuel Cell Stacks by Sub-Zero Operation – An In Situ and Ex Situ Analysis, J. Power Sources 182 (2008) 175–187.
[32] S. Ge, C.Y. Wang, Cyclic Voltammetry Study of Ice Formation in the PEFC Catalyst Layer during Cold Start, J. Electrochem. Soc. 154 (2007) B1399–B1406.
[33] T. Hottinen, O. Himanen, S. Karvonen, I. Nitta, Inhomogeneous Compression of PEMFC Gas Diffusion Layer, J. Power Sources 171 (2007) 113.
[34] V. Mishra, F. Yang, R. Pitchumani, Measurement and Prediction of Electrical Contact Resistance Between Gas Diffusion Layers and Bipolar Plate for Applications to PEM Fuel Cells, ASME J. Fuel Cell Sci. Technol. 1 (2004) 2.
[35] U. Beuscher, Simon J.C. Cleghorn, W.B. Johnson, Challenges for PEM Fuel Cell Membranes, Intl. J. Energy Res. 29 (2005) 1103–1112.
[36] A.K. Srouji, Cold Start Induced Damage to Polymer Electrolyte Fuel Cell, Masters Thesis, The Pennsylvania State University, 2010.
[37] C. Chacko, R. Ramasamy, S. Kim, M. Khandelwal, M.M. Mench, Characteristic Behavior of Polymer Electrolyte Fuel Cell Resistance During Cold Start, J. Electrochem. Soc. 155 (2008) 1145–B1154.
[38] A.A. Pesaran, G. Kim, G.D. Gonder, PEM Fuel Cell Freeze and Rapid Startup Investigation, Milestone Report, NREL/MP 540 (2005) 8760.

[39] M. Khandelwal, M.M. Mench, Direct Measurement of Through-plane Thermal Conductivity and Contact Resistance in Fuel Cell Materials, J. Power Sources 161 (2006) 1106–1115.
[40] K.T. Cho, M.M. Mench, Fundamental Characterization of Evaporative Water Removal from Fuel Cell Diffusion Media, J. Power Sources 195 (2009) 3858–3869.
[41] K.T. Cho, M.M. Mench, Coupled Effects of Flow Field Geometry and Diffusion Media Material Structure on Evaporative Water Removal from Polymer Electrolyte Fuel Cells, International Journal of Hydrogen Energy 35 (2010) 12329–12340.
[42] K.T. Cho, M.M. Mench, Effect of Material Properties on Evaporative Water Removal from Polymer Electrolyte Fuel Cell Diffusion Media, J. Power Sources 195 (2010) 6748–6757.
[43] N. Pekula, K. Heller, P.A. Chuang, A. Turhan, M.M. Mench, J.S. Brenizer, K. Ünlü, Study of Water Distribution and Transport in a Polymer Electrolyte Fuel Cell using Neutron Imaging, Nucl. Instru Meth, in Phys. Res. Section A: Accelerators, Spectrometers, Detectors and Associated Equipment 542 (2005) 134–141.
[44] M. Perry, T. Patterson, J. ONeill, PEM Fuel Cell Freeze Durability and Cold Start Project, US DOE Annual Program Review Proceedings, DOE, Washington, DC, (2007). http://www.eere.energy.gov.
[45] R.L. Fuss, Freeze-Protecting a Fuel Cell by Vacuum Drying, United States patent, 6358637 (2002).
[46] R. Bradean, H. Haas, K. Eggen, C. Richards, T. Vrba, Stack Models and Designs for Improving Fuel Cell Startup from Freezing Temperatures, ECS Trans. 3 (2006) 1159–1168.
[47] A.Z. Weber, J. Newman, Coupled Thermal and Water Management in Polymer Electrolyte Fuel Cells, J. Electrochem. Soc. 153 (2006) A2205.
[48] M. Perry, T. Patterson, J. ONeill, Start and Operation of PEMFC Stacks under Sub-Freezing Conditions, proceedings of 5th International Conference on Fuel Cell Science Engineering and Technology, ASME (2007).
[49] R. Zaffou, H.R. Kunz, J.M. Fenton, Temperature-Driven Water Transport in Polymer Electrolyte Fuel Cells, ECS Trans. 3 (2006) 909–913.
[50] R. Zaffou, J.S. Yi, H.R. Kunz, J.M. Fenton, Temperature-Driven Water Transport through Membrane Electrode Assembly of Proton Exchange Membrane Fuel Cells, Electrochem. Solid-State Lett. 9 (2006) A418–A422.
[51] J.P.G. Villaluenga, B. Seoane, V.M. Barragan, C. Ruiz-Bauza, Thermo-osmosis of Mixtures of Water and Methanol through a Nafion Membrane, J. Membr. Sci. 274 (2006) 116–122.
[52] M. Tasaka, T. Hirai, R. Kiyono, T. Aki, Solvent Transport across Cation-Exchange Membranes under a Temperature Difference and under an Osmotic Pressure Difference, J. Membr. Sci. 71 (1992) 151–159.
[53] M. Tasaka, T. Mizuta, O. Sekiguchi, Mass Transfer through Polymer Membranes due to a Temperature Gradient, J. Membr. Sci. 54 (1990) 191–204.
[54] S. Kim, M.M. Mench, Temperature Gradient Induced Water Transport in Polymer Electrolyte Membranes, ECS Trans., Phoenix meeting (2008).
[55] S. Kim, M.M. Mench, Investigation of Temperature-Driven Water Transport in Polymer Electrolyte Fuel Cell: Thermo-osmosis in Membranes, J. Memb. Sci. 328 (2008) 113–120.
[56] M. Khandelwal, S. Lee, M.M. Mench, Model to Predict Temperature and Capillary Pressure Driven Water Transport in PEFCs After Shutdown, J. Electrochem. Soc. 156 (2009) B703–B715.
[57] M. Kandelwal, M.M. Mench, J. Power Source (2010). in press.
[58] A. Turhan, K. Heller, J.S. Brenizer, M.M. Mench, Passive Control of Liquid Water Storage and Distribution in a PEFC through a Flow-Field Design, J. Power Sources 180 (2008) 773–783.

[59] A. Turhan, S. Kim, M. Hatzell, M.M. Mench, Impact of Channel Wall Hydrophobicity on Through-plane Water Distribution and Flooding Behavior in a Polymer Electrolyte Fuel Cell, Electrochim. Acta 55 (2010) 2734–2745.

[60] M.M. Mench, Advanced Modeling in Fuel Cell Systems: A Review of Modeling Approaches, Proceedings of the 18th World Hydrogen Energy Conference, 2010.

[61] X. Yu, B. Zhou, A. Sobiesiak, Water and Thermal Management for Ballard PEM Fuel Cell Stack, J. Power Sources 147 (2005) 184–195.

[62] M.S. Wilson, J.A. Valerio, S. Gottesfeld, Low Platinum Loading Electrodes for Polymer Electrolyte Fuel Cells Fabricated using Thermoplastic Ionomers, Electrochim. Acta 40 (1995) 355–363.

[63] T. Patterson, R. Balliet, PEM Fuel Cell Freeze Durability and Cold Start Project, DOE Hydrogen Program Annual Progress Report (2006) 910–912.

[64] T. Patterson, J.P. Meyers, PEM Fuel Cell Freeze Durability and Cold Start Project, Proceedings of 2006 DOE Annual Program Review, DOE, Washington DC, 2006.

[65] E.A. Cho, J.J. Ko, H.Y. Ha, S.A. Hong, K.Y. Lee, T.W. Lim, I.H. Oh, Effects of Water Removal on the Performance Degradation of PEMFCs Repeatedly Brought to $< 0°C$, J. Electrochem. Soc. 151 (2004) A661–A665.

[66] M. Oszcipok, M. Zedda, D. Riemann, D. Geckeler, Low Temperature Operation and Influence Parameters on the Cold Start Ability of Portable PEMFCs, J. Power Sources 154 (2006) 404–411.

[67] M. Oszcipok, D. Riemann, U. Kronenwett, M. Kreideweis, M. Zedda, Statistic Analysis of Operational Influences on the Cold Start Behaviour of PEM Fuel Cells, J. Power Sources 145 (2005) 407–515.

[68] R. Mukundan, Y.S. Kim, F. Garzon, B. Pivovar, Freeze/Thaw Effects in PEM Fuel Cells, ECS Trans. 1 (8) (2006) 403–413.

Chapter 7

Experimental Diagnostics and Durability Testing Protocols

Mike L. Perry,[1] Ryan Balliet[2] and Robert M. Darling[1]
[1]*United Technologies Research Center, CT, USA,* [2]*Department of Chemical Engineering, University of California, Berkeley, CA, USA*

1. INTRODUCTION

Recently, a large number of papers have been published on the durability of polymer electrolyte fuel cells (PEFCs). A recent review article [1] included over 500 references on this topic. Books on PEFC durability, including this one, have also begun to appear [2,3]. The emphasis of these papers and chapters is on degradation mechanisms and approaches to decay prevention. Separately, the subject of diagnostics has been covered, mostly in books on fuel cells, with the primary emphasis being on learning what limits their initial performance. The subject of how to decide what causes decay in PEFCs has not been discussed extensively. Determining the causes of degradation can be difficult and therefore one can benefit from following a methodology that can identify likely causes in a logical and efficient manner. This chapter provides guidance on how to use in-cell diagnostics to uncover the causes of decay in a PEFC that have occurred over time.

Determining the root cause of degradation often requires application of numerous investigative techniques, including the destructive analysis of cells and a thorough analysis of components to determine what physical changes have occurred since beginning-of-life (BOL). End-of-life (EOL) analysis can include a variety of sophisticated analytical tools and techniques that are often time consuming and expensive. The effective use of in-cell diagnostics can provide insights into where to look and what to look for during a destructive analysis, thereby saving time and money. This is especially true if a systematic set of diagnostics is used to isolate the components that are responsible for decay.

A PEFC consists of a number of repeating components including a membrane, anode and cathode catalyst layers (CLs), anode and cathode gas diffusion layers (GDLs), bipolar plates, and seals. Therefore, employing a method that can identify the components responsible for performance loss is

Polymer Electrolyte Fuel Cell Degradation. DOI: 10.1016/B978-0-12-386936-4.10007-7

of paramount importance. Additionally, in-cell diagnostics can be used to assess the physical changes that have occurred in degraded components. Finally, one would like to determine the mechanisms responsible for causing these physical changes. This last step is beyond the scope of this chapter because it requires knowledge of the design and a history of the operating conditions. However, if one has already determined what changes have occurred, then it is much easier to ascertain the causes. Therefore, in-cell diagnostics are a logical first step in determining the mechanisms responsible for decay, and their effective use can minimize the speculation that occurs prior to the completion of a root-cause investigation.

This chapter ends with a discussion of accelerated stress tests (ASTs). This is a related topic because the goal of durability testing is often to accelerate a particular decay mechanism or stress specific components within the PEFC. In order to do this effectively, one should understand what conditions cause PEFCs to decay, including what operating parameters (e.g. T, V) have a significant effect on the rate of each of these mechanisms. An overview of the major types of accelerated tests, and the advantages and disadvantages of each, is the focus of the last section.

2. GENERAL COMMENTS ON DIAGNOSTIC TEST PROCEDURES

The purpose of this chapter is to discuss the diagnostic tests that can be performed to uncover the causes of degradation. An extensive set of tests and procedures designed to accomplish this task is discussed. The diagnostic procedures that can actually be applied to a particular fuel cell may be limited if it is integrated into a system. For example, generating a well-defined polarization curve can be challenging on a stack of fuel cells integrated into an automobile. Algorithms for filtering and interpolating data, that necessarily introduce uncertainty, may be needed to create familiar polarization curves. Often one is limited to monitoring decay under repeatable conditions and assessing the health of the membrane on fuel cells operating at customer sites. Evaluations like those described in this chapter may be possible after the stack is removed from service. On the other hand, most of the tests described in this chapter can be done in a laboratory. Frequent, targeted diagnostics during durability testing should be done whenever possible to establish the characteristics and causes of decay.

Two types of tests that are common in laboratories are those done on production-scale cells and those done on cells with smaller active areas. Generally, production-scale cells are operated close to design conditions. Large deviations in test conditions may be inappropriate. For example, doubling the flow in a cell that is designed to drop air pressure by 50 kPa by diluting with nitrogen will yield a pressure drop of approximately100 kPa. The resulting difference in average pressure complicates interpretation of results. More latitude exists when a smaller, or sub-scale, cell is substituted for a production-scale cell. A sub-scale cell can be designed to operate like a differential cell that corresponds to some portion of the active area of a production cell or it can be operated at conditions similar to a large cell. The flow fields can be designed to accommodate either approach. The flows will be large, and utilizations of reactants small, when following the differential approach. Conversely, the flows will be small and the utilizations large when the small cell is operated like the larger production cell. The differential approach can be complicated by water balance. The relative humidity of the incoming gases must be estimated for the portion of the active area of interest, or tests must be run over a range of humidity levels. The overall water balance may be nearly preserved when a small-scale cell is run at high reactant utilizations, but the details of mass transport outside the catalyst layer are unlikely to match a large cell. Many of the diagnostic tests are better suited to the differential approach. However, scaling the transport characteristics of the flow fields and diffusion layers is complicated and should be done carefully. Diagnostics conducted on a combination of production-scale and sub-scale cells can be especially informative.

3. POLARIZATION-CHANGE CURVE

A logical first step in determining the causes of performance loss is to categorize what overpotentials have changed since BOL. As is well known, there are three major types of overpotential in fuel cells, namely; kinetic, ohmic, and mass transport [4]. One can assess how the different overpotentials evolve by plotting the change in voltage versus current density (one may choose to correct the voltage for internal resistance, and track changes in internal resistance separately), as illustrated in Fig. 7.1. This polarization-change plot is constructed by taking the difference between a polarization curve taken at BOL[1] and the latest polarization curve. The difference is taken this way in order to return a positive value. Thus,

$$\Delta V = V_{BOL} - V \tag{7.1}$$

Operating conditions, including dwell times and procedures that recover performance, must be the same for all curves. Measuring polarization curves before and after applying recovery procedures should allow one to separate recoverable and irrecoverable decay.

The limiting cases depicted in Fig. 7.1 were constructed by subtracting voltages calculated with a simple model of polarization. Parameters

---

[1] BOL is herein defined as the peak performance of the cell (i.e. after the break-in period where the cell performance is still increasing), since including the effects of the break-in period makes analyzing degradation more difficult. The cause of performance changes during the break-in period is not the subject of this chapter; however, the process described here could be used to analyze any change in performance, including performance improvements experienced during the break-in period.

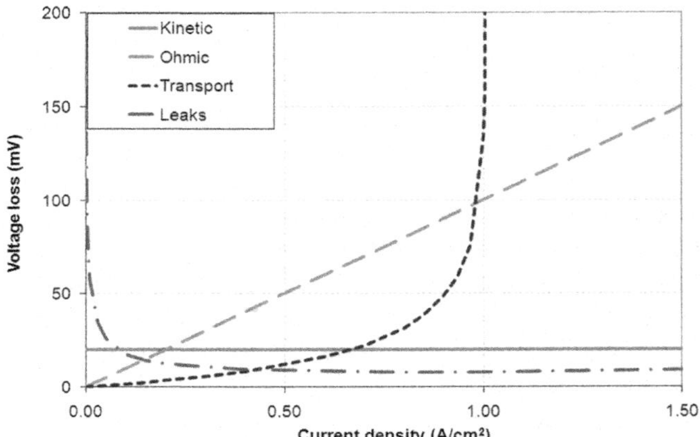

FIGURE 7.1 Limiting cases of polarization-change curves. Polarization-change plots are constructed by taking the difference between a recent polarization curve and one taken at peak performance. Four limiting cases are shown here: 1) Kinetic (solid line), 2) Ohmic (dashed line), 3) Transport (dotted curved line), and 4) Leak (dash and dot curve).

characterizing kinetic, ohmic, and transport were adjusted individually in three separate cases and subtracted from the simulated BOL. The case of increasing leaks is discussed in section 4.4. The voltages were calculated using a simple set of equations that give a good representation of a PEFC polarization curve:

$$V = U - b\log[i/(a\, i_0\, \delta_{CL})] + b\log[1 - i/i_{\lim}] - iR' \tag{7.2}$$

$$U = U^{\theta} + b\log[p_o/p_o^{\theta}] + (RT/2F)\ln[p_H/p_H^{\theta}] \tag{7.3}$$

In the above equations, i is the superficial current density based on the geometric surface area of the membrane. The first term in Eqn 7.2 is a potential intercept that sums the standard thermodynamic potential of the cell, which depends on temperature, and the partial pressure dependencies for the ORR and HOR, as shown in Eqn 7.3. The ORR is assumed to follow Tafel kinetics, first order in oxygen concentration in this formulation. The Tafel slope is related to the cathodic transfer coefficient by the equation:

$$\alpha_c = 2.303RT/b\,F \tag{7.4}$$

Different reaction orders have been reported in the literature for oxygen reduction [5]. The HOR is described by the Nernst equation in Eqn 7.3. The second term in Eqn 7.2 comes from using Tafel kinetics to describe the ORR. The term in the denominator is the product of the exchange current density per unit platinum area, i_0, and the platinum area per unit geometric area, $a\delta_{CL}$. The denominator is often included in the potential intercept term. The dependence of the exchange current density on the partial pressure of oxygen is explicitly stated in Eqn 7.2, however i_0 may also depend on acidity, water concentration, and temperature [6]. The third term is a concentration overpotential for oxygen reduction. The limiting current, i_{lim}, occurs when the oxygen concentration is zero at the cathode catalyst layer and should be approximately proportional to oxygen mole fraction. The final term in Eqn 7.2 treats resistive losses. *R'* is the sum of the resistances of all layers within a cell and contact resistances. All of these terms are further defined and described elsewhere [7]. Equations 7.2 and 7.3 describe the polarization behavior of a typical fuel cell reasonably well and serve as a useful basis for discussion of the qualitative features that Fig. 7.1 is intended to illustrate. These equations are not expected to be quantitatively accurate in all cases. For example, these equations cannot predict a doubling of the Tafel slope caused by ohmic or transport limitations in the cathode catalyst layer. This behavior is often observed in porous electrodes, like those employed in fuel cells, and is discussed in subsequent sections of this chapter.

3.1. Key Limiting Cases of Polarization-Change Curves

Four polarization-change curves are depicted in Fig. 7.1; each represents the shape expected if only one type of overpotential changes. These polarization-change curves are the difference between peak and decayed voltage at different current densities. All other independent variables are held constant.

The first limiting case depicted in Fig. 7.1 is a horizontal line, resulting from an increase in kinetic losses for the ORR. This case is realized by decreasing the denominator of the second term in Eqn 7.2, while maintaining a constant Tafel slope. The voltage difference is:

$$\Delta V = b \log[(a_b \, i_{0,b} \, \delta_{CL,b})/(a \, i_0 \, \delta_{CL})] \tag{7.5}$$

The subscript *b* denotes BOL. In principle, decreases in either the exchange current density or the catalytic surface area, $a\delta_{CL}$, could be responsible. Kinetic losses associated with the ORR should be independent of current density, provided the Tafel slope does not change. Practically, the Tafel slope does not appear to change significantly as platinum catalysts degrade. For the case shown in Fig. 7.1, it was assumed that the catalytically active surface area per unit volume of electrode, *a*, decreased by 50%. Redistribution of platinum during operation may also lead to increases in ohmic and transport losses in the cathode.

The second limiting case is a line that intercepts the origin of the plot. This is the theoretical expectation for losses due solely to an increase in resistance. Thus,

$$\Delta V = i(R' - R'_b) \tag{7.6}$$

For the case shown in Fig. 7.1, it was assumed that the resistance, *R'*, doubled from 100 to 200 m$\Omega \cdot$cm^2.

If increases in mass-transport overpotential are the sole cause of decay, then one would expect a curve that intersects the origin and has a roughly exponential shape:

$$\Delta V = b \log[(1 - i/i_{lim,b})/(1 - i/i_{lim})] \quad (7.7)$$

In this case, the performance loss approaches infinity as the current approaches the new limiting current. In practice, if the current exceeds the limiting current for oxygen reduction, hydrogen evolution will occur at the cathode and the voltage loss will be limited. The original limiting current can no longer be achieved because of the degradation in mass transport. For the case shown in Fig. 7.1, it was assumed that the limiting current, i_{lim}, had been reduced by 50% from 2.010 to 1.005 A/cm^2. Mass transport losses through the depth of the cathode catalyst layer may not display limiting-current behavior because reaction and transport may be parallel, not serial, mathematically. This depends on the nature of the model used to describe transport losses in the catalyst layer. The current in the Tafel term of Eqn 7.2 may be multiplied by an effectiveness factor to address parallel transport losses in the catalyst layer. Because the effectiveness factor is a function of cathode potential, Eqn 7.2 will be implicit in voltage. Effectiveness factors are commonly used to describe reaction and diffusion in catalyst pellets and adoption of this approach to porous fuel cell electrodes has been described elsewhere [8,9].

The fourth limiting case depicted in Fig. 7.1 is observed if leaks are the sole cause of decay. Gas leaks in a cell can increase due to membrane or seal degradation. Electrical shorting, which can also increase with time as electrical contact between anode and cathode components improves due to degradation of the membrane or seals, behaves similarly on this plot. In either case, the impact of leaks on cell performance is greatest at low current densities. This behavior can be readily understood if one thinks of a leak as an additional load imposed on the cell. In the case of an electrical short, this additional load is the current being carried by the short, which means that the measured current is lower than the actual current through the active area of the cell. A gas leak across the membrane is equivalent to an electrical short in the way that it affects polarization and the magnitude of the leak can be expressed as an equivalent current density (i.e. in A/cm^2) [10]. At low current densities this equivalent leak current will be as high, or higher, than the measured current density and the impact on cell performance will be significant. At higher current densities the impact will be relatively small, because the leak current is small compared to the measured current. For the case illustrated in Fig. 7.1, the leak current was increased from 0 to 50 mA/cm^2. The current density in Eqn 7.1 should be replaced by $i + i_x$, where i is the current density in the external circuit and i_x is the leak current density, to account for leaks.

3.2. Analyzing Actual Polarization-Change Curves

PEFC degradation usually results from changes to more than one type of overpotential. For example, an increase in kinetic losses often occurs early in the life of a PEFC, since the small catalyst particles employed to maximize catalyst surface area are prone to dissolve and grow larger with time [1]. This kinetic loss may be accompanied by a transport loss since, at a given current density, the reactant may have to penetrate deeper into the degraded catalyst layer and the rate of reactant transport to each remaining catalyst site must be proportionally greater. In this case, the result would be a combination of kinetic and reactant-transport curves depicted in Fig. 7.1. This would be an approximately exponential curve that intercepts the ordinate at a positive value, which is a common result with actual cell data. In this case, the intercept provides a quantitative estimate of the change in polarization due to increased kinetic overpotential. With real data, which always has some degree of noise, one can use a best fit of the horizontal portion of the curve to provide a better estimate of the intercept value. Similarly, a combination of kinetic and ohmic changes can result in a sloped line with a non-zero intercept.

A polarization-change curve with an intercept of zero is uncommon, because PEFCs typically experience a loss in catalytic area. In other words, one almost always measures a lower open circuit voltage (OCV) as a cell degrades. However, if one measures a very large change in OCV and much smaller changes at higher current densities, then this is good indication that the cell has developed a leak or short. Since OCV measurements can be noisy, it is good practice to also include at least one measurement at a low current density (e.g. 10 mA/cm^2 or less) to make this comparison. As shown in Fig. 7.1, a leak is the only case where the polarization-change curve should exhibit a negative slope. In this case, the magnitude of the leak current can be roughly estimated by extrapolating the vertical portion of the polarization-change curve to intercept the horizontal axis; however, doing this requires a significant number of measurements at low current densities and there are better methods to quantify leaks that are provided in Section 4.4.

The graphical analysis described here is simple and useful; however, it should be conducted with some caution. For example, small decreases in the limiting current may result in a roughly linear function in a polarization-change plot at currents that are a small fraction of the limiting current. One should record data out to the limiting current to obtain the best possible insight. In any case, one should seek to verify the sources of overpotential by conducting the more advanced cell diagnostics described in subsequent sections. However, this simple graphical analysis is a useful first step since it utilizes readily available data and it provides insight into what may be the most fruitful subsequent diagnostics.

4. ISOLATING THE COMPONENTS RESPONSIBLE FOR PERFORMANCE LOSS

This section will outline a methodology to determine what components in a PEFC are most likely responsible for the performance loss. The first step is constructing a polarization-change plot, such as that shown in Fig. 7.1, to make a preliminary diagnosis. Next, in-cell diagnostics targeted at these types of losses are used in order to try to isolate the responsible component(s). Conceptually, the results of these tests can then be used to reduce the amount of *ex-situ* analysis that must be done to determine the nature of the damage responsible for the performance loss.

Using targeted diagnostics to isolate the causes of decay is the subject of this section. The components susceptible to each kind of loss are discussed, along with common mechanisms and specific diagnostic procedures. A summary of this section is provided in Table 7.1, where each row corresponds to one of the first three limiting cases discussed in Section 3. The fourth limiting case, leaks, is summarized in Section 4.4 and Table 7.3. Only the most common mechanisms are included here and they are not discussed in detail because that is the subject of other chapters in this book.

A theme throughout this section is the necessity to run combinations of diagnostics rather than relying on any single experiment. There are several reasons for this approach. First, as discussed in Section 3.2, it is common for a cell to experience more than one type of degradation over the course of a durability test. Furthermore, the symptoms of one decay mechanism can often appear similar to those of another. Finally, a given type of degradation may apply to more than one component within the cell. Combinations of diagnostics are therefore often required to determine with confidence which types of losses are present and the components responsible. Examples of such scenarios will be discussed.

4.1. Catalytic Activity Losses

As shown in Fig. 7.1, voltage loss that is nearly independent of current density may indicate a loss of catalytic surface area or activity. This behavior is only true for reactions that follow simple kinetic expressions with constant Tafel slopes, which is usually true at the cathode of a PEFC. The first row in Table 7.1 corresponds to this case. Because the cell reactions are presumed to happen only within the catalyst layers, this type of loss indicates an issue within either the anode or cathode. Which electrode has the issue will depend on the particular mechanism.

4.1.1. Common Mechanisms

A decrease in catalytic activity can result from a loss of electrochemical surface area (ECSA), which can result from a variety of mechanisms that have

TABLE 7.1 Summary of Recommended In-cell Diagnostics for Different Types of Performance Losses

Type of Loss Observed	Some Possible Contributors and Typical Diagnostic Tests		
	Anode Diffusion Media or Catalyst Layer	Membrane	Cathode Diffusion Media or Catalyst Layer
Catalytic activity Performance loss nearly independent of current density	**Possible causes** – Loss of catalyst area by sintering or dissolution – Contamination by adsorption* **Diagnostics** – Cyclic voltammogram of anode		**Possible causes** – Loss of catalyst area by sintering or dissolution – Contamination by adsorption* – Pt oxide formation* **Diagnostics** – Cyclic voltammogram of cathode – Tafel measurement
Ohmic (ionic or electronic) Performance loss proportional to current density	**Possible causes** – Dryout of ionomer* – Contamination by foreign cations* – Increased contact resistance, intra- or interlayer **Diagnostics** – Electrochemical H_2 pump – Current interrupt – Impedance measurement plus model	**Possible causes** – Dryout of membrane* – Contamination by foreign cations* **Diagnostics** – Impedance measurement of ionic resistance – Electrochemical H_2 pump – Current interrupt	**Possible causes** – Dryout of ionomer* – Contamination by foreign cations* – Increased contact resistance, intra- or interlayer **Diagnostics** – Electrochemical H_2 pump – Current interrupt – Impedance measurement plus model

(Continued)

TABLE 7.1 Summary of Recommended In-cell Diagnostics for Different Types of Performance Losses—cont'd

Type of Loss Observed	Some Possible Contributors and Typical Diagnostic Tests		
	Anode Diffusion Media or Catalyst Layer	Membrane	Cathode Diffusion Media or Catalyst Layer
Reactant mass transfer Performance loss exponential with current density	**Possible causes** – Flooding of diffusion media or catalyst layer* – Reactant channel blockage* – Carbon oxidation **Diagnostics** – Voltage gain, H_2 vs. dilute H_2 – Compare H_2/N_2 with H_2/He – Voltage vs. H_2 utilization – Reactant pressure drop – Impedance measurement plus model		**Possible causes** – Flooding of diffusion media or catalyst layer* – Reactant channel blockage* – Carbon oxidation **Diagnostics** – Voltage gain, O_2 vs. air – Dilute O_2 IV curve – H_2/O_2 IV curve – Voltage vs. O_2 utilization – H_2/Helox IV curve – Reactant pressure drop – Impedance measurement plus model

* *Often substantially reversible.*

been summarized elsewhere [1]. Additionally, the activity of the catalyst may change due to particle-size effects or change in the composition of the surface in the case of alloy catalysts. In addition, potential cycling leads to dissolution of platinum and alloying elements, causing these elements to be redistributed within the cell, often to areas such as the membrane where they can no longer promote the desired reactions [11,12].

A decrease in catalytic activity can also be caused by adsorption of contaminants. Common culprits include CO and H_2S on the anode and NH_3 and SO_X on the cathode [13]. The impact on performance may not be independent of current density in cases where contamination of the catalytic surface is extensive. Poisoning of platinum at the anode by CO, for example, lowers the limiting current for hydrogen oxidation significantly. Oxidation of platinum in the cathode may also lead to decreased activity [14].

Contamination is often reversible. Raising the potential of the anode is often sufficient to remove adsorbed species. Prolonged operation of the cell in the absence of the contaminant has been shown to remove some species that adsorb on the cathode [10]. Raising the potential of the cathode can also remove adsorbed species; this can be accomplished by stopping the fuel cell and allowing the electrodes to approach the reversible potential for oxygen reduction [15,16]. Platinum oxides can be stripped by lowering the cathode potential [17,18].

4.1.2. In-cell Diagnostics

4.1.2.1. Tafel Slope Measurement

The voltage is measured as a function of current density, while temperature, operating pressure, inlet humidification, and reactant utilizations are held constant. Flows are often held constant instead of utilizations on smaller cells. A semi-log plot is then constructed with the overpotential corrected for internal resistance on the ordinate, and current density corrected for shorting and crossover on the abscissa. The slope of the resulting line at low-to-moderate current densities, where mass-transfer effects are negligible, is known as the Tafel slope and can be used to find the apparent charge transfer coefficient, α_c, using the Tafel equation. The measured Tafel slope may be larger than the true kinetic Tafel slope when using a porous electrode due to ohmic and transport losses in the cathode. Transport effects may be compensated for by using a current density that has been corrected for transport losses, $i/(1-i/i_{lim})$, as the abscissa. Mass-transport losses may also be eliminated to a large extent by generating the polarization with oxygen instead of air. An extrapolation of the Tafel slope to zero overpotential gives the exchange-current density, i_o, for a smooth electrode. In the case of a porous electrode it provides the superficial exchange current density, $a\ i_o\ \delta_{CL}$ [4, 19]. Many researchers prefer to use $i_{(0.9V)}$ at specified oxygen partial pressure, temperature, and humidity to avoid extrapolation over orders of magnitude of current density. Measuring polarization with the fuel and air swapped to generate a Tafel plot for oxygen on the normal fuel electrode can be useful for tracking decay at the anode. This approach works best when the anode catalyst is platinum, but may be appropriate for some alloys.

The exchange current density is of particular interest because it provides a measure of catalyst performance. A higher exchange current density indicates greater catalytic activity [20]. Tracking the superficial exchange current density as a function of run time, therefore, is a means of tracking catalyst degradation over the life of the cell. One can measure the ECSA (described in the following section) as a function of time to track the catalytic area, $a\delta_{CL}$, in order to distinguish between losses due to reduction in area and those due to changes in the rate constant. This approach may be imperfect because ECSA requires only protons, not a gaseous reactant, to access the catalytic surface.

4.1.2.2. Cyclic Voltammetry (CV)

Nitrogen, or liquid water, flows over the working electrode while hydrogen, or dilute hydrogen, flows over the counter electrode, which generally also functions as the reference electrode when the test is performed *in-situ*. Temperature, gas humidity, and gas flows are held constant. A potentiostat is used to increase the potential to an upper limit at a constant rate, and then to decrease the potential at a constant rate back to the starting point. Due to the nature of the reactions and the state of the catalyst at different potentials, the shape of the *i-V* characteristic during the forward scan differs from the reverse scan, and when both curves are plotted together they are known as a cyclic voltammogram [21].

Analysis of the various peaks in a CV can yield a range of information, including potentials where charge-transfer reactions occur and the presence of contaminants, as well as changes in the carbon supports. For durability studies, one result of particular interest is the electrochemical surface area (ECSA). This is obtained from a CV by identifying the hydrogen adsorption or desorption peak and then integrating the current passed during the time under this peak to give the total charge passed as hydrogen is adsorbed onto the catalyst surface. The electrochemical surface area can then be estimated from the total charge passed [4]. Tracking ECSA with run time is a means of tracking catalyst degradation by a variety of mechanisms over the life of the cell. Relative to the Tafel method, this method has the advantage that it can be used to study either the anode or the cathode regardless of the composition of the anode catalyst. The ECSA measured in cell may be compared to measurements made in acid flooded electrodes to estimate platinum utilization. Additional information may be extracted by analyzing other peaks of interest on the voltammogram [4].

4.1.2.3. Electrochemical Impedance Spectroscopy (EIS)

The impedance of the cell is measured as a function of current density by superimposing a low-magnitude alternating-current signal over a range of frequencies and measuring the voltage response [18]. Temperature, operating pressure, inlet humidification, and reactant concentrations are held constant.

For a given operating condition, the imaginary component of the impedance is plotted against the real component over the range of frequencies used. Such a figure is called a Nyquist plot [22]. Information regarding cathode mass-transfer resistance, anode and cathode kinetics, and ionic resistance may be extracted from the Nyquist plot with an equivalent-circuit model. It can be difficult to separate and assign parameter values to phenomena that have similar time constants. Alternatively, one can use a physics-based model translated into the frequency domain for the analysis [23]. In any case, comparing EIS results as a function of operating time can provide a means to measure changes in polarization parameters over the life of the cell.

4.2. Ohmic Losses

Voltage decay that is a linear function of current density may be caused by increasing resistance. The second row in Table 7.1 corresponds to this case. Determining the sources of ohmic losses can be challenging. The loss may be due either to changes in the electronic or ionic resistance and may be a result of a change in conductivity within a layer or at an interface between layers. Every layer of the cell is either electronically or ionically conductive, and the catalyst layers must be both.

4.2.1. Common Mechanisms

Ionic conductivity is a strong function of ionomer hydration. Drier conditions result in higher ohmic losses, within both the catalyst layers and the bulk membrane. Operating at higher relative humidity should hydrate the ionomer and reverse this type of loss. Irreversible chemical degradation of the ionomer can also take place with time and this is often accelerated by operation under drier conditions. Although this degradation will impact ionic conductivity, it is usually a highly localized phenomenon that results in loss of structural integrity and leakage through the membrane before having an appreciable impact on cell conductivity.

Contamination by metal cations (such as ferrous ions) may also decrease conductivity [24]. Similar to catalyst contamination, ionic contamination is often reversible. Ionomer is essentially a cation-exchange media, which can be restored to the acid form in analogous manner to renewing a demineralizer bed. For example, operation in the absence of the contaminant with porous-plate cells [25] with a demineralizer bed in the coolant loop is effective at removing ionic contaminants, especially if the cell is operated at high current densities to maximize the flux of clean water passing through the cell provided by the porous plates and electro-osmotic drag. Recovery from cationic contaminants in conventional cells with solid bipolar plates has also been shown, however some cations are removed more easily than others and the recovery mechanism is poorly understood [26].

Increased contact resistance may increase ohmic losses. This may occur if the compressive load relaxes or if gaps develop between the layers of the cell, as can happen if there is significant carbon oxidation [27]. Severe corrosion of carbon catalyst support will also result in other losses, such as increased ohmic and transport losses due to the change in the catalyst layer structure.

4.2.2. In-cell Diagnostics

4.2.2.1. High-frequency Resistance Measurement

The impedance of the cell is measured as a function of current density by applying a low-magnitude, constant-frequency alternating-current signal and measuring the voltage response [16]. Temperature, pressure, inlet humidification, and inlet reactant concentrations are held constant. Flow may vary with current.

The frequency that is used to measure the impedance must be high enough to justify the assumption that the impedance is entirely real and therefore equivalent to the internal resistance of the cell [16]. Using the high-frequency resistance (HFR) method as a function of run time, therefore, is a way to measure changes in resistance over the life of the cell. This method essentially measures the ionic resistance of the membrane and the electronic resistance of the other components, since the current follows the path of least resistance in the catalyst layers and the much larger ionic resistance of these layers is not significant in this measurement. Therefore, changes in ionic resistance of the catalyst layers are not directly measured. However, as described in Section 4.1.2.3, one can obtain additional insight into the sources of polarization, including ohmic losses within electrodes, by using full-frequency impedance spectroscopy combined with a cell model; this is discussed below in Section 4.3.2.2.

4.2.2.2. Electrochemical Hydrogen Pump

The normal fuel electrode is connected to the negative terminal of a power supply while the normal air electrode is connected to the positive terminal. Pure hydrogen flows on the normal air electrode while nitrogen or hydrogen flows on the normal fuel electrode. In this configuration, hydrogen oxidation occurs at what is normally the air electrode while hydrogen evolution occurs at the normal fuel electrode. Voltage is measured as a function of current density while temperature, pressure, hydrogen utilization, and inlet humidification are held constant.

The voltage increases nearly linearly with current density due to the facile hydrogen kinetics and negligible mass-transport resistance [28]. The slope of the resulting voltage vs. current-density curve represents a good approximation to the internal resistance for a properly functioning cell, including both ionic and electronic contributions [29]. Although this measurement does include all the layers of the cell, the current distribution in the normal electrode is probably different when oxygen reduction occurs. Because the reaction profile is not the same, the ohmic contribution of the cathode in fuel-cell mode is not assessed. Periodic electrochemical hydrogen pumps provide a good method to measure changes in resistance with time provided the overpotentials associated with hydrogen reaction and transport are constant and negligible.

When the hydrogen pump is used to measure resistance it is generally not important whether the test is run such that protons move toward the normal fuel or air electrode. That is, the test arrangement described above may be reversed. However, during the measurement electro-osmotic drag tends to dry out the anodic catalyst layer. As a result, when the measurement is performed such that protons move away from the fuel-cell cathode catalyst layer liquid water may be removed, resulting in a significant improvement in performance at high current density once the test is complete. For this reason the hydrogen pump is sometimes used to identify and recover from flooding of the cathode [30].

Hydrogen pumping is useful for diagnosing anode polarization when operating on contaminated hydrogen made by re-forming hydrocarbons. Losses occurring in the anode catalyst layer can be isolated by subtracting resistance measured by current interruption or HFR from the measured voltage, with the assumption that losses in the hydrogen evolving catalyst layer are negligible. The hydrogen pump with reformate or a H_2/N_2 mixture with the appropriate concentration of hydrogen is fed to the anode and neat hydrogen fed to the cathode. It might be necessary to inject a small amount of oxygen into the hydrogen stream to prevent contamination of the cathode.

4.2.2.3. Current Interrupt

The cell voltage is measured as a function of time once current is interrupted, usually for a total period of several milliseconds. Temperature, operating pressure, inlet humidification, and reactant concentrations are held constant.

When current to the external circuit is interrupted, the cell voltage approaches open circuit in a characteristic two-step manner. First, there is a sharp increase that is almost instantaneous. The magnitude of this voltage difference divided by the current density is the internal resistance of the cell (both ionic and electronic). This is followed by a slower rise in voltage that approaches OCV asymptotically with a time constant associated with the activation polarization and capacitance of the cathode. Interrupting the current as a function of time, therefore, is a way to measure changes in resistance over the life of the cell. As with HFR, this method primarily measures the ionic resistance of the membrane and the electronic resistances of the other components, so changes in ionic resistance in the catalyst layer are not assessed.

4.3. Reactant Mass-Transport Losses

Performance loss that is a strongly increasing function of current density may indicate an increase in reactant mass-transport losses. The third row in Table 7.1 corresponds to this case. As with ohmic losses, the cause may be on either the anode or cathode. However, with mass-transport losses it is relatively straightforward to determine which side is causing the problem. On the other hand, it can be difficult to isolate the particular component responsible. Reactant transport losses may be caused by changes in transport resistance either external to the catalyst layer or internal to it, and the symptoms of some types of reactant transport loss may be confused with ionic transport losses. Differentiating between reactant transport losses and ionic transport losses in the catalyst layers can be challenging, and the use of physics-based models can be helpful to distinguish between these two sources of polarization [7,20]. This section will focus on diagnostics that can assist in attributing losses to reactant transport.

4.3.1. Common Mechanisms

Accumulation of liquid water in flow channels and gas diffusion layers may cause reactant transport losses to increase. This may result from changes in material properties, such as hydrophobicity, over time.

Accumulation of liquid water within the catalyst layer related to changes in hydrophobicity may also be a problem. Normal diffusion of gases in the pores of the catalyst layer should decrease as $(1\text{-}S)^r$, where S is the saturation and the exponent r is probably greater than 1.5. Loss of porosity due to the oxidation of carbon support may also diminish gas transport rates in the cathode catalyst layer. The latter mechanism is often observed in localized areas in full-size cells due to non-uniformities in fuel or current distribution during start or normal operation [31].

If reactant transport losses are due to flooding, it is sometimes possible to reverse the loss, at least temporarily, by removing excess water. Many methods for doing so are available. Water may be removed electrochemically (as described in section 4.2.2.2), by operating the cell at high temperature to increase the vapor pressure of the trapped water, operating the cell with drier reactants, or drying the cell out while it is not operating. Whether or not the improvement obtained by these procedures is stable, however, depends on the cause of the flooding. If the flooding is due to an irreversible mechanism such as a change in wettability or pore structure of the catalyst layer, it is likely to return shortly after the cell is brought back to normal operating conditions. If the flooding is due to an excursion to an abnormal operating condition, such as a cold start, it is often substantially reversible and the recovery in performance is persistent [32].

4.3.2. *In-cell Diagnostics*

A good first step when analyzing transport losses is to use the Tafel plot described in Section 4.1.2.1 and estimate the current density where the performance deviates from the kinetic Tafel slope by a specified amount. Because the overpotential is corrected for resistance (i.e. is *iR-free*) and anode polarization is negligible or may be accounted for, this deviation indicates where the overpotential begins to be controlled by transport. Therefore, tracking the current density where this deviation occurs as a function of time gives an indication of whether transport losses are changing over the life of the cell. An alternative method is to note where the current density on the polarization-change curve described in Section 3 deviates from a horizontal line by a specified amount since this indicates where either transport or ohmic changes are becoming significant. The shape of the polarization-change curve, as well as the tests described in this section and Section 4.2, can then help determine the causes of these changes beyond the losses in catalytic activity.

4.3.2.1. Reactant Concentration

The response in voltage to changes in reactant concentrations in either the fuel or air stream is measured. Temperature, pressure, inlet humidification, utilizations or flows, and concentrations on the other electrode are held constant. As an example, the performance on hydrogen and air may be compared to the performance on hydrogen and oxygen. Maintaining the same flow at the different concentrations is desirable, because this makes membrane hydration levels more consistent. This can be difficult to achieve with dilute reactants, like 4% oxygen in nitrogen, because the corresponding utilization on air would be very high. Other strategies to maintain consistent hydration levels, like adjusting temperature while maintaining the same utilization, may be more appropriate. Comparing inlet concentrations that differ by smaller factors like two, instead of the factor of five between air and oxygen, can be helpful from this perspective, as well for other reasons mentioned below.

At low current densities the difference in voltage will primarily be due to the kinetic overpotential for oxygen reduction, which is usually assumed to be first order in oxygen concentration with a transfer coefficient of unity [26]. At high current densities, the difference will also include the overpotential due to mass-transport resistance [17]. Tracking sensitivity to reactant concentration as a function of time, therefore, gives an indication of whether mass-transfer resistance is changing over the life of the cell.

However, as noted above, the major transport resistance within the catalyst layer may be either ohmic or diffusive [33]. For the case of the cathode catalyst layer, proper analysis of Tafel plots taken at different oxygen concentrations can help to distinguish between these possibilities. This is done by determining the reaction order with respect to oxygen concentration within the double-Tafel region. For an electrode dominated by oxygen diffusion limitations, first-order dependence is expected, while if ohmic limitations dominate, half-order dependence is expected [34]. A summary of these different limiting cases is provided in Table 7.2. It is not uncommon for both transport limitations to be significant at high current densities and, in this case, the reaction order with respect to oxygen concentration should be somewhere between one-half and unity. Transport resistance in the catalyst layer may also be significant on length scales different from the thickness of the catalyst layer. Important length scales include the thickness of the ionomer film and agglomerate diameter. Reference [5] reviews effects at different length scales.

One should measure performance on oxygen concentrations that are both higher and lower than the oxygen concentration of interest, in order to determine the reaction order with respect to oxygen concentration at the condition of interest. Most PEFC cathodes designed to operate on air are ionically limited on pure oxygen and are limited by oxygen delivery at low oxygen concentrations (e.g. 4% O_2), so one needs to measure performance at conditions closer to the oxygen concentration of interest to gain insight into what is dominating the cathode performance. For example, to diagnose what is occurring on air, a good procedure is to measure the performance on 'half air' (i.e. 10.5% O_2) and 'double air' (i.e. 42% O_2) in addition to recording the performance on air.

TABLE 7.2 Characteristics of Important Limiting Cases of Cathode Overpotential

Major Sources of Cathode Overpotential	Tafel Slope	Oxygen Reaction Order	Oxygen Gain**
Kinetic only	Single	1	Normal
Kinetic and Diffusion* (oxygen transport)	Double	1	Double
Kinetic and Ohmic (ionic transport)	Double	1/2	Normal

** Through the thickness of the electrode.*
*** Voltage on oxygen minus voltage on air.*

4.3.2.2. Electrochemical Impedance Study

As mentioned in Section 4.1.2.3, one can obtain insight into sources of polarization by combining physics-based models and EIS. Although this can become quite complex, multiple diagnostics can be combined to help separate different sources of polarization. For example, EIS can be conducted at different oxygen concentrations on the cathode to help resolve ohmic losses in the catalyst layer from gas-transport losses throughout the cathode [20]. Alternatively, one can use nitrogen on the cathode to simplify the analysis, since this eliminates faradic reactions throughout most of the cathode [36]. One can also combine these two approaches [37]. In any case, these are all indirect measurements and require models with certain assumptions to interpret the results. The simplest approach to analyzing EIS data is to use equivalent-circuit analysis, but one must be cautious since a good fit to an equivalent-circuit model does not necessarily mean it is correct. However, for the purpose of identifying what qualitative changes have occurred in a cell, this type of relatively simple analysis can be helpful, especially if it is combined with some of the other diagnostics discussed here. Therefore, analyzing EIS data as a function of run time can provide insight into the type of changes that are occurring over the life of the cell.

4.3.2.3. Reactant Utilization

The voltage is measured at a given current density as the utilization of the reactant of interest is varied. Again, temperature, pressure, humidification, and utilization of the other reactant are constant. The utilization is defined as the ratio of moles consumed divided by moles fed, and is the inverse of the stoichiometric ratio. As the utilization increases, the concentration of the reactant in the gas channels decreases, and therefore the driving force for transport of reactants to the electrode decreases. When humidity is high, voltage decreases sharply as utilization and, therefore, reactant transport limitations increase. When humidity is low, the voltage may show a maximum as a function of utilization that corresponds to the best balance between water and reactant transport.

The difference in voltage between low and high utilizations depends on flow patterns and transport characteristics. A cell that has a high mass-transfer resistance will be more sensitive to reactant utilization than a cell with a low mass-transfer resistance. Tracking sensitivity to reactant utilization as a function of run time, therefore, gives an indication of whether mass-transfer resistance is changing over the life of the cell. The utilization response depends on flow field design, since the degree of mixing and flow uniformity are important; therefore, this type of testing is typically only of interest on large cells and stacks.

It should be noted that using this test on the anode can be problematic due to the risk of irreversible damage due to fuel starvation [38,39]. Typically it is not advisable to operate the cell at very high fuel utilizations, but this in turn means that there may only be a few millivolts of difference between the low-utilization and high-utilization points for a hydrogen utilization sweep. Consequently, changes in the sensitivity to utilization may not be definitively observed for some time. One way to overcome this issue is to increase the magnitude of the signal relative to noise through the use of dilute hydrogen.

Generally it is easier to interpret results obtained with varying reactant concentrations than results obtained with varying reactant utilizations for the reasons given in Section 2. However, reactant utilization sweeps can be very useful to detect changes in reactant distribution or local damage to the cell. One can compare reactant utilization results with those obtained at low utilizations and a reactant concentration that is close to the average concentration of the former to look for unusual trends. If local cell damage is suspected, then one can compare results obtained at high utilizations with the reactant fed in opposite directions through the cell (from normal outlet to normal inlet).

4.3.2.4. Reactant Diluent Substitution

The voltage as a function of current density when a given reactant diluent is used is compared with the voltage when a different diluent is used. Temperature, operating pressure, inlet humidification, reactant concentrations, reactant utilizations, and the type diluent used for the other reactant (if any) are held constant. As an example, the performance of a cell operating on hydrogen and air may be compared to that of the cell operating on hydrogen and a mix of oxygen and helium (commonly referred to as 'helox') [26].

The purpose of using a different diluent is to change the gas-phase diffusivity of the reactant. Continuing with the helox example, if the bulk of the cathode mass-transfer resistance is in the gas phase, the performance at high current densities should be significantly better using helox because the diffusion coefficient of oxygen in helium is significantly higher than the diffusion coefficient of oxygen in nitrogen. On the other hand, if diffusion of oxygen through liquid water or ionomer is more important than diffusion in the gas phase then the performances on air and helox will be similar [16]. Tracking sensitivity to the type of diluent used as a function of run time, therefore, is a way to measure changes in the amount of flooding over the life of the cell.

4.3.2.5. Pressure Drop

The pressure drop across the cell is measured as a function of current density or reactant utilization. Temperature, pressure, inlet humidification, and inlet compositions are held constant.

Pressure drop at a given operating condition depends upon the structure of the cell, including the dimensions of the channels, the thickness and porosity of the diffusion media, and the degree of inter-digitation [40]. Pressure drop also depends on the amount of liquid water present in the channels and diffusion media, which depends on physical and chemical properties. It is possible for these properties to change with time, which in turn changes the amount of liquid water present at a given operating condition [16]. Tracking pressure drop as a function of run time, therefore, is a way to detect changes in the amount of liquid water in the cell over time. If the pressure drop does increase with time, it can often be reduced by briefly increasing the flow to a rate sufficient to remove water that has collected. Whether or not the improvement is stable, however, depends on the cause of the flooding.

4.4. Leaks

Performance loss that primarily affects the voltage at low current density may indicate leakage, either of gas crossing through the membrane or as current due to an electrical short from anode to cathode. Table 7.3 corresponds to this case. It is generally possible to distinguish between electrical shorts and reactant crossover.

Neither overboard leakage nor gas-to-coolant leakage are considered here. These types of leaks are generally straightforward to monitor over time, using periodic pressure tests, for example.

4.4.1. Common Mechanisms

Over the course of a durability test, chemical attack may compromise the integrity of the membrane, leading to pinholes through which gases can move [41]. Seal degradation may also lead to gas leakage at the edges of the cell.

Electrical shorts may occur when chemical attack leads to membrane thinning or if fibers puncture the membrane. Degradation of perimeter seals may lead to electrical shorts at edges.

4.4.2. In-cell Diagnostics

4.4.2.1. Electrochemical Leakage Test

Pure hydrogen flows on the counter electrode, which is generally the normal fuel electrode, while nitrogen flows on the working electrode. The counter

TABLE 7.3 Summary of Recommended In-cell Diagnostics for Cells with Leaks

Type of Loss Observed	Some Possible Contributors and Typical Diagnostic Tests Membrane or Seals
Leakage (electrical or reactant) Performance loss primarily at low current densities	**Possible causes** – Short circuit through or around membrane – Reactant leakage due to membrane or seal failure **Diagnostics** – Electrochemical shorting/crossover test – OCV decay time during N_2 purge of cathode – Open circuit voltage response to reactant pressure

electrode generally serves as the reference electrode for convenience. The current density is measured as a function of applied voltage between zero and roughly 0.5 V. This is often called driven-cell mode. Temperature, pressure, and inlet humidification are held constant.

In this configuration, hydrogen oxidation occurs at the working electrode, which is the normal air electrode, while hydrogen evolution occurs at the counter electrode. The hydrogen oxidation rate is limited by the rate of molecular hydrogen transport through the membrane. As the voltage is increased, the current density increases rapidly and then, once the limiting current is reached, increases linearly due to the electrical shorts in the cell. The limiting current is thus a measurement of the crossover current while the slope of the linear region is a measurement of the electrical resistance of the cell. Using the electrochemical leakage test as a function of run time, therefore, is a way to measure changes in both the amount of gas leakage as well as the magnitude of internal short circuits.

To measure electrical shorts by themselves, the same arrangement and procedure may be used, but with nitrogen flowing on both sides of the cell [42].

4.4.2.2. Open Circuit Leakage Tests

In the open circuit voltage (OCV) decay test, the cell is first brought to open circuit using typical reactant flow rates. Temperature, operating pressure, inlet humidification, and inlet reactant concentrations are held constant. The gas on the cathode side is then quickly switched from air to nitrogen, or is simply turned off, and the time required for the cell voltage to drop to a certain defined level (for example, 100 mV) is measured.

Voltage decay time is a strong function of both the amount of reactant leaking and the magnitude of internal short circuits present in the cell. Tracking the amount of time that it takes for the OCV to decay as a function of run time is a way to measure changes in the amount of gas leakage and internal shorting. However, repeatability can be poor due to the sensitivity of the decay time to factors such as hydration, the timing and sequence of gas switching, and reactant pressures.

Another problem with the OCV decay test is that it cannot distinguish between gas leakage and electrical shorting. To separate the two, a separate test, called an OCV pressure sweep, may be performed. Again, the cell is first brought to open circuit using typical reactant flow rates. Temperature, inlet humidification, and inlet reactant concentrations are held constant. The OCV is then measured at different fuel pressures. The air pressure is held constant. Because gas leakage between anode and cathode is a strong function of the pressure difference between the two sides while short-circuit current is not, if the OCV increases significantly when the fuel pressure is increased, a gas leak is present. Tracking the sensitivity of OCV to pressure difference as a function of run time, therefore, is a way to measure changes in the amount of gas leakage over the life of the cell.

5. ACCELERATED TEST PROTOCOLS

The purpose of this section is to discuss accelerated testing of PEFCs in a general way. A detailed analysis of particular accelerated stress tests (ASTs), like those advocated by the US Department of Energy (DOE), is not the objective; nor is a review of the results of such tests. Recent articles of this nature include those of Zhang et al. [43]. Rather, the goals, strengths, and weaknesses of different approaches to accelerated testing are discussed in the context of fuel-cell development. The principles of accelerated testing are discussed in numerous textbooks on reliability [44,45,46].

Accelerated stress testing may be regarded as one method for estimating reliability. Common approaches to estimating the reliability of a product include using predictions based on information in data bases, the assessment of historical or demonstration data, and laboratory simulation of normal operation. The first method probably does not apply to PEFCs at their current stage of development as dramatic differences in degradation rates (from 1 to 210 μV/h) have been reported in the literature [47,48]. Progress has been made in understanding the physical phenomena that cause decay and in ranking their contributions, but quantitative predictive capability has not yet been achieved [1]. Data from previously fielded units is probably the best predictor of performance. Field data can also be used to establish the suitability and acceleration factors of ASTs. Unfortunately, accumulating field data can be expensive and time consuming. DOE's operating time requirements are 5,000 h for transportation and 40,000 h for stationary applications [1,49]. Laboratory simulation of normal operation is probably the most common method for estimating reliability before fielding PEFC-based products. Units can be monitored closely with frequent diagnostic tests, and they can be removed from test for examination without affecting customer satisfaction. However, this approach is time consuming, which can slow product development.

Laboratory tests should include all conditions that may arise during normal operation, while avoiding events that should not occur in the field. Fortunately, in many respects, the decay characteristics of a PEFC stack are independent of the number and active area of the cells. Thus, it is often possible to reduce costs by testing scaled versions of the final product. For example, with representative test protocols, a stack with 20 cells may be an adequate representation of a power plant with 600 cells. Tests on full-scale articles may be required, when transient effects, like starting and stopping, are important [50]. Accumulating meaningful statistics may also be particularly important for some degradation mechanisms, such as membrane damage.

Good ASTs should reduce the time and cost required to develop new products and improve existing products, while also reducing technical risks to acceptable levels. The tests should provide physical insight about the nature of the failure modes likely to occur during operation, and should not cause failures for reasons that would not occur in the field. The tests must be carefully crafted to achieve these goals bearing in mind the application, because different stressors are pertinent to transportation, stationary, and auxiliary applications. Generally speaking, degradation can be accelerated by increasing stress levels above those experienced in the field, or by increasing the frequency of stress application (especially for units that normally operate intermittently) [40]. The first method is of primary interest for PEFCs in transportation and stationary applications, the second method is of interest for auxiliary applications.

ASTs can be categorized as qualitative or quantitative. Qualitative ASTs are useful for screening and ranking new materials; quantitative ASTs can be used to accurately predict the reliability of a product. Large acceleration factors are more appropriate for qualitative tests, because extrapolation over a very large range is likely to lead to inaccuracy. Performance decay related to changes in oxygen transport capability, for example, usually increases with time in a very non-linear way [51]. More than one stress can be applied in an AST. However, physical insight is more difficult to obtain from tests with more than one acceleration factor.

Table 7.4 categorizes conditions that are known to damage PEFCs. A detailed discussion of these mechanisms is beyond the scope of this chapter. Explanation of degradation mechanisms, including known accelerating factors, is covered in other chapters of this book and other texts [1–3]. The extent to which a particular PEFC is exposed to any condition depends upon the application and the design of the system. For example, the magnitude and frequency of load changes imposed upon a fuel cell in an automobile depends on how it is hybridized with energy storage devices. Knowing the levels of stress present during normal operation is critical to constructing an effective testing plan. Similarly, it is important to know the bounds on stressor values. For example, using temperatures above the glass-transition point of a membrane may lead to erroneous conclusions.

Two methods for applying stresses to test articles are considered below. The first method is an accelerated mission; the second method is a series of tests designed to isolate and accelerate individual decay processes. While these two approaches are discussed as alternatives for the sake of exposition, a development program will probably include both types of testing. Redundant testing is a logical risk reduction technique.

Using an accelerated mission that exposes the design, or at least the major components, to all of the conditions expected during operation – such as starting and stopping, load cycling, freezing and thawing, and thermal cycling – is an attractive approach, because it is expected to involve a relatively small amount of testing. Increasing the severity of the conditions to accelerate degradation relative to the field by a known amount may be tricky, however, especially when more than one degradation phenomenon is important. Accelerated mission testing approaches the laboratory simulation technique discussed above if the acceleration factors approach unity. While time savings may be minimal, significant cost savings may be possible by using a scaled test

TABLE 7.4 Summary of Major PEFC Stressors and Decay Mechanisms

STRESS		DECAY MODE				
Type	**Magnitude**	**Carbon Corrosion**	**Platinum Dissolution**	**Membrane Damage***	**Structural Damage****	**Activity Loss*****
Potential	High	X	X			
	Low					
	Cycles	X	X	X		
Humidity	High	X	X			
	Low			X		
	Cycles			X		
Load	High					
	Low	X				
	Cycles	X	X	X		
Temperature	High	X	X	X		
	Low (freezing)				X	
	Cycles				X	
Contaminants	High			X		X

** Mechanical and chemical degradation.*
*** Physical changes, especially to the electrode layers.*
****Activity losses beyond those due to platinum dissolution.*

article. A problem with mission testing is that it can be difficult to explain the failure of a test article exposed to many conditions, making it difficult to implement corrective actions. Another problem is that the results are less meaningful to the wider technical community, because different developers are likely to implement different mission tests.

Another approach is to use a series of tests to accelerate and isolate different decay modes. The ASTs given by the DOE exemplify this approach [52], and these ASTs cover many of the stressors listed in Table 7.4. This approach is expected to require more test articles than a mission approach, but the information provided by the tests should be easier to interpret and will simplify the assessment of mitigation approaches. Ideally, the damage induced by the ASTs should be linked to damage observed in the field by known factors. If this information is not available, it should at least be possible to use the ASTs to screen new components. Two potential issues associated with applying this approach to qualify components are: (1) all sources of decay must be identified and addressed with individual tests, and (2) synergistic effects that may occur in actual operation may not be apparent. Meaningful statistics should be collected for ASTs that are highly accelerated, either by running replicate single cells or multi-cell stacks. The results of an AST may depend upon the active area and details of the flow field and hardware designs and should be extrapolated to different designs cautiously.

One example which shows the difficulty of isolating decay modes involves the degradation of platinum and carbon. Corrosion of carbon generally increases with increasing temperature, voltage, and concentration of water. Platinum catalyst also tends to increase the corrosion rate. The lower voltage limit and cycling profile and frequency are important during potential cycling experiments. Platinum dissolution and carbon corrosion tend to be accelerated by the same variables, but with different sensitivities. Thus, constructing a test to accelerate only one of these corrosion rates is difficult. For conventional Pt/C electrodes, frequent cycling to potentials below 1V tends to emphasize degradation of the metal, while cycling to higher potentials tends to emphasize carbon corrosion.

Another example involves PtCo. PtCo loses less surface area during potential cycling than Pt, and PtCo reduces fluoride loss from the membrane [53]. This example shows that any change in materials should be vetted against all failure modes in order to capture unexpected effects. This comes at the cost of significant additional testing. An example of a synergistic effect that might escape a test plan designed to isolate individual mechanisms involves the influence of platinum in the membrane on membrane degradation. It has been reported that the rate of membrane degradation is increased by the platinum band deposited in the membrane during cycling [54]. The importance of this effect may not be adequately captured in steady-state tests of membrane durability. This particular example is more complicated, as it has been reported that Pt particles deposited within the membrane suppress chemical degradation [55]. The addition of Pt has long been claimed to be an effective mitigation strategy for membrane degradation [56]. Clearly, one must be careful in drawing conclusions based on limited testing, especially when multiple decay mechanisms are occurring simultaneously.

As is evident in Table 7.4, the stressors that are commonly employed to create ASTs are the same factors that one should strive to minimize in fielded products in order to maximize PEFC life [57]. In essence, a key goal of a PEFC developer is to minimize exposure to stressful conditions in fielded products as much as possible. If this is done, then creating ASTs is also easier since one does not have to invoke extreme conditions to enable sufficient acceleration factors. Both the development of PEFCs with long life and the development of effective ASTs require an excellent understanding of all of the possible PEFC decay mechanisms, including the stressors for each of these mechanisms. Many of the conditions that degrade PEFCs have now been discussed in the literature, making it possible to construct qualitative accelerated tests for individual mechanisms with an increasing degree of confidence.

Development of fuel cell products relies on information from all sources. Field trials are extremely valuable as they subject cells to intended and unintended operating conditions. Unintended conditions may arise, for example, if a strategy to mitigate a particular type of decay results in the acceleration of another type of degradation. Field trials help to inform accelerated test protocols in two ways: (1) they identify the actual operating conditions, and (2) they identify why units actually fail. Therefore, regular correlation between field results and laboratory results is essential in the evolution of AST protocols.

NOMENCLATURE AND ABBREVIATIONS

a	specific interfacial area, cm^{-1}
AST	accelerated stress test
b	Tafel slope, mV/dec
BOL	beginning of life
CL	catalyst layer
CV	cyclic voltammogram
DOE	US Department of Energy
ECSA	electrochemical surface area
EIS	electrochemical impedance spectroscopy
EOL	end of life
F	Faraday's constant, 96,487 C/equiv
GDL	gas diffusion layer
HFR	high-frequency resistance
HOR	hydrogen-oxidation reaction
h	hours
i	superficial current density, A/cm^2
i_o	exchange current density for the oxygen-reduction reaction, A/cm^2
i_{lim}	limiting current density, A/cm^2
OCV	open circuit voltage, V
ORR	oxygen-reduction reaction

p	partial pressure of reactant
PEFC	polymer electrolyte fuel cell
U^{θ}	standard cell potential, V
R	universal gas constant, 8.3143 J/mol K
R'	total ohmic resistance, ohm cm^2
S	saturation
T	temperature, °C or K
V	voltage, V

SUBSCRIPTS

b	beginning of life
H	hydrogen
O	oxygen
x	crossing through the membrane

GREEK

α_c	charge transfer coefficient for the oxygen-reduction reaction
δ_{CL}	thickness of catalyst layer, cm

REFERENCES

[1] R. Borup, J. Meyers, B. Pivovar, Y.S. Kim, R. Mukundan, N. Garland, D. Myers, M. Wilson, F. Garzon, D. Wood, P. Zelenay, K. More, K. Stroh, T. Zawodzinski, J. Boncella, J.E. McGrath, M. Inaba, K. Miyatake, M. Hori, K. Ota, Z. Ogumi, S. Miyata, A. Nishikata, Z. Siroma, Y. Uchimoto, K. Yasuda, K. Kimijima, N. Iwashita, Scientific Aspects of Polymer Electrolyte Fuel Cell Durability and Degradation, Chem. Rev. 107 (2007) 3904.

[2] F. Buchi, M. Inaba, T. Schmidt (Eds), Fuel Cell Durability, Springer, New York, 2009.

[3] W. Vielstich, H. Gasteiger, A. Lamm (Eds), Handbook of Fuel Cells Fundamentals, Technology, and Applications, Vol. 5, John-Wiley & Sons, Hoboken NJ, 2009.

[4] M. Mench, Fuel Cell Engines, John Wiley & Sons, Hoboken NJ, 2008.

[5] K.C. Neyerlin, W. Gu, J. Jorne, H.A. Gasteiger, Determination of Catalyst Unique Parameters for the Oxygen Reduction Reaction in a PEMFC, Journal of Electrochemical Society 153 (2006) A1955.

[6] K.C. Neyerlin, W. Gu, J. Jorne, H.A. Gasteiger, Effect of Relative Humidity on Oxygen Reduction Kinetics in a PEMFC, Journal of Electrochemical Society 152 (2005) A1073.

[7] A.Z. Weber, R.M. Darling, J. Newman, Modeling Two-Phase Behavior in PEFCs, Journal of Electrochemical Society 151 (2004) A1715.

[8] A.Z. Weber, J. Newman, Modeling Transport in Polymer-Electrolyte Fuel Cells, Chemical Reviews 104 (2004) 4679.

[9] J. Giner, C. Hunter, The Mechanism of Operation of the Teflon-Bonded Gas Diffusion Electrode: A Mathematical Model, Journal of Electrochemical Society 116 (1969) 1124.

[10] M. Mench, Fuel Cell Engines, John Wiley & Sons, Inc, 2008. 468.

[11] T.W. Patterson (Ed), in Fuel Cell Technology Topical Conference Proceedings, AICHE Spring National Meeting (2002), 2002.

[12] B. Wu, G.E. Gray, T.F. Fuller, PEM fuel cell Pt/C dissolution and deposition in Nafion electrolyte, Electrochemical and Solid-State Letters 10 (2007) B101.

[13] X. Cheng, Z. Shi, N. Glass, L. Zhang, J.J. Zhang, D.T. Song, Z.S. Liu, H.J. Wang, J. Shen, A review of PEM hydrogen fuel cell contamination: Impacts, mechanisms, and mitigation, Journal of Power Sources 165 (2007) 739.

[14] A. Damjanovic, P.G. Hudson, On the kinetics and mechanism of O_2 reduction at oxide film covered Pt Electrodes. 1. Effect of oxide film thickness on kinetics, Journal of the Electrochemical Society 135 (1988) 2269.

[15] T.W. Patterson, M.L. Perry, T. Skiba, P. Yu, T.D. Jarvi, J.A. Leistra, H. Chizawa, T. Aoki, Decontamination procedure for a fuel cell power plant, US Patent 7 (2008) 442. 453.

[16] I.G. Urdampilleta, F.A. Uribe, T. Rockward, E.L. Brosha, B.S. Pivovar, F. Garzon, PEMFC poisioning with H_2S: Dependence on operating conditions, in: T. Fuller (Ed), Proton Exchange Membrane Fuel Cells, 7, 2007, p. 383.

[17] T. Jarvi, T. Patterson, N. Cipollini, J. Hertzberg, M. Perry, Recoverable performance losses in PEM fuel cells, Electrochemical Society Meeting Abstracts, Paris, 2003.

[18] R.M. Darling, J.P. Meyers, Mathematical model of platinum movement in PEM fuel cells, Journal of the Electrochemical Society 152 (2005) A242.

[19] F. Barbir, PEM Fuel Cells: Theory and Practice, Elsevier, 2005.

[20] J. Newman, K.E. Thomas-Alyea, Electrochemical Systems, John Wiley & Sons, New York, 2004.

[21] A.J. Bard, L.R. Faulkner, Electrochemical Methods: Fundamentals and Applications, John Wiley & Sons, New York, 2001.

[22] R. O'Hayre, S.-W. Cha, W. Colella, F. Prinz, Fuel Cell Fundamentals, John Wiley & Sons, New York, 2006.

[23] T.E. Springer, T.A. Zawodzinski, M.S. Wilson, S. Gottesfeld, Characterization of Polymer Electrolyte Fuel Cells Using AC Impedance Spectroscopy, Journal of the Electrochemical Society 143 (1996) 587.

[24] R.M. Darling, J.P. Meyers, Mathematical model of platinum movement in PEM fuel cells, Journal of the Electrochemical Society 152 (2005) A242.

[25] A.Z. Weber, R.M. Darling, Understanding porous water-transport plates in polymer-electrolyte fuel cells, J. of Power Sources 168 (2007) 191.

[26] F. Garzon, 2008 Annual DOE Fuel Cell Program Review (2008). Project ID FC30.

[27] P. Zhou, C.W. Wu, G.J. Ma, Influence of clamping force on the performance of PEMFCs, Journal of Power Sources 163 (2007) 874.

[28] J.M. Sedlak, J.F. Austin, A.B. LaConti, Hydrogen recovery and purification using the solid polymer electrolyte electrolysis cell, International Journal of Hydrogen Energy 6 (1981) 45.

[29] S. Srinivasan, Fuel Cells: From Fundamentals to Applications, Springer, New York, 2006.

[30] P.L. Hagans, G. Resnick, Performance recovery process for PEM fuel cells, US Patent 6 (2004) 709. 777.

[31] C.A. Reiser, L. Bregoli, T.W. Patterson, J.S. Yi, J.D. Yang, M.L. Perry, T.D. Jarvi, A reverse-current decay mechanism for fuel cells, Electrochemical and Solid-State Letters 8 (2005) A273.

[32] C. Chacko, R. Ramasamy, S. Kim, M. Khandelwal, M. Mench, Characteristic behavior of polymer electrolyte fuel cell resistance during cold start, Journal of the Electrochemical Society 155 (2008) B1145.

[33] J. Newman, W. Tiedemann, Porous-electrode theory with battery applications, AICHE Journal 21 (1975) 25.

[34] M.L. Perry, J. Newman, E.J. Cairns, Mass transport in gas-diffusion electrodes: A diagnostic tool for fuel-cell cathodes, Journal of the Electrochemical Society 145 (1998) 5.

[35] M.H. Eikerling, K. Malek, Q. Wang, Catalyst Layer Modeling: Structure, Properties, and Performance, in: J. Zhang (Ed), PEM Fuel Cell Electrocatalysts and Catalyst Layers: Fundamentals and Applications, Springer, New York, 2008.

[36] M.C. Lefebvre, R.B. Martin, P.G. Pickup, Characterization of ionic conductivity profiles within photon exchange membrane fuel cell gas diffusion electrodes by impedance spectroscopy, Electrochemical and Solid-State Letters 2 (6) (1999) 259–261.

[37] Y. Liu, M.W. Murphy, D.R. Baker, W. Gu, C. Ji, J. Jorne, H.A. Gasteiger, Proton conduction and oxygen reduction kinetics in PEM fuel cell cathodes: effects of ionomer-to-carbon ratio and relative humidity, Journal of the Electrochemical Society 156 (2009) B970.

[38] T.W. Patterson, R.M. Darling, Damage to the cathode catalyst of a PEM fuel cell caused by localized fuel starvation, Electrochemical and Solid State Letters 9 (2006) A183.

[39] N. Takeuchi, T.F. Fuller, Modeling and Investigation of Design Factors and Their Impact on Carbon Corrosion of PEMFC Electrodes, Journal of The Electrochemical Society 155 (2008) 770.

[40] J.S. Yi, T.V. Nguyen, Multi-Component Transport in the Porous Electrodes of PEM Fuel Cells with Interdigitated Gas Distributors, J. Electrochem. Society 146 (1999) 38.

[41] W. Vielstich, A. Lamm, H.A. Gasteigher, Handbook of Fuel Cells, John Wiley & Sons, West Sussex, 2003.

[42] E.L. Thompson, J. Jorne, H.A. Gasteiger, Oxygen Reduction Reaction Kinetics in Subfreezing PEM Fuel Cells, Journal of the Electrochemical Society 154 (2007) B783.

[43] S. Zhang, X. Yuan, H. Wang, W. Merida, H. Zhu, J. Shen, S. Wu, J. Zhang, A review of accelerated stress tests of MEA durability in PEM fuel cells, International Journal of Hydrogen Energy 34 (2009) 388.

[44] C.E. Ebeling, An Introduction to Reliability and Maintainability Engineering, McGraw-Hill, 1997.

[45] W.O. Meeker, L.A. Escobar Statistical, Methods for Reliability Data, John Wiley & Sons, New York, 1998.

[46] P.D.T. O'Connor, Practical Reliability Engineering, Third Ed. John Wiley & Sons, New York, 1991.

[47] F.A. de Bruijn, V.A.T. Dam, G.J.M. Janssen, Review: Durability and Degradation Issues of PEM Fuel Cell Components, Fuel Cells 08 (2008) 3. 1.

[48] J. Wu, X.Z. Yuan, J.J. Martin, H. Wang, J. Zhang, J. Shen, S. Wu, W. Merida, A review of PEM fuel cell durability: Degradation mechanisms and mitigation strategies, Journal of Power Sources 184 (2008) 104.

[49] Hydrogen, Fuel Cells & Infrastructure Technologies Program Multi-Year Research, Development and Demonstration Plan, August (2006) (http://www1.eere.energy.gov/hydrogenandfuelcells/mypp/)

[50] M. Perry, T. Patterson, C. Reiser, System Strategies to mitigate carbon corrosion in fuel cells, ECS Trans. 3 (2006) 783.

[51] M.F. Mathias, R. Makharia, H.A. Gasteiger, J.J. Conley, T.J. Fuller, C.J. Gittleman, S.S. Kocha, D.P. Miller, C.K. Mittelsteadt, T. Xie, S.G. Yan, P.T. Yu, Two fuel cell cars in every garage? The Electrochemical Society Interface, Fall (2005) 24.

[52] Appendix C of DOE Solicitation DE-PS36-08GO98009 (2007).

[53] L. Protsailo, Alternative Materials for PEM Fuel Cell Systems, Fuel Cells 2005 Conference, Asilomar, CA, February 2005.

[54] A. Ohma, S. Suga, S. Yamamoto, K. Shinohara, J. Electrochem. Soc. 154 (2007) B757.

[55] E. Endoh, S. Hommura, S. Terazono, H. Widjaja, J. Anzai, Degradation Mechanism of the PFSA Membrane and Influence of Deposited Pt in the Membrane, ECS Trans. 11 (2007) 1083.

[56] M. Watanabe, Solid Polymer Electrolyte Fuel Cell, US Patent 5 (1995) 472. 799A.

[57] M. Perry, R. Darling, S. Kandoi, T. Patterson, C. Reiser, Operating Requirements for Durable PEFC Stacks, in: F. Buchi, M. Inaba, T. Schmidt (Eds), Fuel Cell Durability, Springer, New York, 2009.

Chapter 8

Advanced High Resolution Characterization Techniques for Degradation Studies in Fuel Cells

Feng-Yuan Zhang,* Suresh G. Advani and Ajay K. Prasad
Center for Fuel Cell Research, Department of Mechanical Engineering, University of Delaware, Newark, DE, USA
Current affiliation: Electrochemical Energy Storage and Conversion, Department of Mechanical, Aerospace and Biomedical Engineering, The University of Tennessee, Knoxville, TN, USA

1. INTRODUCTION

Fuel cells directly convert chemical energy into electrical energy with high efficiency and minimal environmental impact, and are promising energy conversion systems for portable, automotive, and stationary power applications. Although significant progress has been made in the past decades in terms of overall performance, power density, and cost reduction, the performance degradation of fuel cells with use continues to be an important barrier to commercialization [1–12]. Fuel cell degradation, which includes catalyst-particle agglomeration/sintering/oxidation, carbon-support corrosion, membrane thinning and decomposition, loss of hydrophobicity, interfacial delamination, and mechanical fatigue, is caused by many factors – such as the materials used, component fabrication methods, fuel cell design, assembly and integration, operating conditions, and the presence of impurities and contaminants in the system. In addition, water/thermal management plays a critical role in fuel cell degradation and performance. Improper water management during fuel cell operation can lead to either membrane dryout or flooding of the gas diffusion and catalyst layers, which can cause deterioration of the chemical/physical/mechanical properties of fuel cell components, including membranes, catalyst layers and gas diffusion media. For example, if the membrane dries out during operation, its strength will deteriorate, and the ionic transport resistance of the membrane and catalyst layer will significantly increase. Furthermore, flooding will block the transport of fuel/oxidant into

Polymer Electrolyte Fuel Cell Degradation. DOI: 10.1016/B978-0-12-386936-4.10008-9

the reaction sites and lead to local starvation and undesirable flow/temperature/current distributions, which can induce carbon-support corrosion, catalyst oxidation, an increase in size of the pore structure and even cracks within the structure. At the same time, the flooding accelerates a number of fuel cell degradation mechanisms, including catalyst agglomeration/migration/sintering, ionomer erosion/decomposition, loss of hydrophobicity, deterioration of gas diffusion layers, and corrosion of the components. Moreover, during operations at sub-zero temperatures (e.g. cold start and freeze/thaw cycles), flooding can cause ice formation in the membrane, catalyst layer, microporous layer, and/or gas diffusion layer, which leads to significant performance decay and material degradation due to local cyclical volume expansion, fatigue stresses and structural damage.

Understanding the degradation mechanisms in fuel cells is a critical area of study which has motivated many efforts to develop and apply advanced high-resolution techniques to characterize and quantify the state of various chemical species and components before, during and after fuel cell operation. In this chapter, we will discuss several *in-situ* and *ex-situ* diagnostic tools for the study of fuel cell degradation, summarize previous and current efforts, and evaluate the potential of each technique to study the degradation of fuel cell materials and components. These include: optical visualization, neutron imaging, electron spectroscopy and microscopy, X-ray techniques, magnetic resonance imaging, and thermal mapping. In addition to the high-resolution characterization techniques reviewed in this chapter, there exists a separate class of electrochemical techniques for evaluating degradation; however, we have not reviewed the latter in order to restrict the scope and length of the chapter.

This chapter is organized as follows. Next in Section 2, we discuss the optical visualization technique which has been widely used to investigate water dynamics in flow channels, diffusion media and catalyst layers. In Section 3, we present a review of thermal neutron imaging which has the ability to quantify water content during fuel cell operation. Section 4 discusses the utility of magnetic resonance imaging (MRI) for studying fuel cell degradation. In Section 5, we summarize how various electron spectroscopy and microscopy techniques have been used for *ex-situ* characterization of material degradation in the gas diffusion layer (GDL) and the catalyst-coated membrane (CCM). In Section 6, the work of various researchers who have used X-ray techniques to probe the state of materials during or after fuel cell operation is presented. Section 7 discusses the utility of thermal mapping as a diagnostic tool for studying fuel cell degradation.

2. OPTICAL VISUALIZATION

It has long been recognized that efficient water management is strongly correlated to the durability of polymer electrolyte membrane fuel cells (PEMFC) [2,6,7,9,11,13,14]. The cathode side of the PEMFC is prone to liquid water accumulation due to electro-osmotic water transport across the membrane and water production from the oxygen reduction reaction (ORR). Efficient removal of excess water is required, not only to promote adequate transport of fuel and oxidant to the electrodes but also to minimize PEMFC degradation. If present, liquid water can cover the catalyst sites and greatly increase the transport resistance of reactant gases across the thickness of the catalyst layer, a phenomenon referred to as 'catalyst layer flooding' [15–21]. Liquid water may also saturate the pores of the gas diffusion layer (GDL), thus blocking the oxygen transport into the catalyst layer (CL), and causing 'GDL flooding' [22–37]. If excess liquid water accumulates in the flow channel, a water lens or band may form inside the channel, thereby clogging and shutting down the oxidizer flow, a situation known as 'channel flooding' [25,33,38–44]. Flooding severely limits the power output of the fuel cell by starving it of reactant gases, which causes undesirable flow/temperature/current distributions and accelerates the carbon-support corrosion, catalyst oxidation, loss of hydrophobicity, increase in size of the pore structure and even cracks within the structure. Hence, flooding is clearly detrimental to fuel cell performance. Thus, more effective water management will not only provide important performance gains, but will also enable significant lifetime enhancement by minimizing the associated degradation mechanisms. Hence there is a need to understand water dynamics during fuel cell operation in order to help develop effective water management strategies. One commonly used approach to investigate water transport is optical visualization in specially designed transparent operational fuel cells, wherein a transparent window is typically used to replace the conventional metallic backing plate. This technique has been employed extensively to visualize liquid water transport in flow channels [25,33,42–79], diffusion media [25,31,33,34,42,51,52,61,71,78,80–90], including GDLs and microporous layers (MPLs), and catalyst layers [19,91–93], which we summarize below.

2.1. Flow Channels

Flow channel design and material properties can influence water management effectiveness in fuel cells. A good design can promote optimal conditions for the membrane, GDL, and catalyst layer and slow down their degradation. The objective is to keep the membrane hydrated, ensure the reactants can access the catalyst sites, and remove products efficiently in a manner that precludes flooding.

Liquid water transport has been investigated in the flow channels of fuel cells using optical visualization. Tuber et al. [45] visualized water transport in the cathode channels of a transparent fuel cell operating at low current densities and room temperature. They found that the surface property of carbon papers directly influences the accumulation of liquid water in the gas channels of the cathode. For operating temperatures of around 30°C, hydrophilic diffusion layers turned out to be very effective with respect to fuel cell performance, and led to less liquid water in the flow channel and reduced performance degradation. Yang et al. [25] visualized liquid water transport under automotive conditions of 0.82 A/cm^2, 70°C, and 2 atm. The authors observed that a liquid film formed on hydrophilic channel walls and caused channel clogging. As a consequence, electrochemical reaction in the channel was shut down leading to a sharp decline in the performance. Zhang et al. [33] made *in-situ* observations of liquid water distribution on the GDL surface and inside the gas channels to investigate the mechanisms of liquid water removal in a transparent operating PEMFC. They found that liquid water was removed from the gas channel by mist flow at high gas velocities, by a steady corner flow at low gas velocities and low water production, and by an annular film/slug flow at low gas velocities and high production of liquid water. They also found that very hydrophilic channel surfaces promote liquid water removal in fuel cells. Recently, Metz et al. [73] studied passive water management based on capillary effects in microstructures. They found that liquid water could be better removed by hydrophilic channels with a tapered cross-section.

Using a fast digital camera in combination with automated image processing, Stumper and Stone [44] revealed a distinct 'wedge-shaped' pattern of liquid water distribution in the direction perpendicular to the channels, which indicated that most of the reactant gas flowed through the middle channels. These results illustrate the feasibility of using flow visualization for the optimization of flow distribution within a fuel cell under two-phase flow conditions. Ous and Arcoumanis [63] investigated the accumulation of water in the cathode/anode serpentine flow channels of a transparent PEMFC. They confirmed that sufficiently high air stoichiometry was capable of extracting all the water from the cathode channels without causing membrane dehydration. Increasing the operating temperature of the cell was also very effective for the water extraction process. Water was absent in the anode flow channels under their operating conditions with high stoichiometry (> 5) and lower humidity ($< 20\%$).

Anode channel flooding has been visualized in an operating fuel cell with an optical window. Spernjak et al. [42] introduced a microporous layer (MPL) on the cathode side which created a pressure barrier for water produced at the catalyst layer and resulted in anode flooding. Liu et al. [62] studied liquid water distribution and pressure drop characteristics in anode and cathode parallel flow channels and reported much higher liquid water content in the cathode flow channels than in the anode flow channels, because water accumulated on the anode side by diffusing across the membrane from the cathode side due to a high water concentration gradient.

Murahashi et al. [74] studied performance decay and two-phase flow in a fuel cell with a single serpentine flow channel for 4,000 h under degradation conditions of low humidity and high CO concentration. They observed the production of water droplets and measured the transition point of liquid water onset. As the performance decayed with time, the transition point to a two-phase flow moved toward the cathode exit, as shown in Fig. 8.1. The circles in each image indicate

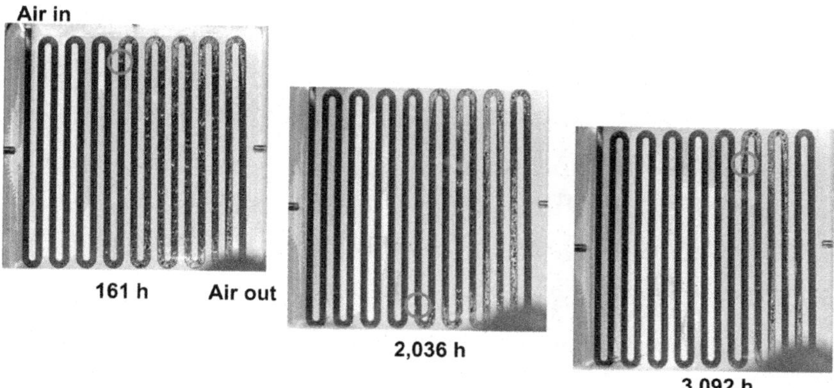

FIGURE 8.1 Photographs of water droplets at 161 h, 2,036 h and at 3,092 h; **O**: the liquid water onset location where a water droplet is first observed in the flow channel. *Reprinted from [74] with permission from Elsevier*

the onset location, where the liquid water is condensed on the gas diffusion layer. They revealed that the degradation of electrodes occurred at the cathode inlet and the anode outlet under low humidity and high CO concentration, which led to a decay in performance. Zhang et al. [75] investigated two-phase flow in horizontal parallel channels. They found that high gas velocities are required to ensure an even flow distribution of gas and liquid in all channels and avoid reactant maldistribution for the two-phase flow regime, which led to flooding or drying in different regions of the active cell area in parallel channels.

2.2. Gas Diffusion Media

Gas diffusion media (GDM), including gas diffusion layers (GDLs) and microporous layers (MPLs), are located between flow channels and catalyst layers in fuel cells. The rates at which the water droplets initiate and grow from the GDL surface are correlated to the degradation of the GDL and its associated components. Yang et al. [25] employed optical visualization to observe that liquid water emerges from the GDL surface in the form of droplets with diameters ranging from a few to hundreds of μm in operating fuel cells. Droplets of water appeared and grew at preferential openings on the GDL surface. Surface tension plays an important role for water droplets on the GDL surface. It was also revealed that water droplets grow intermittently. This result was confirmed by Litster et al. [52] with *ex-situ* visualization of through-plane liquid water transport in the GDL using a fluorescent tracer. They found that water passed slowly through the constrictions of the cross-sections generated by intersecting fibers and expanded rapidly above this cross-section until the fluid interface contacts the next set of fiber intersections. They also revealed that water will preferentially pass through fiber cross-sections featuring the

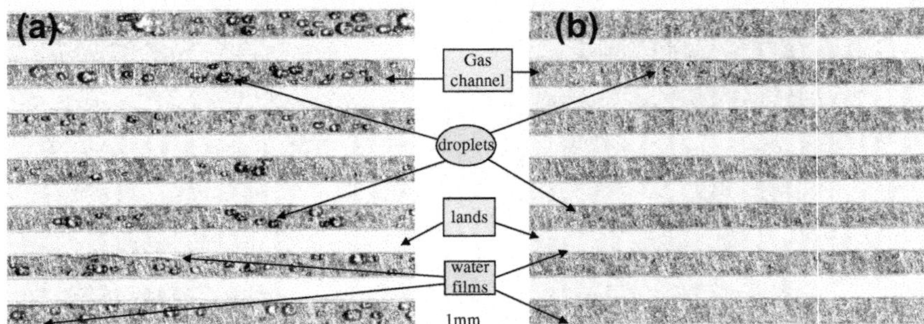

FIGURE 8.2 Liquid water distribution on the GDL at the same section of the test cell after running the fuel cell at 0.8 A/cm^2 for 30 minutes with flow velocities in the channel of (a) 1.43 m/s, and (b) 7.15 m/s. The view of the 7-channel parallel flowfield, close to the outlet, is 14 mm high and 16 mm wide. The flow is from right to left. *Reproduced with permission from [33]. Copyright 2006, The Electrochemical Society*

greatest spacing, as this reduces the capillary pressure resistance. Recently, Gao et al. [84] and Bazylak et al. [81] further studied unstable flow in GDL and observed the change of breakthrough locations and the recession of water pathways upon breakthrough. They found that water flow in PEMFC GDLs was similar to the column flows commonly observed in hydrophobic soils. For better understanding of the transport mechanism of water and droplets in the GDL, Zhang et al. [33] carried out an *in-situ* study with both environmental scanning electron microscopy (ESEM) and optical visualization. The authors revealed that liquid water was driven by capillary forces through the pore pathway to the GDL surface. A critical capillary pressure was needed and was built up due to the continuous water generation in an operating FC. The operating flow rate, GDL hydrophobicity, and capillary force played important roles in the liquid water distribution and droplet emergence and growth on the GDL surface at preferential sites as shown in Fig. 8.2. Most of the pores of the GDL remained free of any liquid due to their hydrophobicity. The liquid water coverage rate on the GDL, as shown in Fig. 8.2(a), was found to be about 11%. Kimball et al. [71] also found that no liquid water flowed through the GDL until a critical hydrostatic pressure head is applied by raising the water reservoir to an appropriate height.

Spernjak et al. [42] examined the effectiveness of various GDL materials and the influence of the MPL on water transport. They found that untreated GDLs with lower hydrophobicity were unable to push the water to the membrane side, which resulted in low ionic conductivity of the membrane. Wet-proofed GDLs managed to expel water in the form of discrete droplets over the entire exposed GDL surface, while leaving the majority of the pores available for gas transport. Adding an MPL increased the pressure barrier, resulting in a well-hydrated membrane. Nam et al. [86] also found that the MPL plays a similar role in water control, and revealed that the MPL reduces the frequency of liquid water breakthrough toward the GDL. They suggested that the MPL should provide a small pore size toward the CL side to reduce the size and saturation level of interfacial water droplets. Nishida et al. [87] studied *in-situ* the effects of GDL thickness and observed fewer droplets on thicker GDLs than on thinner ones. They also revealed that liquid water accumulation inside the GDL increased with its thickness and thus resulted in FC performance reduction and degradation. They confirmed that the thinner GDL provided better water management, which has been one of the key issues for future GDL development [94,95]. Using *ex-situ* visualizations, Bazylak et al. [82] showed that compression caused irreversible damage at the surface of the GDL in the form of fiber and PTFE coating breakage and deformation, resulting in greater proportions of hydrophilic to hydrophobic surface area. They found that compression altered liquid water pathways and favored flow in the compressed areas of the GDL due to both morphological changes and loss of hydrophobicity.

Liquid water removal from the GDL has also been studied with both *ex-situ* and *in-situ* visualization. Droplets are removed from the GDL surface by capillary force, shear drag force, and/or gravity. In most cases, the effect of gravity is negligible due to the very small Bond number. Zhang et al. [33] studied the *in-situ* liquid water removal mechanism from the GDL by introducing the droplet detached diameter (DDD). They observed that when the droplet diameter exceeded the DDD, the droplet was detached from the surface by the drag force followed by a mist flow in the gas channel. When the DDD was larger than the channel dimension, liquid water was only removed by capillary forces and involved interaction with channel walls followed by annular film flow and/or liquid slug flow in the channel. They created a model to correlate the DDD to the mean gas velocity and GDL surface properties in the channel, which could be used in the design of flow channels to reduce flooding and improve performance and material durability. Chen et al. [83] observed *ex-situ* the formation, growth, and removal or instability of water droplets from the GDL of a simulated PEMFC cathode. They found that droplet removal was enhanced by increasing flow channel length or mean gas flow velocity, decreasing channel height or contact angle hysteresis, or making the GDL more hydrophobic. After a water drop detached, Kimball et al. [71] found that it moved freely along the surfaces in the gas flow channel, running into other drops and combining to form liquid slugs.

Kumbur et al. [51] employed simultaneous visualization of both the top and side views of a water droplet to conduct a combined theoretical and experimental *ex-situ* study on the influence of controllable engineering parameters including surface PTFE coverage, channel geometry, droplet chord length and height, and operational air flow rate on liquid droplet deformation and removal at the interface of the GDL and the gas flow channel. They found that the lowest channel height was the most effective for droplet removal in the absence of channel-wall interactions, and higher PTFE loadings of the GDL promoted deformation of the droplet, causing higher contact angle hysteresis. In addition, the influence of PTFE content on contact angle hysteresis was more important in high air flow rate regimes (*Re* $\geq$ 600) than under low air flow conditions. By investigating the detachment of water droplets from porous carbon material surfaces, Theodorakakos et al. [89] found that the droplet shape changes dynamically from its static position until it finally loses contact with the wall surface and is swept away by the air.

The dynamics of the formation, growth and removal of liquid water in/on gas diffusion media are directly associated with the pore morphology and surface properties of GDM as well as operating conditions. The degradation of GDM, which is accelerated due to GDM flooding, will lead to deterioration of the surface chemistry of fibers, pore morphology and loss of hydrophobicity of the GDM surface. In addition, GDM flooding will block the local transport and cause CL starvation, which damages catalyst layers. By monitoring the liquid water dynamics and distribution on GDM, optical visualization provides valuable and time-dependent information about the degradation of GDM and the associated components.

2.3. Catalyst Layers

Catalyst layers (CLs) are critical components within fuel cells as they are the sites at which the electrochemical reaction occurs. Their relevance for water management has become recognized in recent years [16–18,21,96–99]. In addition, the complex and heterogeneous micro-/nano-structure of the CL, coupled with the fact that it is concealed behind the GDL within the fuel cell, makes it a challenging task for optical visualization. Zhang et al. [19] characterized *in-situ* the micro-scale water transport on the catalyst layer by designing and fabricating a catalyst-visible operational transparent fuel cell and developing a micro-visualization system. The optical subsystem was carefully designed with a spatial resolution of 5 μm and a working distance of up to 10 cm, while the temporal resolution was limited by the camera frame rate. As shown in Fig. 8.3, they observed that micro-scale water droplets initially form with diameters ranging from 2–5 μm, which are held on to the site by surface tension as they grow and coalesce due to water production and transport. They observed that the evaporation rate varied spatially over the cathode CL within a scale of less than 50 μm, which indicated that the process of evaporation could be enhanced and controlled after its detailed mechanism was understood. Their observations also revealed four possible pathways of water dynamics on the CL surface and clarified liquid water evolution/transport on the CL surface. Water first condenses and/or flows into the hydrophilic pores and floods the catalyst sites therein. As liquid water continues to accumulate on the membrane side and the hydrophilic pores are flooded, water will eventually penetrate into larger hydrophobic pores when the capillary pressure reaches a critical value, which was confirmed by recent modeling efforts [100–102]. In addition, the

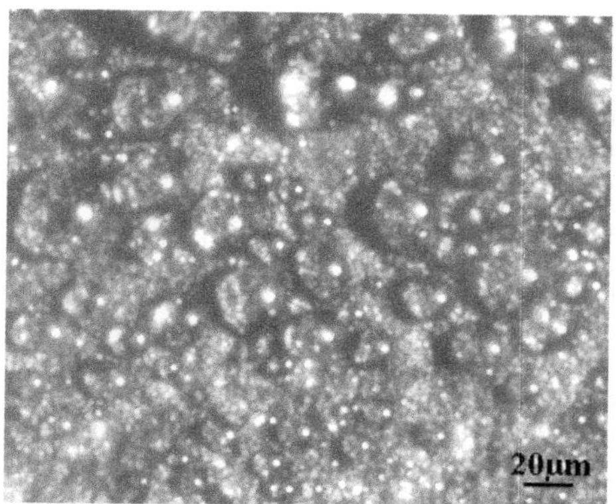

FIGURE 8.3 A typical image of micro droplets on the cathode catalyst layer (CCL) in an operating fuel cell. Initially, water droplets were observed with diameters ranging from 2–5 μm. They are held on to the site by surface tension as they grow and coalesce due to water production and transport. *Reproduced with permission from [19]. Copyright 2007, The Electrochemical Society*

authors proposed to derive *in-situ* the net water transport coefficient across the MEA based on determining the local water accumulation rate on the CL surface from optical visualization and image processing techniques.

Several efforts were also made on cold starts with a catalyst-visible operational transparent fuel cell. Ge and Wang [91] observed that water underwent phase transition from liquid to ice/frost on the cathode CL surface during a sub-zero start-up. They found that water droplets and ice particles on the CL surface differed in shape: the droplets appeared round, while the ice particles were irregular. They revealed that the freezing-point depression of water in the CL was negligibly small in comparison with the large sub-zero temperature range of interest in automotive applications. Even when a PEMFC was operated below the freezing point at $-10°C$, Ishikawa et al. [92] observed liquid water formation on the cathode CL. They revealed that the temperature of the water rose to $0°C$ when the water began to solidify. During cold starts, ice formation in the membrane, catalyst layer, MPL, and GDL would lead to a significant decay in performance, and to material degradation due to local volume expansion, resulting stresses, and structure damage. In addition, it may cause changes in the GDL/CL/membrane pore structure, delamination within the MEA, loss in catalyst electrochemical surface area, and decrease of material strength. Thus, characterizing liquid water dynamics and developing strategies to enhance freeze-tolerance is an important task for optical visualization.

In summary, low-cost, high-resolution optical visualization has proved its utility for both *in-situ* and *ex-situ* investigation of two-phase flow and water transport in flow channels, gas diffusion media and catalyst layers, although its spatial and temporal resolutions are mainly limited by the employed microscope magnification, optical working distance, and camera speed. In addition, the development of image processing techniques has enabled quantitative studies of water transport and distribution, which can be correlated with the degradation of these critical components and their impact on FC performance.

Some of the main challenges for this technique are to preserve a realistic operating environment such as maintaining the same temperature field, as well as surface and interface properties as in the conventional fuel cell. Limitations of optical visualization can be overcome by using other techniques to characterize water liquid/vapor transport in the CCL, GDL and MPL, and the interfacial effects and phase change in fuel cells. One such technique is neutron imaging.

3. NEUTRON IMAGING

Neutron imaging is a non-intrusive and *in-situ* diagnostic tool for spatially resolving liquid water distribution and accumulation inside an operating fuel cell [103–110]. Unlike optical visualization it can directly provide quantitative information on the water content. It relies on the measurement of an attenuated neutron signal when a sample is placed in the path of a neutron beam. The image contrast is obtained due to the relatively high absorption cross-section of hydrogen compared to that of the various component materials used in fuel cells (for example, aluminum, stainless steel and graphite). In addition, since the density of hydrogen atoms in liquid water is much higher than in gaseous hydrogen at standard pressure, this method can be used to detect the presence and amount of liquid water in operational fuel cells, with insensitivity to common fuel cell materials [103–107,111].

Different from direct optical visualization, neutron imaging can detect liquid water in optically opaque regions of interest, such as under the lands and inside GDLs/MEAs. The water content in different parts of the cell can be identified and estimated by the use of masking techniques [104,112]. Recent research has demonstrated the capability of imaging with a spatial resolution of up to 25 μm [113–119], and temporal resolution can be as high as 30 frames/s [120,121]. This method has been widely used for investigating materials [115,122–127], flow channels [41,48,104,112,115,116,118,121,124,125,128–136], and transport phenomena in a single cell [104,111,112,116,122,124,125,127–130,134,135,137–141] as well as in a stack [133,142].

Bellows et al. [103] demonstrated the ability of the neutron imaging technique to measure the response of water content in the membrane with respect to changes in gas humidification levels. The authors measured in-plane water gradient profiles across the Nafion membrane in an operating PEM FC. Mosdale et al. [143] used small-angle neutron imaging (SANI) to measure the water profile across the membrane thickness in a PEM FC, and also to detect regions of the membrane that experienced swelling due to excess water accumulation. They showed that the inversion of the water concentration profile was dependent on whether the H_2 sent to the anode was dry or not. Xu et al. [144] and Gebel et al. [137,138] further applied SANI in a graphite single fuel cell to determine and to quantify the transverse water concentration within a membrane in an operational MEA. In addition, they probed different areas of the active surface in an operational fuel cell. The membrane was found to be drier at the gas inlet than at the gas outlet. These studies suggest that neutron imaging can be a very useful tool for degradation studies. For example, cyclical water accumulation and swelling of the membrane can lead to hygrothermal fatigue with associated failure mechanisms such as catalyst layer delamination. Delamination causes unrecoverable interfacial degradation between the CL and the PEM, including increase of contact resistance, loss of catalytic activity, and development of pinholes, cracks and flooding areas. In addition, the agglomeration of Pt, carbon corrosion, and ionomer dissolution can be affected by water accumulation and high relative humidity.

Geiger et al. [145] reported the gas/liquid two-phase flow patterns in the flow fields with neutron imaging. However, the low spatial resolution and image acquisition rates limited the investigation of the two-phase flow to real time. Satija et al. [104] quantitatively analyzed water dynamics in an operational fuel cell system with in-plane neutron imaging and improved spatial/temporal resolution. They created a three-dimensional water content profile over 2,000 s by using tomography and masking techniques. Kramer et al. [111, 146] revealed a lower water accumulation in fuel cells with a serpentine channel than with an interdigitated channel. Kim et al. [116,139] investigated the distribution and transport of water by varying the flow directions in anode and cathode channels and the differential pressures (100, 200, 300 kPa) using a 3-parallel serpentine single PEMFC. They found that at a given current density the amount of water produced in the fuel cell increased as the partial pressure increased and the water production for the counter-current flow, where the flow directions of anode and cathode were opposite, was more uniform than for the co-current one due to the opposite humidity gradients of fuel and air.

Neutron imaging techniques have been used to quantify the liquid water distribution in a PEMFC under a variety of flow rates, humidities, and currents with paper or cloth GDLs. Kowal et al. [131] noted that the paper GDL held roughly 60% of the total water stored under the lands and the remaining 40% in, or under, the channels. The authors revealed that the paper GDL could hold more water per volume of GDL under the lands than the cloth. Yoshizawa et al. [126] also investigated the effects of GDL type and flow field design on fuel cell performance and found that the carbon-cloth GDL was less influenced by the accumulated water than the carbon-paper GDL. With a serpentine flow field, water accumulated in the corners and the gas bypassed the flow field. These phenomena are the main causes of performance degradation due to local flooding with a serpentine flow field. Chen et al. [141] investigated the effect of cathode inlet RH and stoichiometry on liquid water accumulation and distribution. It was observed that under fully humidified conditions, the water content at the anode GDL decreased with increasing current densities. In addition, at under-saturated operating conditions, the water content in GDLs decreased with increasing flow stoichiometry. It can be inferred from these studies that neutron imaging can play a powerful role in characterizing water dynamics and management and, therefore, in analyzing the related degradation mechanisms and predicting the durability of the MEA.

Considering that the liquid water content in fuel cells is integrated along the beam direction with neutron imaging, several efforts have been made to separately evaluate the effects of liquid water residing in different cell compartments, including three-dimensional (3D) in-plane tomography [104], image processing algorithms [125,134] and modified cell design [141]. Spernjak et al. [112,147] further proposed a method incorporating simultaneous neutron imaging and direct visualization for better understanding and distinguishing water dynamics in both cathode and anode channels. By combining the two techniques, two sets of images of liquid water were obtained in an operational, transparent, PEM FC; the optical data provided a qualitative assessment of water content in the channels, whereas the neutron data provided a quantitative measure of the water content through the thickness of the cell. The authors demonstrated this experimental approach for characterizing the water transport within different flow field configurations and evaluating the water management efficiency of the GDL materials. Pekula et al. [120] reported the tendency for liquid water to accumulate along or under the channel walls at 180° turns, and revealed a slug-flow regime up to current densities of least 1 A/cm^2 in an operating fuel cell. The authors observed anode flow channel blockage at low power, while higher power conditions resulted in more dispersed distribution of liquid droplets. Siegel et al. [148] employed neutron imaging over four continuous days of fuel cell testing as they investigated the accumulation of liquid water in both the anode and cathode gas channels. They observed anode channel flooding followed by a significant voltage decrease even without cathode channel flooding, indicating that anode flooding also led to performance drop and material degradation.

The spatial and temporal resolution of neutron imaging was improved to view the fuel cell from the side, in order to differentiate liquid water between fuel cell components [117–119]. Kim et al. [113,114,149] first investigated the phase-change-induced water transport of polymer electrolyte fuel cell materials subjected to a temperature gradient. They observed that the thermo-osmotic flux of water in three different commercial perfluorinated PEMs is directed from the cold to the hot side of the membrane, and that membrane reinforcement with a PTFE fibril-reinforced composite did not affect the flow direction. The net flux, however, was altered by the reinforcement structure. They revealed that phase-change-induced flow can also be important in PEMFCs. As shown in Fig. 8.4, when the liquid side was colder, the phase-change-induced flow reversed. When the temperature of the gas-phase side was lower, the flow from the hot side to the cold side was augmented by the phase-change-induced flow and the gas-phase side filled with liquid water much more quickly, which indicated that along with capillary flow, the phase-change-induced flow from the MEA to the flow channel played an important role in water transport.

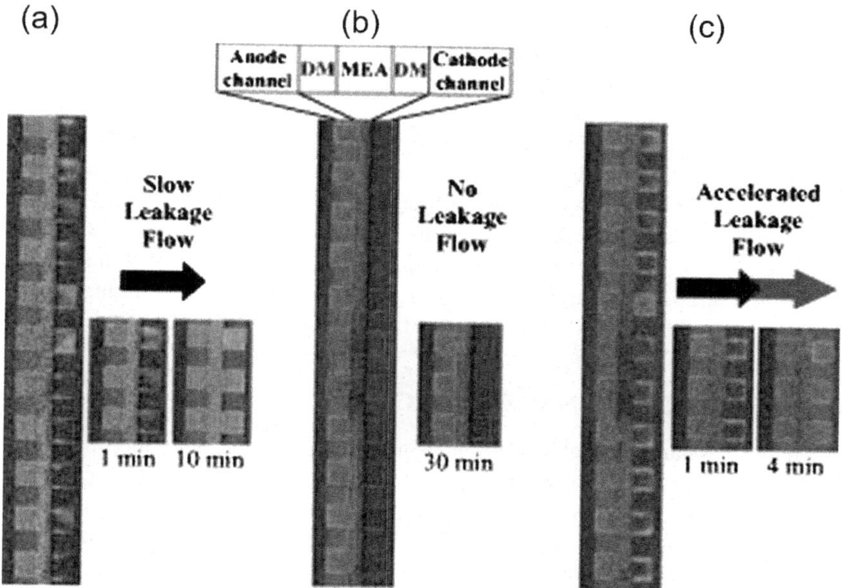

FIGURE 8.4 Neutron image of water distribution in an MEA with hydrophilic DM under different temperature gradients in an operational fuel cell (liquid water in red color): (a) anode/cathode=65°C/65°C, (b) anode/cathode=60°C/70°C, and (c) anode/cathode=70°C/60°C. *Reproduced with permission from [149]. Copyright 2009, The Electrochemical Society*

Recently, Turhan et al. [136] investigated the effect of varying the channel size, land width, and land-to-channel ratio (*L*:*C*) with a counter-flow serpentine/parallel combination design. They found that water content increased with *L*:*C* ratio. Trabold et al. [41] employed neutron imaging to investigate the accumulation of liquid water in a PEMFC with a serpentine flow field. They focused on the cathode side and evaluated the influence of the inlet reactant gas humidity and inlet/outlet header differential pressure. Among several findings, they observed that water tends to accumulate in the 180° bends of both the serpentine anode and cathode flow fields. A higher gas velocity is sufficient to remove liquid water in flow channels even when operating at a current density of 1.0 A/cm^2. Similar results were reported by Owejan et al. [125,134] and they further provide quantitative analysis and revealed that the GDL accumulated liquid water close to the transport limiting point based on the liquid water content, compressed thickness and porosity of the GDL. Complete saturation of the GDL could also influence the durability of the material.

Neutron imaging is a very powerful technique for *in-situ* visualization and quantification of liquid water in an operating PEMFC. In contrast to direct optical visualization, neutron imaging provides the ability to acquire data from inside critical components of conventional fuel cells, such as under the lands, inside the GDL and MPL, and inside the electrode and the membrane. However, this capability has been limited due to the lower spatial and temporal resolution of these techniques. Moreover, its integration times are also insufficient to measure the micro-scale transport of liquid water. This section indicates that despite the potential for neutron imaging to provide unique insights into degradation from sub-optimal water management, very little degradation work has been pursued with neutron imaging to date. Hence there is a good opportunity to apply neutron imaging to degradation studies. Hardware enhancements will enable this technique to continue to improve in terms of the spatial and temporal resolution, and thus shed new light on correlating water management within the MEA to degradation.

4. MAGNETIC RESONANCE IMAGING

Magnetic resonance imaging (MRI), or nuclear magnetic resonance imaging (NMRI), is primarily a non-destructive and non-invasive imaging technique most commonly used in radiology to visualize the internal structure and function of an object. It uses a powerful magnetic field to align the nuclear magnetization of hydrogen atoms in water. In addition, radio frequency (RF) fields are applied to systematically alter the alignment of this magnetization, causing the hydrogen nuclei to produce a rotating magnetic field detectable by the scanner. This signal can also be manipulated by additional magnetic fields to build up enough information to reconstruct an image of the object [150,151]. MRI has been successfully used to explore degradation in PEMFCs by characterizing the membrane [152–162], visualizing water transport [163–169], and assessing new strategies for water management [153,154,158,159,163,166,169]. MRI visualization was achieved with a spatial resolution of up to 8 μm along the through-plane direction in the membrane by using a highly sensitive radio frequency (RF) coil system and a specially designed PEMFC [157,158]. Its temporal resolution has been reported at about 50 s [154].

Minard et al. [170] employed MRI to visualize water transport inside a PEMFC during 11.4 h of continuous operation with a constant load. They revealed the formation of a dehydration front propagated slowly over the surface of the fuel cell membrane, as shown in Fig. 8.5. After traversing the entire PEM surface, channels in the gas manifold began to flood on the cathode side. They demonstrated the power of MRI for visualizing time-dependent water distributions across membranes during PEM fuel cell operation, and highlighted its potential for studying fuel cell degradation.

MRI has revealed that membrane hydration/dehydration is far from homogeneous. Bedet et al. [164] designed a single cell optimized for MRI experiments using a PMMA support with a perimetric gold wire for current collection. Inhomogeneous hydration of the membrane was partly attributed to its swelling within the gas channel. Tsushima at al. [155,169] also observed

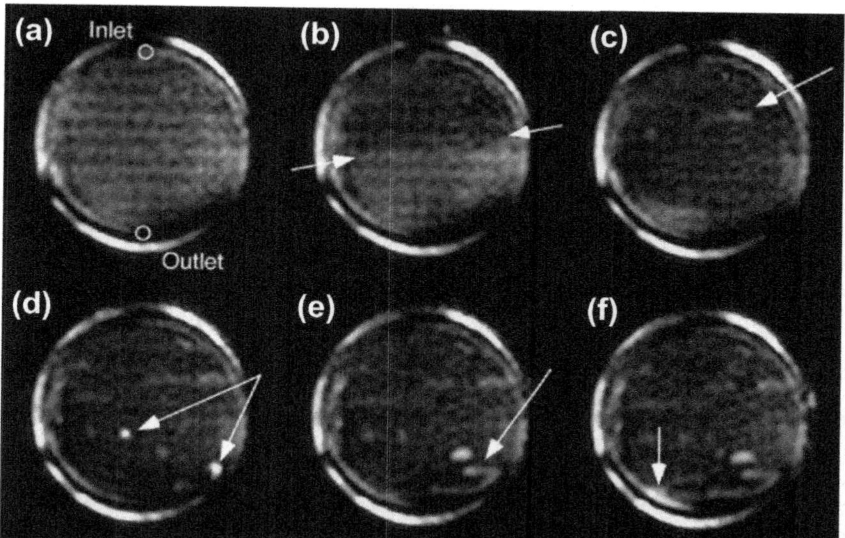

FIGURE 8.5 Selected MR images showing water in an operating fuel cell over an 11.4 h period. Images were acquired at: (a) the start of the experiment and after (b) 66, (c) 164, (d) 290, (e) 473, and (f) 678 min. The fuel and oxidant inlet and outlet positions are labeled and the arrows in (b) highlight the position of the dehydration front that moved from the inlet towards the outlet. The grey scale was chosen to highlight the dehydration front formation (drier regions appear darker) and the actual signal intensity varied by approximately 10%. Regions showing flooding are also highlighted by white arrows. Images were obtained every 128 s and the full set of 320 images was assembled into a continuous movie that is available from the authors. *Reprinted from [170] with permission from Elsevier*

both homogeneous and inhomogeneous distributions of water content within the membrane. They revealed that the rate-determining process of water permeation across the membrane was related to the diffusion process through the membrane and the interfacial transport process across the membrane-gas interface. Teranishi et al. [154,168] further used MRI measurements to determine the water transfer coefficient and maximum water content of a membrane in a fuel cell. They found that the maximum water content was consistent with the measured relationship between water content and membrane swelling, which are directly related to the membrane degradation.

The research with MRI on fuel cells has been limited due to lack of availability and material constraints (the MRI technique requires the subject materials to be nonmagnetic). It is due to the latter reason that the water content in the CL and GDL is difficult to visualize with MRI. More efforts in fuel cell research can be expected with the continuing development of MRI. Recently, Dunbar and Masel [165,166] used MRI to obtain a three-dimensional quantitative water distribution profile using Teflon® flow fields inside an operating fuel cell. They showed the feasibility of MRI for studying water formation, distribution, and flow patterns in fuel cell flow fields as well as in the membrane. Zhang et al. [157,158] obtained high spatial resolution MR images of $<8\ \mu m$ in an image acquisition time of less than 2 min using a novel methodology. These developments of 3D and high-resolution images permit water content measurements under transient conditions, including wetting and drying, and provide more accurate information of degradation of associated components in fuel cells. Despite these studies, it can be concluded that the potential of MRI to analyze cell degradation has not been fully exploited. Future MRI studies are expected to fill this gap.

5. ELECTRON SPECTROSCOPY AND MICROSCOPY

Electron spectroscopy and microscopy are analytical techniques for characterizing materials by probing their atomic and molecular electronic structure, including X-ray photoelectron spectroscopy (XPS), scanning electron microscopy (SEM), and transmission electron microscopy (TEM) [171–173]. XPS (also known as electron spectroscopy for chemical analysis (ESCA)) detects photoelectrons that are ejected from an inner-shell orbital of an atom by X-rays and is frequently used to determine the chemical composition and oxidation state of materials on a selected region of the surface. In SEM, the electrons interact with the atoms that make up the sample's surface and reveal information about its morphology, composition, and other properties. Environmental SEM (ESEM) can be used to observe samples in low-pressure gaseous environments and high relative humidity. In TEM, an image is formed from the interaction of a beam of electrons as it transits through an ultra-thin specimen. TEM is capable of imaging at a significantly higher resolution than SEM. The addition of Si drift detector (SDD) technology to high spatial resolution scanning TEM (STEM) can enable compositional analysis at <2 Å [174–178]. These techniques have been used to study the durability and degradation of electrodes [178–234], membranes [175,196,219,235–264], bipolar plates [265–274], and other components and materials [196,207,275–280].

5.1. X-ray Photoelectron Spectroscopy (XPS)

Schulze et al. [189] studied the change in the distribution of the platinum catalyst in membrane-electrode assemblies during PEMFC operation. Coupled with SEM, the XPS investigation of the electrode, which was electrochemically stressed in an aqueous electrolyte, demonstrated that the platinum movement from anode to cathode is supported by the physical presence of liquid water. The platinum concentration on the anode was found to decrease significantly due to electrochemical stressing and electrical field gradients. Schulze et al. [202] further identified two degradation processes of MEAs based on the results of XPS and SEM:

1. the agglomeration of the platinum catalyst mainly in the cathode, and
2. the disintegration of PTFE and the correlated decrease of hydrophobicity.

They also observed that platinum oxide particles with a size of approximately 50 μm were formed on the backside of the gas diffusion layers (GDLs) of the cathodes during the fuel cell experiments. Shao et al. [199–201] conducted studies on the electrochemical stability of carbon-black-supported and carbon-nanotube-supported Pt (Pt/C and Pt/CNTs) electrodes. They found that the degree of oxidation of carbon black was higher than that of carbon nanotubes, implying that carbon nanotubes were more resistant to electrochemical oxidation than carbon black. Shao et al. [215] used TEM and XPS to investigate the degradation of Pt/C electrocatalysts. They found that Pt/C degrades much faster under potential step conditions than under potential-static holding conditions. Guilminot et al. [244,245] also revealed that aged MEAs exhibited severe degradation of most elements (Pt, C, F), coupled with a dramatic change in Pt particle shape, mean particle size, and density over the carbon substrate.

Recently, XPS was applied to quantitatively analyze catalyst layer degradation by determining both the elemental concentrations and chemical states of C, F, Pt, O and S in PEM FCs. After about 300 hours of fuel cell operation, Zhang et al. [227] reported that the ionomer in the catalyst layer dissolves and/or decomposes as characterized by a decrease of CF_3 and CF_2 species and an increase in oxidized forms of carbon (e.g. C–O and C=O). This was also confirmed by SEM observations of the unused and used catalyst layer surface. The ratio of carbon in fluorinated forms to carbon in graphitic forms decreased from $\approx$ 4.2 to 1.4, as shown in Fig. 8.6. The oxidized states of platinum and carbon were found to be substantially higher for the used samples.

In addition to catalyst layers, the degradation of membranes and other materials have been investigated with XPS. Chen et al. [243] studied the chemical degradation in Nafion® membranes and observed the $(CF_2)_n$ polymer backbone had decomposed after treatment with various Fenton's reagents. Two membrane degradation mechanisms were proposed: degradation initiated by polymer defects; and by cross-linking of sulfonic end groups. To study changes on the surfaces of elastomeric gasket materials before and after exposure to the simulated fuel cell environment over time, Tan et al. [278,279] employed both XPS and attenuated total reflection Fourier transform infrared (ATR-FTIR) spectroscopy. They found that the surface chemistry changed significantly, and that chemical degradation was due to de-crosslinking and chain scission in the backbone. Aragane et al. [281] studied the degradation of the polytetrafluoroethylene (PTFE) material in a phosphoric acid fuel cell (PAFC) using XPS and found that the PTFE began to degrade through oxidation after about 1,000 h of operation.

XPS has also been used to investigate degradation mechanisms in DMFCs and alkaline fuel cells (AFCs). Wang et al. [217,218] evaluated the cathodic Pt

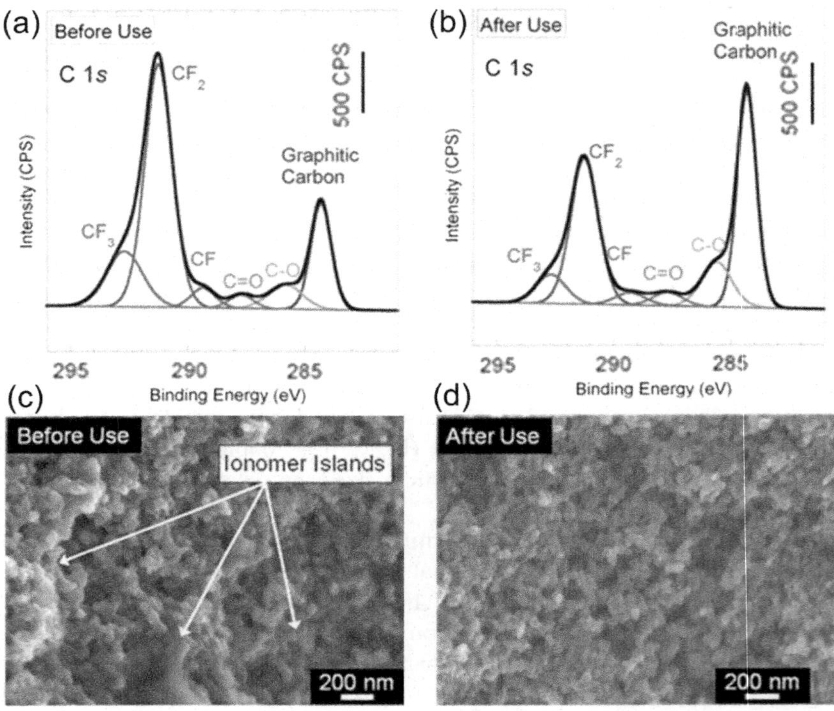

FIGURE 8.6 Results from high resolution XPS spectra of carbon and SEM images of the catalyst layer before and after fuel cell operation. (a) Unused-XPS, (b) used-XPS, (c) unused-SEM, and (d) used-SEM. There were two dominant carbon peaks at the binding energies of 284.3 and 291.3 eV, which are assigned to graphitic carbon and fluorinated carbon in the ionomer, respectively. In addition, minor components consisting of oxidized forms of carbon in the C–O and C=O states were observed at 285.8 and 287.7 eV, respectively. *Reprinted from [227] with permission from Elsevier*

black catalysts in DMFCs prior to and after the life tests with both XPS and XRD. The authors found that the precious metal content gradually decreased with test time, whereas the content of Pt oxides in cathodic catalysts increased. Lai et al. [247] investigated the degradation of the MEA in a DMFC at highly anodic potential over 160 hours with XPS, TEM, electron probe microanalysis (EPMA), and XRD. They showed that the sulfonic acid vanished in the broken anodic layer compared to the original sample and the particle sizes of the catalysts increased due to aggregation. Gülzow et al. [183,186] studied the long-term influence of CO_2-containing reaction gases on electrodes in AFCs. Surprisingly, the XPS characterization of used cathodes exposed to pure and to CO_2-containing oxygen yielded no significant difference.

5.2. Electron Microscopy – SEM and TEM

Electron microscopy has been widely used to study MEA degradation under different operating conditions. Akita et al. [176] reported the change of TEM image patterns of deposited Pt in the membrane phase after 86 hours of accelerated testing at a constant voltage of 1.0 V. Schulze et al. [202] employed a 'dead-ended' hydrogen supply at the anode (100% utilization) with various purging intervals to understand degradation in the MEA. Taniguchi et al. [216] investigated CL degradation by cell reversal with air. Yasuda et al. [261,282] studied electrochemically cycled MEAs. All these investigations presented strong evidence for platinum dissolution/sintering, loss of electrochemical surface area, and diffusion of the dissolved platinum into the membrane. Wang et al. [283] carried out a 2,250 h life test of PEMFC at a current density of 160 mA/cm^2. They revealed that the electrochemically active surface areas of anodic and cathodic catalysts initially increased, and then decreased with the operation time. The particle size of the cathodic Pt/C catalyst was evidently bigger than that of the anodic one. After long-term operation of PEMFCs under various operating and feed conditions, Kim et al. [284] also found using TEM that large quantities of cathode Pt were dissolved when the residual oxygen concentration inside the cathode electrode was high with a low reduction current density.

SEM and TEM have shown that catalyst agglomeration and detachment from the carbon support are among the primary causes of catalyst underutilization. Xie et al. [194] observed catalyst particles detached from the carbon surface in the recast Nafion ionomer network. Cathode catalyst agglomeration occurred mainly in the first 500 h of operation and the rate of agglomeration reduced through the subsequent 500 h, as shown in Fig. 8.7. They observed catalyst migration toward the interface of the cathode catalyst layer and the membrane. Yoda et al. [205] evaluated degradation of Pt catalysts and the carbon support after a daily start and stop (DSS) operation with cyclic voltammetry and TEM. Recently, Mayrhofer et al. [210,211] used TEM to observe the identical locations of a catalyst before and after electrochemical treatment, by utilizing a TEM gold finder grid as a support for a standard Pt fuel cell catalyst. They demonstrated that the observed changes of the catalyst particles such as particle detachment and particle movement/agglomeration were a direct consequence of the applied electrochemical treatment.

Grolleau et al. [285] studied the effect of potential cycling on the structure and activity of Pt nanoparticles dispersed on different carbon supports. TEM results indicated that before potential cycling the mean particle size on the non-oxidized support (Pt/XC72) was a little higher than that on the functionalized one (Pt/XC72$_{HNO3}$), being 2.5 and 2.0 nm, respectively. However, 400 potential cycles led to a greater increase in the particle size on the functionalized support, reaching 5.5 nm compared to 4.0 nm for the non-functionalized one. Matsuoka et al. [209] investigated the degradation of Pt supported on carbon catalysts

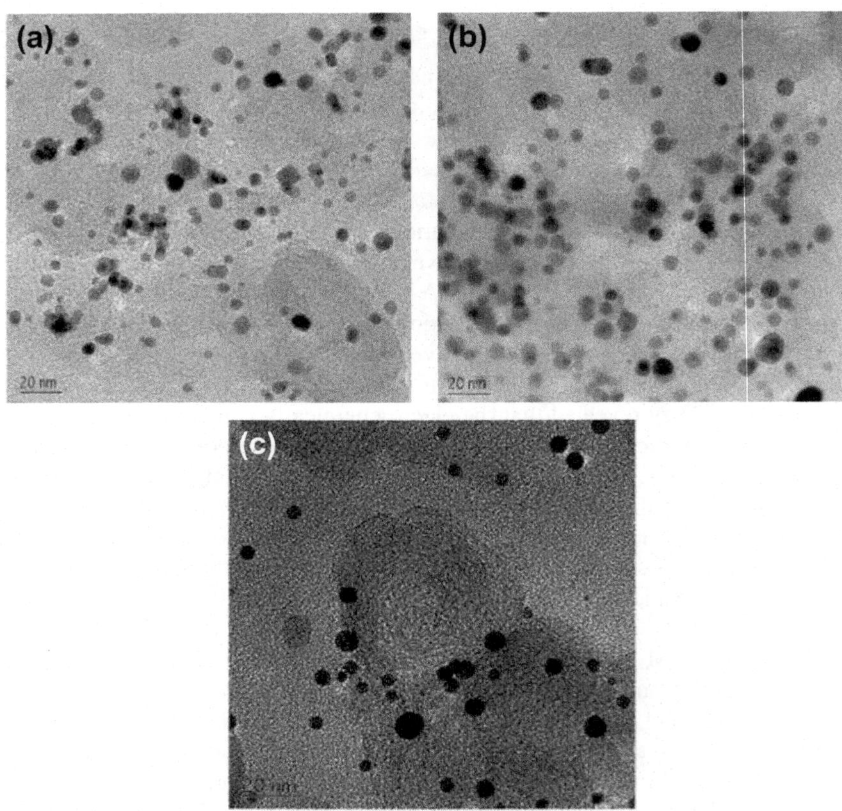

FIGURE 8.7 Pt_3Cr/C cathode catalyst clusters (high magnification TEM) in (a) fresh catalyst layer, (b) catalyst layer after ~500 h of testing, and (c) catalyst layer after 1,000 h of testing. The average Pt_3Cr particle size for the fresh MEA increased from 4-12 nm to 6 to >20 nm (~1.5-2 times) during the initial ~500 h, but increased much less with an additional 500 h of testing of the second MEA. *Reproduced with permission from [194]. Copyright 2005, The Electrochemical Society*

(Pt/C) due to four anion species (Cl^-, F^-, SO_4^{2-}, NO_3^-). TEM and SEM results after 50 h tests also revealed partial Pt dissolution and deposition in the membrane and the formation of a Pt band.

Park et al. [178] investigated the durability behavior of Pt-Ru anode catalysts under virtual DMFC operating conditions at the atomic scale using high-resolution transmission electron microscopy (HR-TEM) coupled with time-of-flight secondary ion mass spectroscopy (TOF-SIMS). They found the Pt-Ru anode catalysts decomposed into small particles via morphological change and cracking. The Pt and Ru ions moved across the solid electrolyte membrane and redeposited at the cathode. Electron spectroscopy and microscopy, including XPS, SEM, ESEM, and TEM, have become more attractive for the *ex-situ* characterization of degradation of fuel cell materials and components. With a surface sensitivity of up to 0.01%, XPS provides quantitative information about elemental variation and changes in chemical composition and oxidation state on selected surfaces of fuel cell materials. The recent development of higher resolution imaging down to sub nm enables SEM/TEM to examine finer details – even as small as a single column of atoms. With these tools, analyses of degradation on the MEA have been widely reported. Multi-dimensional (for example 2D with XPS or 3D with SEM/TEM) and more quantitative exploration of the catalyst layer, electrode assembly, and membrane will provide better understanding of degradation mechanisms in fuel cells.

6. X-RAY TECHNIQUES

X-ray techniques are a family of non-destructive analytical techniques which reveal information about the crystallographic structure, chemical composition, and physical properties of materials and thin films. When a monochromatic beam of X-ray photons impinges on a given specimen, it will either scatter, fluoresce or get absorbed. The coherently scattered photons may undergo subsequent interference leading in turn to the generation of diffraction maxima. These three basic phenomena give rise to the three important X-ray methods: the scattering effect, which is the basis of X-ray diffraction (XRD); the fluorescence effect, for X-ray fluorescence (XRF) spectrometry; and the absorption technique, for X-ray tomography or radiography [286–288]. These techniques have been widely used in the study of PEMFCs [175,239,242,248,283,289–351], DMFCs [160,204,217,218,228,247,352–367], SOFC [220,368–388], PAFCs [389] and alkaline fuel cells [390]. With three-dimensional (3D) computed tomography (CT) coupled with condenser optics and imaging objective lenses, X-ray techniques are capable of resolving down to sub-50 nm scale in fuel cells [326,381,385,391–394].

6.1. X-ray Diffraction

XRD has provided important information about fuel cell degradation by deriving the crystallite size change with a spatial resolution down to the sub-nanometer scale based on the measured peak full width at half maximum (FWHM) of the X-ray diffraction pattern. Borup et al. [305] conducted a single cell durability test by examining the degradation of Pt electrocatalysts, and also studied the effect that fuel cell operating conditions have on electrocatalyst particle size, based on XRD patterns. As shown in Fig. 8.8, the electrocatalysts after testing show a narrowing of the diffraction peaks, which is indicative of larger particles, compared with the fresh catalyst particles. The mean particle size was calculated by XRD patterns from the X-ray wavelength, the angle at the position of the crystal peak, and the width of the diffraction peak at half

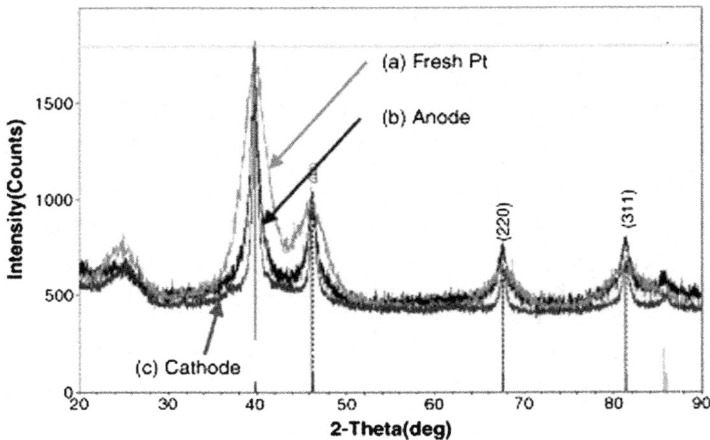

FIGURE 8.8 X-ray diffraction patterns for (a) fresh Pt catalyst, (b) anode Pt catalyst, and (c) cathode Pt catalyst after fuel cell testing at 0.60 V for 3,500 h. *Reprinted from [305] with permission from Elsevier*

height. They found that the cathode catalyst particle size grew from about 1.9 to 3.5 nm after a typical test cycle of 1,200 hours. During the cycles, the cathode potential was swept linearly with time from 0.1 V to a voltage which was varied from 0.8 to 1.5 V. This increase in size of the catalyst particles was greater than the one observed during steady-state testing, in which the particles grew only to 2.6 nm after 900 h and 3.1 nm after 3,500 h. Catalyst particle growth is due to agglomeration which causes a reduction in catalyst surface area and a drop in fuel cell performance. Low relative humidity decreased the platinum particle growth, but substantially increased carbon corrosion and loss. The rate of carbon corrosion of the electrode catalyst layer was also found to increase with increasing potential. Carbon corrosion also leads to a drop in fuel cell performance due to the isolation of catalyst nanoparticles, and the increase in ohmic resistance of the catalyst layer.

Koh et al. [316] correlated structure, composition, and electrochemical behavior of carbon-supported Pt-Co alloy electrocatalysts in PEMFC cathode electrode layers, and established a number of detailed structure-property relationships for this class of materials using synchrotron XRD. They revealed that low-temperature preparation conditions favored the formation of multiple primarily fcc Pt-Co alloy phases, while high-temperature preparation conditions for carbon-supported $Pt_{50}Co_{50}$ catalysts led to the formation of the chemically ordered fct $Pt_{50}Co_{50}$ alloy phase. These studies indicate that X-ray methods provide an improved understanding of the relationships between the crystallographic phase, chemical ordering, and the corrosion stability of fuel cell catalysts within polymer-electrolyte/catalyst composites, leading to more rational designs for durable catalyst materials.

Prasanna et al. [340] studied the effect of fabrication methods on the durability of PEMFCs. Membrane-electrode assemblies (MEAs) were fabricated using a conventional method, a catalyst-coated membrane (CCM) method, and a CCM-hot pressed method. Before and after long-term operation, the physical and chemical characteristics of the MEAs were analyzed using XRD and other methods. They found that the CCM MEA exhibited the lowest degradation rate as well as the best initial performance. It was found that the lower porosity of the CCM-hot pressed MEA increased the probability of water accumulation in the catalytic layers, and also of gas diffusion, leading to quicker performance degradation.

Wang et al. [283] further employed XRD to characterize both anodic and cathodic catalysts before and after lifetime tests. The results showed that the electrochemical surface areas (ECSA) of anodic and cathodic catalysts initially increased, and then decreased with operation time. The particle size of the cathodic Pt/C catalyst was found to be larger than that of the anodic one. They also revealed that the degradation of the cathodic catalyst for oxygen electro-reduction was the main cause of the PEMFC performance decay.

XRD was also used to investigate carbon-supported $Pd_{100-x}Mo_x$ ($0 \leq x \leq 40$) nanoparticles synthesized by a simultaneous thermal decomposition of palladium acetylacetonate and molybdenum carbonyl in an organic solvent (*o*-xylene) in the presence of Vulcan XC-72R carbon, followed by heat treatment up to 900°C in H_2 atmosphere [341]. XRD revealed the formation of single-phase face-centered cubic solid solutions for $0 \leq x \leq 30$ after heat treating at 900°C and the occurrence of a Mo_2C impurity phase for $x = 40$. This also indicted that XRD can be an effective way for studying the degradation of Pd-based alloy catalysts.

Loster et al. [319] developed a technique to separate the individual components from a composite XRD pattern. After extraction of the catalyst signals of the anode and the cathode from the diffraction patterns of an entire MEA, line profile analysis provided particle size distributions of the catalyst particles, and also the total surface area. They found a significant surface loss during operation that was similar to the non-reversible voltage degradation under constant current operation. The electrodes of the identical MEA exposed to different gases showed approximately one-third of the surface loss, suggesting that particle growth may be induced by other mechanisms in addition to fuel cell operation; such as particle agglomeration, a process that is due to the high mobility of the metal particles on the carbon support.

6.2. X-ray Fluorescence Spectrometry

After Kulesza et al. [289] introduced XRF to detect the presence of platinum with very low loadings on a carbon substrate, this method was quickly adopted – as reported in other publications on fuel cells [289,292–294,303,352–354,360–364,387,388,395–402]. Ahn et al. [293] used XRF to investigate the degradation of a counter-flow type 40-cell PEMFC stack. They found a rapid decay in performance after continuous operation for 1,800 h; the degradation of the catalyst and contamination of the polymer electrolyte membrane were the main causes of this sudden decay. Adjemian et al. [303] studied metal-oxide-recast Nafion composite membranes and indicated a specific chemical interaction between polymer sulfonate groups and the metal oxide surface for systems that provide a good elevated-temperature performance (i.e. fuel cell operation above 120°C). Coupled with electrochemical characterizations, they found that composite systems that incorporated either a TiO_2 or a SiO_2 phase produced superior elevated-temperature, low-humidity behavior compared to that of a simple Nafion-based fuel cell.

Piela et al. [354] studied the ruthenium crossover from the Pt-Ru black DMFC anode through the solid electrolyte membrane to the Pt black cathode using XRF and XRD. They revealed that Ru crossover occurred both under abnormal conditions, such as cell reversal resulting in very high anode potentials, and under normal DMFC operating conditions. XRF data also confirmed the presence of Ru in cathode half-MEAs, and in the intermediate membranes placed between the anode and the cathode half-MEAs. The presence of Ru at the cathode has a negative impact on oxygen reduction kinetics, and leads to DMFC performance degradation.

Sachdeva et al. [400] studied Ag^+-loaded Nafion-117 membrane before and after reduction with $NaBH_4$ with XRD and XRF. They indicated that Ag nanoparticles were formed only on the membrane surface exposed to $NaBH_4$ solution. The authors discovered that the original nanostructure of the Nafion-117 membrane was regenerated and the cluster-channel network in the membrane remained intact even after completion of the formation of Ag nanoparticles. This work also demonstrated an interesting possibility of using XRF to investigate membrane degradation due to metal ion contaminants.

6.3. X-ray Absorption Technique

X-ray absorption techniques have long been used for imaging purposes due to their ability to 'see through' the surface and capture information about the internals of a material. The X-ray intensity of the signal is attenuated as it travels through the material. The transmitted signal provides information based on the variation of absorption within the material. With increasing computing power, three-dimensional (3D) computed tomography (CT) now provides more valuable insights into fuel cells [124,135,289,291,294,297, 307,310,326,333,343,346,348,349,381,385,391–394,401,403–405]. In addition, X-ray CT instruments using advanced detectors, condenser optics and imaging objective lenses are capable of reaching sub-50 nm resolution in fuel cells [326,381,385,391–394]. Izzo et al. [381] visualized the actual porous microstructure of an SOFC and achieved a spatial resolution of 42.7 nm. As shown in Fig. 8.9, they identified the solid and porous regions, respectively.

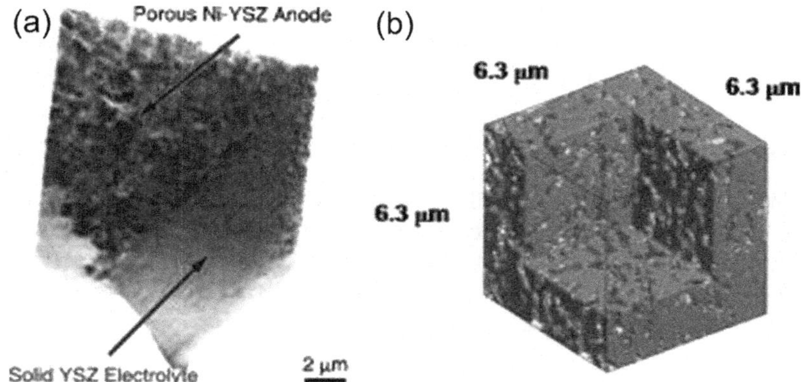

FIGURE 8.9 (a) A frame from the processed volume movie, where the material is black and void/empty space is white due to the selection of a threshold value that yielded a porosity matching experimental measurement for the sample, and (b) cut-away isosurface plot of the binary representative volume data. *Reproduced with permission from [381]. Copyright 2008, The Electrochemical Society*

Here, the outer shell was the solid YSZ electrolyte layer, while the inner layer was the porous Ni-YSZ anode. Similarly, Griesser et al. [403] also used 3D, computer aided, X-ray CT to detect the variation in thickness of the electrolyte as well as the homogeneity in thickness of the electrodes deposited. They revealed that gas diffusion through the electrode layer could become a problem when the thickness of the electrode layer was too high. On the other hand, if the layers were too thin, it could lead to a high electrical series resistance of the electrode.

Coupled with microtomography or high-resolution synchrotron radiography, X-ray absorption techniques were recently employed for the investigation of water transport and water management in fuel cells [124,135, 310,349,404–407]. Sinha et al. [310] used X-ray microtomography to obtain 3D images of liquid water distribution in a gas diffusion layer (GDL) during gas purge with a resolution of $10 \times 10 \times 13.4$ μm. They demonstrated that the liquid water distribution could be quantified at the component level. Using high-resolution synchrotron X-ray radiography, Hartnig et al. [124] studied the evolution of water clusters and the transport of liquid water from the catalyst layer through the gas diffusion layer (GDL) to the gas channels in a PEMFC. They obtained spatial and time resolutions of up to 3 μm and 5 s, respectively, and separately quantified the liquid water content in the respective components. They revealed the diffusion barrier in the interfacial area of the CL, MPL and the adjacent GDL, and confirmed that liquid water existed on the CL surface, which was directly observed with optical visualization by Zhang et al. [408]. The presence of liquid water flooding decreases PEMFC performance, and accelerates a number of fuel cell degradation mechanisms, including corrosion.

Thus, a better understanding of the liquid water transport phenomena will provide significant performance gains, as well as lifetime enhancement, in the fuel cell.

7. THERMAL MAPPING

Thermal management of fuel cells has long been recognized as an important factor in minimizing fuel cell degradation. Heat is always generated due to various loss mechanisms inherent in the operation of fuel cells, and inadequate thermal management can accelerate fuel cell degradation. Heat generation can include entropic heat of electrochemical reactions, crossover, and ohmic resistances, as well as latent heat release due to water condensation and interconnection resistances [8,109,409,410]. The degradation mechanisms associated with poor thermal management include thermal stresses on various components, local dehydration/thinning of membranes, agglomeration/sintering of catalysts, increase of interfacial resistance, and flooding in fuel cells. Spatial temperature variations can occur when these heat-generating processes occur preferentially in different parts of the fuel cell stack. Therefore, understanding the temperature distribution and thermal effects on fuel cell components and the overall system is critical in optimizing the performance and durability of fuel cells. In recent years, there have been a number of studies on temperature distribution and measurement in the active area of a fuel cell using thermal infrared (IR) cameras [129,411–417], novel film temperature sensors [418–424], thermocouples [425,426], and micro- and millimeter-scale sensors [427]. Thermal imaging systems can reach a measurement accuracy of up to 0.1°C and spatial resolution of 0.5 mm [412].

Wang et al. [411] developed an IR-transparent PEMFC incorporating a barium fluoride window on the anode side. They measured the temperature at certain points along the flow channel with an accuracy of better than ±0.3°C in the range of 20–80°C. They found that both the temperatures and non-uniformity of temperature distribution over the surface increase with the current density. Hakenjos et al. [413] further designed a PEMFC for combined measurements of current, temperature and water distribution on the cathode side. The back plate of the cathode flow field was a zinc selenide optical window which is transparent to infrared (IR) as well as visible wavelengths. They also found that large current densities lead to elevated temperatures on the cathode side. Moreover, the flooded areas showed higher temperatures than non-flooded areas. Shimoi et al. [416] also visualized the temperature field in a PEMFC with thermal IR imaging and found that a local temperature maximum in the membrane might exceed the membrane design limit, which would accelerate the membrane's physical degradation and ultimately result in membrane failure.

Thermal investigation is particularly important when a fuel cell is exposed to sub-zero temperatures since fuel cells undergo severe degradation during

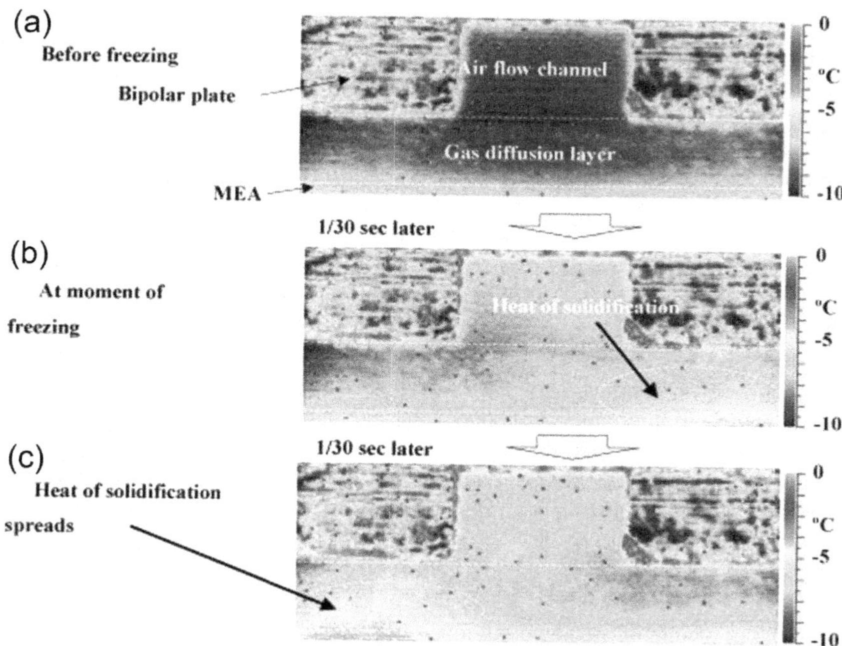

FIGURE 8.10 Infrared images inside PEMFC cross-section: (a) before freezing, (b) at moment of freezing, and (c) 1/30 s after freezing. *Reprinted from [417] with permission from Elsevier*

cold starts as well as cycling between sub- and above-zero temperatures. The possible degradation mechanisms include ice formation and removal, loss of protonic conductivity, delamination of the MEA, dissolution/thinning of membrane, loss in catalyst electrochemical surface area, and decrease of material strength. Ishikawa et al. [417] investigated the phenomenon of water freezing in operating PEMFCs with thermal infrared and visible images. They found that super-cooled water was generated on the gas diffusion layer (GDL) surface and that water freezing occurred at the interface between the GDL and MEA at the moment that the cell performance deteriorates. In addition, they clarified that the heat of solidification was released at the GDL/MEA interface at the moment when super-cooled water froze inside the PEMFC cross-section, as shown in Fig. 8.10. They revealed that ice formation between the GDL/MEA interface blocked air access leading to a drop in cell performance.

To measure the temperature distribution within an operating PEMFC, Mench et al. [418] proposed a thin-film thermal sensor which was placed inside the MEA between two sheets of Nafion. He et al. [419] further developed the thin-film thermal sensor, which is around 16 μm thick and is fabricated using micro-fabrication techniques. They revealed an electrolyte temperature increase of 1.5°C in real-time when operating the fuel cell at 0.2 V and a current density of 0.19 A/cm^2. Lee et al [422] demonstrated the feasibility of embedding 2 μm thick flexible thin film sensors within the MEA of micro-fuel cells to measure the local temperature and humidity, and found that the maximum temperature difference between MEA and the outer surface of a bipolar plate is 5.7°C.

The non-uniform temperature distribution within an operating fuel cell and its change with operating conditions and operating time, including cold starts and repeated cycling, are critical to the fuel cell performance and durability, since they are directly related to thermal stresses on various components and the associated degradation. In addition, an improved understanding of the thermal effects in fuel cells will provide more accurate guidelines for the improvement of cell design and operation, proper water/thermal management and validation of numerical models. A knowledge of the temperature field is critical for understanding liquid/vapor transport, phase change, and interfacial effects. Although research to date on thermal effects and temperature distribution is limited due to cost and technical challenges, future investigations are needed to better understand their effects on performance degradation and durability.

8. SUMMARY AND OUTLOOK

Degradation is one of the key challenges facing fuel cells researchers that must be minimized in order to facilitate large-scale commercialization. Various efforts with advanced high-resolution characterization techniques have been made to understand the underlying degradation mechanisms and improve fuel cell durability. In this chapter, we have discussed the most relevant *in-situ* and *ex-situ* diagnostic tools for fuel cell degradation studies and categorized them as: optical visualization, neutron imaging, electron spectroscopy and microscopy, X-ray techniques, magnetic resonance imaging, and thermal mapping. Table 8.1 summarizes the principal features of all the techniques reviewed here. This chapter reveals that while a few techniques have been successfully employed for studying degradation to date, there exists a significant opportunity to apply some powerful, yet under-utilized, tools for this purpose.

Optical visualization has proved to be an efficient and low-cost way to investigate water management and its associated degradation in PEMFCs. By directly visualizing liquid water formation/transport/removal in flow fields, diffusion media and catalyst layers from macro to microscale, it provides direct information about water distribution and surface degradation of fuel cell materials and components. Coupled with quantitative analysis, this method can easily be used to derive changes in water production/transport, to capture phase transients, and to monitor the structure and morphology of materials in an operating fuel cell. In addition, this technique can also assist in the investigation of thermal management in combination with the thermal mapping technique. Neutron imaging and magnetic resonance imaging are very powerful *in-situ* techniques to visualize and quantify liquid water in an operating PEMFC.

TABLE 8.1 Summary of High-Resolution Diagnostic Techniques for Degradation Studies in Fuel Cells

Method	Applications	Spatial/Temporal Resolution	Advantages	Challenges
Optical visualization	Water management; Material/component degradation	5 μm/0.03 s	*In-situ* and *ex-situ*; nonintrusive; 2D mapping	Limited spatial and temporal resolution; requires component substitution and modification with optical transparent part; only surface observation possible; quantitative information is limited
Neutron imaging	Water management	25 μm/0.05 s	*In-situ*; nonintrusive; 2D mapping; can use conventional fuel cell; subsurface observation possible; quantitative information	Limited availability of technique; spatial resolution and temporal resolution
Magnetic resonance imaging	Water management	8 μm/50 s	*In-situ*; nonintrusive; 2D mapping; subsurface observation possible; quantitative information	Limited availability; trade-off between spatial and temporal resolution; nonmagnetic materials needed
Electron spectroscopy and microscopy	Material/component degradation	Sub nm/(*ex-situ*)	2D mapping; high spatial resolution	*In-situ* studies are difficult; quantitative information is limited

(*Continued*)

TABLE 8.1 Summary of High-Resolution Diagnostic Techniques for Degradation Studies in Fuel Cells—cont'd

Method	Applications	Spatial/Temporal Resolution	Advantages	Challenges
X-ray techniques	Material/component degradation; Water management	1 nm/5 s	*In-situ* and *ex-situ*; nonintrusive; 2D/3D; high spatial resolution; quantitative information possible	Low temporal resolution; *in-situ* only possible for X-ray absorption; cannot probe materials with high atomic number
Thermal mapping	Thermal/water management;	500 μm/0.03 s	*In-situ* and *ex-situ*; nonintrusive; 2D mapping; quantitative information is possible; temporal resolution can be easily improved	IR transparent materials needed; certain working distance needed; component substitution and modification; spatial resolution and temporal resolution

Unlike direct optical visualization, these methods are able to obtain sub-surface information about transport phenomena occurring inside the material, or under certain components within the fuel cell, such as under the lands, inside the GDL and MPL and inside the electrode and membrane. The unique insight provided by these techniques, however, is somewhat limited by their lower spatial and temporal resolutions. Furthermore, only nonmagnetic materials are suitable for MRI techniques, which prevents the study of water transport and degradation in carbon diffusion media and electrodes.

Electron spectroscopy and microscopy, including X-ray photoelectron spectroscopy (XPS), scanning electron microscopy (SEM), and transmission electron microscopy (TEM), have become more attractive for *ex-situ* characterization of material degradation. With a surface sensitivity of up to 0.01%, XPS provides information about element variation including changes in their chemical composition and state of oxidation on a selected material surface in fuel cells. The recent development of higher resolution SEM/TEM (down to sub nm) has made it possible to examine fine details – even as small as a single column of atoms. Compositional analyses of degradation of catalyst alloys and the particle structure at the nm scale using these tools have been reported. Further quantitative investigations of material degradation are expected in the future with these electron techniques.

In general, X-ray techniques reveal degradation information in terms of the crystal size, chemical composition, and the internal structure of materials and components in fuel cells. X-ray diffraction (XRD) has been widely used to detect *ex-situ* catalyst degradation by deriving the crystal size change with a spatial resolution down to sub nm levels. Coupled with three-dimensional (3D) computed tomography (CT), the X-ray absorption technique has obtained an *in-situ* spatial resolution of 3 μm. It is capable of reaching sub-50 nm resolution in fuel cells by using advanced detectors, condenser optics and imaging objective lenses. It has also shown potential for the *in-situ* investigation of MEA degradation, liquid transport and two-phase flow in the diffusion media and electrodes.

The understanding of degradation mechanisms in fuel cells is still quite limited. A given degradation mechanism in a material, component or system usually results from multiple causes, including water/thermal management, operating conditions and environment, and impurities and contamination. Degradation mechanisms include catalyst-particle agglomeration/sintering/oxidation, carbon-support corrosion, membrane thinning and decomposition, loss of hydrophobicity, interfacial delamination, and mechanical fatigue. In addition, improper water management during fuel cell operation can lead to either membrane dryout or flooding of gas diffusion and catalyst layers which can cause deterioration of chemical/physical/mechanical properties of fuel cell components, including membranes, catalyst layers and gas diffusion media. For example, if the membrane dries out during operation its strength will deteriorate, and the ionic transport resistance of the membrane and catalyst layer will significantly increase. Furthermore, flooding will block/reduce the transport of fuel/oxidant into the reaction sites and lead to local starvation and undesirable flow/temperature/current distributions, which can induce carbon-support corrosion, catalyst oxidation, increase in size of the pore structure and even cracks within the structure. At the same time, flooding accelerates a number of fuel cell degradation mechanisms, including catalyst agglomeration/migration/sintering, ionomer erosion/decomposition, loss of hydrophobicity, deterioration of gas diffusion layers, and corrosion of the components. Moreover, during fuel cell sub-zero operations (e.g. cold start and freeze/thaw cycles), flooding can cause ice formation in the membrane, catalyst layer, microporous layer, and gas diffusion layer, which leads to significant performance decay and material degradation due to local cyclical volume expansion, fatigue stresses and structural damage.

Intensive efforts are required to establish a quantitative correlation between the degradation of a single material arising from multiple pathways to performance loss, and a quantitative link from individual component degradation to the system failure. High-resolution diagnostic tools, such as those described in this chapter, have a prominent role in performing the innovative and comprehensive analyses that are necessary. In addition, measurements obtained with hybrid diagnostic methods by simultaneously combining two or more techniques can provide critical insights that are not possible by applying the same methods individually. For example, the degradation of the catalyst layer is correlated to the change in size of the catalyst/carbon particle, its chemical state, composition and distribution, ionomer durability, heat and liquid water product transport, and interfacial effects. Multiple techniques with their respective strengths must be jointly leveraged to shed light on these complex and interlinked processes. Various combinations of the diagnostic techniques discussed in this chapter could be developed to create a comprehensive understanding of fuel cell degradation in the future.

REFERENCES

[1] J. Larminie, A. Dicks, Fuel cell systems explained, Second Ed., J. Wiley, Chichester, West Sussex, 2003. xxii, p. 406.

[2] W. Vielstich, A. Lamm, H. Gasteiger, Handbook of Fuel Cells: Fundamentals, Technology and Applications, Wiley, New York, 2003.

[3] N. Demirdoven, J. Deutch, Hybrid cars now, fuel cell cars later, Science 305 (5686) (2004) 974–976.

[4] B. Du, Q.H. Guo, R. Pollard, D. Rodriguez, C. Smith, J. Elter, PEM fuel cells: Status and challenges for commercial stationary power applications, Jom 58 (8) (2006) 45–49.

[5] F.A. de Bruijn, V.A.T. Dam, G.J.M. Janssen, Durability and degradation issues of PEM fuel cell components, Fuel Cells 8 (1) (2008) 3–22.

[6] M.M. Mench, Fuel cell engines, John Wiley & Sons, Hoboken, N.J., 2008, xi, p. 515.

[7] T.J. Schmidt, Proton exchange fuel cell durability, Springer, New York, 2008.

[8] W. Schmittinger, A. Vahidi, A review of the main parameters influencing long-term performance and durability of PEM fuel cells, Journal of Power Sources 180 (1) (2008) 1–14.

[9] J.F. Wu, X.Z. Yuan, J.J. Martin, H.J. Wang, J.J. Zhang, J. Shen, S.H. Wu, W. Merida, A review of PEM fuel cell durability: Degradation mechanisms and mitigation strategies, Journal of Power Sources 184 (1) (2008) 104–119.

[10] Z.G. Yang, Recent advances in metallic interconnects for solid oxide fuel cells, International Materials Reviews 53 (1) (2008) 39–54.

[11] N. Yousfi-Steiner, P. Mocoteguy, D. Candusso, D. Hissel, A. Hernandez, A. Aslanides, A review on PEM voltage degradation associated with water management: Impacts, influent factors and characterization, Journal of Power Sources 183 (1) (2008) 260–274.

[12] S.S. Zhang, X.Z. Yuan, H.J. Wang, W. Merida, H. Zhu, J. Shen, S.H. Wu, J.J. Zhang, A review of accelerated stress tests of MEA durability in PEM fuel cells, International Journal of Hydrogen Energy 34 (1) (2009) 388–404.

[13] C. Wang, Fundamental models for fuel cell engineering, Chem. Rev. 104 (10) (2004) 4727–4766.

[14] A.Z. Weber, J. Newman, Coupled thermal and water management in polymer electrolyte fuel cells, Journal of the Electrochemical Society 153 (12) (2006) A2205–A2214.

[15] C. Marr, X. Li, Composition and performance modeling of catalyst layer in a proton exchange membrane fuel cell, Journal of Power Sources 77 (1) (1999) 17–27.

[16] M. Eikerling, Water management in cathode catalyst layers of PEM fuel cells, Journal of the Electrochemical Society 153 (2006) E58.

[17] H. Meng, A three-dimensional PEM fuel cell model with consistent treatment of water transport in MEA, Journal of Power Sources 162 (1) (2006) 426–435.

[18] N. Djilali, Computational modelling of polymer electrolyte membrane (PEM) fuel cells: Challenges and opportunities, Energy 32 (4) (2007) 269–280.

[19] F.Y. Zhang, D. Spernjak, A.K. Prasad, S.G. Advani, In situ characterization of the catalyst layer in a polymer electrolyte membrane fuel cell, Journal of the Electrochemical Society 154 (2007) B1152.

[20] Y. Wang, X. Feng, Analysis of Reaction Rates in the Cathode Electrode of Polymer Electrolyte Fuel Cell I. Single-Layer Electrodes, Journal of the Electrochemical Society 155 (2008) B1289.

[21] Y. Wang, X. Feng, Analysis of the Reaction Rates in the Cathode Electrode of Polymer Electrolyte Fuel Cells, Journal of the Electrochemical Society 156 (2009) B403.

[22] J. Baschuk, X. Li, Modelling of polymer electrolyte membrane fuel cells with variable degrees of water flooding, Journal of Power Sources 86 (1–2) (2000) 181–196.

[23] T. Berning, N. Djilali, A 3D, multiphase, multicomponent model of the cathode and anode of a PEM fuel cell, Journal of the Electrochemical Society 150 (2003) A1589.

[24] A. Weber, J. Newman, Modeling transport in polymer-electrolyte fuel cells, Chemical reviews 104 (2004) 4679–4726.

[25] X.G. Yang, F.Y. Zhang, A.L. Lubawy, C.Y. Wang, Visualization of liquid water transport in a PEFC, Electrochemical and Solid State Letters 7 (11) (2004) A408–A411.

[26] J. Benziger, J. Nehlsen, D. Blackwell, T. Brennan, J. Itescu, Water flow in the gas diffusion layer of PEM fuel cells, Journal of Membrane Science 261 (1–2) (2005) 98–106.

[27] G. Lin, T. Van Nguyen, Effect of thickness and hydrophobic polymer content of the gas diffusion layer on electrode flooding level in a PEMFC, Journal of the Electrochemical Society 152 (2005) A1942.

[28] U. Pasaogullari, C. Wang, K. Chen, Two-phase transport in polymer electrolyte fuel cells with bilayer cathode gas diffusion media, Journal of the Electrochemical Society 152 (2005) A1574.

[29] Y. Cai, J. Hu, H. Ma, B. Yi, H. Zhang, Effect of water transport properties on a PEM fuel cell operating with dry hydrogen, Electrochimica Acta 51 (28) (2006) 6361–6366.

[30] J. Kowal, A. Turhan, K. Heller, J. Brenizer, M. Mench, Liquid water storage, distribution, and removal from diffusion media in PEFCS, Journal of the Electrochemical Society 153 (2006) A1971.

[31] H. Yamada, T. Hatanaka, H. Murata, Y. Morimoto, Measurement of flooding in gas diffusion layers of polymer electrolyte fuel cells with conventional flow field, Journal of the Electrochemical Society 153 (2006) A1748.

[32] L. You, H. Liu, A two-phase flow and transport model for PEM fuel cells, Journal of Power Sources 155 (2) (2006) 219–230.

[33] F.Y. Zhang, X.G. Yang, C.Y. Wang, Liquid water removal from a polymer electrolyte fuel cell, Journal of the Electrochemical Society 153 (2) (2006) A225–A232.

[34] E. Kumbur, K. Sharp, M. Mench, On the effectiveness of Leverett approach for describing the water transport in fuel cell diffusion media, Journal of Power Sources 168 (2) (2007) 356–368.

[35] S. Park, J. Lee, B. Popov, Effect of PTFE content in microporous layer on water management in PEM fuel cells, Journal of Power Sources 177 (2) (2008) 457–463.

[36] Y. Wang, Modeling of two-phase transport in the diffusion media of polymer electrolyte fuel cells, Journal of Power Sources 185 (2008) 261–271.

[37] H Meng, Multi-dimensional liquid water transport in the cathode of a PEM fuel cell with consideration of the micro-porous layer (MPL), International Journal of Hydrogen Energy 34 (13) (2009) 5488–5497.

[38] D. Natarajan, T. Van Nguyen, Three-dimensional effects of liquid water flooding in the cathode of a PEM fuel cell, Journal of Power Sources 115 (1) (2003) 66–80.

[39] P. Nguyen, T. Berning, N. Djilali, Computational model of a PEM fuel cell with serpentine gas flow channels, Journal of Power Sources 130 (1–2) (2004) 149–157.

[40] A. Su, F. Weng, C. Hsu, Y. Chen, Studies on flooding in PEM fuel cell cathode channels, International Journal of Hydrogen Energy 31 (8) (2006) 1031–1039.

[41] T. Trabold, J. Owejan, D. Jacobson, M. Arif, P. Huffman, In situ investigation of water transport in an operating PEM fuel cell using neutron radiography: Part 1–Experimental method and serpentine flow field results, International Journal of Heat and Mass Transfer 49 (25–26) (2006) 4712–4720.

[42] D. Spernjak, A.K. Prasad, S.G. Advani, Experimental investigation of liquid water formation and transport in a transparent single-serpentine PEM fuel cell, Journal of Power Sources 170 (2) (2007) 334–344.

[43] H. Li, Y. Tang, Z. Wang, Z. Shi, S. Wu, D. Song, J. Zhang, K. Fatih, H. Wang, A review of water flooding issues in the proton exchange membrane fuel cell, Journal of Power Sources 178 (1) (2008) 103–117.

[44] J. Stumper, C. Stone, Recent advances in fuel cell technology at Ballard, Journal of Power Sources 176 (2) (2008) 468–476.

[45] K. Tüber, D. Pócza, C. Hebling, Visualization of water buildup in the cathode of a transparent PEM fuel cell, Journal of Power Sources 124 (2) (2003) 403–414.

[46] J. Martin, P. Oshkai, N. Djilali, Flow structures in a U-Shaped fuel cell flow channel: Quantitative visualization using particle image velocimetry, Journal of Fuel Cell Science and Technology 2 (1) (2005) 70–80.

[47] K. Sugiura, M. Nakata, T. Yodo, Y. Nishiguchi, M. Yamauchi, Y. Itoh, Evaluation of a cathode gas channel with a water absorption layer/waste channel in a PEFC by using visualization technique, Journal of Power Sources 145 (2) (2005) 526–533.

[48] T.A. Trabold, Minichannels in polymer electrolyte membrane fuel cells, Heat Transfer Engineering 26 (3) (2005) 3–12.

[49] Y.H. Cai, J. Hu, H.P. Ma, B.L. Yi, H.M. Zhang, Effects of hydrophilic/hydrophobic properties on the water behavior in the micro-channels of a proton exchange membrane fuel cell, Journal of Power Sources 161 (2) (2006) 843–848.

[50] P.A.C. Chang, J. St-Pierre, J. Stumper, B. Wetton, Flow distribution in proton exchange membrane fuel cell stacks, Journal of Power Sources 162 (1) (2006) 340–355.

[51] E.C. Kumbur, K.V. Sharp, M.M. Mench, Liquid droplet behavior and instability in a polymer electrolyte fuel cell flow channel, Journal of Power Sources 161 (1) (2006) 333–345.

[52] S. Litster, D. Sinton, N. Djilali, Ex situ visualization of liquid water transport in PEM fuel cell gas diffusion layers, Journal of Power Sources 154 (1) (2006) 95–105.

[53] H. Ma, H. Zhang, J. Hu, Y. Cai, B. Yi, Diagnostic tool to detect liquid water removal in the cathode channels of proton exchange membrane fuel cells, Journal of Power Sources 162 (1) (2006) 469–473.

[54] H.P. Ma, H.M. Zhang, J. Hu, Y.H. Cai, B.L. Yi, Diagnostic tool to detect liquid water removal in the cathode channels of proton exchange membrane fuel cells, Journal of Power Sources 162 (1) (2006) 469–473.

[55] K. Sugiura, T. Yodo, M. Yamauchi, K. Tanimoto, Visualization of electrolyte volatile phenomenon in DIR-MCFC, Journal of Power Sources 157 (2) (2006) 739–744.

[56] Z.G. Zhan, J.S. Xiao, D.Y. Li, M. Pan, R.Z. Yuan, Effects of porosity distribution variation on the liquid water flux through gas diffusion layers of PEM fuel cells, Journal of Power Sources 160 (2) (2006) 1041–1048.

[57] Z.G. Zhan, J.S. Xiao, M. Pan, R.Z. Yuan, Characteristics of droplet and film water motion in the flow channels of polymer electrolyte membrane fuel cells, Journal of Power Sources 160 (1) (2006) 1–9.

[58] S. Ge, C.Y. Wang, Liquid water formation and transport in the PEFC anode, Journal of the Electrochemical Society 154 (10) (2007) B998–B1005.

[59] T.H. Ha, H.S. Kim, K.D. Min, Oxygen concentration in the cathode channel of PEM fuel cell using gas chromatograph, International Journal of Automotive Technology 8 (1) (2007) 119–126.

[60] R.S. Jayashree, M. Mitchell, D. Natarajan, L.J. Markoski, P.J. Kenis, Microfluidic hydrogen fuel cell with a liquid electrolyte, Langmuir 23 (13) (2007) 6871–6874.

[61] S. Litster, C.R. Buie, T. Fabian, J.K. Eaton, J.G. Santiago, Active water management for PEM fuel cells, Journal of the Electrochemical Society 154 (10) (2007) B1049–B1058.

[62] X. Liu, H. Guo, F. Ye, C. Ma, Water flooding and pressure drop characteristics in flow channels of proton exchange membrane fuel cells, Electrochimica Acta 52 (11) (2007) 3607–3614.

[63] T. Ous, C. Arcoumanis, Visualisation of water droplets during the operation of PEM fuel cells, Journal of Power Sources 173 (1) (2007) 137–148.

[64] F.M. Wang, J.E. Steinbrenner, C.H. Hidrovo, T.A. Kramer, E.S. Lee, S. Vigneron, C.H. Cheng, J.K. Eaton, K.E. Goodson, Investigation of two-phase transport phenomena in microchannels using a microfabricated experimental structure, Applied Thermal Engineering 27 (10) (2007) 1728–1733.

[65] C. Xu, T.S. Zhao, A new flow field design for polymer electrolyte-based fuel cells, Electrochemistry Communications 9 (3) (2007) 497–503.

[66] X. Zhu, P.C. Sui, N. Djilali, Dynamic behaviour of liquid water emerging from a GDL pore into a PEMFC gas flow channel, Journal of Power Sources 172 (1) (2007) 287–295.

[67] F. Barreras, A. Lozano, L. Valino, R. Mustafa, C. Marin, Fluid dynamics performance of different bipolar plates – Part I. Velocity and pressure fields, Journal of Power Sources 175 (2) (2008) 841–850.

[68] A. Bazylak, J. Heinrich, N. Djilali, D. Sinton, Liquid water transport between graphite paper and a solid surface, Journal of Power Sources 185 (2) (2008) 1147–1153.

[69] J. Cho, H.S. Kim, K. Min, Transient response of a unit proton-exchange membrane fuel cell under various operating conditions, Journal of Power Sources 185 (1) (2008) 118–128.

[70] H.S. Kim, K. Min, Experimental investigation of dynamic responses of a transparent PEM fuel cell to step changes in cell current density with operating temperature, Journal of Mechanical Science and Technology 22 (11) (2008) 2274–2285.

[71] E. Kimball, T. Whitaker, Y.G. Kevrekidis, J.B. Benziger, Drops, slugs, and flooding in polymer electrolyte membrane fuel cells, Aiche Journal 54 (5) (2008) 1313–1332.

[72] A. Lozano, L. Valino, F. Barreras, R. Mustata, Fluid dynamics performance of different bipolar plates – Part II. Flow through the diffusion layer, Journal of Power Sources 179 (2) (2008) 711–722.

[73] T. Metz, J. Viertel, C. Muller, S. Kerzenmacher, N. Paust, R. Zengerle, P. Koltay, Passive water management for fuel-cells using capillary microstructures, J. Micromech. Microeng 18 (104007) (2008) 104007.

[74] T. Murahashi, H. Kobayashi, E. Nishiyama, Combined measurement of PEMFC performance decay and water droplet distribution under low humidity and high CO, Journal of Power Sources 175 (1) (2008) 98–105.

[75] L. Zhang, H. Bi, D. Wilkinson, J. Stumper, H. Wang, Gas-liquid two-phase flow patterns in parallel channels for fuel cells, Journal of Power Sources 183 (2) (2008) 643–650.

[76] N. Akhtar, A. Qureshi, J. Scholta, C. Hartnig, M. Messerschmidt, W. Lehnert, Investigation of water droplet kinetics and optimization of channel geometry for PEM fuel cell cathodes, International Journal of Hydrogen Energy 34 (7) (2009) 3104–3111.

[77] I.S. Hussaini, C.Y. Wang, Visualization and quantification of cathode channel flooding in PEM fuel cells, Journal of Power Sources 187 (2) (2009) 444–451.

[78] Z. Lu, S.G. Kandlikar, C. Rath, M. Grimm, W. Domigan, A.D. White, M. Hardbarger, J.P. Owejan, T.A. Trabold, Water management studies in PEM fuel cells, Part II: Ex situ investigation of flow maldistribution, pressure drop and two-phase flow pattern in gas channels, International Journal of Hydrogen Energy 34 (8) (2009) 3445–3456.

[79] L.J. Yu, W.C. Chen, M.J. Qin, G.P. Ren, Experimental research on water management in proton exchange membrane fuel cells, Journal of Power Sources 189 (2) (2009) 882–887.

[80] A. Bazylak, Liquid water visualization in PEM fuel cells: A review, International Journal of Hydrogen Energy 34 (9) (2009) 3845–3857.

[81] A. Bazylak, D. Sinton, N. Djilali, Dynamic water transport and droplet emergence in PEMFC gas diffusion layers, Journal of Power Sources 176 (1) (2008) 240–246.

[82] A. Bazylak, D. Sinton, Z.S. Liu, N. Djilali, Effect of compression on liquid water transport and microstructure of PEMFC gas diffusion layers, Journal of Power Sources 163 (2) (2007) 784–792.

[83] K. Chen, M. Hickner, D. Noble, Simplified models for predicting the onset of liquid water droplet instability at the gas diffusion layer/gas flow channel interface, International Journal of Energy Research 29 (12) (2005) 1113–1132.

[84] B. Gao, T.S. Steenhuis, Y. Zevi, J.Y. Parlange, R.N. Carter, T.A. Trabold, Visualization of unstable water flow in a fuel cell gas diffusion layer, Journal of Power Sources 190 (2) (2009) 493–498.

[85] Q. Liao, X. Zhu, X.Y. Zheng, Y.D. Ding, Visualization study on the dynamics of CO_2 bubbles in anode channels and performance of a DMFC, Journal of Power Sources 171 (2) (2007) 644–651.

[86] J.H. Nam, K.J. Lee, G.S. Hwang, C.J. Kim, M. Kaviany, Microporous layer for water morphology control in PEMFC, International Journal of Heat and Mass Transfer 52 (11–12) (2009) 2779–2791.

[87] K. Nishida, T. Murakami, S. Tsushima, S. Hirai, Microscopic visualization of state and behavior of liquid water in a gas diffusion layer of PEFC, Electrochemistry 75 (2) (2007) 149–151.

[88] R.P. Ramasamy, E.C. Kumbur, M.M. Mench, W. Liu, D. Moore, M. Murthy, Investigation of macro- and micro-porous layer interaction in polymer electrolyte fuel cells, International Journal of Hydrogen Energy 33 (13) (2008) 3351–3367.

[89] A. Theodorakakos, T. Ous, A. Gavaises, J.M. Nouri, N. Nikolopoulos, H. Yanagihara, Dynamics of water droplets detached from porous surfaces of relevance to PEM fuel cells, Journal of Colloid and Interface Science 300 (2) (2006) 673–687.

[90] J. Wu, X. Zi Yuan, H. Wang, M. Blanco, J. Martin, J. Zhang, Diagnostic tools in PEM fuel cell research: Part II Physical/chemical methods, International Journal of Hydrogen Energy 33 (6) (2008) 1747–1757.

[91] S.H. Ge, C.Y. Wang, Characteristics of subzero startup and water/ice formation on the catalyst layer in a polymer electrolyte fuel cell, Electrochimica Acta 52 (14) (2007) 4825–4835.

[92] Y. Ishikawa, T. Morita, K. Nakata, K. Yoshida, M. Shiozawa, Behavior of water below the freezing point in PEFCs, Journal of Power Sources 163 (2) (2007) 708–712.

[93] A. Li, S. Chan, N. Nguyen, Anti-flooding cathode catalyst layer for high performance PEM fuel cell, Electrochemistry Communications 11 (4) (2009) 897–900.

[94] F.Y. Zhang, A.K. Prasad, S.G. Advani, Investigation of a copper etching technique to fabricate metallic gas diffusion media, Journal of Micromechanics and Microengineering 16 (11) (2006) N23–N27.

[95] F.Y. Zhang, S.G. Advani, A.K. Prasad, Performance of a metallic gas diffusion layer for PEM fuel cells, Journal of Power Sources 176 (1) (2008) 293–298.

[96] Q. Ye, T. Van Nguyen, Three-dimensional simulation of liquid water distribution in a PEMFC with experimentally measured capillary functions, Journal of the Electrochemical Society 154 (2007) B1242.

[97] Songprakorp, R., Investigation of transient phenomena of proton exchange membrane fuel cells. 2008.

[98] J. Maruyama, M. Umemura, M. Inaba, A. Tasaka, I. Abe, Carbon Surface Oxidation by Short-Term Ozone Treatment for Modeling Long-Term Degradation of Fuel Cell Cathodes, Journal of the Electrochemical Society 156 (2009) A181.

[99] J. Nam, K. Lee, G. Hwang, C. Kim, M. Kaviany, Microporous layer for water morphology control in PEMFC, International Journal of Heat and Mass Transfer 52 (11–12) (2009) 2779–2791.

[100] K. Jiao, B. Zhou, Effects of electrode wettabilities on liquid water behaviours in PEM fuel cell cathode, Journal of Power Sources 175 (1) (2008) 106–119.

[101] M. Eikerling, K. Malek, and Q. Wang, Catalyst Layer Modeling: Structure, Properties and Performance. book chapter in PEM Fuel Cells Catalysts and Catalyst Layers-Fundamentals and Applications, Ed. JJ Zhang, 2008 381–446.

[102] J. Liu, M. Eikerling, Model of cathode catalyst layers for polymer electrolyte fuel cells: The role of porous structure and water accumulation, Electrochimica Acta 53 (13) (2008) 4435–4446.

[103] R.J. Bellows, M.Y. Lin, M. Arif, A.K. Thompson, D. Jacobson, Neutron imaging technique for in situ measurement of water transport gradients within Nafion in polymer electrolyte fuel cells, Journal of the Electrochemical Society 146 (3) (1999) 1099–1103.

[104] R. Satija, D.L. Jacobson, M. Arif, S.A. Werner, In situ neutron imaging technique for evaluation of water management systems in operating PEM fuel cells, Journal of Power Sources 129 (2) (2004) 238–245.

[105] D. Hussey, D. Jacobson, M. Arif, P. Huffman, R. Williams, J. Cook, New neutron imaging facility at the NIST, Nuclear Inst. and Methods in Physics Research, A 542 (1–3) (2005) 9–15.

[106] Du, B., G. Wang, J. Elter, R. Pollard, D. Jacobson, D. Hussey, M. Arif, and G. Eisman. Applications of Neutron Radiography in PEM fuel cell Research and Development. 2006.

[107] K.J. Coakley, D.S. Hussey, Feasibility of single-view coded source neutron transmission tomography, Measurement Science & Technology 18 (11) (2007) 3391–3398.

[108] H. Li, Y.H. Tang, Z.W. Wang, Z. Shi, S.H. Wu, D.T. Song, J.L. Zhang, K. Fatih, J.J. Zhang, H.J. Wang, Z.S. Liu, R. Abouatallah, A. Mazza, A review of water flooding issues in the proton exchange membrane fuel cell, Journal of Power Sources 178 (1) (2008) 103–117.

[109] J.F. Wu, X.Z. Yuan, H.J. Wang, M. Blanco, J.J. Martin, J.J. Zhang, Diagnostic tools in PEM fuel cell research: Part II – Physical/chemical methods, International Journal of Hydrogen Energy 33 (6) (2008) 1747–1757.

[110] A. Bazylak, Liquid water visualization in PEM fuel cells: A review, International Journal of Hydrogen Energy 34 (2009) 3845–3857.

[111] D. Kramer, J.B. Zhang, R. Shimoi, E. Lehmann, A. Wokaun, K. Shinohara, G.G. Scherer, In situ diagnostic of two-phase flow phenomena in polymer electrolyte fuel cells by neutron imaging Part A. Experimental, data treatment, and quantification, Electrochimica Acta 50 (13) (2005) 2603–2614.

[112] D. Spernjak, S.G. Advani, A.K. Prasad, Simultaneous Neutron and Optical Imaging in PEM Fuel Cells, Journal of the Electrochemical Society 156 (1) (2009) B109–B117.

[113] S. Kim, A. Heller, M. Hatzell, M. Mench, D. Hussey, D. Jacobson, High Resolution Neutron Imaging of Temperature-Driven Flow in Polymer Electrolyte Fuel Cells, 98, Transactions American Nuclear Society, 2008. 39.

[114] S. Kim, M. Mench, Investigation of temperature-driven water transport in polymer electrolyte fuel cell: Thermo-osmosis in membranes, Journal of Membrane Science 328 (1–2) (2009) 113–120.

[115] D.S. Hussey, D.L. Jacobson, M. Arif, J.P. Owejan, J.J. Gagliardo, T.A. Trabold, Neutron images of the through-plane water distribution of an operating PEM fuel cell, Journal of Power Sources 172 (1) (2007) 225–228.

[116] T. Kim, J. Kim, C. Sm, S. Lee, Y. Son, M. Kim, Experimental Approaches for Water Discharge Characteristics in PEMFC Using Neutron Imaging Technique at Conrad, HMI, Nuclear Engineering and Technology 41 (1) (2009) 135–142.

[117] M. Hickner, N. Siegel, K. Chen, D. Hussey, D. Jacobson, M. Arif, Understanding liquid water distribution and removal phenomena in an operating PEMFC via neutron radiography, Journal of the Electrochemical Society 155 (2008) B294.

[118] M.A. Hickner, N.P. Siegel, K.S. Chen, D.S. Hussey, D.L. Jacobson, M. Arif, In situ high-resolution neutron radiography of cross-sectional liquid water profiles in proton exchange membrane fuel cells, Journal of the Electrochemical Society 155 (4) (2008) B427–B434.

[119] R. Mukundan, J. Davey, T. Rockward, J. Spendelow, B. Pivovar, D. Hussey, D. Jacobson, M. Arif, R. Borup, Imaging of Water Profiles in PEM Fuel Cells using Neutron Radiography: Effect of Operating Conditions and GDL Composition, ECS Trans. 11 (1) (2007) 411–422.

[120] N. Pekula, K. Heller, P. Chuang, A. Turhan, M. Mench, J. Brenizer, K. Ünlü, Study of water distribution and transport in a polymer electrolyte fuel cell using neutron imaging, Nuclear Inst. and Methods in Physics Research, A 542 (1–3) (2005) 134–141.

[121] D.J. Ludlow, C.M. Calebrese, S.H. Yu, C.S. Dannehy, D.L. Jacobson, D.S. Hussey, M. Arif, M.K. Jensen, G.A. Eisman, PEM fuel cell membrane hydration measurement by neutron imaging, Journal of Power Sources 162 (1) (2006) 271–278.

[122] R. Fluckiger, S.A. Freunberger, D. Kramer, A. Wokaun, G.G. Scherer, F.N. Buchi, Anisotropic, effective diffusivity of porous gas diffusion layer materials for PEFC, Electrochimica Acta 54 (2) (2008) 551–559.

[123] J.T. Gostick, M.A. Ioannidis, M.W. Fowler, M.D. Pritzker, Pore network modeling of fibrous gas diffusion layers for polymer electrolyte membrane fuel cells, Journal of Power Sources 173 (1) (2007) 277–290.

[124] C. Hartnig, I. Manke, R. Kuhn, S. Kleinau, J. Goebbels, J. Banhart, High-resolution in-plane investigation of the water evolution and transport in PEM fuel cells, Journal of Power Sources 188 (2) (2009) 468–474.

[125] J.P. Owejan, T.A. Trabold, J. Gagliardo, D.L. Jacobson, R.N. Carter, D.S. Hussey, M. Arif, Voltage instability in a simulated fuel cell stack correlated to cathode water accumulation, Journal of Power Sources 171 (2) (2007) 626–633.

[126] K. Yoshizawa, K. Ikezoe, Y. Thsaki, D. Kramer, E.H. Lehmann, G.G. Scherer, Analysis of gas diffusion layer and flow-field design in a PEMFC using neutron radiography, Journal of the Electrochemical Society 155 (3) (2008) B223–B227.

[127] J.B. Zhang, D. Kramer, R. Shimoi, Y. Ona, E. Lehmann, A. Wokaun, K. Shinohara, G.G. Scherer, In situ diagnostic of two-phase flow phenomena in polymer electrolyte fuel cells by neutron imaging Part B. Material variations, Electrochimica Acta 51 (13) (2006) 2715–2727.

[128] P. Boillat, D. Kramer, B.C. Seyfang, G. Frei, E. Lehmann, G.G. Scherer, A. Wokaun, Y. Ichikawa, Y. Tasaki, K. Shinohara, In situ observation of the water distribution across a PEFC using high resolution neutron radiography, Electrochemistry Communications 10 (4) (2008) 546–550.

[129] J. Diep, D. Kiel, J. St-Pierre, A. Wong, Development of a residence time distribution method for proton exchange membrane fuel cell evaluation, Chemical Engineering Science 62 (3) (2007) 846–857.

[130] C. Hartnig, I. Manke, N. Kardjilo, A. Hilger, M. Gruenerbel, J. Kaczerowski, J. Banhart, W. Lehnert, Combined neutron radiography and locally resolved current density measurements of operating PEM fuel cells, Journal of Power Sources 176 (2) (2008) 452–459.

[131] J.J. Kowal, A. Turhan, K. Heller, J. Brenizer, M.M. Mench, Liquid water storage, distribution, and removal from diffusion media in PEFCS, Journal of the Electrochemical Society 153 (10) (2006) A1971–A1978.

[132] X.G. Li, I. Sabir, J. Park, A flow channel design procedure for PEM fuel cells with effective water removal, Journal of Power Sources 163 (2) (2007) 933–942.

[133] D.A. McKay, J.B. Siegel, W. Ott, A.G. Stefanopoulou, Parameterization and prediction of temporal fuel cell voltage behavior during flooding and drying conditions, Journal of Power Sources 178 (1) (2008) 207–222.

[134] J.P. Owejan, T.A. Trabold, D.L. Jacobson, M. Arif, S.G. Kandlikar, Effects of flow field and diffusion layer properties on water accumulation in a PEM fuel cell, International Journal of Hydrogen Energy 32 (17) (2007) 4489–4502.

[135] J. St-Pierre, PEMFC in situ liquid-water-content monitoring status, Journal of the Electrochemical Society 154 (7) (2007) B724–B731.

[136] A. Turhan, K. Heller, J.S. Brenizer, M.M. Mench, Passive control of liquid water storage and distribution in a PEFC through flow-field design, Journal of Power Sources 180 (2) (2008) 773–783.

[137] G. Gebel, O. Diat, Neutron and X-ray scattering: Suitable tools for studying ionomer membranes, Fuel Cells 5 (2) (2005) 261–276.

[138] G. Gebel, O. Diat, S. Escribano, R. Mosdale, Water profile determination in a running PEMFC by small-angle neutron scattering, Journal of Power Sources 179 (1) (2008) 132–139.

[139] T. Kim, C. Sim, M. Kim, Research on water discharge characteristics of PEM fuel cells by using neutron imaging technology at the NRF, HANARO, Applied Radiation and Isotopes 66 (5) (2008) 593–605.

[140] C. Ziegler, T. Heilmann, D. Gerteisen, Experimental study of two-phase transients in PEMFCs, Journal of the Electrochemical Society 155 (4) (2008) B349–B355.

[141] Y. Chen, H. Peng, D. Hussey, D. Jacobson, D. Tran, T. Abdel-Baset, M. Biernacki, Water distribution measurement for a PEMFC through neutron radiography, Journal of Power Sources 170 (2) (2007) 376–386.

[142] I. Manke, C. Hartnig, M. Grunerbel, J. Kaczerowski, W. Lehnert, N. Kardjilov, A. Hilger, J. Banhart, W. Treimer, M. Strobl, Quasi-in situ neutron tomography on polymer electrolyte membrane fuel cell stacks, Applied Physics Letters 90 (18) (2007).

[143] R. Mosdale, G. Gebel, M. Pineri, Water profile determination in a running proton exchange membrane fuel cell using small-angle neutron scattering, Journal of Membrane Science 118 (2) (1996) 269–277.

[144] F. Xu, O. Diat, G. Gebel, A. Morin, Determination of Transverse Water Concentration Profile Through MEA in a Fuel Cell Using Neutron Scattering, Journal of the Electrochemical Society 154 (2007) B1389.

[145] A. Geiger, A. Tsukada, E. Lehmann, P. Vontobel, A. Wokaun, G. Scherer, In situ investigation of two-phase flow patterns in flow fields of PEFC's using neutron radiography, Fuel Cells 2 (2) (2002).

[146] D. Kramer, E. Lehmann, G. Frei, P. Vontobel, A. Wokaun, G.G. Scherer, An on-line study of fuel cell behavior by thermal neutrons, Nuclear Instruments & Methods in Physics Research Section a-Accelerators Spectrometers Detectors and Associated Equipment 542 (1–3) (2005) 52–60.

[147] D. Spernjak, A.K. Prasad, S.G. Advani, In Situ Comparison of Water Content and Dynamics in Parallel, Single-Serpentine, and Interdigitated Flow Fields of PEM Fuel Cells, Journal of Power Sources 195 (11) (2010) 3553–3568.

[148] J. Siegel, D. McKay, A. Stefanopoulou, D. Hussey, D. Jacobson, Measurement of Liquid Water Accumulation in a PEMFC with Dead-Ended Anode, Journal of the Electrochemical Society 155 (2008) B1168.

[149] S. Kim, M.M. Mench, Investigation of Temperature-Driven Water Transport in Polymer Electrolyte Fuel Cell: Phase-Change-Induced Flow, Journal of the Electrochemical Society 156 (3) (2009) B353–B362.

[150] P.T. Callaghan, Principles of nuclear magnetic resonance microscopy, Clarendon Press, Oxford University Press, Oxford [England], New York, 1991, xvii, p. 492.

[151] J.B. Lambert, E.P. Mazzola, Nuclear magnetic resonance spectroscopy: an introduction to principles, applications, and experimental methods, Pearson/Prentice Hall, Upper Saddle River, N.J., 2004, xiv, 341, p. 8.

[152] S. Tsushima, S. Hirai, K. Kitamura, M. Yamashita, S. Takasel, MRI application for clarifying fuel cell performance with variation of polymer electrolyte membranes: Comparison of water content of a hydrocarbon membrane and a perfluorinated membrane, Applied Magnetic Resonance 32 (1–2) (2007) 233–241.

[153] K.W. Feindel, S.H. Bergens, R.E. Wasylishen, The use of H-1 NMR microscopy to study proton-exchange membrane fuel cells, Chemphyschem 7 (1) (2006) 67–75.

[154] K. Teranishi, S. Tsushima, S. Hirai, Study of the effect of membrane thickness on the performance of polymer electrolyte fuel cells by water distribution in a membrane, Electrochemical and Solid State Letters 8 (6) (2005) A281–A284.

[155] S. Tsushima, K. Teranishi, S. Hirai, Magnetic resonance imaging of the water distribution within a polymer electrolyte membrane in fuel cells, Electrochemical and Solid State Letters 7 (9) (2004) A269–A272.

[156] S. Tsushima, S. Takita, S. Hirai, N. Kubo, K. Aotani, Magnetic Resonance Imaging of a Polymer Electrolyte Membrane under Water Permeation, Experimental Heat Transfer 22 (1) (2009) 1–11.

[157] Z. Zhang, J. Martin, J. Wu, H. Wang, K. Promislow, B.J. Balcom, Magnetic resonance imaging of water content across the Nafion membrane in an operational PEM fuel cell, Journal of Magnetic Resonance 193 (2) (2008) 259–266.

[158] Z.H. Zhang, A.E. Marble, B. MacMillan, K. Promislow, J. Martin, H.J. Wang, B.J. Balcom, Spatial and temporal mapping of water content across Nafion membranes under wetting and drying conditions, Journal of Magnetic Resonance 194 (2) (2008) 245–253.

[159] A.K. Sahu, G. Selvarani, S.D. Bhat, S. Pitchumani, P. Sridhar, A.K. Shukla, N. Narayanan, A. Banerjee, N. Chandrakumar, Effect of varying poly(styrene sulfonic acid) content in poly(vinyl alcohol)-poly(styrene sulfonic acid) blend membrane and its ramification in hydrogen-oxygen polymer electrolyte fuel cells, Journal of Membrane Science 319 (1-2) (2008) 298–305.

[160] A.K. Sahu, G. Selvarani, S. Pitchumani, P. Sridhar, A.K. Shukla, N. Narayanan, A. Banerjee, N. Chandrakumar, PVA-PSSA membrane with interpenetrating networks and its methanol crossover mitigating effect in DMFCs, Journal of the Electrochemical Society 155 (7) (2008) B686–B695.

[161] T.A. Zawodzinski, T.E. Springer, F. Uribe, S. Gottesfeld, Characterization of Polymer Electrolytes for Fuel-Cell Applications, Solid State Ionics 60 (1–3) (1993) 199–211.

[162] T.A. Zawodzinski, C. Derouin, S. Radzinski, R.J. Sherman, V.T. Smith, T.E. Springer, S. Gottesfeld, Water-Uptake by and Transport through Nafion(R) 117 Membranes, Journal of the Electrochemical Society 140 (4) (1993) 1041–1047.

[163] S. Ha, Z. Dunbar, R. Masel, Magnetic resonance imaging (MRI): A new tool for fuel cell research, Abstracts of Papers of the American Chemical Society 230 (2005) U1638–U1639.

[164] J. Bedet, G. Maranzana, S. Leclerc, O. Lottin, C. Moyne, D. Stemmelen, P. Mutzenhardt, D. Canet, Magnetic resonance imaging of water distribution and production in a 6 cm^2 PEMFC under operation, International Journal of Hydrogen Energy 33 (12) (2008) 3146–3149.

[165] Z. Dunbar, R.I. Masel, Quantitative MRI study of water distribution during operation of a PEM fuel cell using Teflon (R) flow fields, Journal of Power Sources 171 (2) (2007) 678–687.

[166] Z.W. Dunbar, R.I. Masel, Magnetic resonance imaging investigation of water accumulation and transport in graphite flow fields in a polymer electrolyte membrane fuel cell: Do defects control transport? Journal of Power Sources 182 (1) (2008) 76–82.

[167] K.W. Feindel, S.H. Bergens, R.E. Wasylishen, Use of hydrogen-deuterium exchange for contrast in H–1 NMR microscopy investigations of an operating PEM fuel cell, Journal of Power Sources 173 (1) (2007) 86–95.

[168] K. Teranishi, S. Tsushima, S. Hirai, Analysis of water transport in PEFCs by magnetic resonance imaging measurement, Journal of the Electrochemical Society 153 (4) (2006) A664–A668.

[169] S. Tsushima, K. Teranishi, K. Nishida, S. Hirai, Water content distribution in a polymer electrolyte membrane for advanced fuel cell system with liquid water supply, Magnetic Resonance Imaging 23 (2) (2005) 255–258.

[170] K.R. Minard, V.V. Viswanathan, P.D. Majors, L.Q. Wang, P.C. Rieke, Magnetic resonance imaging (MRI) of PEM dehydration and gas manifold flooding during continuous fuel cell operation, Journal of Power Sources 161 (2) (2006) 856–863.

[171] R.F. Egerton, Electron energy-loss spectroscopy in the electron microscope, Plenum Press, New York, 1986, xii, p. 410.

[172] J.T. Yates, Experimental innovations in surface science: a guide to practical laboratory methods and instruments, AIP Press, New York, 1998, Springer. xv, p. 904.

[173] D. Briggs, J.T. Grant, Surface analysis by Auger and X-ray photoelectron spectroscopy, IM Publications, Chichester, West Sussex, U.K, 2003, xi, p. 899.

[174] More, K., L. Allard, H. Meyer, and S. Reeves, Microstructural Characterization of PEM Fuel Cell MEAs, in DOE FY 2009 Progress Report, D.F.P. Report, Editor. 2009.

[175] R. Borup, J. Meyers, B. Pivovar, Y.S. Kim, R. Mukundan, N. Garland, D. Myers, M. Wilson, F. Garzon, D. Wood, P. Zelenay, K. More, K. Stroh, T. Zawodzinski, J. Boncella, J.E. McGrath, M. Inaba, K. Miyatake, M. Hori, K. Ota, Z. Ogumi, S. Miyata, A. Nishikata, Z. Siroma, Y. Uchimoto, K. Yasuda, K.I. Kimijima, N. Iwashita, Scientific aspects of polymer electrolyte fuel cell durability and degradation, Chemical Reviews 107 (10) (2007) 3904–3951.

[176] T. Akita, T. Hiroki, S. Tanaka, T. Kojima, M. Kohyama, A. Iwase, F. Hori, Analytical TEM observation of Au-Pd nanoparticles prepared by sonochemical method, Catalysis Today 131 (1–4) (2008) 90–97.

[177] K. More, I. Milton, VA 3 Microstructural Characterization of PEM Fuel Cell MEAs. 2006 DOE Hydrogen Review, FC27.

[178] G.S. Park, C. Pak, Y.S. Chung, J.R. Kim, W.S. Jeon, Y.H. Lee, K. Kim, H. Chang, D. Seung, Decomposition of Pt-Ru anode catalysts in direct methanol fuel cells, Journal of Power Sources 176 (2) (2008) 484–489.

[179] J. Li, Catalyst Layer Degradation, Diagnosis and Failure Mitigation. Chapter 23. In the book of PEM fuel cell electrocatalysts and catalyst layers, Springer (2008).

[180] K. Tomantschger, R. Findlay, M. Hanson, K. Kordesch, S. Srinivasan, Degradation Modes of Alkaline Fuel-Cells and Their Components, Journal of Power Sources 39 (1) (1992) 21–41.

[181] Y. Kiros, S. Schwartz, Long-term hydrogen oxidation catalysts in alkaline fuel cells, Journal of Power Sources 87 (1–2) (2000) 101–105.

[182] D. Boxall, G. Deluga, E. Kenik, W. King, C. Lukehart, Rapid synthesis of a Pt1Ru1/Carbon nanocomposite using microwave irradiation: A DMFC anode catalyst of high relative performance, Chem. Mater 13 (3) (2001) 891–900.

[183] E. Gulzow, M. Schulze, G. Steinhilber, Investigation of the degradation of different nickel anode types for alkaline fuel cells (AFCs), Journal of Power Sources 106 (1–2) (2002) 126–135.

[184] T. Ralph, M. Hogarth, Catalysis for low temperature fuel cells, Platinum Metals Review 46 (1) (2002) 3–13.

[185] E. Antolini, Formation of carbon-supported PtM alloys for low temperature fuel cells: a review, Materials Chemistry & Physics 78 (3) (2003) 563–573.

[186] E. Gulzow, M. Schulze, Long-term operation of AFC electrodes with CO_2 containing gases, Journal of Power Sources 127 (1–2) (2004) 243–251.

[187] W.K. Hu, X.P. Gao, Y. Kiros, E. Middelman, D. Noreus, Zr-based AB(2)-type hydrogen storage alloys as dual catalysts of gas-diffusion electrodes in an alkaline fuel cell, Journal of Physical Chemistry B 108 (26) (2004) 8756–8758.

[188] A. Schulze, E. Gulzow, Degradation of nickel anodes in alkaline fuel cells, Journal of Power Sources 127 (1–2) (2004) 252–263.

[189] M. Schulze, A. Schneider, E. Gulzow, Alteration of the distribution of the platinum catalyst in membrane–electrode assemblies during PEFC operation, Journal of Power Sources 127 (1–2) (2004) 213–221.

[190] N. Wagner, M. Schulze, E. Gulzow, Long term investigations of silver cathodes for alkaline fuel cells, Journal of Power Sources 127 (1–2) (2004) 264–272.

[191] Y.L. Liu, C.G. Jiao, Microstructure degradation of an anode/electrolyte interface in SOFC studied by transmission electron microscopy, Solid State Ionics 176 (5–6) (2005) 435–442.

[192] K. More, K. Reeves, D. Blom, Microstructural Characterization of PEM Fuel Cell MEAs, Oak Ridge National Laboratory, Oak Ridge, TN, 2005.

[193] D. Waldbillig, A. Wood, D.G. Ivey, Electrochemical and microstructural characterization of the redox tolerance of solid oxide fuel cell anodes, Journal of Power Sources 145 (2) (2005) 206–215.

[194] J. Xie, D.L. Wood, K.L. More, P. Atanassov, R.L. Borup, Microstructural changes of membrane electrode assemblies during PEFC durability testing at high humidity conditions, Journal of the Electrochemical Society 152 (5) (2005) A1011–A1020.

[195] H. Chhina, S. Campbell, O. Kesler, An oxidation-resistant indium tin oxide catalyst support for proton exchange membrane fuel cells, Journal of Power Sources 161 (2) (2006) 893–900.

[196] H. Kim, S.J. Shin, Y.G. Park, J. Song, H.T. Kim, Determination of DMFC deterioration during long-term operation, Journal of Power Sources 160 (1) (2006) 440–445.

[197] J. Li, Y. Liang, Q.C. Xu, X.Z. Fu, J.Q. Xu, J.D. Lin, D.W. Liao, Synthesis and characterization of sub-10 nm Platinum Hollow Spheres as electrocatalyst of direct methanol fuel cell, J Nanosci Nanotechnol 6 (4) (2006) 1107–1113.

[198] S.T. Liu, K. Takahashi, K. Fuchigami, K. Uematsu, Hydrogen production by oxidative methanol reforming on Pd/ZnO: Catalyst deactivation, Applied Catalysis A-General 299 (2006) 58–65.

[199] Y. Shao, G. Yin, Y. Gao, P. Shi, Durability Study of Pt/ C and Pt/ CNTs Catalysts under Simulated PEM Fuel Cell Conditions, Journal of the Electrochemical Society 153 (2006) A1093.

[200] Y.Y. Shao, G.P. Yin, Y.Z. Gao, Study of the electrochemical stability of Pt/C and Pt/CNTs electrodes, Acta Chimica Sinica 64 (16) (2006) 1752–1756.

[201] Y.Y. Shao, G.P. Yin, J. Zhang, Y.Z. Gao, Comparative investigation of the resistance to electrochemical oxidation of carbon black and carbon nanotubes in aqueous sulfuric acid solution, Electrochimica Acta 51 (26) (2006) 5853–5857.

[202] M. Schulze, N. Wagner, T. Kaz, K.A. Friedrich, Combined electrochemical and surface analysis investigation of degradation processes in polymer electrolyte membrane fuel cells, Electrochimica Acta 52 (6) (2007) 2328–2336.

[203] Y. Shao-Horn, W.C. Sheng, S. Chen, P.J. Ferreira, E.F. Holby, D. Morgan, Instability of supported platinum nanoparticles in low-temperature fuel cells, Topics in Catalysis 46 (3–4) (2007) 285–305.

[204] L.S. Sarma, C.H. Chen, G.R. Wang, K.L. Hsueh, C.P. Huang, H.S. Sheu, D.G. Liu, J.F. Lee, B.J. Hwang, Investigations of direct methanol fuel cell (DMFC) fading mechanisms, Journal of Power Sources 167 (2) (2007) 358–365.

[205] T. Yoda, H. Uchida, M. Watanabe, Effects of operating potential and temperature on degradation of electrocatalyst layer for PEFCs, Electrochimica Acta 52 (19) (2007) 5997–6005.

[206] S.M. Choi, J.H. Kim, J.Y. Jung, E.Y. Yoon, W.B. Kim, Pt nanowires prepared via a polymer template method: Its promise toward high Pt-loaded electrocatalysts for methanol oxidation, Electrochimica Acta 53 (19) (2008) 5804–5811.

[207] H.S. Choo, T. Kinumoto, M. Nose, K. Miyazaki, T. Abe, Z. Ogumi, Electrochemical oxidation of highly oriented pyrolytic graphite during potential cycling in sulfuric acid solution, Journal of Power Sources 185 (2) (2008) 740–746.

[208] Y.S. Chung, C. Pak, G.S. Park, W.S. Jeon, J.R. Kim, Y. Lee, H. Chang, D. Seung, Understanding a degradation mechanism of direct methanol fuel cell using TOF-SIMS and XPS, Journal of Physical Chemistry C 112 (1) (2008) 313–318.

[209] K. Matsuoka, S. Sakamoto, K. Nakato, A. Hamada, Y. Itoh, Degradation of polymer electrolyte fuel cells under the existence of anion species, Journal of Power Sources 179 (2) (2008) 560–565.

[210] K.J.J. Mayrhofer, S.J. Ashton, J.C. Meier, G.K.H. Wiberg, M. Hanzlik, M. Arenz, Non-destructive transmission electron microscopy study of catalyst degradation under electrochemical treatment, Journal of Power Sources 185 (2) (2008) 734–739.

[211] K.J.J. Mayrhofer, J.C. Meier, S.J. Ashton, G.K.H. Wiberg, F. Kraus, M. Hanzlik, M. Arenz, Fuel cell catalyst degradation on the nanoscale, Electrochemistry Communications 10 (8) (2008) 1144–1147.

[212] J. Peron, Y. Nedellec, D.J. Jones, J. Roziere, The effect of dissolution, migration and precipitation of platinum in Nafion®-based membrane electrode assemblies during fuel cell operation at high potential, Journal of Power Sources 185 (2) (2008) 1209–1217.

[213] I. Roche, E. Chainet, J. Vondrak, M. Chatenet, Durability of carbon-supported manganese oxide nanoparticles for the oxygen reduction reaction (ORR) in alkaline medium, Journal of Applied Electrochemistry 38 (9) (2008) 1195–1201.

[214] M.S. Saha, R.Y. Li, X.H. Sun, High loading and monodispersed Pt nanoparticles on multiwalled carbon nanotubes for high performance proton exchange membrane fuel cells, Journal of Power Sources 177 (2) (2008) 314–322.

[215] Y.Y. Shao, R. Kou, J. Wang, V.V. Viswanathan, J.H. Kwak, J. Liu, Y. Wang, Y.H. Lin, The influence of the electrochemical stressing (potential step and potential-static holding) on the degradation of polymer electrolyte membrane fuel cell electrocatalysts, Journal of Power Sources 185 (1) (2008) 280–286.

[216] A. Taniguchi, T. Akita, K. Yasuda, Y. Miyazaki, Analysis of degradation in PEMFC caused by cell reversal during air starvation, International Journal of Hydrogen Energy 33 (9) (2008) 2323–2329.

[217] Z.B. Wang, H. Rivera, X.P. Wang, H.X. Zhang, P.M. Feng, E.A. Lewis, E.S. Smotkin, Catalyst failure analysis of a direct methanol fuel cell membrane electrode assembly, Journal of Power Sources 177 (2) (2008) 386–392.

[218] Z.B. Wang, Y.Y. Shao, P.J. Zuo, X.P. Wang, G.P. Yin, Durability studies of unsupported Pt cathodic catalyst with working time of direct methanol fuel cells, Journal of Power Sources 185 (2) (2008) 1066–1072.

[219] J. Wu, X. Yuan, J. Martin, H. Wang, J. Zhang, J. Shen, S. Wu, W. Merida, A review of PEM fuel cell durability: Degradation mechanisms and mitigation strategies, Journal of Power Sources 184 (1) (2008) 104–119.

[220] M.J. Zhi, X.Q. Chen, H. Finklea, I. Celik, N.Q.Q. Wu, Electrochemical and microstructural analysis of nickel–yttria-stabilized zirconia electrode operated in phosphorus-containing syngas, Journal of Power Sources 183 (2) (2008) 485–490.

[221] C.Y. Du, M. Chen, X.Y. Cao, G.P. Yin, P.F. Shi, A novel CNT@SnO_2 core-sheath nanocomposite as a stabilizing support for catalysts of proton exchange membrane fuel cells, Electrochemistry Communications 11 (2) (2009) 496–498.

[222] Y. Gu, J. St-Pierre, A. Joly, R. Goeke, A. Datye, P. Atanassov, Aging Studies of Pt/Glassy Carbon Model Electrocatalysts, Journal of the Electrochemical Society 156 (2009) B485.

[223] M.S. Saha, Y. Chen, R. Li, X. Sun, Enhancement of PEMFC performance by using carbon nanotubes supported Pt-Co alloy catalysts, Asia-Pacific Journal of Chemical Engineering 4 (1) (2009) 12–16.

[224] A. Sarkar, A.V. Murugan, A. Manthiram, Low cost Pd-W nanoalloy electrocatalysts for oxygen reduction reaction in fuel cells, Journal of Materials Chemistry 19 (1) (2009) 159–165.

[225] K. Sato, H. Abe, T. Misono, K. Murata, T. Fukui, M. Naito, Enhanced electrochemical activity and long-term stability of Ni-YSZ anode derived from NiO-YSZ interdispersed composite particles, Journal of the European Ceramic Society 29 (6) (2009) 1119–1124.

[226] S. Uhlenbruck, T. Moskalewicz, N. Jordan, H.J. Penkalla, H.P. Buchkremer, Element interdiffusion at electrolyte-cathode interfaces in ceramic high-temperature fuel cells, Solid State Ionics 180 (4–5) (2009) 418–423.

[227] F.Y. Zhang, S.G. Advani, A.K. Prasad, M. Boggs, S. Sullivan, T. Beebe, Quantitative characterization of catalyst layer degradation in PEM fuel cells by X-ray photoelectron spectroscopy, Electrochimica Acta 54 (16) (2009) 4025–4030.

[228] W.M. Chen, G.Q. Sun, X.S. Zhao, P.C. Sun, S.H. Yang, Q. Xin, Degradation of electrocatalysts performance in direct methanol fuel cells, Chemical Journal of Chinese Universities-Chinese 28 (5) (2007) 928–931.

[229] R. Borup, J. Davey, F. Garzon, D. Wood, P. Welch, K. More, Polymer Electrolyte Membrane (PEM) Fuel Cell Durability, Materials Research Highlight, LALP-06–108, 2006.

[230] More, K., VII. I. 2 Microstructural Characterization of Polymer Electrolyte Membrane Fuel Cell (PEMFC) Membrane Electrode Assemblies (MEAs). 2005 DOE Review Report, vii_i_2.

[231] E. Ding, K. More, T. He, Preparation and characterization of carbon-supported PtTi alloy electrocatalysts, Journal of Power Sources 175 (2) (2007) 794–799.

[232] P. Ferreira, Y. Shao-Horn, D. Morgan, R. Makharia, S. Kocha, H. Gasteiger, Instability of Pt/C Electrocatalysts in Proton Exchange Membrane Fuel Cells, Journal of the Electrochemical Society 152 (2005) A2256.

[233] P. Moloney, C. Huffman, M. Springer, O. Gorelik, P. Nikolaev, E. Sosa, S. Arepalli, and L. Yowell, PEM fuel cell electrodes using Single Wall Carbon Nanotubes. Mater. Res. Soc. Symp. Proc. 2006, 885,75.

[234] T. Takeguchi, Y. Anzai, R. Kikuchi, K. Eguchi, W. Ueda, Preparation and Characterization of CO-Tolerant Pt and Pd Anodes Modified with SnO Nanoparticles for PEFC, Journal of the Electrochemical Society 154 (2007) B1132.

[235] M.M. Nasef, H. Saidi, M.A. Yarmo, Surface investigations of radiation grafted FEP-g-polystyrene sulfonic acid membranes using XPS, Journal of New Materials for Electrochemical Systems 3 (4) (2000) 309–317.

[236] M.M. Nasef, H. Saidi, M.A. Yarmo, Cation exchange membranes by radiation-induced graft copolymerization of styrene onto PFA copolymer films. IV. Morphological investigations using X-ray photoelectron spectroscopy, Journal of Applied Polymer Science 77 (11) (2000) 2455–2463.

[237] B. Tazi, O. Savadogo, Parameters of PEM fuel-cells based on new membranes fabricated from Nafion®, silicotungstic acid and thiophene, Electrochimica Acta 45 (25–26) (2000) 4329–4339.

[238] M.M. Nasef, H. Saidi, Post-mortem analysis of radiation grafted fuel cell membrane using X-ray photoelecton spectroscopy, Journal of New Materials for Electrochemical Systems 5 (3) (2002) 183–189.

[239] C.D. Huang, K.S. Tan, H.Y. Lin, K.L. Tan, XRD and XPS analysis of the degradation of the polymer electrolyte in H_2—O_2 fuel cell, Chemical Physics Letters 371 (1–2) (2003) 80–85.

[240] R. Souzy, B. Ameduri, Functional fluoropolymers for fuel cell membranes, Progress in Polymer Science 30 (6) (2005) 644–687.

[241] B. Bae, H. Ha, D. Kim, Nafion®-graft-polystyrene sulfonic acid membranes for direct methanol fuel cells, Journal of Membrane Science 276 (1–2) (2006) 51–58.

[242] M.H. Woo, O. Kwon, S.H. Choi, M.Z. Hong, H.W. Ha, K. Kim, Zirconium phosphate sulfonated poly (fluorinated arylene ether)s composite membranes for PEMFCs at 100-140°C, Electrochimica Acta 51 (27) (2006) 6051–6059.

[243] C. Chen, G. Levitin, D. Hess, T. Fuller, XPS investigation of Nafion® membrane degradation, Journal of Power Sources 169 (2) (2007) 288–295.

[244] E. Guilminot, A. Corcella, M. Chatenet, F. Maillard, F. Charlot, G. Berthome, C. Iojoiu, J. Sanchez, E. Rossinot, E. Claude, Membrane and active layer degradation upon PEMFC steady-state operation, Journal of the Electrochemical Society 154 (2007) B1106.

[245] E. Guilminot, A. Corcella, M. Chatenet, F. Maillard, F. Charlot, G. Berthome, C. Iojoiu, J.Y. Sanchez, E. Rossinot, E. Claude, Membrane and active layer degradation upon PEMFC steady-state operation – I. Platinum dissolution and redistribution within the MEA, Journal of the Electrochemical Society 154 (11) (2007) B1106–B1114.

[246] F. Zaragoza-Martin, D. Sopena-Escario, E. Morallon, C.S.M. de Lecea, Pt/carbon nanofibers electrocatalysts for fuel cells Effect of the support oxidizing treatment, Journal of Power Sources 171 (2) (2007) 302–309.

[247] C.M. Lai, J.C. Lin, K.L. Hsueh, C.P. Hwang, K.C. Tsay, L.D. Tsai, Y.M. Peng, On the accelerating degradation of DMFC at highly anodic potential, Journal of the Electrochemical Society 155 (8) (2008) B843–B851.

[248] J.H. Chen, M.L. Zhai, M. Asano, L. Huang, Y. Maekawa, Long-term performance of polyetheretherketone-based polymer electrolyte membrane in fuel cells at 95°C, Journal of Materials Science 44 (14) (2009) 3674–3681.

[249] L. Xiong, A. Manthiram, Nanostructured Pt–M/C (M= Fe and Co) catalysts prepared by a microemulsion method for oxygen reduction in proton exchange membrane fuel cells, Electrochimica Acta 50 (11) (2005) 2323–2329.

[250] K. Lim, H. Oh, H. Kim, Use of a carbon nanocage as a catalyst support in polymer electrolyte membrane fuel cells, Electrochemistry Communications 11 (6) (2009) 1131–1134.

[251] H. Chhina, D. Susac, S. Campbell, O. Kesler, Transmission Electron Microscope Observation of Pt Deposited on Nb-Doped Titania, Electrochemical and Solid-State Letters 12 (2009) B97.

[252] P. Nicholson, S. Zhou, G. Hinds, A. Wain, A. Turnbull, Electrocatalytic activity mapping of model fuel cell catalyst films using scanning electrochemical microscopy, Electrochimica Acta 54 (19) (2009) 4525–4533.

[253] R. Kou, Y. Shao, D. Wang, M. Engelhard, J. Kwak, J. Wang, V. Viswanathan, C. Wang, Y. Lin, Y. Wang, Enhanced activity and stability of Pt catalysts on functionalized graphene sheets for electrocatalytic oxygen reduction, Electrochemistry Communications (2009).

[254] Z. Peng, H. Yang, PtAu bimetallic heteronanostructures made by post-synthesis modification of Pt-on-Au nanoparticles, Nano Research 2 (5) (2009) 406–415.

[255] F. Kadirgan, S. Beyhan, T. Atilan, Preparation and characterization of nano-sized Pt–Pd/C catalysts and comparison of their electro-activity toward methanol and ethanol oxidation, International Journal of Hydrogen Energy 34 (10) (2009) 4312–4320.

[256] V. Sethuraman, B. Lakshmanan, J. Weidner, Quantifying desorption and rearrangement rates of carbon monoxide on a PEM fuel cell electrode, Electrochimica Acta (2009).

[257] T. Madden, D. Weiss, N. Cipollini, D. Condit, M. Gummalla, S. Burlatsky, V. Atrazhev, Degradation of Polymer-Electrolyte Membranes in Fuel Cells, Journal of the Electrochemical Society 156 (2009) B657.

[258] J. Chen, M. Zhai, M. Asano, L. Huang, Y. Maekawa, Long-term performance of polyetheretherketone-based polymer electrolyte membrane in fuel cells at 95°C, Journal of Materials Science 44 (14) (2009) 3674–3681.

[259] P. Ferreira-Aparicio, M. Folgado, L. Daza, High surface area graphite as alternative support for proton exchange membrane fuel cell catalysts, Journal of Power Sources (2008).

[260] K. More, VI 2 Microstructural Characterization of PEM Fuel Cell Membrane Electrode Assemblies, FY 2004 Progress Report on DOE Hydrogen Program (2004).

[261] K. Yasuda, A. Taniguchi, T. Akita, T. Ioroi, Z. Siroma, Platinum dissolution and deposition in the polymer electrolyte membrane of a PEM fuel cell as studied by potential cycling, Physical Chemistry Chemical Physics 8 (6) (2006) 746–752.

[262] H. Nakajima, S. Nomura, T. Sugimoto, S. Nishikawa, I. Honma, High temperature proton conducting organic/inorganic nanohybrids for polymer electrolyte membrane, Journal of the Electrochemical Society 149 (2002) A953.

[263] E. Guilminot, A. Corcella, F. Charlot, F. Maillard, M. Chatenet, Detection of Pt Ions and Pt Nanoparticles Inside the Membrane of a Used PEMFC, Journal of the Electrochemical Society 154 (2007) B96.

[264] S. Zhang, X. Yuan, H. Wang, W. Mérida, H. Zhu, J. Shen, S. Wu, J. Zhang, A review of accelerated stress tests of MEA durability in PEM fuel cells, International Journal of Hydrogen Energy 34 (1) (2009) 388–404.

[265] H. Wang, M. Brady, K. More, H. Meyer, J. Turner, Thermally nitrided stainless steels for polymer electrolyte membrane fuel cell bipolar plates Part 2: Beneficial modification of passive layer on AISI446, Journal of Power Sources 138 (1–2) (2004) 79–85.

[266] M. Brady, H. Wang, and J. Turner, Surface modified stainless steels for PEM fuel cell bipolar plates. 2005, Google Patents.

[267] M. Brady, B. Yang, H. Wang, J. Turner, K. More, M. Wilson, F. Garzon, The formation of protective nitride surfaces for PEM fuel cell metallic bipolar plates, JOM 58 (8) (2006) 50–57.

[268] I. Paulauskas, M. Brady, H. Meyer, R. Buchanan, L. Walker, Corrosion behavior of CrN, Cr_2N and p phase surfaces on nitrided Ni-50Cr for proton exchange membrane fuel cell bipolar plates, Corrosion Science 48 (10) (2006) 3157–3171.

[269] J. Turner, H. Wang, and M. Brady, Corrosion Protection of Metallic Bipolar Plates for Fuel Cells (Presentation). 2006. DOE Review Report, vii_d.

[270] S. Anandan, Y. Ikuma, K. Kakinuma, K. Niwa, Synthesis and Characterization of a Highly Crystalline Novel Mesoporous C- and N-Codoped TiO_2 Nanophotocatalyst, Nano 3 (5) (2008) 367–372.

[271] M. Brady, K. Weisbrod, I. Paulauskas, R. Buchanan, K. More, H. Wang, M. Wilson, F. Garzon, L. Walker, Preferential thermal nitridation to form pin-hole free Cr-nitrides to protect proton exchange membrane fuel cell metallic bipolar plates, Scripta Materialia 50 (7) (2004) 1017–1022.

[272] M. Brady, P. Tortorelli, K. More, E. Payzant, B. Armstrong, H. Lin, M. Lance, F. Huang, M. Weaver, Coating and near-surface modification design strategies for protective and functional surfaces, Materials and Corrosion 56 (11) (2005).

[273] Y. Hung, H. Tawfik, D. Mahajan, Durability and characterization studies of polymer electrolyte membrane fuel cell's coated aluminum bipolar plates and membrane electrode assembly, Journal of Power Sources 186 (1) (2009) 123–127.

[274] B. Du, Q. Guo, Z. Qi, L. Mao, R. Pollard, J. Elter, 12 Materials for Proton Exchange Membrane Fuel Cells, Materials for the Hydrogen Economy (2008) 251.

[275] M. Schulze, M. Lorenz, N. Wagner, E. Gulzow, XPS analysis of the degradation of Nafion, Fresenius Journal of Analytical Chemistry 365 (1–3) (1999) 106–113.

[276] M. Schulze, T. Knori, A. Schneider, E. Gulzow, Degradation of sealings for PEFC test cells during fuel cell operation, Journal of Power Sources 127 (1–2) (2004) 222–229.

[277] M. Schulze, C. Christenn, XPS investigation of the PTFE induced hydrophobic properties of electrodes for low temperature fuel cells, Applied Surface Science 252 (1) (2005) 148–153.

[278] J.Z. Tan, Y.J. Chao, J.W. Van Zee, W.K. Lee, Degradation of elastomeric gasket materials in PEM fuel cells, Materials Science and Engineering A-Structural Materials Properties Microstructure and Processing 445 (2007) 669–675.

[279] J.Z. Tan, Y.J. Chao, M. Yang, C.T. Williams, J.W. Van Zee, Degradation Characteristics of Elastomeric Gasket Materials in a Simulated PEM Fuel Cell Environment, Journal of Materials Engineering and Performance 17 (6) (2008) 785–792.

[280] J. Nam, K. Lee, G. Hwang, C. Kim, M. Kaviany, Microporous layer for water morphology control in PEMFC, International Journal of Heat and Mass Transfer (2009).

[281] J. Aragane, H. Urushibata, Xps and Fe-Sem Analysis of Platinum Electrocatalysts in a Phosphoric-Acid Fuel-Cell, Nippon Kagaku Kaishi(9) (1995) 736–742.

[282] K. Yasuda, A. Taniguchi, T. Akita, T. Ioroi, Z. Siroma, Characteristics of a platinum black catalyst layer with regard to platinum dissolution phenomena in a membrane electrode assembly, Journal of the Electrochemical Society 153 (8) (2006) A1599–A1603.

[283] Z.B. Wang, P.J. Zuo, X.P. Wang, J. Lou, B.Q. Yang, G.P. Yin, Studies of performance decay of Pt/C catalysts with working time of proton exchange membrane fuel cell, Journal of Power Sources 184 (1) (2008) 245–250.

[284] L. Kim, C.G. Chung, Y.W. Sung, J.S. Chung, Dissolution and migration of platinum after long-term operation of a polymer electrolyte fuel cell under various conditions, Journal of Power Sources 183 (2) (2008) 524–532.

[285] C. Grolleau, C. Coutanceau, F. Pierre, J. Léger, Effect of potential cycling on structure and activity of Pt nanoparticles dispersed on different carbon supports, Electrochimica Acta 53 (24) (2008) 7157–7165.

[286] R. Jenkins, An introduction to X-ray spectrometry, Heyden, London, New York, 1974, xi, p. 163.

[287] R. Jenkins, American Society for Nondestructive Testing, X-ray diffraction and fluorescence. The Nondestructive testing handbook on radiography and radiation testing section = 7, American Society for Nondestructive Testing, Columbus, Ohio, 1983, p 30.

[288] M.F. Guerra, An overview on the ancient goldsmith's skill and the circulation of gold in the past: the role of X-ray based techniques, X-ray Spectrometry 37 (4) (2008) 317–327.

[289] P.J. Kulesza, W.Y. Lu, L.R. Faulkner, Cathodic Fabrication of Platinum Microparticles Via Anodic-Dissolution of a Platinum Counterelectrode - Electrocatalytic Probing and Surface-Analysis of Dispersed Platinum, Journal of Electroanalytical Chemistry 336 (1–2) (1992) 35–44.

[290] M.S. Wilson, F.H. Garzon, K.E. Sickafus, S. Gottesfeld, Surface-Area Loss of Supported Platinum in Polymer Electrolyte Fuel-Cells, Journal of the Electrochemical Society 140 (10) (1993) 2872–2877.

[291] P.G. Allen, S.D. Conradson, M.S. Wilson, S. Gottesfeld, I.D. Raistrick, J. Valerio, M. Lovato, In-situ Structural Characterization of a Platinum Electrocatalyst by Dispersive-X-ray Absorption-Spectroscopy, Electrochimica Acta 39 (16) (1994) 2415–2418.

[292] C. Roth, M. Goetz, H. Fuess, Synthesis and characterization of carbon-supported Pt-Ru-WOx catalysts by spectroscopic and diffraction methods, Journal of Applied Electrochemistry 31 (7) (2001) 793–798.

[293] S.Y. Ahn, S.J. Shin, H.Y. Ha, S.A. Hong, Y.C. Lee, T.W. Lim, I.H. Oh, Performance and lifetime analysis of the kW-class PEMFC stack, Journal of Power Sources 106 (1–2) (2002) 295–303.

[294] E.M. Crabb, M.K. Ravikumar, Y. Qian, A.E. Russell, S. Maniguet, J. Yao, D. Thompsett, M. Hurford, S.C. Ball, Controlled modification of carbon supported platinum electrocatalysts by Mo, Electrochemical and Solid State Letters 5 (1) (2002) A5–A9.

[295] J. Roziere, D.J. Jones, Non-fluorinated polymer materials for proton exchange membrane fuel cells, Annual Review of Materials Research 33 (2003) 503–555.

[296] X. Cheng, L. Chen, C. Peng, Z.W. Chen, Y. Zhang, Q.B. Fan, Catalyst microstructure examination of PEMFC membrane electrode assemblies vs. time, Journal of the Electrochemical Society 151 (1) (2004) A48–A52.

[297] S. Mukerjee, R.C. Urian, S.J. Lee, E.A. Ticianelli, J. McBreen, Electrocatalysis of CO tolerance by carbon-supported PtMo electrocatalysts in PEMFCs, Journal of the Electrochemical Society 151 (7) (2004) A1094–A1103.

[298] D.C. Papageorgopoulos, M.P. de Heer, M. Keijzer, J.A.Z. Pieterse, F.A. de Bruijn, Non-alloyed carbon-supported PtRu catalysts for PEMFC applications, Journal of the Electrochemical Society 151 (5) (2004) A763–A768.

[299] Z.G. Shao, P. Joghee, I.M. Hsing, Preparation and characterization of hybrid Nafion-silica membrane doped with phosphotungstic acid for high temperature operation of proton exchange membrane fuel cells, Journal of Membrane Science 229 (1–2) (2004) 43–51.

[300] J.H. Tian, F.B. Wang, Z.Q. Shan, R.J. Wang, J.Y. Zhang, Effect of preparation conditions of Pt/C catalysts on oxygen electrode performance in proton exchange membrane fuel cells, Journal of Applied Electrochemistry 34 (5) (2004) 461–467.

[301] C. Wu, F. Wu, Y. Bai, B.L. Yi, H.M. Zhang, Cobalt boride catalysts for hydrogen generation from alkaline $NaBH_4$ solution, Materials Letters 59 (14–15) (2005) 1748–1751.

[302] Y. Xu, J.H. Tian, C. Zhang, Z.Q. Shan, Preparation by inorganic colloid method and characterization of Pt/C catalysts, Chinese Journal of Inorganic Chemistry 21 (10) (2005) 1475–1478.

[303] K.T. Adjemian, R. Dominey, L. Krishnan, H. Ota, P. Majsztrik, T. Zhang, J. Mann, B. Kirby, L. Gatto, M. Velo-Simpson, J. Leahy, S. Srinivasant, J.B. Benziger, A.B. Bocarsly, Function and characterization of metal oxide-naflon composite membranes for elevated-temperature H_2/O_2 PEM fuel cells, Chemistry of Materials 18 (9) (2006) 2238–2248.

[304] N. Benker, C. Roth, M. Mazurek, H. Fuess, Synthesis and characterisation of ternary Pt/Ru/Mo catalysts for the anode of the PEM fuel cell, Journal of New Materials for Electrochemical Systems 9 (2) (2006) 121–126.

[305] R.L. Borup, J.R. Davey, F.H. Garzon, D.L. Wood, M.A. Inbody, PEM fuel cell electrocatalyst durability measurements, Journal of Power Sources 163 (1) (2006) 76–81.

[306] M. Cai, M.S. Ruthkosky, B. Merzougui, S. Swathirajan, M.P. Balogh, S.H. Oh, Investigation of thermal and electrochemical degradation of fuel cell catalysts, Journal of Power Sources 160 (2) (2006) 977–986.

[307] M. Mazurek, N. Benker, C. Roth, T. Buhrmester, H. Fuess, Electrochemical impedance and X-ray absorption spectroscopy (EXAFS) as in-situ methods to study the PEMFC anode, Fuel Cells 6 (1) (2006) 16–20.

[308] Y.Y. Shao, G.P. Yin, Y.Z. Gao, P.F. Shi, Durability study of Pt/C and Pt/CNTs catalysts under simulated PEM fuel cell conditions, Journal of the Electrochemical Society 153 (6) (2006) A1093–A1097.

[309] Z.G. Shao, H.F. Xu, M.Q. Li, I.M. Hsing, Hybrid Nafion–inorganic oxides membrane doped with heteropolyacids for high temperature operation of proton exchange membrane fuel cell, Solid State Ionics 177 (7–8) (2006) 779–785.

[310] P. Sinha, P. Halleck, C. Wang, Quantification of liquid water saturation in a PEM fuel cell diffusion medium using X-ray microtomography, Electrochemical and Solid-State Letters 9 (2006) A344.

[311] K. Suarez-Alcantara, A. Rodriguez-Castellanos, R. Dante, O. Solorza-Feria, $Ru_xCr_ySe_z$ electrocatalyst for oxygen reduction in a polymer electrolyte membrane fuel cell, Journal of Power Sources 157 (1) (2006) 114–120.

[312] H.D. Wang, X.Q. Sun, Y. Ye, S.L. Qiu, Radiation induced synthesis of Pt nanoparticles supported on carbon nanotubes, Journal of Power Sources 161 (2) (2006) 839--842.

[313] Y.F. Zhai, H.M. Zhang, J.W. Hu, B.L. Yi, Preparation and characterization of sulfated zirconia(SO_4^{2-}/ZrO_2)/Nafion composite membranes for PEMFC operation at high temperature/low humidity, Journal of Membrane Science 280 (1–2) (2006) 148–155.

[314] H.W. Zhang, B.K. Zhu, Y.Y. Xu, Composite membranes of sulfonated poly(phthalazinone ether ketone) doped with 12-phosphotungstic acid (H3PW12O40) for proton exchange membranes, Solid State Ionics 177 (13–14) (2006) 1123–1128.

[315] G.W. Chen, S.H. Li, Q. Yuan, Pd-Zn/Cu-Zn-Al catalysts prepared for methanol oxidation reforming in microchannel reactors, Catalysis Today 120 (1) (2007) 63–70.

[316] S. Koh, J. Leisch, M.F. Toney, P. Strasser, Structure-activity-stability relationships of Pt-Co alloy electrocatalysts in gas-diffusion electrode layers, Journal of Physical Chemistry C 111 (9) (2007) 3744–3752.

[317] G. Liu, H. Zhang, J.W. Hu, Novel synthesis of a highly active carbon-supported $Ru_{85}Se_{15}$ chalcogenide catalyst for the oxygen reduction reaction, Electrochemistry Communications 9 (11) (2007) 2643–2648.

[318] G. Liu, H.M. Zhang, M.R. Wang, H.X. Zhong, J. Chen, Preparation, characterization of ZrO_xN_y/C and its application in PEMFC as an electrocatalyst for oxygen reduction, Journal of Power Sources 172 (2) (2007) 503–510.

[319] M. Loster, D. Balzar, K.A. Friedrich, J. Garche, X-ray line profile analysis of nanoparticles in proton exchange membrane fuel cell electrodes, Journal of Physical Chemistry C 111 (26) (2007) 9583–9591.

[320] S. Ma, Q. Chen, F.H. Jogensen, P.C. Stein, E.M. Skou, F-19 NMR studies of Nafion (TM) ionomer adsorption on PEMFC catalysts and supporting carbons, Solid State Ionics 178 (29–30) (2007) 1568–1575.

[321] B. Moreno, E. Chinarro, J.L.G. Fierro, J.R. Jurado, Synthesis of the ceramic-metal catalysts (PtRuNi-TiO_2) by the combustion method, Journal of Power Sources 169 (1) (2007) 98–102.

[322] B. Moreno, E. Chinarro, J.C. Perez, J.R. Jurado, Combustion synthesis and electrochemical characterisation of Pt-Ru-Ni anode electrocatalyst for PEMFC, Applied Catalysis B-Environmental 76 (3–4) (2007) 368–374.

[323] D.G. Tong, X. Han, W. Chu, H. Chen, X.Y. Ji, Preparation of mesoporous Co-B catalyst via self-assembled triblock copolymer templates, Materials Letters 61 (25) (2007) 4679–4682.

[324] P.E. Tsiakaras, PtM/C (M = Sn, Ru, Pd, W) based anode direct ethanol-PEMFCs: Structural characteristics and cell performance, Journal of Power Sources 171 (1) (2007) 107–112.

[325] Y. Wang, D.O. Northwood, An investigation into TiN-coated 316L stainless steel as a bipolar plate material for PEM fuel cells, Journal of Power Sources 165 (1) (2007) 293–298.

[326] F.C. Wu, C.C. Wan, Y.Y. Wang, L.D. Tsai, K.L. Hsueh, Improvement of Pt-catalyst dispersion and utilization for direct methanol fuel cells using silane coupling agent, Journal of the Electrochemical Society 154 (6) (2007) B528–B532.

[327] J.H. Zeng, Z.L. Zhao, J.Y. Lee, P.K. Shen, S.Q. Song, Do magnetically modified PtFe/C catalysts perform better in methanol electrooxidation? Electrochimica Acta 52 (11) (2007) 3673–3679.

[328] Y.F. Zhai, H.M. Zhang, D.M. Xing, Z.G. Shao, The stability of Pt/C catalyst in H_3PO_4/PBI PEMFC during high temperature life test, Journal of Power Sources 164 (1) (2007) 126–133.

[329] A. Bayrakceken, A. Smirnova, U. Kitkamthorn, M. Aindow, L. Turker, I. Eroglu, C. Erkey, Pt-based electrocatalysts for polymer electrolyte membrane fuel cells prepared by supercritical deposition technique, Journal of Power Sources 179 (2) (2008) 532–540.

[330] C. Bi, H.M. Zhang, Y. Zhang, X.B. Zhu, Y.W. Ma, H. Dai, S.H. Xiao, Fabrication and investigation of SiO_2 supported sulfated zirconia/Nafion (R) self-humidifying membrane for proton exchange membrane fuel cell applications, Journal of Power Sources 184 (1) (2008) 197–203.

[331] Y.Z. Chen, Z.P. Shao, N.P. Xu, Ethanol steam reforming over Pt catalysts supported on $Ce_xZr_{1-x}O_2$ prepared via a glycine nitrate process, Energy & Fuels 22 (3) (2008) 1873–1879.

[332] A.R. dos Santos, M. Carmo, A. Oliveira-Neto, E.V. Spinace, J.G.R. Poco, C. Roth, H. Fuess, M. Linardi, Electrochemical and impedance spectroscopy studies in H_2/O_2 and methanol/O_2 proton exchange membrane fuel cells, Ionics 14 (1) (2008) 43–51.

[333] A.C. Garcia, V.A. Paganin, E.A. Ticianelli, CO tolerance of PdPt/C and PdPtRu/C anodes for PEMFC, Electrochimica Acta 53 (12) (2008) 4309–4315.

[334] E. Guilminot, R. Gavillon, M. Chatenet, S. Berthon-Fabry, A. Rigacci, T. Budtova, New nanostructured carbons based on porous cellulose: Elaboration, pyrolysis and use as platinum nanoparticles substrate for oxygen reduction electrocatalysis, Journal of Power Sources 185 (2) (2008) 717–726.

[335] O.E. Haas, S.T. Briskeby, O.E. Kongstein, M. Tsypkin, R. Tunold, B.T. Borresen, Synthesis and characterisation of RuxTix-1O2 as a catalyst support for polymer electrolyte fuel cell, Journal of New Materials for Electrochemical Systems 11 (1) (2008) 9–14.

[336] T. Jian-Hua, G. Peng-Fei, Z. Zhi-Yuan, L. Wen-Hui, S. Zhony-Qiang, Preparation and performance evaluation of a Nafion-TiO_2 composite membrane for PEMFCs, International Journal of Hydrogen Energy 33 (20) (2008) 5686–5690.

[337] N.D. Nam, J.G. Kim, Electrochemical Behavior of CrN Coated on 316L Stainless Steel in Simulated Cathodic Environment of Proton Exchange Membrane Fuel Cell, Japanese Journal of Applied Physics 47 (8) (2008) 6887–6890.

[338] K. Oishi, O. Savadogo, New Method of Preparation of Catalyzed Gas Diffusion Electrode for Polymer Electrolyte Fuel Cells Based on Ultrasonic Direct Solution Spray Reaction, Journal of New Materials for Electrochemical Systems 11 (4) (2008) 221–227.

[339] N. Patel, R. Fernandes, G. Guella, A. Kale, A. Miotello, B. Patton, C. Zanchetta, Structured and nanoparticle assembled Co-B thin films prepared by pulsed laser deposition: A very efficient catalyst for hydrogen production, Journal of Physical Chemistry C 112 (17) (2008) 6968–6976.

[340] M. Prasanna, E.A. Cho, T.H. Lim, I.H. Oh, Effects of MEA fabrication method on durability of polymer electrolyte membrane fuel cells, Electrochimica Acta 53 (16) (2008) 5434–5441.

[341] A. Sarkar, A.V. Murugan, A. Manthiram, Synthesis and characterization of nanostructured Pd-Mo electrocatalysts for oxygen reduction reaction in fuel cells, Journal of Physical Chemistry C 112 (31) (2008) 12037–12043.

[342] K. Suarez-Alcantara, O. Solorza-Feria, Kinetics and PEMFC performance of $Ru_xMo_ySe_z$ nanoparticles as a cathode catalyst, Electrochimica Acta 53 (15) (2008) 4981–4989.

[343] S. Takenaka, H. Matsumori, H. Matsune, E. Tanabe, M. Kishida, High durability of carbon nanotube-supported Pt electrocatalysts covered with silica layers for the cathode in a PEMFC, Journal of the Electrochemical Society 155 (9) (2008) B929–B936.

[344] X.G. Yang, Y. Tabuchi, F. Kagami, C.Y. Wang, Durability of membrane electrode assemblies under polymer electrolyte fuel cell cold-start cycling, Journal of the Electrochemical Society 155 (7) (2008) B752–B761.

[345] J. Zhang, PEM fuel cell electrocatalysts and catalyst layers: fundamentals and applications, Springer, London, 2008, xxi, p. 1137.

[346] S.M. Choi, J.S. Yoon, H.J. Kim, S.H. Nam, M.H. Seo, W.B. Kim, Electrochemical benzene hydrogenation using PtRhM/C (M = W, Pd, or Mo) electrocatalysts over a polymer electrolyte fuel cell system, Applied Catalysis A-General 359 (1–2) (2009) 136–143.

[347] Y. Hung, H. Tawfik, D. Mahajan, Durability and characterization studies of polymer electrolyte membrane fuel cell's coated aluminum bipolar plates and membrane electrode assembly, Journal of Power Sources 186 (1) (2009) 123–127.

[348] H. Niwa, K. Horiba, Y. Harada, M. Oshima, T. Ikeda, K. Terakura, J. Ozaki, S. Miyata, X-ray absorption analysis of nitrogen contribution to oxygen reduction reaction in carbon alloy cathode catalysts for polymer electrolyte fuel cells, Journal of Power Sources 187 (1) (2009) 93–97.

[349] L.G.S. Pereira, V.A. Paganin, E.A. Ticianelli, Investigation of the CO tolerance mechanism at several Pt-based bimetallic anode electrocatalysts in a PEM fuel cell, Electrochimica Acta 54 (7) (2009) 1992–1998.

[350] H.J. Yu, L.J. Yang, L. Zhu, X.Y. Jian, Z. Wang, L.J. Jiang, Anticorrosion properties of Ta–coated 316L stainless steel as bipolar plate material in proton exchange membrane fuel cells, Journal of Power Sources 191 (2) (2009) 495–500.

[351] D. Zhao, B.L. Yi, H.M. Zhang, H.M. Yu, L. Wang, Y.W. Ma, D.M. Xing, Cesium substituted 12–tungstophosphoric ($Cs_xH_{3-x}PW_{12}O_{40}$) loaded on ceria-degradation mitigation in polymer electrolyte membranes, Journal of Power Sources 190 (2) (2009) 301–306.

[352] Nitani, H., T. Ono, Y. Honda, A. Koizumi, T. Nakagawa, T. Yamamoto, H. Daimon, and Y. Kurobe, EXAFS Study on Nanosized PtRu Catalyst for Direct Methanol Fuel Cell. Materials Research Society Symposium Proceedings, 2005, Vol. 900, pages 283–288.

[353] A. Arico, V. Baglio, A. Di Blasi, E. Modica, P. Antonucci, V. Antonucci, Analysis of the high-temperature methanol oxidation behaviour at carbon-supported Pt–Ru catalysts, Journal of Electroanalytical Chemistry 557 (2003) 167–176.

[354] P. Piela, C. Eickes, E. Brosha, F. Garzon, P. Zelenay, Ruthenium crossover in direct methanol fuel cell with Pt-Ru black anode, Journal of the Electrochemical Society 151 (12) (2004) A2053–A2059.

[355] L.H. Jiang, G.Q. Sun, S.L. Wang, G.X. Wang, Q. Xin, Z.H. Zhou, B. Zhou, Electrode catalysts behavior during direct ethanol fuel cell life-time test, Electrochemistry Communications 7 (7) (2005) 663–668.

[356] W.M. Chen, G.Q. Sun, J.S. Guo, X.S. Zhao, S.Y. Yan, J. Tian, S.H. Tang, Z.H. Zhou, Q. Xin, Test on the degradation of direct methanol fuel cell, Electrochimica Acta 51 (12) (2006) 2391–2399.

[357] M.K. Jeon, K.R. Lee, K.S. Oh, D.S. Hong, J.Y. Won, S. Li, S.I. Woo, Current density dependence on performance degradation of direct methanol fuel cells, Journal of Power Sources 158 (2) (2006) 1344–1347.

[358] J.S. Guo, G.Q. Sun, Z.M. Wu, S.G. Sun, S.Y. Yan, L. Cao, Y.S. Yan, D.S. Su, Q. Xin, The durability of polyol-synthesized PtRu/C for direct methanol fuel cells, Journal of Power Sources 172 (2) (2007) 666–675.

[359] Z.X. Liang, T.S. Zhao, C. Xu, J.B. Xu, Microscopic characterizations of membrane electrode assemblies prepared under different hot-pressing conditions, Electrochimica Acta 53 (2) (2007) 894–902.

[360] H. Nitani, T. Nakagawa, H. Daimon, Y. Kurobe, T. Ono, Y. Honda, A. Koizumi, S. Seino, T.A. Yamamoto, Methanol oxidation catalysis and substructure of PtRu bimetallic nanoparticles, Applied Catalysis A-General 326 (2) (2007) 194–201.

[361] V. Baglio, A. Di Blasi, C. D'Urso, V. Antonucci, A. Aricò, R. Ornelas, D. Morales-Acosta, J. Ledesma-Garcia, L. Godinez, L. Arriaga, Development of Pt and Pt-Fe Catalysts Supported on Multiwalled Carbon Nanotubes for Oxygen Reduction in Direct Methanol Fuel Cells, Journal of the Electrochemical Society 155 (2008) B829.

[362] V. Baglio, A. Di Blasi, C. D'Urso, V. Antonucci, A.S. Arico, R. Ornelas, D. Morales-Acosta, J. Ledesma-Garcia, L.A. Godinez, L.G. Arriaga, L. Alvarez-Contreras, Development of Pt and Pt-Fe catalysts supported on multiwalled carbon nanotubes for oxygen reduction in direct methanol fuel cells, Journal of the Electrochemical Society 155 (8) (2008) B829–B833.

[363] A. Castro Luna, A. Bonesi, W. Triaca, V. Baglio, V. Antonucci, A. Aricò, Pt–Fe cathode catalysts to improve the oxygen reduction reaction and methanol tolerance in direct methanol fuel cells, Journal of Solid State Electrochemistry 12 (5) (2008) 643–649.

[364] A.M.C. Luna, A. Bonesi, W.E. Triaca, V. Baglio, V. Antonucci, A.S. Arico, Pt-Fe cathode catalysts to improve the oxygen reduction reaction and methanol tolerance in direct methanol fuel cells, Journal of Solid State Electrochemistry 12 (5) (2008) 643–649.

[365] B. Mecheri, A. D'Epifanio, E. Traversa, S. Licoccia, Sulfonated polyether ether ketone and hydrated tin oxide proton conducting composites for direct methanol fuel cell applications, Journal of Power Sources 178 (2) (2008) 554–560.

[366] Z.B. Wang, X.P. Wang, P.J. Zuo, B.Q. Yang, G.P. Yin, X.P. Feng, Investigation of the performance decay of anodic PtRu catalyst with working time of direct methanol fuel cells, Journal of Power Sources 181 (1) (2008) 93–100.

[367] J. Zhang, G.P. Yin, Y.Y. Shao, Z.B. Wang, Q.Z. Lai, Alteration of the membrane electrode assemblies during direct methanol fuel cell activation process, Rare Metal Materials and Engineering 37 (3) (2008) 476–479.

[368] Y.Y. Huang, K. Ahn, J.M. Vohs, R.J. Gorte, Characterization of Sr-doped $LaCoO_3$-YSZ composites prepared by impregnation methods, Journal of the Electrochemical Society 151 (10) (2004) A1592–A1597.

[369] C. Chervin, R.S. Glass, S.M. Kauzlarich, Chemical degradation of $La_{1-x}S_xMnO_3/Y_2O_3$-stabilized ZrO_2 composite cathodes in the presence of current collector pastes, Solid State Ionics 176 (1–2) (2005) 17–23.

[370] M. Kumar, M.A. Kulandainathan, I.A. Raj, R. Chandrasekaran, R. Pattabiraman, Electrical and sintering behaviour of $Y_2Zr_2O_7$ (YZ) pyrochlore-based materials-the influence of bismuth, Materials Chemistry and Physics 92 (2–3) (2005) 295–302.

[371] S. Larrondo, M.A. Vidal, B. Irigoyen, A.F. Craievich, D.G. Lamas, I.O. Fabregas, G.E. Lascalea, N.E.W. de Reca, N. Amadeo, Preparation and characterization of Ce/Zr mixed oxides and their use as catalysts for the direct oxidation of dry CH_4, Catalysis Today (2005) 107–108, 53–59.

[372] B.D. Madsen, S.A. Barnett, Effect of fuel composition on the performance of ceramic-based solid oxide fuel cell anodes, Solid State Ionics 176 (35–36) (2005) 2545–2553.

[373] S.P. Simner, M.D. Anderson, G.G. Xia, Z. Yang, L.R. Pederson, J.W. Stevenson, SOFC performance with Fe-Cr-Mn alloy interconnect, Journal of the Electrochemical Society 152 (4) (2005) A740–A745.

[374] B. Huang, X.F. Ye, S.R. Wang, H.W. Nie, J. Shi, Q. Hu, J.Q. Qian, X.F. Sun, T.L. Wen, Performance of Ni/ScSZ cermet anode modified by coating with $Gd_{0.2}Ce_{0.8}O_2$ for an SOFC running on methane fuel, Journal of Power Sources 162 (2) (2006) 1172–1181.

[375] X.J. Chen, Q.L. Liu, S.H. Chan, N.P. Brandon, K.A. Khor, Sulfur tolerance and hydrocarbon stability of $La_{0.75}Sr_{0.25}Cr_{0.5}Mn_{0.5}O_3/Gd_{0.2}Ce_{0.8}O_{1.9}$ composite anode under anodic polarization, Journal of the Electrochemical Society 154 (11) (2007) B1206–B1210.

[376] C.M. Grgicak, J.B. Giorgi, Improved performance of Ni- and Co-YSZ anodes via sulfidation to NiS- and CoS-YSZ. Effects of temperature on electrokinetic parameters, Journal of Physical Chemistry C 111 (42) (2007) 15446–15455.

[377] B. Huang, S.R. Wang, R.Z. Liu, X.E. Ye, H.W. Nie, X.E. Sun, T.L. Wen, Performance of $La_{0.75}Sr_{0.25}Cr_{0.5}Mn_{0.5}O_3$-delta perovskite-structure anode material at lanthanum gallate electrolyte for IT-SOFC running on ethanol fuel, Journal of Power Sources 167 (1) (2007) 39–46.

[378] M. Lang, P. Szabo, Z. Ilhan, S. Cinque, T. Franco, G. Schiller, Development of solid oxide fuel cells and short stacks for mobile application, Journal of Fuel Cell Science and Technology 4 (4) (2007) 384–391.

[379] D. Marrero-Lopez, D. Perez-Coll, J.C. Ruiz-Morales, J. Canales-Vazquez, M.C. Martin-Sedeno, P. Nunez, Synthesis and transport properties in $La_{2-x}A_xMo_2O_{9-\delta}$ ($A = Ca^{2+}$, Sr^{2+}, Ba^{2+}, K^+) series, Electrochimica Acta 52 (16) (2007) 5219–5231.

[380] A. D'Epifani, E. Fabbri, E. Di Bartolomeo, S. Licoccia, E. Traversa, Design of $BaZr_{0.8}Y_{0.2}O_3$-delta protonic conductor to improve the electrochemical performance in intermediate temperature solid oxide fuel cells (IT-SOFCs), Fuel Cells 8 (1) (2008) 69–76.

[381] J.R. Izzo, A.S. Joshi, K.N. Grew, W.K.S. Chiu, A. Tkachuk, S.H. Wang, W.B. Yun, Nondestructive reconstruction and analysis of SOFC anodes using X-ray computed tomography at sub-50 nm resolution, Journal of the Electrochemical Society 155 (5) (2008) B504–B508.

[382] D.J. Jan, C.T. Lin, C.F. Ai, Structural characterization of $La_{0.67}Sr_{0.33}MnO_3$ protective coatings for solid oxide fuel cell interconnect deposited by pulsed magnetron sputtering, Thin Solid Films 516 (18) (2008) 6300–6304.

[383] C. Lee, J. Bae, Oxidation-resistant thin film coating on ferritic stainless steel by sputtering for solid oxide fuel cells, Thin Solid Films 516 (18) (2008) 6432–6437.

[384] Z. Zeng, K. Natesan, S.B. Cai, Characterization of oxide scale on alloy 446 by X-ray nanobeam analysis, Electrochemical and Solid State Letters 11 (1) (2008) C5–C8.

[385] W.K.S. Chiu, A.S. Joshi, K.N. Grew, Lattice Boltzmann model for multi-component mass transfer in a solid oxide fuel cell anode with heterogeneous internal reformation and electrochemistry, European Physical Journal-Special Topics 171 (2009) 159–165. 232.

[386] C. Johnson, N. Orlovskaya, A. Coratolo, C. Cross, J. Wu, R. Gemmen, X. Liu, The effect of coating crystallization and substrate impurities on magnetron sputtered doped $LaCrO_3$ coatings for metallic solid oxide fuel cell interconnects, International Journal of Hydrogen Energy 34 (5) (2009) 2408–2415.

[387] D. La Rosa, A. Sin, M. Faro, G. Monforte, V. Antonucci, A. Aricò, Mitigation of carbon deposits formation in intermediate temperature solid oxide fuel cells fed with dry methane by anode doping with barium, Journal of Power Sources 193 (1) (2009) 160–164.

[388] N. Laorodphan, P. Namwong, W. Thiemsorn, M. Jaimasith, A. Wannagon, T. Chairuangsri, A low silica, barium borate glass-ceramic for use as seals in planar SOFCs, Journal of Non-Crystalline Solids 355 (1) (2009) 38–44.

[389] M. Watanabe, K. Tsurumi, T. Mizukami, T. Nakamura, P. Stonehart, Activity and Stability of Ordered and Disordered Co-Pt Alloys for Phosphoric-Acid Fuel-Cells, Journal of the Electrochemical Society 141 (10) (1994) 2659–2668.

[390] S.J.J. Lue, J.Y. Chen, J.M. Yang, Crystallinity and stability of poly(vinyl alcohol)-fumed silica mixed matrix membranes, Journal of Macromolecular Science Part B-Physics 47 (1) (2008) 39–51.

[391] M. Feser, J. Gelb, H. Chang, H. Cui, F. Duewer, S.H. Lau, A. Tkachuk, W. Yun, Sub-micron resolution CT for failure analysis and process development, Measurement Science & Technology 19 (9) (2008).

[392] A. Tkachuk, F. Duewer, H.T. Cui, M. Feser, S. Wang, W.B. Yun, X-ray computed tomography in Zernike phase contrast mode at 8 keV with 50-nm resolution using Cu rotating anode X-ray source, Zeitschrift Fur Kristallographie 222 (11) (2007) 650–655.

[393] L. Helfen, T. Baumbach, P. Cloetens, J. Baruchel, Phase-contrast and holographic computed laminography, Applied Physics Letters 94 (10) (2009).

[394] Z.W. Hu, F. De Carlo, Noninvasive three-dimensional visualization of defects and crack propagation in layered foam structures by phase-contrast microimaging, Scripta Materialia 59 (10) (2008) 1127–1130.

[395] H. Bolivar, S. Izquierdo, R. Tremont, C. Cabrera, Methanol oxidation at Pt/MoO_x/$MoSe_2$ thin film electrodes prepared with exfoliated $MoSe_2$, Journal of Applied Electrochemistry 33 (12) (2003) 1191–1198.

[396] C.K. Rhee, M. Wakisaka, Y.V. Tolmachev, C.M. Johnston, R. Haasch, K. Attenkofer, G.Q. Lu, H. You, A. Wieckowski, Osmium nanoislands spontaneously deposited on a Pt(111) electrode: an XPS, STM and GIF-XAS study, Journal of Electroanalytical Chemistry 554 (2003) 367–378.

[397] K. Hayashi, N. Furuya, Preparation of gas diffusion electrodes by electrophoretic deposition, Journal of the Electrochemical Society 151 (2004) A354.

[398] N. Lakshmi, N. Rajalakshmi, K.S. Dhathathreyan, Functionalization of various carbons for proton exchange membrane fuel cell electrodes: analysis and characterization, Journal of Physics D-Applied Physics 39 (13) (2006) 2785–2790.

[399] M. Nagao, A. Takeuchi, P. Heo, T. Hibino, M. Sano, A. Tomita, A Proton-Conducting In-Doped SnPO Electrolyte for Intermediate-Temperature Fuel Cells, Electrochemical and Solid-State Letters 9 (2006) A105.

[400] A. Sachdeva, S. Sodaye, A.K. Pandey, A. Goswami, Formation of silver nanoparticles in poly(perfluorosulfonic) acid membrane, Analytical Chemistry 78 (20) (2006) 7169–7174.

[401] S. Stoupin, E.H. Chung, S. Chattopadhyay, C.U. Segre, E.S. Smotkin, Pt and Ru X-ray absorption spectroscopy of PtRu anode catalysts in operating direct methanol fuel cells, Journal of Physical Chemistry B 110 (20) (2006) 9932–9938.

[402] Z. Ma, Q. Liu, Z.M. Cui, S.W. Bian, W.G. Song, Parallel array of Pt/polyoxometalates composite nanotubes with stepwise inside diameter control and its application in catalysis, Journal of Physical Chemistry C 112 (24) (2008) 8875–8880.

[403] S. Griesser, G. Buchinger, T. Raab, D.P. Claassen, D. Meissner, Characterization of fuel cells and fuel cell systems using three-dimensional X-ray tomography, Journal of Fuel Cell Science and Technology 4 (1) (2007) 84–87.

[404] I. Manke, C. Hartnig, M. Grunerbel, W. Lehnert, N. Kardjilov, A. Haibel, A. Hilger, J. Banhart, H. Riesemeier, Investigation of water evolution and transport in fuel cells with high resolution synchrotron X-ray radiography, Applied Physics Letters 90 (17) (2007).

[405] A. Vabre, S. Legoupil, F. Buyens, O. Gal, R. Riva, O. Gerbaux, A. Memponteil, Metallic foams characterization with X-ray microtomography using Medipix2 detector, Nuclear Instruments & Methods in Physics Research Section a-Accelerators Spectrometers Detectors and Associated Equipment 576 (1) (2007) 169–172.

[406] S.J. Lee, N.Y. Lim, S. Kim, G.G. Park, C.S. Kim, X-ray imaging of water distribution in a polymer electrolyte fuel cell, Journal of Power Sources 185 (2) (2008) 867–870.

[407] T. Mukaide, S. Mogi, J. Yamamoto, A. Morita, S. Koji, K. Takada, K. Uesugi, K. Kajiwara, T. Noma, In situ observation of water distribution and behaviour in a polymer electrolyte fuel cell by synchrotron X-ray imaging, Journal of Synchrotron Radiation 15 (2008) 329–334.

[408] F.Y. Zhang, D. Spernjak, A.K. Prasad, S.G. Advani, In situ characterization of the catalyst layer in a polymer electrolyte membrane fuel cell, Journal of the Electrochemical Society 154 (11) (2007) B1152–B1157.

[409] S.G. Kandlikar, Z.J. Lu, Thermal management issues in a PEMFC stack - A brief review of current status, Applied Thermal Engineering 29 (7) (2009) 1276–1280.

[410] A. Faghri, Z. Guo, Challenges and opportunities of thermal management issues related to fuel cell technology and modeling, International Journal of Heat and Mass Transfer 48 (19–20) (2005) 3891–3920.

[411] M.H. Wang, H. Guo, C.F. Ma, Temperature distribution on the MEA surface of a PEMFC with serpentine channel flow bed, Journal of Power Sources 157 (1) (2006) 181–187.

[412] D.J.L. Brett, P. Aguiar, R. Clague, A.J. Marquis, S. Schottl, R. Simpson, N.P. Brandon, Application of infrared thermal imaging to the study of pellet solid oxide fuel cells, Journal of Power Sources 166 (1) (2007) 112–119.

[413] A. Hakenjos, H. Muenter, U. Wittstadt, C. Hebling, A PEM fuel cell for combined measurement of current and temperature distribution, and flow field flooding, Journal of Power Sources 131 (1–2) (2004) 213–216.

[414] G. Ju, K. Reifsnider, X.Y. Huang, Infrared thermography and thermoelectrical study of a solid oxide fuel cell, Journal of Fuel Cell Science and Technology 5 (3) (2008).

[415] M.G. Santarelli, P. Leone, M. Cali, G. Orsello, Experimental analysis of the voltage and temperature behavior of a solid oxide fuel cell generator, Journal of Fuel Cell Science and Technology 4 (2) (2007) 143–153.

[416] R. Shimoi, M. Masuda, K. Fushinobu, Y. Kozawa, K. Okazaki, Visualization of the membrane temperature field of a polymer electrolyte fuel cell, Journal of Energy Resources Technology-Transactions of the Asme 126 (4) (2004) 258–261.

[417] Y. Ishikawa, H. Harnada, M. Uehara, M. Shiozawa, Super-cooled water behavior inside polymer electrolyte fuel cell cross-section below freezing temperature, Journal of Power Sources 179 (2) (2008) 547–552.

[418] M. Mench, D. Burford, and T. Davis. In situ temperature distribution measurement in an operating polymer electrolyte fuel cell. Proceedings of IMECE'03, 2003 ASME International Mechanical Engineering Congress & Exposition Washington DC, November 16– 21 (2003).

[419] S.H. He, M.M. Mench, S. Tadigadapa, Thin film temperature sensor for real-time measurement of electrolyte temperature in a polymer electrolyte fuel cell, Sensors and Actuators A-Physical 125 (2) (2006) 170–177.

[420] C.I. Lee, H.S. Chu, Effects of temperature on the location of the gas-liquid interface in a PEM fuel cell, Journal of Power Sources 171 (2) (2007) 718–727.

[421] C.Y. Lee, C.L. Hsieh, G.W. Wu, Novel method for measuring temperature distribution within fuel cell using microsensors, Japanese Journal of Applied Physics Part 1-Regular Papers Brief Communications & Review Papers 46 (5A) (2007) 3155–3158.

[422] C.Y. Lee, W.J. Hsieh, G.W. Wu, Embedded flexible micro-sensors in MEA for measuring temperature and humidity in a micro-fuel cell, Journal of Power Sources 181 (2) (2008) 237–243.

[423] C.Y. Lee, R.D. Huang, C.W. Chuang, Novel integration approach for in situ monitoring of temperature in micro-direct methanol fuel cell, Japanese Journal of Applied Physics Part 1-Regular Papers Brief Communications & Review Papers 46 (10A) (2007) 6911–6914.

[424] C.Y. Lee, S.J. Lee, Y.C. Hu, W.P. Shih, W.Y. Fan, C.W. Chuang, Real Time Monitoring of Temperature of a Micro Proton Exchange Membrane Fuel Cell, Sensors 9 (3) (2009) 1423–1432.

[425] M. Adzic, M.V. Heitor, D. Santos, Design of dedicated instrumentation for temperature distribution measurements in solid oxide fuel cells, Journal of Applied Electrochemistry 27 (12) (1997) 1355–1361.

[426] M. Wilkinson, M. Blanco, E. Gu, J.J. Martin, D.P. Wilkinson, J.J. Zhang, H. Wang, In situ experimental technique for measurement of temperature and current distribution in proton exchange membrane fuel cells, Electrochemical and Solid State Letters 9 (11) (2006) A507–A511.

[427] T. Fabian, R. O'Hayre, F.B. Prinz, J.G. Santiago, Measurement of temperature and reaction species in the cathode diffusion layer of a free-convection fuel cell, Journal of the Electrochemical Society 154 (9) (2007) B910–B918.

Chapter 9

Computational Modeling Aspects of PEFC Durability

Yu Morimoto* and Shunsuke Yamakawa
Toyota Central R&D Labs. Inc., Nagakute, Aichi, Japan

1. INTRODUCTION

This entire book is dedicated to various aspects of the degradation of polymer electrolyte fuel cells. This last chapter describes an approach to this issue using computer modeling. This approach is a very effective and powerful tool, for two main reasons. Firstly, degradation issues in PEFCs are very complex. Therefore, observable phenomena are only the final results of multi-step chemical and thermodynamic reactions. The influences of PEFC operational conditions, such as temperature, pressure and humidification, on the final degradation frequently seem inconsistent. This can happen because a PEFC operational condition that can accelerate one step may decelerate another. Therefore, a total quantitative understanding is very difficult without help from computer modeling studies which describe the balances of all elemental steps and their dependencies on PEFC operational conditions. Secondly, the elemental steps themselves sometimes are not observable, hence their nature often cannot be experimentally defined. A quantum chemical approach can be an effective tool for these problems.

In this chapter, the two modeling approaches, chemical engineering macroscopic modeling and molecular-level quantum chemical microscopic modeling for computational studies on PEFC durability and degradation are first reviewed. Secondly, recent works on macroscopic degradation of platinum catalysts in PEFCs are presented.

2. SIGNIFICANT LITERATURE

2.1. Macroscopic Models of Chemical Membrane Degradation

Membrane degradation has been a focal issue, since it results not only in gradual performance loss but also sudden failure by gas cross

Polymer Electrolyte Fuel Cell Degradation. DOI: 10.1016/B978-0-12-386936-4.10009-0

leakage. The basic mechanisms of the chemical degradation of the membrane are:

1. PFSA membrane is chemically decomposed by peroxide radical attack,
2. Peroxide radicals are generated on Pt when both H_2 and O_2 exist.

There are, however, unanswered questions such as:

1. Is the peroxide radical generated directly on Pt or through hydrogen peroxide with help from transition metal ions?
2. On which platinum (in the cathode, anode or Pt band) is the peroxide radical and/or hydrogen peroxide generated?

In spite of the low level of the understanding of the degradation mechanism, there has been one model published for the chemical membrane degradation [1]. Figure 9.1 shows the domain for the model. The assumed phenomena in this study are:

1. hydrogen crossover from the anode to cathode,
2. peroxide radical formation on the cathode/membrane interface,
3. membrane degradation caused by the radical attack,
4. fluoride formation and transportation,
5. membrane thinning, and
6. OCV change by hydrogen crossover increase and electrochemical Pt area loss.

While there is a good agreement in fluoride emission, the model seems to need various refinements, and further advancement in the understanding of the mechanism.

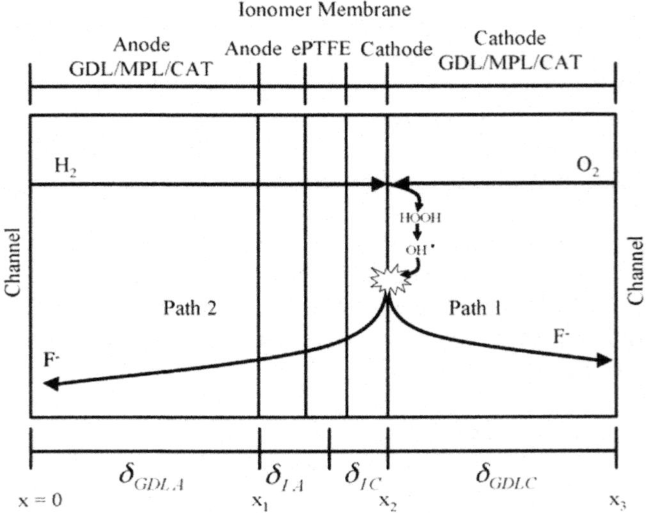

FIGURE 9.1 Domains for chemical degradation model of reference [1]. *Reproduced by permission of The Electrochemical Society.*

2.2. Microscopic Models of Membrane Degradation

As indicated previously, the mechanism of peroxide radical formation is not fully understood. Atrazhev et al. [2] proposed that the peroxide radical can be formed directly on a Pt surface – as shown in Fig. 9.2 – instead of through hydrogen peroxide, using a density function theory calculation. This mechanism allows peroxide radicals to be formed on the cathode, whose potential is too high to form hydrogen peroxide, by the two electron reduction of oxygen. This could explain some experimental results which suggest radical formation on the cathode [3,4,5].

Chemical degradation mechanisms of polymer membranes were studied quantum-chemically for hydrocarbon membranes [6] and for perfluorinated membranes [7]. The latter showed a possible new decomposition mechanism of peroxide radical attack at undissociated SO_3H side chain terminals, as shown in Fig. 9.3. Hydrogen radicals, which can be formed by the reaction between a hydrogen molecule and a peroxide radical, may attack any position on the polymer, in addition to the previously proposed mechanism [8] of peroxide radical attack at any remaining, non-fluorinated, main chain terminals, which would be followed by unzipping reactions.

2.3. Macroscopic Models of Mechanical Membrane Degradation

Membrane failure means a large gas cross-leakage through a pinhole or tear. It is believed to be initiated chemically and to be finished mechanically: i.e. a pinhole or tear is formed by mechanical stress on a chemically degraded membrane. The major source of mechanical stress is RH cycling because

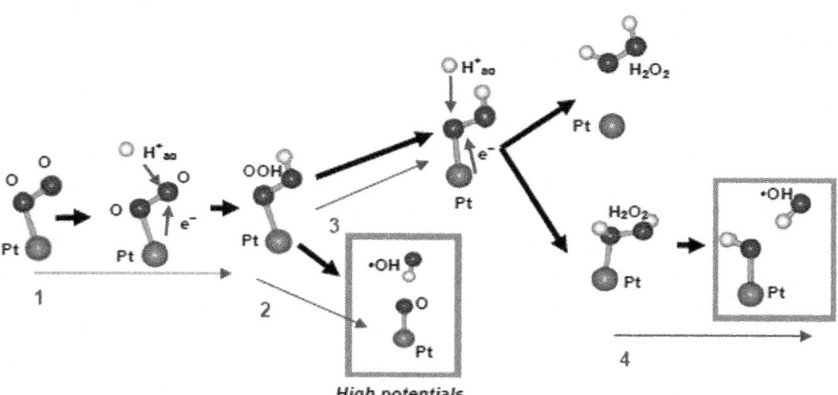

FIGURE 9.2 Direct peroxide formation scheme [2]. *Reproduced by permission of The Electrochemical Society.*

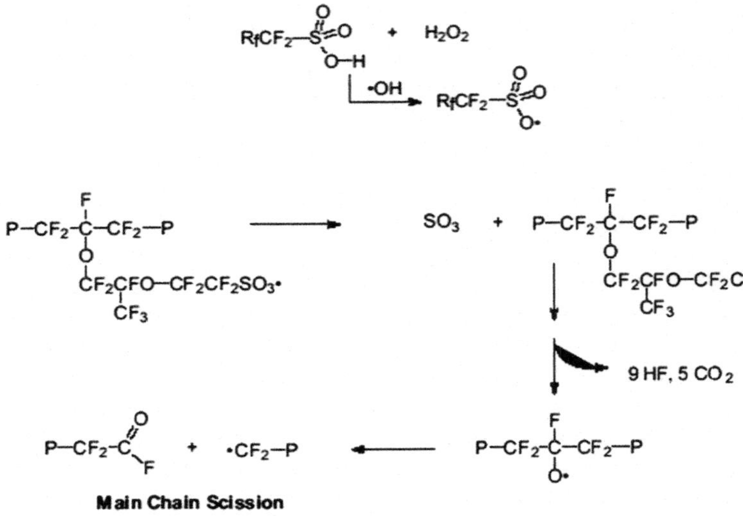

FIGURE 9.3 Peroxide attack on undissociated sulfonic acid end group to form sulfonyl radical and decomposition propagation scheme from sulfonyl radical to main chain scission [7]. *Reproduced by permission of The Electrochemical Society.*

a polymer electrolyte membrane, whether perfluorinated or hydrocarbon, swells at a high RH and shrinks at a low RH. X. Huang et al. [9] employed a finite element code to predict the tensile strain caused by RH cycling using experimental stress-strain data, and showed that a large strain is induced at the corner of a MEA, as shown in Fig. 9.4.

2.4. Macroscopic Models for the Mechanical Degradation of Catalyst Layer and Interface

The catalyst layer has a microstructure of ionomer, Pt-deposited carbon, and pores and this can be changed by various means. Since fuel cell performance is significantly influenced by this microstructure, changes in it are important. Rong et al. [10,11] presented an FEM model to describe debonding and delamination between ionomer and C/Pt agglomerate during humidity and temperature cycles. Figure 9.5 shows their model for the contact between ionomer and catalyst powder, and Fig. 9.6 shows their microstructure model of a catalyst layer. They estimated the strain in the ionomer phase as caused by humidity cycles, predicted the initiation time of delamination between the ionomer and carbon and showed that these mechanical changes can happen, and can affect fuel cell performance.

Delamination can also be an issue when water in the catalyst layer freezes. He et al. [12,13,14] presented a 1D transient model for this phenomenon.

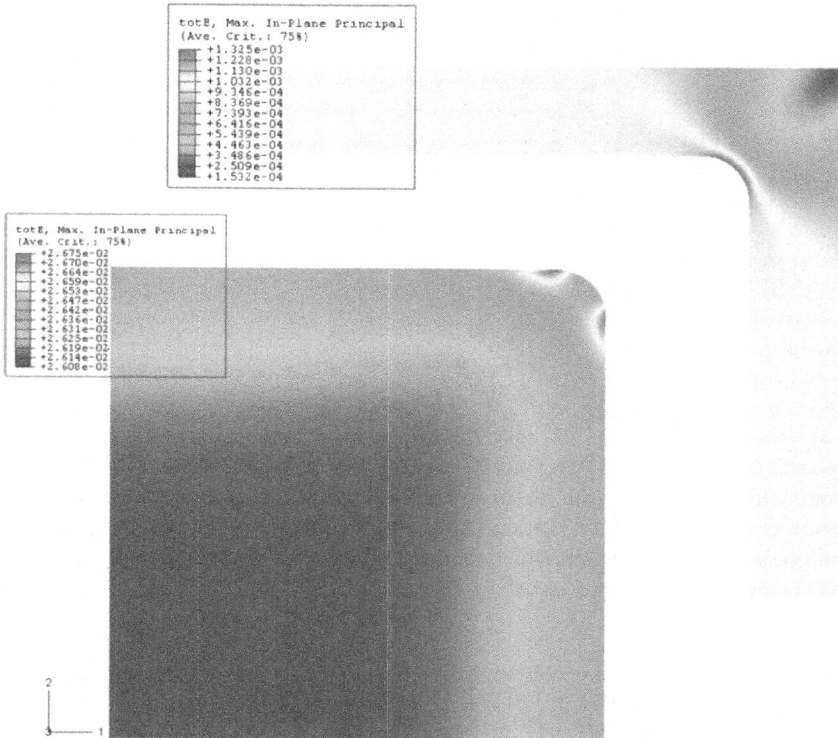

FIGURE 9.4 Model predicted distribution of maximum principal tensile strain in the MEA and edge seals as a result of RH variation from 75 to 0% [9].

Figure 9.7 shows the locations of ice lens formation. The authors showed that ice lens formation between the catalyst layer and the electrolyte membrane, or between the diffusion medium and the catalyst, can cause unrecoverable damage.

2.5. Models of Contamination

Various types of contamination can be introduced into a PEFC from fuel and air flow, and by the corrosion or decomposition of fuel cell materials. Although CO is a common contaminant in the fuel flow, and its poisoning is a serious issue for reformate-fueled FCs, this is a performance issue rather than a degradation issue, since CO is a standard component in a reformate, and such poisoning is recoverable. H_2S in the fuel flow, however, is a cumulative poison that is not easily removed, and therefore contamination by it can be regarded as a degradation issue. This issue was modeled by Shah and Walsh [15]. Their model assumed that H_2S is adsorbed on Pt in the anode, forming Pt-S and predicted

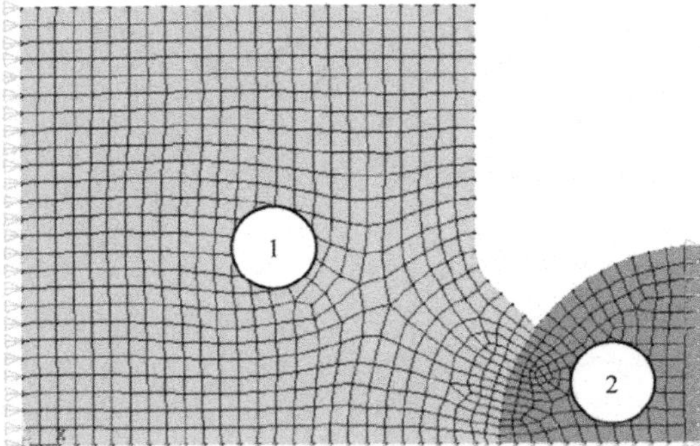

FIGURE 9.5 Model region for the contact between ionomer and catalyst powder; dark gray domain (domain 2) represents the C/Pt agglomerate and the light gray domain (domain 1) represents ionomer [10].

general dependency on temperature and relative humidity. Cation contamination, which could happen through introduction of salt-containing air, was modeled by Weber and Delacourt [16], who considered potassium contamination of the electrolyte, and its transportation causing the proton concentration at the cathode to become zero.

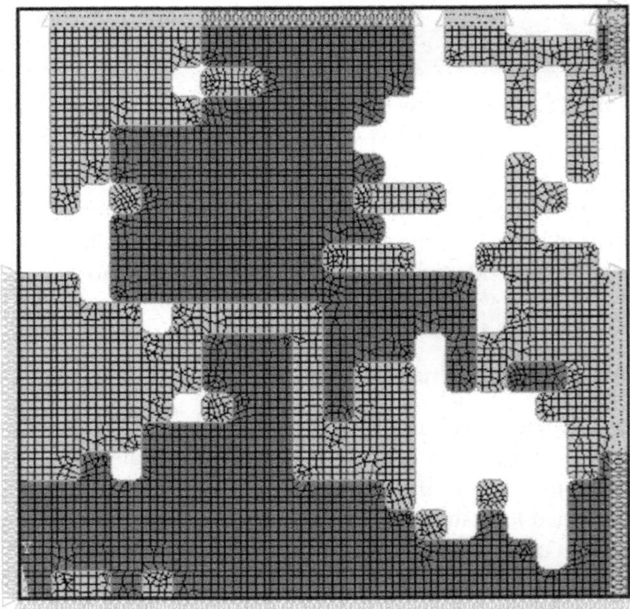

FIGURE 9.6 Model region for the contact between ionomer and catalyst powder; the dark gray domain represents the C/Pt agglomerate and the light gray domain represents ionomer [11].

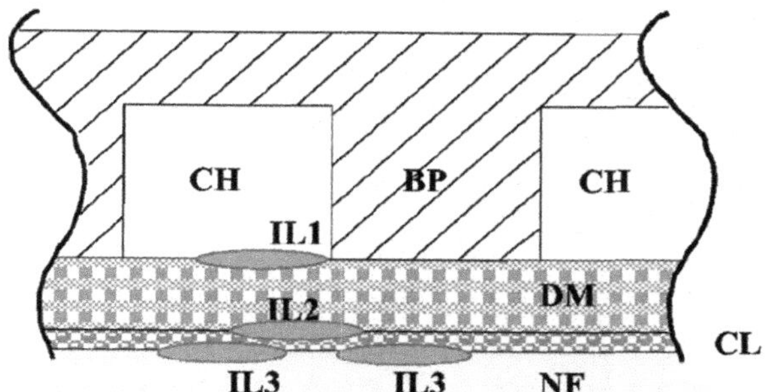

FIGURE 9.7 Illustration showing locations of ice lenses [13]. *Reproduced by permission of The Electrochemical Society.*

2.6. Macroscopic Models of Carbon Corrosion

The mechanism of rapid carbon corrosion during start-stop was first presented in a simple model by Reiser et al. [17], in which carbon in the cathode catalyst layer is quickly corroded when hydrogen occupies only a part of the anode as shown in Fig. 9.8. A detailed mathematical model was subsequently developed by Meyers and Darling [18] to quantitatively study the influence of the operating conditions on the degree of the degradation, not only for start-stop but also for local anode flooding and hydrogen maldistribution. Figure 9.9 shows simulated, transient, carbon corrosion current density during start-up. Fuller and Gray expanded this model to 2D [19] to

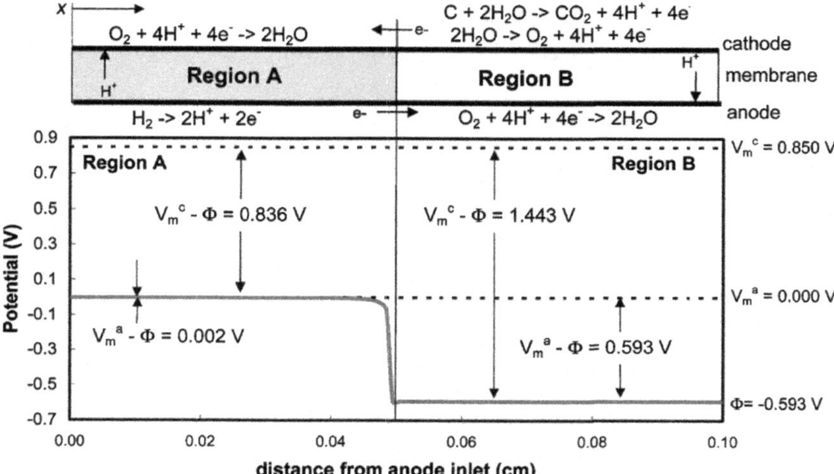

FIGURE 9.8 Potential distributions along anode flow path during reverse current conditions [17]. *Reproduced by permission of The Electrochemical Society.*

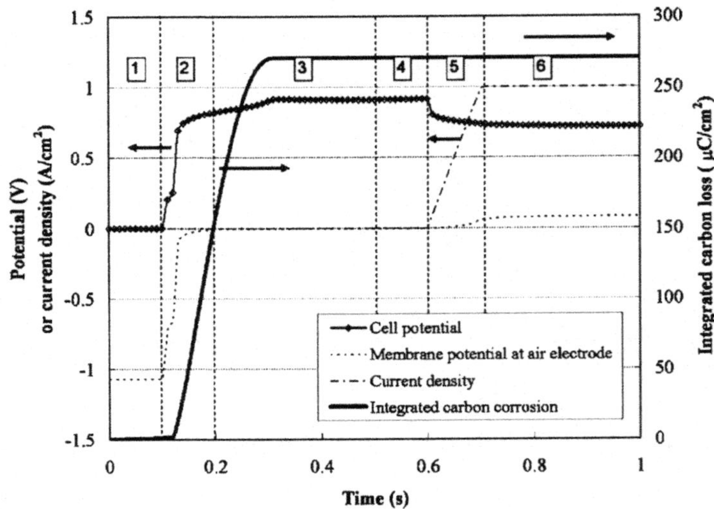

FIGURE 9.9 Time trace of a fuel into air start. Step 1: Air on both electrodes. Step 2: Fuel is introduced to anode. Step 3: Fuel on anode, stagnant air on cathode. Step 4: Flowing air on cathode. Step 5: Load is gradually increased from open circuit to 1 A/cm^2. Step 6: Current is maintained at 1 A/cm^2. Note that as fuel is introduced, the carbon corrosion rate increases sharply and is sustained as long as there is a fuel-air front in the cell. The corrosion rate then drops sharply again, after the fuel-air front passes through the cell [18]. *Reproduced by permission of The Electrochemical Society.*

simulate the potential distribution in the membrane and both the electrodes as shown in Fig. 9.10.

A 3D CFD model was developed by employing these concepts [20]. Various mitigation techniques, such as using a catalyst with high oxygen evolution

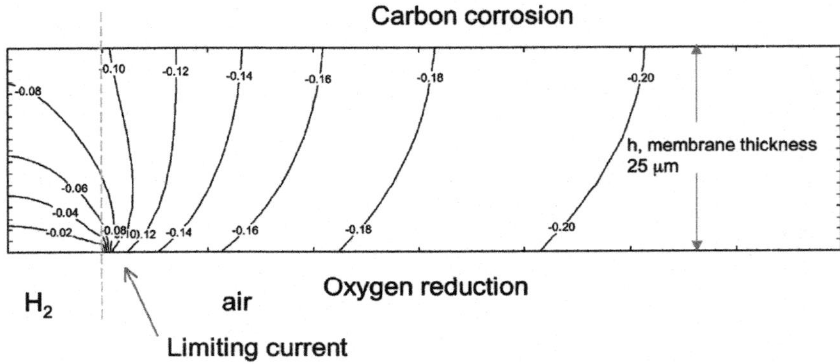

FIGURE 9.10 Simulated 2D potential distribution for partial fuel starvation [19]. *Reproduced by permission of The Electrochemical Society.*

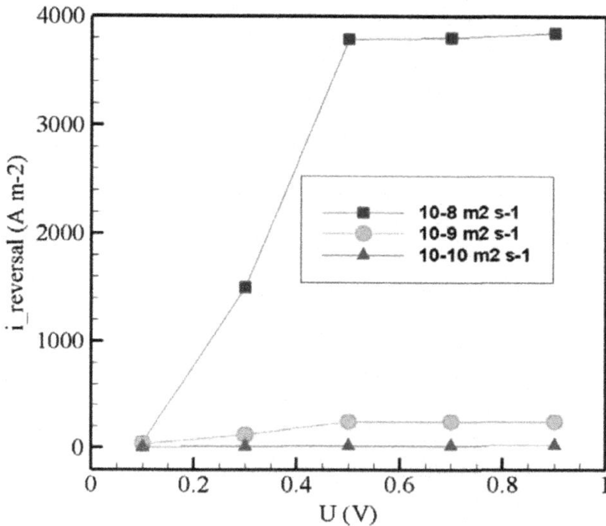

FIGURE 9.11 Average reversal current density as a function of O_2 diffusivity in membrane [21]. *Reproduced by permission of The Electrochemical Society.*

reaction (OER) activity, a membrane of low O_2 diffusivity, corrosion-resistant carbon support and high proton conductivity in the catalyst layer, were evaluated through a 2D model [21]. Figure 9.11 shows the effect of O_2 diffusivity on a membrane.

Franco and Gerard [22] integrated a 1D carbon corrosion model into a fuel cell performance model by estimating the thickness of catalyst layers, the surface area of Pt and the contact resistance between the cathode and a GDL.

2.7. Microscopic Models on Platinum Dissolution

There are only a few reports of microscopic model studies for platinum dissolution. Tian and Anderson [23] showed, by a first principle calculation, that Co is oxidized before Pt on a Pt-Co alloy, and that Pt edge sites are more easily to be oxidized than Pt on the terrace. The latter result suggests that Pt dissolution is initiated at the edge site. Zhou [24] carried out a molecular dynamic (MD) simulation for a charged Pt surface in contact with Nafion containing water. The simulation suggested that Pt disintegration was caused by the interaction between Pt and the sulfonic acid terminal of Nafion, and by an instantaneous temperature rise at the interface of up to 2,000 K. This temperature rise, however, seems too high to be realistic.

2.8. Macroscopic Models of Catalyst Degradation

The loss of area of electrochemically active Pt is one of the most serious degradation symptoms in PEFCs. A similar phenomenon was known in phosphoric acid fuel cells. Bett et al. [25] fitted the surface area decay curve and suggested that Ostwald ripening contributes to the surface area loss. (The term Ostwald ripening is generally used for a phenomenon in which a component species separates, dissolves, or evaporates from smaller particles, and then transfers through a medium and attaches, deposits, or condenses onto larger particles. It results in the disappearance of smaller particles and growth of larger particles.) Their idea suggests that the component species in this case is Pt atom, which transfers onto the surface of carbon.

In PEFCs, the first mathematical model was presented by Darling and Meyers [26]. They consider Pt ion dissolution and its diffusion in the ionomer phase, rather than Pt atom separation and transfer. The phenomana included in the models are electrochemical Pt dissolution into the ionomer, Pt oxide formation and the chemical dissolution of Pt oxide. The rate of Pt dissolution was described as a function dependent on the coverage of Pt oxide and the Pt particle size, using a modified Butler-Volmer equation as follows:

$$r_1 = k_1\theta_{vac}\left[\exp\left(\frac{\alpha_{a,1}n_1F}{RT}(\Phi_1-\Phi_2-U_1)\right) - \left(\frac{C_{Pt^{2+}}}{C_{Pt^{2+},ref}}\right)\exp\left(-\frac{\alpha_{c,1}n_1F}{RT}(\Phi_1-\Phi_2-U_1)\right)\right] \tag{9.1}$$

$$U_1 = U_1^{\theta} - \frac{\alpha_{Pt}M_{Pt}}{2F\rho_{Pt}r} \tag{9.2}$$

The platinum oxide formation reaction rate is described as:

$$r_2 = k_2\left[\exp\left(-\frac{\omega\theta_{PtO}}{RT}\right)\exp\left(\frac{\alpha_{a,2}n_2F}{RT}(\Phi_1-\Phi_2-U_2)\right) - \theta_{PtO}\left(\frac{C_{H^+}^2}{C_{H^+,ref}^2}\right)\exp\left(-\frac{\alpha_{c,2}n_2F}{RT}(\Phi_1-\Phi_2-U_2)\right)\right] \tag{9.3}$$

$$U_2 = U_2^{\theta} + \frac{\alpha_{PtO}M_{PtO}}{2F\rho_{PtO}r} - \frac{\alpha_{Pt}}{2F\rho_{Pt}r} \tag{9.4}$$

Figure. 9.12 shows simulated PtO coverage at a potential sweep. The same authors [27] then expanded their model to describe Pt movement in a fuel cell. Their new model includes Pt ion diffusion through the electrolyte membrane and the anode ionomer, and simulations were carried out assuming a bimodal particle-size distribution. The results successfully predicted that electrochemically active area loss is much quicker during potential cycling than potential

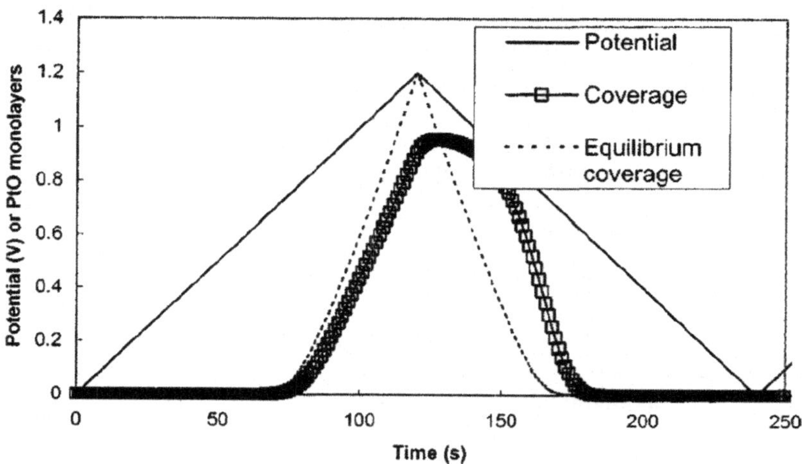

FIGURE 9.12 Comparison of equilibrium oxide coverage to predicted oxide coverage [25]. *Reproduced by permission of The Electrochemical Society.*

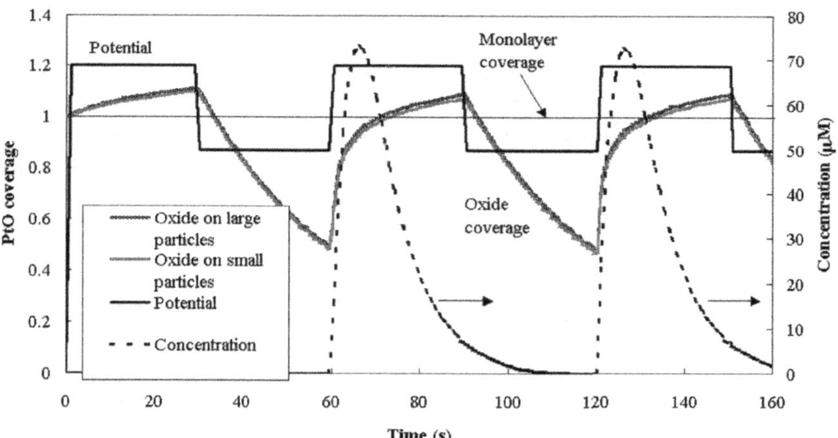

FIGURE 9.13 Oxide coverage and soluble platinum concentration during square-wave cycling from 0.87 to 1.2 V with a period of 60s [26]. *Reproduced by permission of The Electrochemical Society.*

holding at any given potential. Figure 9.13 shows the Pt concentration change during potential cycling. Franco and Tembely [28] utilized a similar Pt dissolution/deposition model and simulated the electrochemical impedance spectroscopic responses based on ionomer/Pt interface. Bi and Fuller [29] added the phenomenon of Pt-band formation as a result of the chemical reduction of diffused Pt ion by cross-over hydrogen at a specific plane in the membrane. Holby et al. [30,31,32] used a similar model, and simulated electrochemically active area loss assuming various particle size distributions for the initial condition. They concluded that the initial particle size distribution significantly affects electrochemically active area loss, and that hydrogen crossover accelerates this loss – presumably because hydrogen reduces Pt ion to Pt metal.

3. OUR RECENT APPROACHES TOWARD MACROSCOPIC MODELS OF CATALYST DEGRADATION

3.1. Simplified Model

We (the authors of this chapter) [33] used a TEM observation-determined particle size distribution and applied a simple Pt dissolution/diffusion/deposition model, in which the dissolution/deposition is treated like the vaporization/condensation of a vapor/liquid using a Kelvin type equation and Pt ion diffusion equation. The dissolution/deposition rate was set to be proportional to the difference between the saturated Pt ion concentration, determined as a function of particle size, and the actual concentration.

3.2. Integrated Model

We have recently upgraded the simplified model to the integrated model, using the modified Butler-Volmer equation of Darling and Meyers [26]. The parameters for Pt oxide formation and reduction were determined to reproduce the cyclic voltammograms measured with various scan rates. The parameters for Pt dissolution and deposition were determined, in order to reproduce the total surface area reduction that had been measured experimentally. The diffusion coefficient of Pt ions was determined to reproduce the Pt mass loss rate. We assumed that Pt ions transported into the membrane were chemically reduced, and formed a Pt band at a position 1 μm away from the catalyst layer/membrane interface. The simulation results are compared with the results of an experiment using the conditions shown in Table 9.1.

Table 9.2 shows the parameters used for this simulation by the integrated model, in comparison with the other reference parameters [26, 29]. To obtain a good fit with the cyclic voltammograms for Pt oxide formation and reduction, the rate constant for the Pt oxide formation/reduction reaction needed to be set to a higher value than that estimated by Darling and Meyers [26]. Furthermore, a much faster Pt dissolution/deposition rate was also suggested.

4. RESULTS AND DISCUSSION

The particle size distribution was simulated by the simplified model in various positions in a catalyst layer from the electrolyte membrane side to the GDL side, in which the initial size distribution was determined by TEM observation and particle size distribution change was simulated. Resulting platinum size distributions before and after degradation at positions near the membrane and

TABLE 9.1 Condition Values of Voltage Cycling

Condition	Value
Lower voltage limit (V)	0.72
Upper voltage limit (V)	0.955
Holding time at upper / lower voltage (s)	4
Sweep rate (V/s)	0.235
Temperature (°C)	80
Thickness of cathode catalyst layer (μm)	10
Cathode Pt loading (mg/cm^2)	0.4
Ionomer / carbon weight ratio	0.75
Initial state of mean Pt particle diameter (nm)	4

near the GDL are shown in Fig. 9.14. Near the membrane, particles smaller than 5nm quickly diminish, and even larger particles decrease in size. Near the GDL, in spite of the similar relative particle size distribution, larger particles show actual growth in size. These simulation results show that relative particle size distribution measurement does not give the overall picture of catalyst degradation unless it accompanies a quantitative analysis of platinum distribution.

TABLE 9.2 Fitting Parameter Values

Parameter	Fitted Value	Reference Value
Pt dissolution reaction rate constant (mol/cm^{2}s)	5×10^{-10} ~ 3×10^{-9} *)	3.4×10^{-13} [25], 3×10^{-10} [28]
Pt oxidation reaction rate constant (mol/cm^{2}s)	6×10^{-10}	1.36×10^{-11} [25], 7×10^{-10} [28]
Pt ions diffusion coefficient (cm^2/s)	2×10^{-6}	1×10^{-6} [25], 1.5×10^{-9} [28]

*) *Values in this range were applicable according to supposed effective PtO coverage. Best-fitted value is still rather vague.*

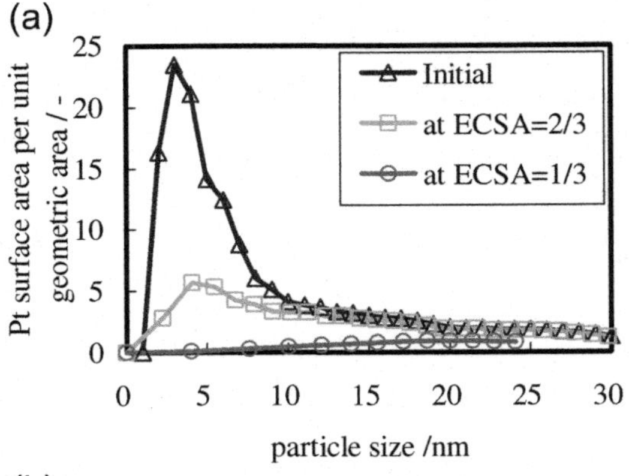

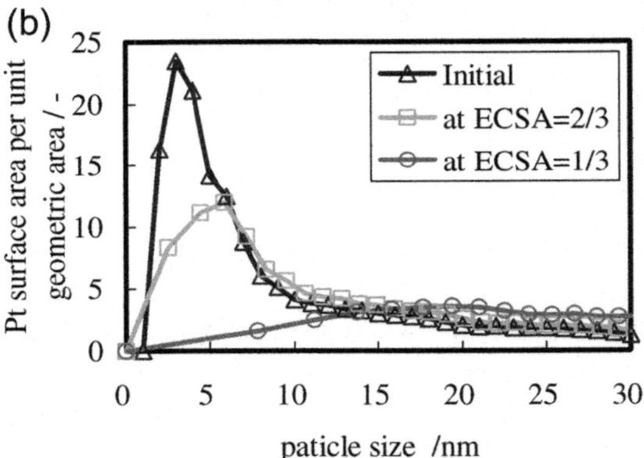

FIGURE 9.14 Distribution of Pt surface area per unit geometric area over particle size at the initial state and at the instances when the total Pt surface area becomes 2/3 and 1/3 of the initial state by potential cycling. (a) at the interface with the electrolyte membrane. (b) at the interface with the GDL.

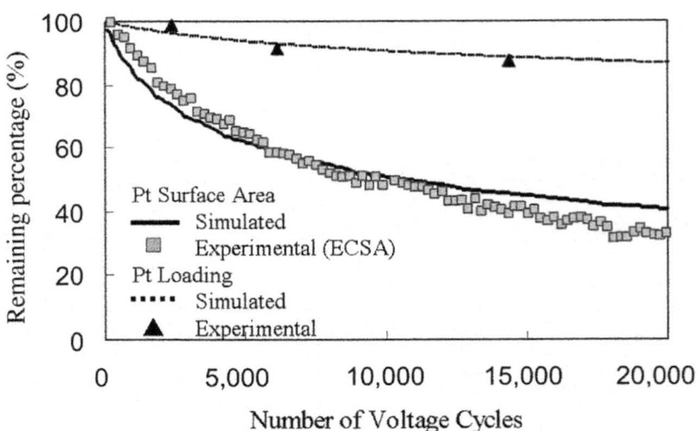

FIGURE 9.15 Platinum surface area and loading losses during voltage cycling in the range of 0.72–0.955 V at 80°C under fully humidified H_2(anode)-N_2(cathode). ECSA was estimated from the total charge of hydrogen deposition reaction on Pt by CV. Pt Loading was estimated by using EPMA.

Using the integrated model with the parameters shown in Table 9.2, the particle size distribution is simulated as a function of time and position, and the total surface area and total Pt loading can be calculated.

Figure 9.15 shows that the experimental and simulated reduction trends of ECSA and total mass (loading) of platinum are in good agreement. The much smaller loading loss than the ECSA loss shows that the ECSA loss is mainly caused by the mechanism of Ostwald ripening through Pt ion diffusion, rather than by Pt ion effluence to the membrane. Simulated and experimental particle size distributions are shown in Fig. 9.16 for positions near the GDL and near the membrane after the potential cycling. At both the positions, particles smaller

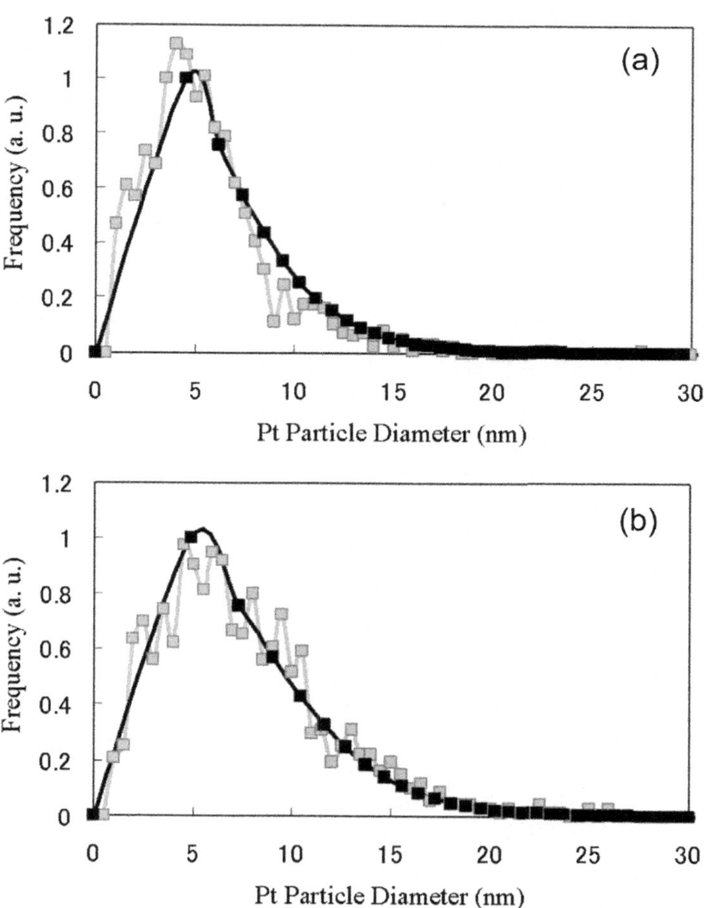

FIGURE 9.16 Size distribution of platinum particles suffering voltage cycles. Corresponding ECSA remaining is 40%. In the two figures, (a) and (b) show the particle distributions near the membrane and gas diffusion layer, respectively. Filled black and gray squares denote results of simulation and experimental values measured from TEM micrographs, respectively.

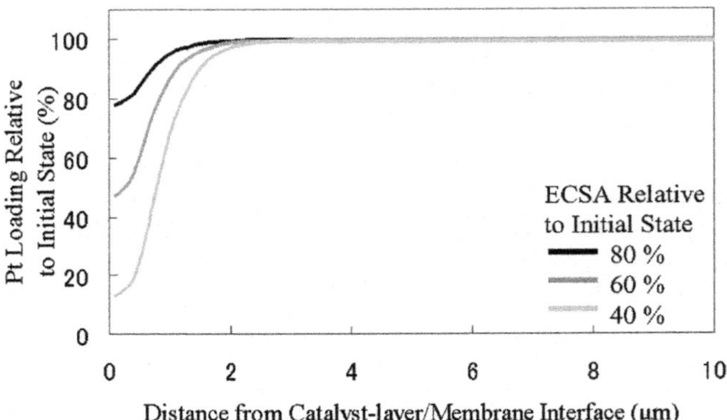

FIGURE 9.17 Percentage of Pt loading relative to initial state at positions away from the catalyst layer/membrane interface. Lines of different gradations correspond to different ECSA remaining.

than 5 nm, which are initially dominant in number, diminish, but larger particles are more prominent in the GDL side. The simulated platinum mass distribution shown in Fig. 9.17 indicates that mass loss is seen only near the membrane. These results clearly indicate the significance of Ostwald ripening for the platinum catalyst degradation.

5. SUMMARY AND FUTURE CHALLENGES

While computational modeling has been widely used in PEFC technological development for performance enhancement, its application to degradation issues has only just begun. This chapter provided a concise review and our studies were presented. As stated previously, a wide variety of degradation studies have been performed, and used for understanding the phenomena and improving the durability of PEFC cells. However, there are a lot of areas needing further study, in order to understand what is really happening, and to plan mitigation strategies.

Areas that would benefit from further research include:

1. Macroscopic model of chemical degradation of membrane: this model should include the influence of formation of Pt band and its effect.
2. Performance prediction modeling, including the effects of membrane degradation, the effect of membrane thinning, and also the catalytic effects of decomposition products.
3. Microscopic Pt dissolution model: this model should elucidate the particle size effect on Pt dissolution, because the particle size term in the modified Butler-Volmer equation by Darling and Meyers [26], which we used in our study also, has not been validated theoretically or experimentally.
4. Performance prediction model including the catalyst degradation: this model should include Pt surface area loss, particle size distribution change,

and decomposition products if membrane decomposition happens simultaneously.

As the wider commercialization of PEFCs comes closer in automobile and static applications, durability issues become increasingly important, because PEFCs with low-cost constitutions such as low Pt loading and hydrocarbon electrolytes need to be operated under harsher conditions, such as high temperature and low humidity. Computational modeling studies in this area will increase in importance, and will play a crucial role in tackling this challenge for years to come.

ACKNOWLEDGMENTS

The authors are deeply grateful to Professor M. Mench for his invitation and warm encouragement. Our gratitude also goes to our colleagues in Toyota Central R&D Labs. Inc., especially Dr S. Hyodo for his support for modeling studies and Mr T. Takeshita, Dr H. Murata, and Mr T. Hatanaka for their experimental contribution and fruitful discussions.

NOMENCLATURE

C_i	concentration of species i, mol/cm^3
F	Faraday's constant, 96,487 C/equiv
k_i	rate constant for reaction i in the forward direction, mol/cm^2 s
M_{Pt}	molecular weight of Pt, 195 g/mol
M_{PtO}	molecular weight of PtO, g/mol
N_i	number of electrons in reaction i
r	particle radius, cm
r_i	rate of reaction i, mol/cm^2 s
R	universal gas constant, J/mol K
t	time, s
T	temperature, K
U_i	thermodynamically reversible potential for reaction i, V

GREEK

$\alpha_{a,i}$	anodic transfer coefficient for reaction i
$\alpha_{c,i}$	cathodic transfer coefficient for reaction i
Φ_1	solid-phase potential, V
Φ_2	membrane-phase potential, V
θ_{PtO}	fraction of platinum surface covered by PtO
θ_{Vac}	fraction of platinum surface not covered by PtO
ρ_{Pt}	density of platinum, 21.0 g/cm^3
ρ_{PtO}	density of platinum oxide, 14.1 g/cm^3
σ_{Pt}	surface tension, J/cm^2
ω	PtO-PtO interaction parameter, J/mol

REFERENCES

[1] S. Kundu, M. Fowler, L.C. Simon, R. Abouatallah, N. Beydokhti, Degradation analysis and modeling of reinforced catalyst coated membranes operated under OCV conditions, J. Power Sources 183 (2008) 619.

[2] V. Atrazhev, E. Timokhina, S.F. Burlatsky, V. Sultanov, T. Madden, M. Gummalla, Direct Mechanism of OH Radicals Formation in PEM Fuel Cells, ECS Trans. 6 (25) (2008) 69.

[3] N. Miyake, M. Wakizoe, E. Honda, T. Ohta, High Durability of Asahi Kasei Aciplex Membrane, ECS Trans. 1 (8) (2006) 249.

[4] N. Hasegawa, T. Asano, T. Hatanaka, M. Kawasumi, Y. Morimoto, Degradation of Perfluorinated Membranes Having Intentionally Formed Pt-Band, ECS Trans. 16 (2) (2008) 1713.

[5] V.O. Mittal, H.R. Kunz, J.M. Fenton, Membrane Degradation Mechanisms in PEMFCs, J. Electrochem. Soc. 154 (2007) B652.

[6] A. Panchenko, DFT investigation of the polymer electrolyte membrane degradation caused by OH radicals in fuel cells, J. Membrane Sciences 278 (2006) 269.

[7] F.D. Coms, The Chemistry of Fuel Cell Membrane Chemical Degradation, ECS Trans. 16 (2) (2008) 235.

[8] D.E. Curtin, R.D. Losenberg, T.J. Henry, P.C. Tangeman, M.E. Tisack, J. Power Sources 131 (2004) 41.

[9] X. Huang, R. Solasi, Y. Zou, M. Feshler, K. Reifsnider, D. Condit, S. Burlatsky, T. Madden, Mechanical Endurance of Polymer Electrolyte Membrane and PEM Fuel Cell Durability, J. Polymer Science Part B-Polymer Physics 44 (2006) 2346.

[10] F. Rong, C. Huang, Z.-S. Liu, D. Song, Q. Wang, Microstructure changes in the catalyst layers of PEM fuel cells induced by load cycling Part I. Mechanical model, J. Power Sources 175 (2008) 699.

[11] F. Rong, C. Huang, Z.-S. Liu, D. Song, Q. Wang, Microstructure changes in the catalyst layers of PEM fuel cells induced by load cycling Part II. Simulation and understanding, J. Power Sources 175 (2008) 712.

[12] S. He, M.M. Mench, Degradation of Polymer-Electrolyte Membranes in Fuel Cells I. Experimental, J. Electrochem. Soc.,153 A1724 (2006).

[13] S. He, S.H. Kim, M.M. Mench, One-Dimensional Transient Model for Frost Heave in Polymer Electrolyte Fuel Cells II. Physical Model, J. Electrochem. Soc. 154 (2007) B1024.

[14] S. He, S.H. Kim, M.M. Mench, 1D Transient Model for Frost Heave in PEFCs III. Heat Transfer, Microporous Layer, and Cycling Effects, J. Electrochem. Soc. 154 (2007) B1227.

[15] A.A. Shah, F.C. Walsh, A model for hydrogen sulfide poisoning in proton exchange membrane fuel cells, J. Power Sources 185 (2008) 287.

[16] A.Z. Weber, C. Delacourt, Mathematical Modeling of Cation Contamination in a Proton-exchange Membrane, Fuel Cells 6 (2008) 459.

[17] C.A. Reiser, L. Bregoli, T.W. Patterson, J.S. Yi, J.D. Yang, M.L. Perry, T.D. Jarvi, A Reverse-Current Decay Mechanism for Fuel Cells, Electrochem. Solid-State Lett. 8 (2005) A273.

[18] J.P. Meyers, R.M. Darling, Model of Carbon Corrosion in PEM Fuel Cells, J. Electrochem. Soc. 153 (2006) A1432.

[19] T. Fuller, G. Gray, Carbon Corrosion Induced by Partial Hydrogen Coverage, ECS Trans. 1 (8) (2006) 345.

[20] K. Jain, A. Gidwani, S. Kumar, J.V. Cole, CFD Study of Carbon Corrosion in PEM Fuel Cells, ECS Trans. 16 (2) (2008) 1323.

[21] J. Hu, P.C. Sui, N. Djilali, S. Kumar, Modelling and Simulations on Mitigation Techniques for Carbon Oxidation Reaction Caused by Local Fuel Starvation in a PEMFC, ECS Trans. 16 (2) (2008) 1313.

[22] A.A. Franco, M. Gerard, Multiscale Model of Carbon Corrosion in a PEFC: Coupling with Electrocatalysis and Impact on Performance Degradation, J. Electrochem. Soc. 155 (2008) B367.

[23] F. Tian, A.B. Anderson, Theoretical Study of Early Steps in Corrosion of Pt and Pt/Co Alloy Electrodes, J. Physical Chem. C 112 (2009) 18566.

[24] X.Y. Zhou, Degradation of Pt Catalysts in PEFCs: A New Perspective from Molecular Dynamic Modeling, Electrochem. Solid-State Lett. 11 (2006) B59.

[25] J.A.S. Bett, K. Kinoshita, P. Stonehart, Crystallite growth of platinum dispersed on graphitized carbon black II. Effect of liquid environment, J. Catalysis 41 (1976) 124.

[26] R.M. Darling, J.P. Meyers, Kinetic Model of Platinum Dissolution in PEMFCs, J. Electrochem. Soc. 150 (2003) A1523.

[27] R.M. Darling, J.P. Meyers, Mathematical Model of Platinum Movement in PEM Fuel Cells, J. Electrochem. Soc. 152 (2005) A242.

[28] A.A. Franco, M. Tembely, Transient Multiscale Modeling of Aging Mechanisms in a PEFC Cathode, J. Electrochem. Soc. 154 (2007) B712.

[29] W. Bi, T. Fuller, Modeling of PEM fuel cell Pt/C catalyst degradation, J. Power Sources 178 (2008) 188.

[30] E. Holby, Y.S. -Horn, D. Morgan, 211th ECS meeting Abstract #907, The Electrochemical Society, Pennington, NJ, 2007.

[31] E. Holby, Y.S. -Horn, A. Sheng, D. Morgan, 212th ECS meeting Abstract #391, The Electrochemical Society, Pennington, NJ, 2007.

[32] E. Holby, Y.S. -Horn, A. Sheng, D. Morgan, 214th ECS meeting Abstract #798, The Electrochemical Society, Pennington, NJ, 2008.

[33] T. Takeshita, H. Murata, T. Hatanaka, Y. Morimoto, Analysis of Pt Catalyst Degradation of a PEFC Cathode by TEM Observation and Macro Model Simulation, ECS Trans. 16 (2) (2008) 367.

Index

A

B

C

D

E

F

G

N

Q

R

S

T

U

V

W

X

Printed in the United States
By Bookmasters